The Palgrave Handbook of Child Mental Health

The Palgrave Handbook of Child Mental Health

Discourse and Conversation Studies

Edited by

Michelle O'Reilly
Senior Lecturer, University of Leicester, UK

Jessica Nina Lester
Assistant Professor, Indiana University, USA

palgrave
macmillan

First published 2015 by
PALGRAVE MACMILLAN

Palgrave Macmillan in the UK is an imprint of Macmillan Publishers Limited,
registered in England, company number 785998, of Houndmills, Basingstoke,
Hampshire RG21 6XS.

Palgrave Macmillan in the US is a division of St Martin's Press LLC,
175 Fifth Avenue, New York, NY 10010.

Palgrave Macmillan is the global academic imprint of the above companies
and has companies and representatives throughout the world.

Palgrave® and Macmillan® are registered trademarks in the United States,
the United Kingdom, Europe and other countries.

ISBN 978–1–137–42830–1

This book is printed on paper suitable for recycling and made from fully
managed and sustained forest sources. Logging, pulping and manufacturing
processes are expected to conform to the environmental regulations of the
country of origin.

A catalogue record for this book is available from the British Library.

Library of Congress Cataloging-in-Publication Data
The Palgrave handbook of child mental health / [edited by] Michelle O'Reilly,
    Senior Lecturer, University of Leicester, UK, Jessica Nina Lester,
    Assistant Professor, Indiana University, USA.
       pages   cm
    ISBN 978–1–137–42830–1 (hardback)
    1. Child psychiatry—Handbooks, manuals, etc.   I. O'Reilly, Michelle.
    II. Lester, Jessica Nina.
    RJ499.P33 2015
    618.92'8914—dc23   2015013091

Contents

Part I   The Place of Conversation/Discourse Analysis in Child Mental Health

Part II   A Critical Approach to Child Mental Health

Tables and Figures

Tables

Figures

Preface

Introduction: Aims and objectives of the book

The Palgrave Handbook of Child Mental Health: Discourse and Conversation Studies is a co-edited volume that brings together a range of applied and theoretical perspectives that examine the conversational practices of children diagnosed with mental health disorders, their parents/families, and their practitioners. The volume includes many contributions that illustrate the ways in which *language as action* is able to assist us in understanding the discursive practices that surround child mental health.

Indeed, mental health research with children and adults has been typically dominated by biological and genetic research driven by a desire for outcomes, but importantly, qualitative research is becoming more accepted (Peters, 2010). For the integration of science and practice to be accepted, the field of mental health needs more process-oriented research that focuses on questions of 'why' and/or 'how' (Rhodes, 2011). This is important for improvements to service provision, the development of new and existing interventions, updates to policy, and understanding how to communicate effectively with children and families (Jensen, 2004).

One of the clear advantages of qualitative research in the field of mental health is that it questions the taken-for-granted language that surrounds it. The World Health Organization (2005) defined mental health broadly as including 'a broad perspective ranging from emotional well-being to mental illness and disorder' (p. 12). In this volume, we aim to take a broad orientation to the construct of 'child mental health', while also assuming that the very meaning is bounded within the discursive practices that make possible identifying, labelling, and treating children as 'disordered'. We recognise that the majority of research in the area of child mental health has been positioned within a discourse of disease and deficit, with metaphors of cure and intervention frequently employed. Notably, methodologies employed to study child mental health have primarily represented disorders as scientific facts to be understood through a positivist lens, and thereby they have frequently evaded the institutional histories and discursive practices that have contributed to the very production of mental health disorders.

In this volume, therefore, we aim to provide a more critical understanding of child mental health by presenting a range of contributions that take up critical, discursive, and/or conversation analytic approaches to the study and discussion of child mental health. While there is some diversity in the

theoretical positions of our contributors, many adopt a broadly social constructionist or closely related theoretical perspective. The contributors move away
from viewing mental illness as an objective truth and, instead, reintroduce the
relevance of language in constructing and deconstructing the assumptions that
surround the diagnosis and treatment of childhood mental health disorders
(Fee, 2000). Thus, the volume illustrates the importance of examining child
mental health from a different perspective, one that assumes that psychiatric
categories are made real through both written and spoken language. Similar to
Harper (1995), throughout this volume, we argue that a discourse/conversation
analysis (DA/CA) approach is useful for the study of mental health.

Why a handbook now?

The World Health Organization (2001) reported that 450 million people worldwide have mental health disorders, with a 20% global prevalence of childhood
mental health disorders. From attention deficit disorder to autism spectrum
disorder, there are a plethora of diagnostic labels used to describe the identified
and everyday performances of children, with many such labels being linked to
diagnostic and treatment protocols. Diagnostic manuals, including the *International Classification of Diseases* manual (currently being updated) and *Diagnostic
and Statistical Manual of Mental Disorders* (recently updated), provide diagnosticians with particular frameworks and language that guide clinical practice and
assist them in making sense of particular behaviours. Further, the meanings
attributed to particular behaviours say much about an individual's past, present,
and, to some extent, determine their future (Crowe, 2000). Thus, how child
mental health is conceived of matters as its very construction has consequences
whether intended or not.

Yet, to date, there is no single volume that has brought together diverse
works focused on child mental health from DA/CA perspectives. Further, there
is relatively little writing that explicitly seeks to bridge the gap between the
everyday work of clinicians and DA/CA perspectives. Recognising these gaps in
the conversation, we explicitly sought to develop a handbook that offered theoretical, empirical, and perhaps most uniquely, clinically relevant insights at the
intersection of child mental health and understandings of *language as action*.
Acknowledging that child mental health is of global concern, we position this
Handbook as both timely and central to ongoing conversations around the
everyday, discursive practices that make the identification and treatment of
children possible.

Child mental health: A critical view and the turn to language

The turn to language (or the 'discursive turn') in child mental health and
other disciplines reflects both challenges to traditional paradigmatic views

on the world and how it works. In this turn towards language, language has become viewed as constitutive of social life, constructing minds and identities rather than simply representative of inner, mental workings. This turn towards language has been noted in broader discussions of disability, with disability studies scholars long pointing to the culturally and socially contingent nature of disability (Oliver, 1983, 1993). While disability theorists have been less apt to consider psychiatric disabilities, their perspectives on disability as being bound within cultural practices are useful for generating new approaches to analyses. Weinberg (1997) and others have noted that mental health disorders have commonly been positioned as ahistorical, non-human objects that exist outside of the social realm. With the turn to language, and the linguistic paradigm more specifically, there has been a call for greater attention to how language creates reality. In child mental health practice, the implications of this turn towards language is that clinicians no longer need to view mental health illness as a discovered truth, but rather as something that is socially constructed (Walker, 2006).

DA and CA are particularly useful methodological perspectives for understanding the implications of this turn to language, and how such a turn may offer new perspectives on child mental health practices. Similar to Grue (2011), we recognise that disability, and more specifically child mental health, is underexplored with discourse and conversation analytic approaches, and thus there is ample room to generate great understanding of the implications of the turn to language. Both DA and CA, rooted in the hermeneutic research tradition, have a close affiliation with the 'discursive turn' (Tseliou, 2013). The discursive turn, or turn to language, is tightly connected with constructionist epistemological perspectives, which we explore in greater detail next.

Social constructionism

Within qualitative research more generally, there is a broad range of accepted and useful theoretical and epistemological frameworks that guide the work of researchers (see O'Reilly & Kiyimba, 2015, for an overview). Arguably, although slightly contentious, all qualitative research is hermeneutic (Rennie, 1999), and within this hermeneutic perspective, there are a range of positions including those that focus on *language as action*. The aim of this volume is to draw attention to a range of evidence in child mental health from critical, discourse, and/or conversation analytic perspectives, and thus we consider here the broad theoretical framework that many of our contributors broadly subscribe to – social constructionism.

CA and, broadly speaking, all forms of DA share a social constructionist viewpoint in the sense that they see language as context-bound, functional, and constructive (Wetherell, Taylor, & Yates, 2001). Thus, while there are theoretical, epistemological, and ontological differences between these various approaches to analysis (particularly the different types of DA), all of these analytic approaches broadly argue that their view of the world is constructed through language and that language is inherently action based.

Importantly, there is no single position of social constructionism, and in terms of philosophy, social constructionism is considered a loose assembly of a range of approaches that include deconstructionism, critical psychology, post-structuralism, and DA (including discursive psychology) (Burr, 2003). Gergen (2009) argued that there are four key assumptions of social constructionism, which include

1. a radical doubt in the taken-for-granted world;
2. knowledge is viewed as historically, culturally, and socially specific;
3. knowledge is not fundamentally dependent on empirical validity, rather it is sustained by social process; and
4. explanations and descriptions of phenomena are never 'neutral', rather they constitute social actions that serve to sustain certain patterns to the exclusion of others.

Importantly, there are some key distinctions to make at this juncture. First, there are broadly two types of constructionist theory, which include both micro- and macro-social constructionism (Gubrium & Holstein, 2008). Gubrium and Holstein illustrated that microforms of social constructionism tend to be concerned with the microstructures of language, and therefore research underpinned by this position tends to focus on talk, situated interaction, and local culture. However, macro-social constructionism is more broadly concerned with the role that linguistic and social structures have in terms of shaping the social world. Therefore, research grounded in a macro-social constructionist perspective may focus on broader, more macro discourses circulating in society that function to generate and sustain inequities, for instance.

Second, social constructionism should not be confused with social construc-tivism (O'Reilly & Kiyimba, 2015), for these two perspectives hold unique and slightly differing presuppositions. This is a common and frequent error made in the literature, one that we suggest has important implications. Where social constructionism gives emphasis to language and narrative, social construc-tivism is more pertinent in science, technology, and mathematics (Gubrium & Holstein, 2008) and arose in cognitive and developmental psychology (Young & Collin, 2004). Thus, social constructivism often shares the commitment of pos-itivism to a dualist epistemology, whereas social constructionism is culturally and historically specific and emphasises that language constitutes, rather than reflects, reality (Young & Collin, 2004). With social constructionism under-girding DA and CA perspectives (Tseliou, 2013), it is critical to attend to these differences, as they have informed and shaped the development of these analytic approaches.

History and landscape of DA and CA perspectives

The different types of DA and CA have a rich history that grew from the critical turn to language. To contextualise the chapters included in this volume, we provide a brief but important overview of both DA and CA.

A brief history and landscape of DA

The use of DA spans several disciplines, and there is considerable variability regarding the ways in which discourse is conceptualised and analysed (Georgaca, 2014). Generally speaking, however, there are two core, overarching strands of DA. First, there is the post-structuralist or Foucauldian strand. This strand focuses on the socially available discourses that people draw upon when they present their views and experiences, which ultimately shape their experiences and form their subjectivity (Georgaca, 2014). In addition, this strand tends to be macro-social constructionist and has a concern with issues of power and inequality. The second strand is reflected in discursive psychology, which focuses on discursive practices, aligns with the principles of CA, and explores how individuals construct a view of themselves as credible (Willig, 2008). However, an important similarity across the approaches is an agreement that language is central to the meanings of the human world (Spong, 2010).

DA means different things to different researchers (Wodak & Meyer, 2009). It tends to be used as an umbrella term for various analytic approaches to language, and within the tradition of DA, there are several different types including traditional, Bhaktian, Foucauldian, critical, and discursive psychology (O'Reilly & Kiyimba, 2015). While broadly sharing the theoretical social constructionist foundation to their work, they do differ in their epistemological and ontological assumptions. For example, some types of DA are concerned with investigating the role of discourse in the production of power within a particular social structure, to see how discourse can sustain and legitimise social inequality (see Wooffitt, 2005, for a full discussion). These discourse analysts take an explicit socio-political position and focus on the role of discourse in the production of the dominance of elite groups and institutions as they are reproduced by the talk or text and given a sense of legitimacy (van Dijk, 2008). Other forms of DA (such as discursive psychology (DP)) align more closely with the epistemological and ontological positions of CA (described in the following section), which in turn informs their data collection methods and assumptions. These types of DA favour the collection of naturally occurring data due to a belief that any speech act produced can only be analysed meaningfully with reference to the situated nature of the talk (Edwards & Potter, 1992).

A brief history and landscape of CA

Broadly speaking, CA is the study of social interaction. It is a practice designed to examine language in interaction in terms of how the turns of talk are designed to perform a particular social action (Antaki, 2011), and this has been conceptualised as 'talk-in-interaction' (Drew & Heritage, 1992). Thus, while CA is interested in language, its object of study is in the interactional organisation of social activities, specifically in the sense that the production of talk and the sense obtained by utterances are not seen in terms of the structure of language but rather in the practical social accomplishment (Hutchby & Wooffitt, 2008).

CA is a specific and important research technique that examines what people actually do and say, as opposed to what they report that they do and say (McCabe, 2006). The methods of CA are data-driven, and researchers employing this technique favour the use of naturally occurring data as the basis of their work (Mazeland, 2006). Naturally occurring data are those data that would be available even if the researcher had not been born (Potter, 2004); in other words, these are material that would exist regardless of the intervention of the researcher in the everyday setting or institutional context. Naturally occurring data are favoured as they capture the actual interaction and retain the situated nature of the conversation, which have the potential to illustrate how participants orient to their setting without the abstraction of the researchers' agenda (Potter, 2004).

These decisions are steeped in the rich history of CA and its allegiances with other sociological approaches. CA was originally pioneered by Harvey Sacks alongside his colleagues, Gail Jefferson and Emanuel Schegloff. The development of CA was influenced by Goffman's observation of people in interaction and Garfinkel's ethnomethodology and reflected a fusion of these approaches in the creation of an empirical method to explore how people produce social order (McCabe, 2006). Sacks was interested in the organisation of social interaction by investigating mundane talk in naturally occurring environments and the availability of what was then new technology, which allowed for the audio-recording of talk so that a more detailed analysis might take place (Seedhouse, 2004).

In a contemporary research culture, there has been a distinction made between pure CA and applied CA. Pure CA provides an analysis of commonplace and mundane conversations, whereas applied CA investigates institutional settings (McCabe, 2006). In other words, pure CA explores the institution of interaction as an item in its own right, and applied CA explores the management of social institutions in interaction (Heritage, 2004). Importantly, there is some overlap between pure and applied CA, but the distinction is helpful. Notably, applied CA constitutes a large volume of the CA literature

(Antaki, 2011), and indeed many of the chapters in this volume take data from institutional settings to offer insights into the area of child mental health. However, the application of CA to institutional talk is not typically related to solving problems of institutions, but rather it explores how institutions manage to carry out their institutional work successfully (Antaki, 2011).

The transcription system and its symbols

There are many debates regarding the role of transcription in research, but there has been some general agreement that transcription is an active process and that the transcription conventions should reflect the data and their purpose (Lapadat, 2000). While not all forms of DA have widely accepted conventions for transcription, CA does and some types of DA also rely on this system. For CA, transcription is seen as a core analytical activity and is seen as the first step in developing a deeper understanding of the communicative process (Roberts & Robinson, 2004). The transcription conventions are thus an integral and essential part of the approach.

Throughout the history of CA (and for some forms of DA), there has been a growing attention to detail (Heritage & Atkinson, 1984), and the Jefferson system was specifically designed for CA to reflect the analytic stance of the approach (Jefferson, 2004). This system, named the 'Jefferson system of transcription', is a well-developed set of symbols that includes a high level of detail, with the transcript being a representation of the speech (Edwards & Potter, 1992). The key conventions are thus designed to build intuitively on familiar forms of literary notation (Hepburn & Bolden, 2013). The symbols are designed to illustrate how words and phrases were sounded out, and thus in transcribing using the Jefferson conventions, transcriptionists do not correct for grammar or pronunciation (Hepburn & Bolden, 2013). However, the detailed transcription system can make it difficult for those less familiar with it to read through a transcript; so some familiarity can be helpful (Mazeland, 2006). Furthermore, the process of transcribing in such a detailed way is a time-consuming one for analysts with estimates being offered at one hour to transcribe one minute of data (Roberts & Robinson, 2004). Many of the main symbols in the system can be seen in Table 1, which are also included in some of the chapters in this volume.

Relevance of DA and CA to child mental health

As this volume reflects, there is a growing and important body of work that takes a discourse and/or conversation analytic approach to the study of child mental health. This orientation stems from a general questioning and challenging of the traditional biomedical perspective on health and illness. DA and CA are very useful methods to look at the language of mental health and its related interactions in the field (Harper, 1995; McCabe, 2006). This is because

Table 1 Jefferson transcription symbols

Symbol	Explanation
(.)	A full stop inside brackets denotes a micro pause, a notable pause but of no significant length.
(0.2)	A number inside brackets denotes a timed pause. This is a pause long enough to time and subsequently show in transcription.
[ ]	Square brackets denote a point where overlapping speech occurs.
><	Arrows surrounding talk show that the pace of the speech has quickened.
<>	Arrows in this direction show that the pace of the speech has slowed down.
( )	Space between brackets denotes that the words spoken here were too unclear to transcribe.
(( ))	Double brackets with a description inserted denote some contextual information where no symbol of representation was available.
<u>Under</u>	When a word or part of a word is underlined, it denotes a raise in volume or emphasis.
↑	When an upward arrow appears, it means there is a rise in intonation.
↓	When a downward arrow appears, it means there is a drop in intonation.
⟶	An arrow like this denotes a particular sentence of interest to the analyst.
CAPITALS	Where capital letters appear, it denotes that something was said loudly or even shouted.
Hum(h)our	When a bracketed 'h' appears it means that there was laughter within the talk.
=	The equal sign represents latched speech, a continuation of talk.
:::	Colons appear to represent elongated speech, a stretched sound.
¿	A weaker rising intonation is indicated by an inverted question mark.
?	Strongly rising terminal intonation.

psychiatric categories are produced through and within language (Harper, 1995), and the mere notions of sanity and insanity, for instance, are typifications that begin with observation and interaction (Roca-Cuberes, 2008). Thus work from the traditions of DA and/or CA have helped to reframe conceptualisations of mental illness and its management by shifting emphasis from the biomedical to interpersonal and socio-cultural, as well as the ways in which we understand the knowledge and practice of disciplines as social practices (Georgaca, 2014).

These approaches have been instrumental also in helping us understand the institutional practices of mental health by bringing new perspectives to therapy, counselling, and psychiatric care. This is useful because language-based analyses are congruent with therapy, as both take place through language and both focus on meaning-making (McLeod, 2001). While these methods do not inform practitioners on how to conduct therapy effectively, they show how the clients and therapists create the therapeutic process and environment together and exactly how they do this (Streeck, 2010). Further, both DA and CA offer evidence that is tangible, empirical, and useful for the examination of therapeutic change (Strong, Busch, & Couture, 2008), enabling a different view on what actually happens in therapeutic services by looking in detail at the interactions between practitioners and their clients (Spong, 2010).

Of course, DA and CA studies explore mental health and therapy slightly differently. DA, as we have noted, is a diverse field. These studies have tended to focus on the interactional construction of the client's problem and the process of solution through clinical dialogue, as well as looking at the role of wider discourses in shaping the problems that clients bring to therapy (Georgaca, 2014). Some forms of DA have specifically focused on the broader social structures, power relations, and meanings in therapy, and they have explored meaning-making beyond the therapy room to look at the extent of relationships between therapy and external discourses (Spong, 2010). CA research, however, has tended to examine how clinical processes are interactionally constituted in the course of therapy and often have looked at the role of the therapist in shaping the interaction so that it complies with the institutional format of therapy (Georgaca, 2014). Nonetheless, despite the differences, it is becoming increasingly recognised that conversational evidence is a useful resource for enriching practice, and the audio- or video-recording of actual practices offers rich opportunities to see how outcomes are shaped by therapeutic dialogue (Strong et al., 2008) and offer practical recommendations for professionals (see e.g. Parker & O'Reilly, 2012).

Structure of the book

The book is structured to reflect the critical emphasis of the contributors to the field of child mental health, as well as to showcase the contemporary work that is being conducted from a discourse or conversation analytic perspective. This emphasis is greatly enhanced by the range of contributions from practising clinical professionals working in the field of mental health, all of whom offer important insights into the world of mental health practice. To illustrate the various developments in the field of child mental health in relation to DA and CA, the book is organised into six different parts:

1. Part I of the book, 'The Place of Conversation/Discourse Analysis in Child Mental Health', provides contextual information related to the overall aims of this volume. In this part, the contributors consider the value of using discourse or conversation analytic perspectives for the study of mental health. The scope of this part is greatly enhanced by the contributions

from a consultant child and adolescent psychiatrist (Karim) and a consultant clinical psychologist (Kiyimba).

2. Part II of the book, 'A Critical Approach to Child Mental Health', draws together contributions that take a critical view of traditional biomedical discourses. This part of the book benefits from the views of experts in the field who are well known for their critical positions. This part is greatly enhanced from the insights of practising professionals with critical insights into the world of mental health (including Timimi, Roy-Chowdhury, Strong, and Sesma-Varquez).

3. Part III of the book, 'The Social Construction of Normal/Abnormal', illustrates social constructionist views of normality and pathology and considers how these ideas are reified in discourse. This part also benefits from the contributions of trained practitioners who offer insight into this area of mental health and language (see e.g. Karim).

4. Part IV of the book, 'Situating and Exploring Child Mental Health Difficulties', considers the issue of problem construction and understandings of the mental health difficulties experienced by children and adolescents. Within this part, there is a focus on particular types of disorders and how they are constructed through discursive and conversation devices. As with the other parts, this is enhanced by the contributions of practising professionals (including Dallos, Pitt, Karim, and Howard).

5. Part V of the book, 'Managing Problem Behaviour', focuses on some of the techniques employed by professionals and/or parents to influence and manage the symptoms of the mental health disorders of the children. Similar to the other parts, the discussions within this are greatly enriched by the contributions of practising professionals (including Muskett and Schäfer).

6. Part VI of the book, 'Child Mental Health Practice', focuses on areas of psychiatric and therapeutic practice. This explores discourses of therapy and counselling and considers specifically help-seeking behaviours of children and their families. Within this part, contributions from those working in the field of mental health are also included (including Kiyimba).

The book is organised in a way as to guide the reader through some of the core issues raised by DA and CA studies, as well as related theoretical discussions. Each chapter attends to the practical implications and recommendations of the work for areas such as clinical practice, educational settings, or social work and social care. This practical focus is facilitated by a 'practice highlights' box at the end of each chapter, along with recommended reading lists for those who wish to explore related concepts beyond this volume. We also provide a glossary at the end of the volume to explain some of the more technical terms used within the chapters. For ease of reference, we provide here a clear and succinct abstract for each chapter, which includes a useful summary of the contents and messages from each contributor.

Chapter 1: The Value of Conversation Analysis for the Study of Children's Mental Health

Alessandra Fasulo

This chapter describes the different ways in which research in CA can expand our understanding of children's mental health and support professionals' work in the area. It reviews existing conversational research in the domains of children's spontaneous communication, the interactional functioning of psychological assessments, talk in consultation and intervention settings, and, more briefly, the effect that parents' mental health conditions may have on children. For each domain, Fasulo reports examples from key studies, explains the relevant methodological issues, and illustrates how a conversation analytic approach can change the way we view and define children's mental health problems, as well as the psychological answers to them. Finally, she points out lines of research that are still open for investigation and potential areas for collaboration with practitioners.

Chapter 2: The Value of Conversation Analysis: A Child Psychiatrist's Perspective

Khalid Karim

Research into child mental health is positioned in medical thinking, and while medical sociology has been influential, the influence of CA in this area has only emerged recently. Karim comes from a rare position, as a practising child and adolescent psychiatrist and someone engaged in partnership with conversation analytic researchers. The chapter provides an overview of child mental health research from scholars in the field, and Karim reflexively develops a more personal argument regarding how this type of 'evidence' might be of use to those who practise child psychiatry. In the chapter, Karim critically reflects on the research done in fields such as psychiatry, counselling, therapy, and child helplines to explore how the method has particular potential to advance these fields.

Chapter 3: The Value of Discourse Analysis: A Clinical Psychologist's View

Nikki Kiyimba

As a practising clinical psychologist and experienced discourse analyst, Kiyimba offers a refreshingly candid and practical appraisal of how DA can provide a workable and desirable solution to the problem of seeking to better understand the therapeutic process in child mental health. She skilfully expounds on the importance of examining the seeming mysteries of what makes a good therapeutic alliance by taking a systematic qualitative analytic stance. Kiyimba argues that in doing so, therapeutic effectiveness can be enhanced by understanding the discourse and constructs that are representative and meaningful to clients. Pragmatically, Kiyimba concludes by arguing that good, quality analysis is best tied to good clinical applicability through the development of meaningful academic–clinical partnerships.

Chapter 4: Looking or Spotting: A Conversation Analytic Perspective on Interaction between a Humanoid Robot, a Co-present Adult, and a Child with an ASC

Paul Dickerson and Ben Robins

Dickerson and Robins examine video data of interactions involving a humanoid robot, a child with an autistic spectrum condition, and a co-present adult. These data are analysed from a conversation analytic perspective, with particular attention to the ways in which examining sequences of interaction may provide a markedly different reading of the data to that derived from an analysis guided by some form of checklist. The analysis presented here suggests that spotting behaviours in isolation can be a blunt instrument, which relies on and reproduces generalised conceptions of the child's pathology, rather than opening up an examination of how the child's behaviour may relate to the specific sequence of interaction. Some of the clinical implications of these findings are considered.

Chapter 5: ADHD: Three Competing Discourses

Adam Rafalovich

Rafalovich presents a DA of the attention deficit hyperactivity disorder (ADHD) phenomenon from three major perspectives, which includes the medical, psychodynamic, and sociological viewpoints. On the medical front, ADHD has an important presence stretching back to the late 19th century, which has focused upon the increasing interest of medicine in childhood deviance, academic performance, and life chances of children. The psychodynamic viewpoint essentially argues that patterns of disruptive childhood behaviour emanate from interactive dynamics within the family and a lack of psychological well-being in children. From the sociological view, ADHD is a product of a variety of social forces that have created a type of perfect storm within which the ADHD diagnosis has gained momentum over time, and therefore an increasing legitimacy. In providing a DA of ADHD, the chapter focuses upon the various conversations that have given rise to ADHD and raised its salience in lay, clinical, and academic circles.

Chapter 6: Discourses on Children's Mental Health: A Critical Review

Tom Strong and Monica Sesma-Vazquez

This chapter presents a critical review of how children's 'mental health, 'mental illness', and 'well-being' have been socially constructed at different moments in history. Strong and Sesma-Vazquez examine different conceptualisations of 'childhood well-being' and tensions across, within, and between the most prevalent contemporary and historical discourses in theory, popular media, and the research literatures regarding children's mental health, mental illness, and well-being. They consider how controversies and differences arise and are sometimes resolved within each of these contexts. Against this backdrop of controversies and contested constructions of childhood well-being, the authors identify practices that stabilise dominant discourses of childhood well-being at cultural, institutional, and familial levels of interaction. They invite readers to critically reflect upon how uses of language can optimise responsiveness to children while avoiding forms of 'discursive capture'.

Chapter 7: Child Mental Health: A Discourse Community

Stephen Gilson and Elizabeth DePoy

In recent times, discursive analytic epistemologies have become central tools within the interdisciplinary field of disability studies. Theorising disability as linguistic artefact has been potent in wrestling atypical embodiment away from its medical deficit prison and repositioning it as a socially and culturally constructed phenomenon. Gilson and DePoy assert that delimiting communication to discourse, conversation, and humans interacting with humans is, however, incomplete. The authors propose that meaning-making and its productions in cultural expectations of development, normalcy, health, and illness is a complex interactive discourse community in which language, object, materiality, and interpretation dance in synergy. They conclude the chapter by illustrating the potency of this analytic framework for rethinking and redesigning not only responses to the mentally disabled child but to reconceptualising childhood disability itself such that profound diversity depth is nurtured within contemporary cultures.

Chapter 8: The Social Construction of Attention Deficit Hyperactivity Disorder

Sami Timimi and Lewis Timimi

Timimi and Timimi examine the social construction of ADHD by looking at its history, its cross-cultural aspects, and its construction through commonly available Internet searches. The diversity of findings contrasts to its narrow construction in developing public discourses. They illustrate the contrast between attempts to identify commonalities, in order to 'squeeze' diverse cultural beliefs and practices around children and child development into simplistic Western-ised categories like ADHD, and more anthropologically informed research that explores these diversities. Thematic content analysis of commonly cited web pages on ADHD is used to understand the type of discourses members of the UK public are likely to encounter when carrying out their own searches on this topic.

Chapter 9: Moral Evaluations in Repertoires of ADHD

Mary Horton-Salway and Alison Davies

Horton-Salway and Davies' critical discursive psychology analysis demonstrates how the category of ADHD is produced in media representations and accounts of parents who have a child with a diagnosis of ADHD. The chapter identifies biological and psychosocial explanations of ADHD that are used to construct moral evaluations about parents and their children and position them in neg-ative ways. Dominant media representations align ADHD with bad parenting, and parents manage a dilemma of positioning either their children as biologi-cally abnormal and in need of medication or themselves as failing families in need of social interventions. The chapter discusses how parents manage these repertoires in order to represent themselves as a competent parenting team.

Chapter 10: Leaving Melancholia: Disruptive Mood Dysregulation Disorder

Valerie Harwood

Harwood focuses on a new child depressive disorder described in *DSM-5* (American Psychiatric Association, 2013). Termed 'disruptive mood dysregulation disorder', this is a depression filled with energy and agitation. This child disorder departs from the most commonly understood character of depression: lifelessness and low energy. The chapter draws on two striking lit-erary and artistic figures, Melancholia and Orestes, to examine this change. Melancholia is arguably familiar to us with a characteristic immobile and downward-looking figure. Orestes, on the other hand, is a figure that at times might remind us of melancholia and, at others, is startlingly energetic and agi-tated. The chapter makes the case that conceptual changes are occurring in how 21st-century depression is understood and raises the question: Will tantrums in children come to be interpreted as signs for the potential of adult depression?

Chapter 11: Why Does a Systemic Psychotherapy 'Work'?

Sim Roy-Chowdhury

Among psychological treatments often used with children, systemic psy-chotherapy occupies an unusual, if not a unique, position by virtue of the fact that its theoretical base leans heavily upon a social constructionist epistemology. Roy-Chowdhury explores the question of the means by which such a therapy might be evaluated that is congruent with its own epistemological assumptions. In so doing, some answers are provided to the question of what a social constructionist psychotherapy looks like in practice. In choosing DA as a vehicle for the evaluation, the adequacy of a research method, itself resting upon social constructionist foundations, could also be tested in answering research questions that have become the preserve of quantitative studies. The findings are discussed in relation to systemic psychotherapy theory and the means by which a psychotherapy that does not embrace the notion of an objective, non-contingent reality can be put into practice.

Chapter 12: Red Flags: The Social Construction of a Symptom

David C. Giles

In this chapter, Giles explores the contribution that online materials have made to the construction of symptoms in autism in conjunction with, and sometimes in contradiction to, the clinical literature. Symptoms are typically represented as 'red flag' behaviours for parents to be concerned about, though these are sometimes hard to distinguish from typically eccentric behaviours of young children. Nevertheless, the lack of scientific evidence does not prevent some of these behaviours from acquiring symptom status, and Giles' analysis looks at the ways in which this is worked up by lay and professional discourse online to create a 'scare' around autism that has parents posting videos of their children on YouTube in order to demonstrate the 'signs' to watch out for.

Chapter 13: The Production of the 'Normal Child': Exploring Co-constructions of Parents, Children, and Therapists

Charlotte Brownlow and Andrea Lamont-Mills

Brownlow and Lamont-Mills explore how positions of 'normal' and 'abnormal' are discursively co-constructed between parent, child, and therapist in real-life clinical interactions. Drawing on clinical consultation sessions, the chapter charts the development of shared understandings of what is 'normal' for children and explores how some productions of 'normality' are more privileged over others. The differing levels of power within the triadic clinical relationship are explored, and the techniques drawn on by members to resist and endorse particular positions are considered. The authors argue that the context of clinical consultations are important for the exploration of the negotiation of positionings, not least given the ready framework provided by the *DSM* for abnormality/normality constructions and the problematisation of abnormality that pervades most psychological encounters.

Chapter 14: Should Autism Be Classified as a Mental Illness/Disability? Evidence from Empirical Work

Michelle O'Reilly, Khalid Karim, and Jessica Nina Lester

There has been great tension regarding whether autism is actually a disability, whether it truly constitutes a mental disorder. In the chapter, O'Reilly and colleagues explore some of the arguments regarding whether autism is or is not misclassified. To achieve this, the authors explore the medical ways of thinking about autism and consider the alternative ideas that have been presented over time. Most specifically, they explore the contribution of social constructionism in challenging the positivist positioning of autism. Using empirical evidence from previous work, they consider the ways in which biomedical thinking has come under criticism and present some of the potential counterarguments to the idea that autism is a disability. The chapter also presents arguments from the autistic community, including those voices which advocate that the condition can be disabling.

Chapter 15: Subjectivity in Autistic Language: Insights on Pronoun Atypicality from Three Case Studies

Laura Sterponi, Kenton de Kirby, and Jennifer Shankey

In this chapter, we examine a language phenomenon typically associated with autism: pronoun reversal and avoidance. These features are traditionally considered pathognomonic of the condition and symptomatic of underlying psychological abnormalities, notably an impaired interpersonal perspective-taking. We report on findings from a comparative case study of the person reference repertoire of three 6-year-old children with autism. Both typical and atypical forms are considered in their spontaneous deployment in everyday verbal interaction with family members and tutors. Our findings suggest that multiple explanatory components, notably contextual, pragmatic, and psychological factors, may result in pronominal atypicality in autism. In particular, we show that (1) pronoun reversal/avoidance may be related to specific conversational contexts rather than being solely the by-product of underlying difficulties inherent to the autistic condition. (2) The child with autism deploys language to address the very difficulties that partly manifest in his communicative behaviour.

Chapter 16: Normative Development and the Autistic Child

Lindsay O'Dell and Charlotte Brownlow

In this chapter, the authors draw on data from online discussion forums for people with autism, parents of people with autism, and professionals to explore the concept of normative development and the comparisons of children with autism to this benchmark. Drawing on a position of critical developmental psychology, the authors interrogate the reification of a universal and 'natural' trajectory of child development and the production of the 'normal' child against which children with autism are measured and seen to be deficient. O'Dell and Brownlow demonstrate the power of developmental and clinical discourse within discussions of autism and 'normality'. The authors further postulate an alternative discourse of development that serves to challenge the taken-for-granted 'natural' trajectory of child development and the place that autism occupies within this construction.

Chapter 17: A Conversation Analysis of the Problem Presentation Phase of Initial Assessment Appointments in a Child and Adolescent Mental Health Service

Victoria Stafford and Khalid Karim

Stafford and Karim focus on the problem presentation phase of initial assessment appointments in a Child and Adolescent Mental Health Service (CAMHS). The variety of ways children present their reason for attendance is explored using CA, as well as how these reasons are responded to by clinicians. Stafford and Karim demonstrate that children present their reason for attendance at CAMHS in three ways: (1) they give a technical mental health reason for their attendance; (2) they give a lay explanation of the difficulties they are experiencing; and (3) they respond with variations of 'I don't know', indicating they have insufficient understanding to give an explanation. These responses are discussed in terms of their clinical relevance, with emphasis on the application of successful child-centred practice.

Chapter 18: The Discursive Construction of Problem Behaviours of Children with Autism in Therapy

Jessica Nina Lester and Michelle O'Reilly

Lester and O'Reilly illustrate the ways in which 'problematic' and 'acceptable' behaviours are socially constructed within the interactions between therapists and children diagnosed with autism spectrum disorders (ASD). They draw upon audio and video data of therapy sessions designed to teach children with ASD how to behave in 'socially appropriate' ways. With a broad focus on how 'appropriate' social behaviours were negotiated and constructed within this setting, they attend to how the lexical items (i.e. *Superflex* and the *unthinkables*) functioned to define and mark what counted as an appropriate behaviour, who was accountable for changing the 'inappropriate' behaviour, and what or who was responsible for causing inappropriate behaviour. They conclude by highlighting the implications for clinicians who engage in the delicate task of naming a child's behaviour 'appropriate' or 'inappropriate'.

Chapter 19: Name-Calling by a Child with Asperger's Syndrome

Johanna Rendle-Short, Ray Wilkinson, and Susan Danby

Rendle-Short, Wilkinson, and Danby show how social interaction is directly relevant to maintaining friendships, mental health and well-being, and supportive peer relations. Using CA, the chapter focuses on conversational participants' pursuit of affiliation and intimacy from a language as action perspective. It focuses on the use of derogatory naming practices by a 10-year-old girl diagnosed with Asperger's syndrome. The analysis shows how derogatory address terms, part of a wider pattern of behaviour evident in this child's interaction, result in behaviour that might be thought of as impolite or lacking in restraint. It also illustrates how a single case study can draw attention to the context-specific nature of interaction when working with children with Asperger's syndrome. The chapter contributes to our understanding of the difficulty in pinpointing, with precision and with clear evidence, what counts as a 'social interaction difficulty' due to the context-specific nature of interaction.

Chapter 20: Attachment Processes and Eating Disorders in Families: Research and Clinical Implications

Rudi Dallos and Sarah Pitt

This chapter offers an analysis of the development of eating disorders in families which are drawn from the authors' clinical and research experience. It is argued that anorexia is connected to the complex and frequently negative attachment experiences that many parents have experienced in their childhoods. Paradoxically, many parents are attempting to offer a better emotional experience for their own children, but given the lack of their own experience of comfort and protection, parents can find that their good intentions can become thwarted. A case study is presented which draws on clinical research with a family and attempts to illustrate how parents' attachment experiences shape the family dynamics in terms of their understandings of their problems and the resulting family dynamics. Implications for clinical work are drawn together from this illustrative case study and the author's previous clinical experience and research with eating disorders.

Chapter 21: Using Discourse Analysis to Study Online Forums for Young People Who Self-Harm

Janet Smithson

In the chapter, Smithson provides a DA to study talk by young people on online self-harm forums. The norms and assumptions of forum talk, what they understand and accept as appropriate forms of support and advice, and which aspects of their behaviour and relationships with parents and healthcare professionals problematise are investigated. Data from three sites are compared: an existing, moderated forum; a research-led forum; and an unmoderated, 'pro' self-harm site. Analysis focuses on three aspects of the forum talk. Young people put effort into providing peer support at any hour, often sensibly, therefore providing a safety net at critical times. Self-harming is routinely normalised, with triggers, and reactions and relationships with professionals and parents being constructed as the problems to address.

Chapter 22: Using Conversation Analysis for Understanding Children's Talk about Traumatic Events

Amanda Bateman, Susan Danby, and Justine Howard

Bateman, Danby, and Howard provide a theoretical overview of literature that uses CA to explore children's interactions related to trauma and associated mental health matters. The relatively new approach of using CA to understand trauma reveals the importance of talk in the process of recovery, as well as *how* the participants co-construct talk about traumatic experiences. The chapter explores literature using a CA approach to investigate children's trauma talk with professionals, as well as literature specifically discussing children's talk about their traumatic experiences with people who are not qualified therapists or psychiatrists. They conclude by calling for more research using a CA approach for investigating children's traumatic experiences due to the insight it provides into each child's personal sense-making of traumatic events with a range of people.

Chapter 23: Food, Eating, and 'Eating Disorders': Analysing Adolescents' Discourse

Maxine Woolhouse and Katy Day

Woolhouse and Day argue for the merits of feminist discourse analytic approaches to adolescents and 'eating disorders'. They first critique psychological theories, arguing these to be limited in their failure to examine cultural discourses and socio-political conditions that shape adolescents' eating practices. Moving on to review feminist discursive research on adolescents and eating, they reveal how eating disordered identities and practices are tied to culturally available discourses. Considering the implications of feminist discursive research, Woolhouse and Day conclude by advocating politically informed interventions rather than conventional psychomedical treatments.

Chapter 24: Presuming Communicative Competence with Children with Autism: A Discourse Analysis of the Rhetoric of Communication Privilege

Jessica Nina Lester

Lester explores the ways in which parents and therapists negotiate the meaning of communication with minimally verbal and non-verbal children with autism.

Findings from a discursive psychology study of the interactions of children with autism and their parents and therapists are presented. Lester provides a detailed analysis of three representative extracts from interviews with parents and therapists of children with autism and two representative extracts from audio and video data of therapy sessions with children with autism. More particularly, Lester discusses two discursive patterns: (1) the positioning of non-verbal behaviours as meaningful and (2) co-constructing legitimate communication. The chapter concludes by discussing how privileging non-normative ways of communicating, in everyday clinical practices, makes evident the tensions between the goal of assisting children to become more functional and able to navigate the world outside of therapy setting and the goal of and commitment to self-expression.

Chapter 25: Parents' Resources for Facilitating the Activities of Children with Autism at Home

Monica Ramey and John Rae

Ramey and Rae aim to contribute to our understanding of parents' resources for supporting children with autism in taking part in joint activities at home. In order to examine such interactions between children and their parents, they use CA. The data are drawn from three families, and the authors distinguish different forms of parental participation. Fundamentally the parent's task, as in much social interaction, is in deciding how to co-participate with the unfolding events. The analysis considers resources for (1) directing the child's attention (2) prompting and pursuing responsive actions and (3) relating to task-related contingencies and opportunities. In these settings, these events are material activities that have their own structures – orientation to this is a lively concern in these interactions.

Chapter 26: Managing and Normalising Emotions and Behaviour: A Conversation Analytic Study of ADHD Coaching

Louise Bradley and Carly W. Butler

Bradley and Butler analyse the interactional practices that underpin the delivery of a cognitive-behavioural programme for children with a diagnosis of ADHD. Focusing on the methods and practices used to deliver the programme, the chapter examines the normal and ordinary ways in which the children's emotions and experiences are constructed, how the children are treated as *experts* to co-produce shared knowledge and understandings of ADHD, and how cognitive tools are implemented and made available for the children to use in their everyday lives to manage their emotions and behaviours. As well as looking at the interactional practices, issues of self-concept and identity are considered to address these under-researched concerns within the ADHD literature.

Chapter 27: Interlocutor Influence on the Communication Behaviours Associated with Selective Mutism

Hanna Schäfer and Tom Muskett

In this chapter, Schäfer and Muskett present a CA case study on a child diagnosed with selective mutism. Turn-by-turn analysis of participant talk and embodied action indicates that the child's interactional conduct varies strikingly depending upon whether an expectation to speak was established through immediately preceding interlocutor turns. This finding provides empirical support for the conjecture that interlocutor conduct shapes the 'problematic' behaviours associated with selective mutism, and it implies that the child may experience a highly context-specific anxiety associated with particular forms of interactional demands. As is discussed, this observation may usefully contribute to treatment planning for diagnosed individuals and has implications for the classification of selective mutism within psychiatric nosologies.

Chapter 28: 'I'm Happy with Who I Am': A Discursive Analysis of the Self-Characterisation Practices of Boys in 'Behaviour' Schools

Linda J. Graham

Previous research has found that children and young people with behavioural disorders tend to exhibit positive illusory bias in their perceptions of self, which is believed to interfere with their ability to modify negative traits. While extensive, this research risks exacerbating already impoverished and limiting views of these young people. It also neglects to consider the powerful role that language plays in shaping perception. In this chapter, Graham draws on post-structural theories of language and DA to examine the self-characterisation practices of 33 boys attending special schools for students with disruptive behaviour. Findings suggest that rather than reflecting illusory bias, positive self-characterisations may instead reflect who the young person wants to be and the image of themselves that they wish others had.

Chapter 29: Therapeutic Vision: Eliciting Talk about Feelings in Child Counselling for Family Separation

Ian Hutchby

In this chapter, Hutchby describes how CA can be applied to study interactions between counsellors and young children. He draws upon a British corpus of counselling sessions for young children (4–12 years) experiencing family separation. In the chapter, Hutchby explores the means by which counsellors manage their 'professional vision' in discussing children's responses to problematic family circumstances. The analysis focuses on the ways that conversational sequences are designed to foreground the child's experiential response to a particular discursive scenario, while also encoding a model of the child as autonomous, socially and cognitively competent, speaking subject. A range of sequence types are considered in relation to managing talk about feelings in this environment.

Chapter 30: Parents' Resistance of Anticipated Blame through Alignment Strategies: A Discursive Argument for Temporary Exclusion of Children from Family Therapy

Nikki Kiyimba and Michelle O'Reilly

The literature on family therapy highlights that there is a general consensus that it is preferable to include children in family therapy sessions. In this chapter, Kiyimba and O'Reilly present a case for the deliberate exclusion of children in the initial stages of a series of therapeutic sessions. The purpose of temporary exclusion is for building alignments between adult parties to facilitate the inclusion of children in later sessions. We work with naturally occurring family therapy data as the basis of our argument. Data used illustrate that where unreflective inclusion occurs, parents typically position the child as the problem in a negative and derogatory way, they 'gossip' about their children as if they are absent, and talk about topics which could be considered developmentally inappropriate for children to hear.

Chapter 31: Parentification: Counselling Talk on a Helpline for Children and Young People

Susan Danby, Jakob Cromdal, Johanna Rendle-Short, Carly W. Butler, Karin Osvaldsson, and Michael Emmison

This chapter investigates counselling interactions where young clients talk about their experiences of taking on family responsibilities normatively associated with parental roles. In research counselling literature, practices where relationships in families operate so that there is a reversal of roles, with children managing the households and caring for parents and siblings, is described as parentification. 'Parentification' is used in the counselling literature as a clinician/researcher term, which we 'respecify' (Garfinkel, 1991) by beginning with an investigation of young clients' own accounts of being an adult or parent and how counsellors orient to these accounts. In addition to providing understandings of how young people propose accounts of their experiences of adult–child role reversal, the chapter contributes to understanding how children and young people use the resources of counselling helplines, as well as how counsellors can communicate effectively with children and young people.

Chapter 32: *'And You? What Do You Think Then?'* Taking Care of Thought and Reasoning in Intellectual Disability

Marilena Fatigante, Saverio Bafaro, and Margherita Orsolini

Fatigante, Bafaro, and Orsolini present a case study of the interactions between a target adolescent, Davide, with a diagnosis of mild intellectual disability, and adult therapists in assessment and therapeutic contexts. The study assumes a Deweyan approach to reflective thinking and analyses how the child's reasoning abilities display in conversation. Their analyses show that the adult–child discourse in these settings (1) enhances the child's argumentative abilities and (2) solicits the child to engage in a more positive consideration of how his mind works. The conclusions target the importance of providing children and adolescents with intellectual disability – who less likely engage in rich and complex social interaction – and opportunities to participate in conversation, as the natural site for developing sense-making tools and coping with self-ascription of incompetence.

Chapter 33: 'You Just Have to Be Cheerful Really': Children's Accounts of Ordinariness in Trauma Recovery Talk

Joyce Lamerichs, Eva Alisic, and Marca Schasfoort

The chapter by Lamerichs, Alisic, and Schasfoort examines children's accounts of ordinariness during psychological research interview about trauma recovery. Furnished with mentions of 'just', it explores how children's tellings work to resist the specific category bound implications of the interviewer's questions: that on the basis of their experience with a traumatic occurrence the child has special knowledge to draw upon and give advice to a friend in similar circumstances. The chapter demonstrates how mentions of 'just' construct the advice children eventually provide as nothing special and as the kind of universally knowable, generic suggestions everybody would offer. It is important that professionals are aware that children's 'just'-prefaced answers are related to question format and the identity membership categories that are made relevant in their questions.

References

American Psychiatric Association (2013). *Diagnostic and statistical manual of mental disorders* (5th edition). Washington, DC: APA.

Antaki, C. (2011). Six kinds of applied conversation analysis. In C. Antaki (Ed.), *Applied conversation analysis: Intervention and change in institutional talk* (pp. 1–14). Basingstoke: Palgrave Macmillan.

Burr, V. (2003). *Social constructionism* (2nd edition). London: Routledge.

Crowe, M. (2000). Constructing normality: A discourse analysis of the DSM-IV. *Journal of Psychiatric and Mental Health Nursing, 7,* 69–77.

Drew, P., & Heritage, J. (1992). Analyzing talk at work: An introduction. In D. Drew & J. Heritage (Eds.), *Talk at work: Interaction in institutional settings* (pp. 3–65). Cambridge: Cambridge University Press.

Edwards, D., & Potter, J. (1992). *Discursive psychology*. London: Sage.

Fee, D. (2000). The broken dialogue: Mental illness as discourse and experience. In D. Fee (Ed.), *Pathology and the postmodern: Mental illness as discourse and experience* (pp. 1–17). London: Sage.

Garfinkel, H. (1991). Respecification: Evidence for locally produced, naturally account-able phenomena of order*, logic, reason, meaning, method, etc. in and as of the essential haecceity of immortal ordinary society, (I) – an announcement of studies. In G. Button (Ed.), *Ethnomethodology and the human sciences* (pp. 10–19). Cambridge: Cambridge University Press.

Georgaca, E. (2014). Discourse analytic research on mental distress: A critical overview. *Journal of Mental Health, 23*(2), 55–61.

Gergen, K. (2009). *An invitation to social constructionism* (2nd edition). Thousand Oaks, CA: Sage.

Grue, J. (2011). Discourse analysis and disability: Some topics and issues. *Discourse and Society, 22*(5), 532–546.

Gubrium, J., & Holstein, J. (2008). The constructionist mosaic. In J. Holstein & J. Gubrium (Eds.), *Handbook of constructionist research* (pp. 3–12). New York: Guildford.

Harper, D. (1995). Discourse analysis and 'mental health'. *Journal of Mental Health, 4,* 347–357.

Hepburn, A., & Bolden, G. (2013). The conversation analytic approach to transcription. In T. Stivers & J. Sidnell (Eds.), *The Blackwell handbook of conversation analysis* (pp. 57–76). Oxford: Blackwell.

Heritage, J. (2004). *Garfinkel and ethnomethodology*. Cambridge: Polity Press.

Heritage, J., & Atkinson, J. M. (1984). Introduction. In J. M. Atkinson & J. Heritage, (Eds.), *Structures of social action: Studies in conversation analysis* (pp. 1–16). Cambridge: Cambridge University Press.

Hutchby, I., & Wooffitt, R. (2008). *Conversation analysis* (2nd edition). Cambridge: Polity Press.

Jefferson, G. (2004). Glossary of transcript symbols with an introduction. In G. H. Lerner (Ed.), *Conversation analysis: Studies from the first generation* (pp. 13–31). Amsterdam: John Benjamins.

Jensen, L. E. (2004). Mental health care experiences: Listening to families. *American Psychiatric Nurses Association, 10*(33), 33–41.

Lapadat, J. (2000). Problematizing transcription: Purpose, paradigm and quality. *International Journal of Social Research Methodology, 3*(3), 203–219.

Mazeland, H. (2006). Conversation analysis. In K. Brown (Ed.), *Encyclopaedia of language and linguistics* (2nd edition, Vol. 3, pp. 153–162). Oxford: Elsevier Science.

McCabe, R. (2006). Conversation analysis. In M. Slade & S. Priebe (Eds.), *Choosing methods in mental health research: Mental health research from theory to practice* (pp. 24–46). Hove: Routledge.

McLeod, J. (2001). *Qualitative research in counseling and psychotherapy*. London: Sage.

Oliver, M. (1983). *Social work with disabled people*. London: Macmillan.

Oliver, M. (1993). Re-defining disability: A challenge to research. In J. Swain, V. Finkelstein, S. French, & M. Oliver (Eds.), *Disabling barriers-enabling environments* (pp. 61–68). London: Sage.

O'Reilly, M., & Kiyimba, N. (2015). *Advanced qualitative research: A guide to contemporary theoretical debates.* London: Sage.

Parker, N., & O'Reilly, M. (2012). 'Gossiping' as a social action in family therapy: The pseudo-absence and pseudo-presence of children. *Discourse Studies, 14*(4), 1–19.

Peters, S. (2010). Qualitative research methods in mental health. *Evidence-Based Mental Health, 13*(2), 35–40.

Potter, J. (2004). Discourse analysis as a way of analysing naturally occurring talk. In D. Silverman (Ed.), *Qualitative research: Theory, method and practice* (2nd edition) (pp. 200–221). London: Sage.

Rennie, D. (1999). *Using qualitative methods in psychology.* Thousand Oaks, CA: Sage.

Rhodes, P. (2011). Why clinical psychology needs process research: An examination of four methodologies. *Clinical Child Psychology, 17*(4), 495–504.

Roberts, F., & Robinson, J. (2004). Interobserver agreement on first-stage conversation analytic transcription. *Health Communication Research, 30*(3), 376–410.

Roca-Cuberes, C. (2008). Membership categorisation and professional insanity ascription. *Discourse Studies, 10*(4), 543–570.

Seedhouse, P. (2004). Conversation analysis methodology. *Language Learning 54*(s1), 1–54.

Spong, S. (2010). Discourse analysis: Rich pickings for counsellors and therapists. *Counselling and Psychotherapy Research, 10*(1), 67–74.

Streeck, U. (2010). A psychotherapist's view of conversation analysis. In A. Peräkylä, C. Antaki, S. Vehvilainen, & I. Leudar (Eds.), *Conversation analysis and psychotherapy* (pp. 173–187). Cambridge: Cambridge University Press.

Strong, T., Busch, R., & Couture, S. (2008). Conversational evidence in therapeutic dialogue. *Journal of Marital and Family Therapy, 34*(3), 388–405.

Tseliou, E. (2013). A critical methodological review of discourse and conversation analysis studies of family therapy. *Family Process, 52*(4), 653–672.

Van Dijk, T. (2008). *Discourse and power.* Basingstoke: Palgrave Macmillan.

Walker, M. T. (2006). The social construction of mental illness and its implications for the recovery model. *International Journal of Psychosocial Rehabilitation, 10*(1), 71–87.

Weinberg, D. (1997). The social construction of non-human agency: The case of mental disorder. *Social Problems, 44*(2), 217–234.

Wetherell, M., Taylor, S., & Yates, S. (Eds.) (2001). *Discourse theory and practice: A reader.* London: Sage.

Willig, C. (2008). *Introducing qualitative research in psychology: Adventures in theory and method* (2nd edition). Maidenhead: McGraw-Hill/Open University Press.

Wodak, R., & Meyer, M. (2009). *Methods of critical discourse analysis* (2nd edition). London: Sage.

Wooffitt, R. (2005). *Conversation analysis and discourse analysis: A comparative and critical introduction.* London: Sage.

World Health Organisation (2001). *International classification of functioning, disability and health: Final draft.* Retrieved December 16, 2014 from http://www.who.int/icidh.

World Health Organization (2005). *Child and adolescent atlas: Resources for child and adolescent mental health.* Geneva: World Health Organization.

Young, R., & Collin, A. (2004). Introduction: Constructivism and social constructionism in the career field. *Journal of Vocational Behavior, 64*, 373–388.

Recommended reading

- Antaki, C. (2011). Six kinds of applied conversation analysis. In C. Antaki (Ed.), *Applied conversation analysis: Intervention and change in institutional talk* (pp. 1–14). Basingstoke: Palgrave Macmillan.
- Georgaca, E. (2014). Discourse analytic research on mental distress: A critical overview. *Journal of Mental Health, 23*(2), 55–61.
- O'Reilly, M., & Kiyimba, N. (2015). *Advanced qualitative research: A guide to contemporary theoretical debates.* London: Sage.
- Potter, J. (2004). Discourse analysis as a way of analysing naturally occurring talk. In D. Silverman (Ed.), *Qualitative research: Theory, method and practice* (2nd edition, pp. 200–221). London: Sage.
- Spong, S. (2010). Discourse analysis: Rich pickings for counsellors and therapists. *Counselling and Psychotherapy Research, 10*(1), 67–74.

Acknowledgements

We would like to extend our gratitude to several people who have supported us during the development of this Handbook. We are particularly thankful to Claire Bone who has offered comments on technical aspects and undertaken some of the proofreading for us. We would also like to acknowledge Palgrave for supporting us throughout this process, particularly Nicola Jones, Libby Forrest, and Eleanor Christie, who encouraged us throughout. We also thank our partners, families, and colleagues who have supported us throughout the process.

More specifically, the quality of the material in this book has been assured through the peer-review process and the support of a range of experts in mental health and/or DA and CA. Each chapter was subject to external feedback from either academics or clinical practitioners, and their useful insights helped the authors to improve and develop their chapters prior to inclusion. We are grateful to these reviewers for the time they took out of their busy schedules to do this task for us:

Alvina Ali is a consultant child and adolescent psychiatrist at Leicestershire NHS Partnership Trust, UK.

Tim Auburn is a Senior Lecturer at the Faculty of Health and Human Sciences, University of Plymouth, UK.

Claire Bone is an Assistant Psychologist with Cygnet Healthcare, Derbyshire, UK.

John Cromby is a Senior Lecturer in Psychology, in the School of Management at the University of Leicester, UK.

Sushie Dobbinson is a speech and language therapist and forensic specialist at the Humber Centre, UK.

Hannah Dostal is Assistant Professor of Literacy Education at the University of Connecticut, USA.

Kathy Evans is Assistant Professor of Special Education at Eastern Mennonite University, USA.

Rachael Gabriel is Assistant Professor of Reading Education at the University of Connecticut, USA.

Hilary Gardner is Senior Lecturer in Human Communication Sciences at the University of Sheffield, UK.

Andrea Garner is a PhD candidate in the Centre for Health Initiatives at the University of Wollongong, Australia.

Eugenie Georgaca is an Assistant Professor in the Faculty of Philosophy at the Aristotle University of Thessaloniki, Greece.

Elizabeth Hale is a chartered health psychologist at Russells Hall Hospital, Dudley, UK.

T.A. McDonald is a PhD candidate at the University of Wisconsin, USA.

Pallab Majumder is a consultant child and adolescent psychiatrist at Nottingham Healthcare NHS Trust, UK.

Peter Muntigl is an Assistant Professor in the Department of Linguistics at Gent University, Belgium, and an Adjunct Professor in the Faculty of Education at Simon Fraser University, Canada.

Ottar Ness is an Associate Professor at Buskerud and Vestfold University College, Norway.

Emily Nussbaum is an Assistant Professor at the College of Education, University of Colorado, USA.

Alison Pilnick is Professor of Language, Medicine and Society at the University of Nottingham, UK.

Marco Pino is a Research Fellow in the School of Health Sciences at the University of Nottingham, UK.

Elizabeth Price is a special education programme administrator in Southwest Washington, USA.

Emma Rich is a Senior Lecturer in the Department of Education at the University of Bath, UK.

William (Bill) Rocque is an Associate Professor in the Department of Sociology and Anthropology at the University of Redlands, USA.

Pablo Ronzoni is a consultant child and adolescent psychiatrist at the 2gether Foundation Trust, Hereford, UK.

Kathryn Runswick-Cole is Senior Research Fellow in Disability Studies and Psychology at Manchester Metropolitan University, UK.

Chloe Shaw is a Research Associate at the Institute for Women's Health, University College London, UK.

Trini Stickle is a PhD candidate in the Department of English Language and Linguistics at the University of Wisconsin-Madison, USA.

Wyke Stommel is Assistant Professor of Language and Communication at Radbound University, Nijmegen, the Netherlands.

Laura Thompson is a researcher at the Unit for Social and Community Psychiatry, Queen Mary University of London, UK.

Eleftheria Tseliou is an Assistant Professor at the Department of Early Childhood Education, University of Thessaly, Greece.

Panos Vostanis is a consultant child and adolescent psychiatrist at Leicestershire NHS Partnership Trust and a professor at the University of Leicester, UK.

Sally Wiggins is a Senior Lecturer at the School of Psychological Sciences and Health, University of Strathclyde, UK.

Val Williams is Head of Norah Fry Research Centre, School for Policy Studies, University of Bristol, UK.

One reviewer preferred to remain anonymous.

Contributors

Editors

Jessica Nina Lester is Assistant Professor of Inquiry Methodology at Indiana University, USA. She teaches research methods courses, and her research focuses on the study and development of qualitative methodologies. She situates her research within discourse studies and disability studies, with a particular focus on education and mental health contexts. She has also published books related to qualitative methodologies and research practices, including *Digital Tools for Qualitative Research* (co-authored).

Michelle O'Reilly is a senior lecturer at Greenwood Institute of Child Health, University of Leicester, UK. She also provides research support to practising clinical professionals working for Leicestershire NHS Partnership Trust. Her research interests are broadly in the areas of child mental health, psychiatric research, family therapy, and qualitative methods. She has published many papers and books in this area, including *Doing Mental Health Research with Children and Adolescents: A Guide to Qualitative Methods* (co-authored).

Contributors

Eva Alisic directs the Trauma Recovery Lab at Monash University in Melbourne, Australia. The lab uses a variety of methods (e.g. epidemiological, meta-analytic, psychometric, observational, and qualitative) to unravel the mechanisms that underlie children's and families' responses to potentially traumatic events. For example, with the Electronically Activated Recorder that works on iPods, the team audio samples parent–child communication shortly after trauma and studies how these interactions relate to children's long-term well-being. A key paper of the lab is 'Building Child Trauma Theory from Longitudinal Studies: A Meta-Analysis'.

Saverio Bafaro is a clinical psychologist. His interests focus on the analysis of interaction in institutional contexts, such as therapeutic interactions with developmentally disabled children and medical consultations. Currently, he is practising in the field of clinical psychology for psychosomatic disturbances. He wrote on medicine and multiculturality on the 2012/2013 special issue of *Salute e Società* (Franco Angeli).

Amanda Bateman is Senior Lecturer in Early Childhood Education and is a member of the Early Years Research Centre at the University of Waikato, New Zealand. She has published in the area of early childhood peer interactions and teacher–child interactions using conversation analysis and membership categorisation analysis. She has led two funded projects and an international collaborative project investigating the impact of the Christchurch earthquakes on the children living there.

Louise Bradley holds a PhD from the Department of Social Sciences at Loughborough University, UK. Her thesis, titled 'Supportive Practices in Interactions with Children that Promote a Positive Sense of Self', uses discursive psychology and conversation analysis to examine the ordinary conversational practices that professionals use to support children with social, emotional, and behavioural difficulties.

Charlotte Brownlow is a senior lecturer in the School of Psychology, Counselling and Community at the University of Southern Queensland, Australia. Her research interests focus on understandings of diversity and difference and the impacts that constructions of these have on the crafting of individual identities, particularly for individuals identifying as being on the autism spectrum.

Carly W. Butler is a Senior Lecturer in Social Psychology in the Department of Social Sciences at Loughborough University, UK. Her research involves using ethnomethodological and conversation analytic methods to examine naturally occurring social interaction, with a focus on children's play and interaction, family interaction, and helpline interaction. She is the author of *Talk and Social Interaction in the Playground* and founding editor of the forthcoming journal *Research on Children's Social Interaction* (first issue in 2016).

Jakob Cromdal is Professor of Educational Practice in the Department of Social and Welfare Studies, Linköping University, Sweden. His research focuses on talk and social interaction among children and youth in a variety of mundane and institutional settings, including playgrounds, classrooms, detention homes, and calls to the emergency services.

Rudi Dallos is Professor and Research Director of the Clinical Psychology training programme at the University of Plymouth, UK. He practises as a family therapist and consults to a range of clinical settings, including child sexual abuse, couples therapy, and work with trauma. He has written a number of papers specifically relating to eating disorders, and his most recent book is *Attachment and Family Therapy* (with Pat Crittenden).

Susan Danby is Professor of Early Childhood at Queensland University of Technology (QUT), Australia, and Health, Wellbeing and Happiness Program Leader at Children and Youth Research Centre at QUT. She researches social interaction in home and school settings, helpline talk, childhood studies, and qualitative methodologies, including ethnomethodology and conversation analysis. She recently co-edited *Disputes in Everyday Life: Social and Moral Orders of Children and Young People* (2012).

Alison Davies holds a PhD from The Open University, UK, and now works as an academic consultant in psychology. Her thesis, titled 'It's a Problem with the Brain: A Discursive Analysis of Parents' Constructions of ADHD', takes a critical discursive psychological approach to the talk of parents of children with an ADHD diagnosis. She works as a consultant and a therapist, and currently her research focuses on narratives of women with ADHD.

Katy Day is a Senior Lecturer in Psychology at Leeds Beckett University, UK. A critical social and health psychologist, she is interested in the use of post-structuralist feminist theory to understand gender- and health-related phenomena. In recent years, her publications in the areas of gender, food, and eating have appeared in *Feminism & Psychology, Journal of Health Psychology*, and *Journal of Gender Studies*, including the article 'Starving in Cyberspace: A Discourse Analysis of Pro-eating Disorder Websites' as well as contributing chapters to a number of edited texts.

Elizabeth DePoy is Professor of Interdisciplinary Disability Studies and Social Work at the University of Maine, USA. Alongside Stephen Gilson, she has published extensively in the fields of disability studies and design. Her most recent book, co-authored with Gilson, *Design and Branding Disability*, proposes disability as a construction of design and branding and calls for the use of five principles to guide contemporary design and market strategies to provoke social change. Responding to technological hegemony, Elizabeth is now integrating post-humanism into her work.

Paul Dickerson is a Principal Lecturer in Psychology at the University of Roehampton, UK, where he has taught qualitative research and social psychology for more than 20 years. He is the author of *Social Psychology: Traditional and Critical Perspectives* (2012). His research principally draws on conversation analysis and is particularly concerned with gaining a better understanding of interactions in which one or more of the participants might be described as having a communicative impairment.

Michael Emmison is an honorary reader and research fellow in the School of Social Science, University of Queensland, Australia. He has worked in several fields, principally language and social interaction, visual research methods, and the sociology of culture. His current research is primarily concerned with the analysis of talk in institutional settings and targeting activities such as help-seeking and advice-giving on a variety of telephone helplines. Previous publications include the co-edited collection *Calling for Help: Language and Social Interaction in Telephone Helplines* (2005).

Alessandra Fasulo is a Senior Lecturer in Psychology at the University of Portsmouth, UK. With a focus on interactional and embodied processes, she has done research on children in the family and school settings, learning disabilities, psychotherapy, and the workplace. She is also interested in narrative and autobiographical discourse. She has co-edited the book *Marked Identities: Narrating Lives between Social Labels and Individual Biographies* (with R. Piazza, 2014, Palgrave Macmillan).

Marilena Fatigante is an Assistant Professor of Social Psychology at the University of Rome La Sapienza, Italy. Her interests range from the analysis of social interaction in both ordinary (e.g. family) and institutional (e.g. doctor–patient psychotherapy) domains to ethical aspects of ethnography and qualitative research. Recent publications include 'Information Giving and Enactment of Consent in Written Consent Forms and in Participants' Talk Recorded in a Hospital Setting' in *Human Studies* (2014).

David C. Giles is Reader in Media Psychology at the University of Winchester, UK. He is the author of *Psychology of the Media* (Palgrave Macmillan, 2010) and has published numerous articles on the analysis of online discussion, particularly in the field of mental health. His central research interest is on the media and mental health, as well as being interested in psychological research methods more generally.

Stephen Gilson is Professor of Interdisciplinary Disability Studies and Social Work at the University of Maine. He has published extensively in the fields of disability studies and design. His most recent book is *Design and Branding Disability* (co-authored with Elizabeth DePoy). In this book the authors propose that disability is a construction of branding and design and they call for the use of a range of principles to guide contemporary design and market strategies to promote social change. Responding to technological hegemony, Stephen is now applying post-humanism into his research on advancing human flourishing.

Linda J. Graham is a principal research fellow at the Faculty of Education at QUT. She is the editor of the *International Journal on School Disaffection* and editor-in-chief of *The Australian Educational Researcher*. Her work focuses on institutional contributions to disruptive student behaviour and the improvement of responses to children who are difficult to teach. Currently, she is leading a prospective longitudinal study of children's school-liking, language, learning, and behaviour in the early years of school.

Valerie Harwood works at the School of Education, University of Wollongong, Australia, where she researches issues that impact inclusion and exclusion from education. In 2013, she was awarded a prestigious Australian Research Council's Future Fellowship to pursue research on educational inclusion and early childhood. Her most recent book, co-authored with Julie Allan, is *Psychopathology at School: Theorising Mental Disorders in Education*. Her earlier book *Diagnosing Disorderly Children* (2005) was awarded runner-up for the UK Times Higher Education NASEN Academic Book Award (2006).

Mary Horton-Salway is an academic consultant in psychology who has led modules in social psychology and discourse analysis at The Open University and published on discourse and the social psychology of health, illness, and disability. Her publications include 'Gendering ADHD: A Discursive Analysis of UK Newspaper Stories' in *Journal of Health Psychology*, published online on 1 October 2012.

Justine Howard is an associate professor and postgraduate programme manager at the Centre for Children and Young People's Health and Wellbeing, Swansea University, UK. She is a chartered psychologist specialising in child development and is trained in developmental and therapeutic play. She has published many books, book chapters, and journal articles focusing on the role of play in children's health, care, and development. She is regularly invited to speak on these topics both nationally and internationally. Her most recent books are *The Essence of Play* and *Play Therapy Today*.

Ian Hutchby is Professor of Sociology at the University of Leicester, UK. His research interests are in language and social interaction in institutional, healthcare, and mediated settings. He specialises in conversation analysis and has published a wide range of books and papers, applying this method to the understanding of lay–professional, adult–child, and conflictual communication. He is the author of *The Discourse of Child Counselling* (2007).

Khalid Karim is both a consultant child and adolescent psychiatrist at Leicestershire NHS Partnership Trust and a senior teaching fellow at the

University of Leicester. He has extensive experience of working with children who have mental health problems and specialises in children with autism. He recently co-authored the book *A Practical Guide to Mental Health Problems in Children with Autistic Spectrum Disorder: 'It's Not Just Their Autism!'* (with Alvina Ali and Michelle O'Reilly).

Kenton de Kirby is a PhD candidate at the Graduate School of Education, University of California, Berkeley, USA. His research interests are varied, encompassing mathematics education, culture–cognition relations, neuropsychology, autism, and language-in-social interaction. His publications likewise extend to each of these topics, which include a co-authored monograph and essays in *Autism* and *Wiley Interdisciplinary Reviews: Cognitive Science*.

Nikki Kiyimba is both a chartered clinical psychologist working for the Cheshire and Wirral Partnership NHS Trust and a senior lecturer at Chester University, UK. She has experience with working with children in care, children of prison inmates, and children at risk of school exclusion. Her research interests are in using discourse and conversation analysis to study therapeutic interactions with adults and children. She is a co-author (as Nikki Parker) of *Doing Mental Health Research with Children and Adolescents: A Guide to Qualitative Methods*.

Joyce Lamerichs is a Lecturer in Health Communication at the VU University in Amsterdam. She uses conversation analysis and discursive psychology to study the interactional practices of health professionals and their clients in different (mediated) settings. She aims to feed the results from her work back to professional's interactional practices and has developed the Discursive Action Method as a stepwise approach for diagnosis and reflection. She has co-authored the chapter 'Reflecting on Your Own Talk: The Discursive Action Method at Work' in the edited book *Applied Conversation Analysis*.

Andrea Lamont-Mills is a senior lecturer in the School of Psychology, Counselling, and Community at the University of Southern Queensland and a registered psychologist. Using discursive psychology, her research focuses on the discourses used in real-world clinical and clinical training settings. In particular, she researches how taken-for-granted therapeutic activities are achieved, co-constructed, and situated through client–health professional interactions.

Tom Muskett is a senior lecturer at Leeds Beckett University, UK. He is a qualified speech and language therapist with a particular interest in applying conversation analysis to clinically relevant data. His current research aims to re-examine mainstream accounts of communication development and disorder by combining interactional analysis with insights from diverse fields, including critical psychology and psychiatry, clinical linguistics, and social studies of disability and childhood.

Lindsay O'Dell is Director of Post Graduate Studies in the Faculty of Health and Social Care at The Open University. Her research interests focus on children and young people who are in some way 'different', including neurological difference, young carers, and language brokers. She is a co-editor of the journal *Children & Society* and recently co-authored ' "Hard Wired at the Factory"? Autism as a Form of Biological Citizenship' in *Critical Autism Studies: Enabling Inclusion, Defending Difference* (edited by M. Orsini and J. Davidson, 2012).

Margherita Orsolini is Professor of Developmental Psychology at the University of Rome La Sapienza and Director of the Centre for the Prevention and Treatment of Learning Disabilities, Italy. Her research focuses on the development of language, the role of adult–child interaction in the development of discourse and cognition, and the assessment and treatment of cognitive impairments in mild intellectual disability. She has edited *Quandoimparare è piùdifficile: Dallavalutazioneall'intervento* (When learning is more difficult: From assessment to intervention) (2011).

Karin Osvaldsson is an Associate Professor of Child Studies at Linköping University, Sweden. She holds a PhD in Child Studies from Linköping University in 2002. Working within the main framework of ethnomethodology and discursive psychology, she is engaged in research on identity and social interaction in various settings, including detention homes for troubled youth, emergency rescue services, and Internet and telephone counselling organisations.

Sarah Pitt is a clinical psychologist within Avon and Wiltshire Partnership Trust in Bristol in a medium-secure unit for women and a psychiatric intensive care unit for women. She qualified as a clinical psychologist in 2012 at the University of Plymouth. She has experience working within an eating disorders service and has an interest in family processes and anorexia. She has previously published research conducted in cognitive remediation and anorexia.

John Rae is Reader in Psychology and Director of the Centre for Research in Cognition, Emotion and Interaction in the Department of Psychology, University of Roehampton, London. His empirical research concerns talk and body movement in social interaction. In addition to everyday interactions, he is interested in interactions involving persons with a challenged capacity (e.g. people with aphasia or with an autism spectrum disorder) and in tele-mediated interactions and computer-mediated communications.

Adam Rafalovich is an Associate Professor of Sociology at Pacific University, Forest Grove, Oregon. He holds a PhD from the University of British Columbia, Canada, which was a social analysis of the response of adult authorities to children suspected of having attention deficit hyperactivity disorder. This informed his book *Framing ADHD children* (2004). He has published widely in the area of medical sociology, with articles appearing in the *Journal for the Theory of Social Behaviour, Deviant Behavior, Sociology of Health and Illness*, and *Symbolic Interaction*.

Monica Ramey completed her BSc in Psychology at the University of Roehampton, UK, in 2013. She is currently seeking to complete an MSc in Applied Behaviour Analysis at The Tizard Centre, University of Kent, UK. She aims to conduct a work-based learning project applying research-based methods of observing, recording, and analysing behaviours with individuals with autism spectrum disorders (ASD), to develop positive behaviour support plans enhancing learning and quality of life. Her main areas of interest are social interactions and learning in individuals with ASD.

Johanna Rendle-Short is a senior lecturer at the Australian National University. She utilises the methodology of conversation analysis or talk-in-interaction within a variety of contexts, including language and learning, media studies, and children and adults who are communicatively impaired. She is particularly interested in how children with Asperger's syndrome or high-functioning autism communicate with those around them, both at home and in the school environment.

Ben Robins is a senior research fellow at the University of Hertfordshire, UK. He completed his PhD in 2005, focusing on assistive technology for children with autism. His publications have won several best conference paper awards. His research started with the AUROA Project and continued in the FP6/7 European projects IROMEC and ROBOSKIN, investigating the potential use of robots as therapeutic/educational tools, encouraging basic communication and social interaction skills in children with autism. Recently, he was programme co-chair, committee member, and special session organiser in several international conferences and has been an invited speaker internationally.

Sim Roy-Chowdhury is a clinical psychologist and a systemic psychotherapist. He is a Clinical Director within East London Foundation Trust and an Honorary Senior Lecturer with the Doctoral Programme in Clinical Psychology at the University of East London. His PhD research analysed family therapy sessions using discourse analysis. He is interested in developing an understanding of subjectivity and intersubjectivity which takes into account wider social, political, and cultural contexts.

Hanna Schäfer is a speech and language therapist working in a private practice in Hamburg, Germany. She completed an undergraduate degree in speech and language therapy at the Hogeschool van Arnhem en Nijmegen, University of Applied Sciences, the Netherlands, and holds an MSc in Human Communication Sciences from the University of Sheffield. Her research focus is on interactions involving persons with communication difficulties with a particular research interest in conversation analysis.

Marca Schasfoort is a part-time Lecturer in Conversation Analysis at the VU University, Amsterdam. In her teaching she focuses on talk in institutional settings. She is also an independent established coach and trainer in several organisations in the Netherlands.

Monica Sesma-Vazquez is a professor at Universidad de las Americas, Mexico City, and also holding a postdoctoral fellowship at the University of Calgary, Canada. She is a psychotherapist and supervisor interested in collaborative and social constructionist practices (principally in therapy, education, supervision, and research). Her focuses are on how therapists, counsellors, and psychiatrists understand the influences of psychiatric/diagnostic culture and 'expert' discourses on clients, as well as how they respond to client's psychiatric/psychological understandings in therapeutic spaces.

Jennifer Shankey (1988–2014) was a research associate at the Graduate School of Education, University of California, Berkeley. Her research interests centred on the communication and social interaction of children with autism. She complemented her academic endeavours with clinical work with those children. Her publications include co-authored papers published in *Autism* and *Journal of Child Language*.

Janet Smithson is a social psychologist and a senior lecturer at the School of Psychology, University of Exeter, UK. She has worked on a variety of national and European funded research projects. Her research interests include gender and discourse, work-life practices and policies, life course transitions, qualitative methodologies, and Internet-mediated communication. She is currently studying mediation and alternative dispute resolution after family breakdown, as well as interaction on online forms for young people who self-harm.

Victoria Stafford is a PhD student at the University of Leicester. Her research interests include mental health, child mental health interactions, and qualitative methods. Her project is a conversation analytic study that examines child mental health assessments and family interactions. She is supervised by Michelle O'Reilly and Ian Hutchby and her article has appeared in *Clinical Child Psychology and Psychiatry* on children's accounts of attending the mental health service.

Laura Sterponi is an associate professor at the Graduate School of Education, University of California, Berkeley. She has developed a research programme that is centrally concerned with the role of language and literacy practices in children's development and education. Her studies have examined communicative practices in both typical and atypical children. Her work on autism aims to illuminate the interactional matrix of key features of autistic communication, such as echolalia and pronominal reversal/avoidance. Her work has appeared in *Autism, Human Development, Discourse Studies, Linguistics & Education*, and *Journal of Child Language*.

Tom Strong is a professor, couple and family therapist, and counsellor-educator at the University of Calgary, who researches and writes on the collaborative, critically informed, and practical potentials of discursive approaches to psychotherapy. With Andy Lock, he has co-authored *Social Constructionism: Sources and Stirrings in Theory and Practice* (2010) and co-edited *Discursive Perspectives in Therapeutic Practice* (2012). His current research focuses on medicalising tensions in counsellor education.

Lewis Timimi is a student at The Priory Academy LSST in Lincoln, UK. The chapter that he has co-authored with Sami Timimi in this volume includes research he conducted as part of an Extended Project Question (EPQ) that he recently completed.

Sami Timimi is a consultant child and adolescent psychiatrist and a visiting professor at the University of Lincoln, UK. He writes from a critical psychiatry perspective on topics relating to mental health and has published more than 100 articles, including in leading journals, and many book chapters. His publications include *Naughty Boys: Anti-Social Behaviour, ADHD and the Role of Culture* (Palgrave Macmillan, 2005) and *The Myth of Autism: Medicalising Men's and Boys' Social and Emotional Competence* (Palgrave Macmillan, 2010, co-authored). He has co-edited three books, including, with Carl Cohen, *Liberatory Psychiatry: Philosophy, Politics and Mental Health* (2008).

Ray Wilkinson is Professor of Human Communication at the University of Sheffield, UK. His work focuses on the analysis of everyday social interactions, particularly of people with communication disorders. In addition to investigating the impact of communication disorders on interactions and aspects of relationships and social identity constituted through these interactions, his work has used such analyses as the basis for planning, implementing, and evaluating training programmes which aim to improve everyday social interaction.

Maxine Woolhouse is Senior Lecturer in Psychology at Leeds Beckett University, UK. Her research and teaching focuses mainly on critical social and health psychology and qualitative research methods. In particular, she is interested in discursive approaches to understand how gender and social class intersect to shape and inform so-called 'normal' and 'disordered' eating and body management practices. This is reflected in the co-authored article ' "Cos Girls Aren't Supposed to Eat like Pigs Are They?" Young Women Negotiating Gendered Discursive Constructions of Food and Eating'.

Abbreviations

ABA	Applied Behavioural Analysis
ADHD	Attention Deficit Hyperactivity Disorder
ADOS	Autism Diagnostic Observation Schedule
ANT	Attachment Narrative Therapy
APA	American Psychological Association/American Psychiatric Association
ASC	Autistic Spectrum Condition
ASD	Autistism Spectrum Disorder
CA	Conversation Analysis
CAMHS	Child and Adolescent Mental Health Services
CDA	Critical Discourse Analysis
DA	Discourse Analysis
DAM	Discursive Action Model
DMDD	Disruptive Mood Dysregulation Disorder
DP	Discursive Psychology
DSM	*Diagnostic and Statistical Manual of Mental Disorders*
GMC	General Medical Council
GP	General Practitioner (the general family doctor in the United Kingdom)
MH	Mental Health
NICE	National Institute of Clinical Excellence
NHS	National Health Service
OCD	Obsessive Compulsive Disorder
PIB	Positive Illusory Bias
PTSD	Post-Traumatic Stress Disorder
RAPID	Reasoning and Problem Solving for Inattentive Detectives
RCT	Randomised Controlled Trial
SAPOL	South Australian Police
SCIB	Sexual Crime Investigation Branch
SLT	Speech and Language Therapist
SM	Selective Mutism
SPICC	Sequentially Planned Integrative Counselling for Children

SRP	Standard Relational Pair
TCU	Turn Construction Unit
UK	United Kingdom
USA	United States of America
WHO	World Health Organization
ZPD	Zone of Proximal Development

Part I

The Place of Conversation/Discourse Analysis in Child Mental Health

1

The Value of Conversation Analysis for the Study of Children's Mental Health

Alessandra Fasulo

Introduction

Children's mental health is a growing concern for both healthcare institutions and academic research, driven by the recognition that mental health issues also affect the younger members of society and that the incidence of these problems is increasing (www.mentalhealth.org.uk). Children's mental health problems can include anxiety, conduct disorder, post-traumatic stress disorder, and depression; however, diagnoses are always advanced with caution as children are highly responsive to changes in their environment and their reactions can be extremely variable and fluid. Intellectual disabilities are also often considered under the umbrella of children's mental health, partly because of the impact that the disability itself may have on the child's psychological well-being. In the following, I will provide a review of research studies conducted within a conversation analysis (CA) framework across the whole range of children's mental health and intellectual disabilities, trying to illustrate how and why CA can be a useful methodological approach for the study of these fields.

CA and the study of interaction

CA is a discipline that works at capturing and describing the details of social interaction. It was developed originally within sociology by Harvey Sacks and his colleagues as an inquiry into the coordination of social conduct; its focus is not language per se or the contents exchanged through talk, but how people know *when* to speak, *what* to say, and *how* to say it in relation to what the other speakers are doing.

Although it has developed a distinct set of methodological practices, CA is in the first place a theoretical approach to social interaction. It considers inter-action as the site where 'social facts' – family, institutions, identities – are produced; these are local and emergent accomplishments, and there is no set of formalised rules that can predict or direct the way people behave *in situ*. CA also assumes that talk-in-interaction is a separate system from written language and its grammar, but that, being highly ordered, it can be investigated for its systematic properties as well. The CA approach is now shared across the human and social sciences, with linguists, anthropologists, and, to a lesser extent, psychologists adopting this framework for the investigations of human conduct from their own disciplinary perspective (for a review of the relation of CA to different disciplines, see Sidnell & Stivers, 2012).

The strength of CA is that the analysis is rooted in what speakers themselves make of the talk of their fellow interactants; for the interpretation of a turn-at-talk, the analyst looks at the design and content of the successive conversational moves, which display the co-participants' understanding of the previous ones (Sacks, 1992). This is a radical innovation compared to previous models for the analysis of language, most of which had looked at sentences in isolation and had tried to identify regularity in the way fixed grammatical structures mapped onto function (Maynard, 2012). On the contrary, CA has revealed that form alone cannot determine use (e.g. a question can be asked with a sentence in declarative form, i.e. a non-interrogative form, and sentences with interrogative form can be used for purposes other than asking, such as criticising; see Levinson, 1983); the position within a sequence of turns is what makes a certain form apt and its purpose clear, therefore the sequence is the unit the analyst has to consider.

Typically, CA starts with what has been called 'unmotivated inquiry', that is to say just searching through data for any new interactional phenomenon that might be there. As Sacks (1992) said, what is to be found in naturalistic interaction cannot be easily imagined, as our conscious representations of how language works have only a remote relationship with what actually happens. Evidence is collected through detailed analysis of sequences, considering all components, that is, words as well as any hearable sound, features of speech such as intonation and pace, silences, and gestures. These features then enter the analysis by means of a specific transcription notation (Jefferson, 2004). Single cases can be sufficient to illustrate a phenomenon; however, more often, after a practice is identified, a 'collection' of similar instances is gathered across different data sets of it in order to get a more secure grip on the phenomenon and its interpretation. Integral to this process is the practice of 'data sessions' in which a researcher shares data with other conversation analysts in a free discussion setting, so that alternative or divergent interpretations can arise.

CA relies on the fact that the main concern of participants in an interaction is to achieve intersubjectivity, namely to align their understandings of what is happening moment-by-moment and to fulfil each other's expectations about what is going to happen next. Speakers are accountable for failures in providing the conditions for intersubjectivity, and conversational practices are finely tuned to this necessity. Research has shown, for example, that turn-beginnings are shaped so as to facilitate the anticipation of what will come next, that speakers use a variety of techniques to smoothe turn transition, and that a spectrum of repair moves can be put in place when the alignment is lost (Atkinson & Heritage, 1984; Schegloff, 2007; ten Have, 1999). The characteristics of the CA approach described above represent a significant advantage in the study of people with communicative impairments or other forms of difficulties, as the analysis starts from the actual interaction and focuses on the resources participants draw upon to achieve intersubjectivity, independently of whether there are violations of given linguistic rules or the deployment of unusual communicative means.

The extensive knowledge developed in CA about the functioning of ordinary conversation has been used as a benchmark for the study of a variety of activity settings, including tribunals, schools, high-technology environments, healthcare, and psychotherapy (Antaki, Vehviläinen, & Leudar, 2008; Heath & Luff, 2000; Heritage & Clayman, 2011; Llewellyn & Hindmarsh, 2010). Comparing practices found in specific institutions or groups with analogous ones used in ordinary conversation (Drew, 2003) helps understanding core aspects of those social settings and possible sources of tensions and misunderstandings and it can therefore be particularly useful for the study of institutions and settings involving children with difficulties.

Conversation Analysis of children's interaction

CA looks at interactions involving children with the same unpresuming gaze it bestows on conversations with adult participants: in other words, what qualifies an interaction as specifically parent–child or child–child should be apparent in the features of the interaction itself (Forrester, 2010). Furthermore, developmental explanations for children's ways of talking are not favoured over more contingent accounts of children's situation and interests;[1] the analysis can show instead how the position and status of children are created through interactional practices in the family and other social environments.

Wootton (1997), studying family interactions around his young daughter between ages 1 and 3, illustrates how conversational competence is formed: the appearance of novel language practices in the child's talk could mostly be related to her recent interactions with family members. As Wootton explains, 'there is a predominant sense that the child is making developmental headway by actively assembling various orders of sense in the world to which she is being exposed' (1997, p. 2). The general lesson of his series of studies on this child is that children learn language as part of their means to operate successfully in their world and are therefore sifting and combining socio-cultural and linguistic competence as part of the same process; at the same time, the studies demonstrate that through conversational means children can adjust very quickly to different contexts and interactional formats. Another major finding of Wootton's work is that children have a paramount interest in intersubjectivity, made dramatically visible by the daughter's distress at not being understood by adults, whereas being able to participate in the social life around her and effectively direct the actions of others were major motivations for the child to engage and experiment with language.

Very recently, CA research has started to look at interactions with infants at a very early age (Berducci, 2010; Rączaszek-Leonardi, Nomikou, & Rohlfing, 2013). It has been long known from developmental studies that infant–caregiver exchanges rely on a turn-taking organisation since the first weeks after birth (Kaye & Brazelton, 1971), and that infants can be active participants in routine activities based on sequential patterns (Reddy, 2008; Trevarthen, 1979). By applying CA methodology and terminology to early interactions, researcher have now started to build a continuum with what we know of conversational organisation in general, shedding light on how intersubjectivity and the principles of sequentiality begin to be established in the first months of life.

On the whole, CA studies with children show that applying a rigorous sequential analysis to children's spontaneous interaction gives access to their sense-making procedures and sheds light on behaviours that may otherwise seem unjustified and incomprehensible. This also creates a favourable terrain for the study of children with difficulties. In the following, I will review existing conversational research in four general domains relevant to the understanding of children and mental health. The first is the study of children's communication per se, exploring not only how particular mental health conditions can affect speech but also what kind of interactional dynamics are generated around the perceived problem. The second domain concerns psychological assessment as the set of institutional practices devoted to defining a child's mental health condition and/or disability. The third is the study of consultation and intervention settings, in which professionals interact with children according to different psychological theories and approaches. A fourth domain that will be discussed more briefly, as less interactional research has been done about it, is the conversational environment of children whose parents suffer from mental health conditions.

Interaction between children with difficulties and their caregivers

A main focus of conversational research in this area has been the study of conditions involving language impairments. Autism[2] is probably the most represented condition (as the chapters in this volume reveal; see e.g. Ramey & Rae, Chapter 25, this volume; Rendle-Short, Danby, & Wilkinson, Chapter 19, this volume) followed by Williams syndrome (Tarlin, Perkins, & Stojanovik, 2006), Down syndrome (Peskett & Wootton, 1985), and other types of linguistic or pragmatic impairment.

Research on clinical infants is still rather limited, but promising studies are looking into early differences in the behaviour of small children with a suspected diagnosis of autism: Esposito, Nakazawa, Venuti, and Bornstein (2013) found that the crying of 13-month-old infants with autism has shorter and fewer utterances (i.e. cry emissions between in-breaths) than those of typically developing children and is heard as expressing more distress; as infants' cry utterances orient caregivers' responses (Berducci, 2010), Esposito et al.'s findings support social learning hypotheses for at least part of the difficulties of children with autism, namely that initial altered behaviours can limit their opportunities of learning communicative and interactional skills (Trevarthen, Aitken, Papoudi, & Robarts, 1998).

A study by Wells, Corrin, and Local (2008) is a particularly good example of an interactional approach to the study of children with difficulties. The authors perform a phonetic analysis comparing the conversational prosody of a typically developing child, a child with specific language impairment, and a child with severe autism. By focusing on different aspects of prosody (design, placement, and focus), they are able to show that the child with autism, who had the most severe difficulties and was almost unintelligible, retained those prosodic features more conducive to successful turn-taking and 'sacrificed' the focus system which is less central (and, in fact, is not even present in all languages). As the authors say:

> Because the method does not assume that unusual prosody is a direct reflex of an underlying processing deficit, it also enables the exploration of compensatory mechanisms.
>
> (Wells et al., 2008, p. 149)

The passage above emphasises that the aim of this kind of research is neither to identify behaviours that may be caused by a faulty neurological substratum, nor to draw the dividing line between children with difficulties and their typically developing peers. By setting aside the deficit as the main focus, children's unusual speech behaviours may be revealed as solutions to communicative obstacles rather than manifestation of the underlying condition.

The study also proves that it is worth considering children's talk in its sequential context: widening the observational angle to include speech partners makes children's verbal behaviour a lot easier to decipher, and helps identifying its interactional orientation. Research on echolalia in autism offers ample evidence of this. Echoes, that is, repetitions of previously memorised content or formulaic phrases by children with autism, were interpreted as signs of withdrawal until sequential analyses showed that they actually indicate interactional engagement. Children with autism find it difficult to improvise newly formed utterances at the pace required by conversation, so they may use this type of talk to fill a conversational slot; by performing slight adjustments in form or prosody, they shape echoes into adaptive responses to the other's turn (Local & Wootton, 1995; Sterponi & Shankey, 2014; Tarplee & Barrow, 1999; Wootton, 1999). The following extract shows such use of repetition for the achievement of 'progressivity' (Schegloff, 2007), namely a well-paced flow of turns between conversants. Aaron, a 5 years and 10 months old boy with high-functioning autism, is playing with his mother on the parents' bed; at the bed's bottom there is the blanket where they let their dog (Yachi) sit (the full sequence is analysed in Sterponi & Fasulo, 2010).

```
1   Mother:    [OH MY GO:SH. LOOK what you did
               ((with mock terrified voice))
2   Aaron:     Hehe [heh hehe hehuhe ((laughs))
3   Mother:         [YOU GOT ON THE BAD BED ((tickling Aaron))
4   Aaron:     Hehe hiih ((rolling on bed))
5   Mother:    You're gonna get bug bites. From Yachi.
6   Aaron:  →  He ↑huhu ((laughs)) (.) or else?
7   Mother:    Or e:lse you're gonna be covered in bug bites
               ((tickling Aaron's bare legs))
8   Aaron:     Hehe hehu hehu ((laughs))
9              no seat on the [bug bed,
10  Mother:                   [Yach- Yachi's bug blanket.
11  Aaron:  →  I'll sit on Yachi's bug bed or else?
12  Mother:    You're gonna get covered with bug bites.
13             ((tickling Aaron's legs))
14  Aaron:     Ha ha ha ha hu ha
               ((laughs with pillow on mouth x))
15             you'll be covered in bug bites.
16  Mother:    Yeah.
```

In line 6, Aaron produces his characteristic echo, the question 'or else', in response to Mom's playful scary scenario; the question gives the floor back to her and acts as an invitation to go on playing. She takes up the invitation and continues with a variant of the former menacing line. The boy laughs again and starts proffering a kind of prohibition, partially repeating what his mother had said in a previous turn (line 9). The mother then completes Aaron's turn in overlap and the boy picks up her words to construct a future event, to which he attaches again the question 'or else' (line 11). The formulaic question appears effective one more time in making his mother continue playing: she tickles him and reiterates the bug line, making Aaron laugh and repeat her threat (lines 14–15).

Even in this brief sequence, it is apparent that the child is fully engaged in the interaction and mobilises the resources he has available to keep the exchange going: questions have the advantage of giving the floor back to the other speaker (Sacks, Schegloff, & Jefferson, 1974) while fulfilling one's interactional duty. A question like 'or else' is not very specific, so it can be bent for use in a wide variety of conversational contexts; furthermore, it may trigger jokes and invention in a complacent interlocutor. The formulaic, repetitive contribution of the child does signal that he has a problem, but the analysis shows that it is not a random nonsensical contribution.

Focusing on the functional aspects of children's atypical communication might help build a new grammar for atypical language uses. Wootton (2002) poses the question of whether interactional approaches have the power to find independent explanations for the characteristics of a condition, different from those provided by deficit-oriented approaches. While this may only be ascertained in the future through prolonged and concerted efforts, research is already mapping alternative language structures and advancing hypotheses on how interactional problems can generate cumulative effects in children's competence across the developmental trajectory. For example, Wootton himself (2002) suggests that the observed scarcity of initiations (conversational moves that open a new sequence) in 'pragmatically unusual' children deprives them of the information that can be gained through repairs and other type of responses about the adequacy of their turns. The atypically low frequency of initiatives, whichever the cause, Wootton argues, may determine the more severe language impairments observed at later stages and that are often interpreted, instead, as directly caused by neurophysiological anomalies. In other words, exploring impaired communication in interaction and comparing atypical children behaviour with that of their typical counterparts – the comparative method being central to the CA approach – can build better theories on the nature of children's problems and devise strategies to compensate for the lack of communicative experience children may suffer from.

Looking at children and caregivers' interaction can help identifying ways of supporting children's abilities. The mother we have just seen interacting with Aaron, for example, demonstrated good practices in letting go of linguistic norms in the production of turns and using language as an instrument for play, mirroring the son's utterances and building up on his echoic contributions in a way that promoted his participation. Brouwer et al. (2011), looking at conversations between children with autism and Down syndrome and their parents, list three main types of positive parental verbal behaviour: 'incorporation' –

incorporating the action of the child in a subsequent one; 'pursuance' – pursuing a relevant action/response that has not yet been delivered by the child, and 'go-along' – going along with the trajectory that the action of the child suggests, whether or not it fits the specific conversational locus. Brouwer et al. argue that such practices should inform to a larger degree than they presently do the communication of professionals with children with difficulties, as well as that of parents.

In order to build support and provide for the enrichment of children's communicative repertoires, interventions can also be directed to the peer group. Ochs, Kremer-Sadlik, Solomon, and Sirota (2001) compared peer interaction of autistic children in classrooms where classmates had received explanations on autism with interaction in classes where the classmates did not know about the child's condition. In the first situation there were more episodes of positive inclusion, as well as fewer episodes of negative exchanges. Schuler (2003) reports that training classmates of autistic children to treat the latter conversational contributions as relevant resulted in a significant increase in the quality and quantity of adequate contributions, showing that facilitating engagement and 'accepting' autistic children's talk can affect not only their motivation to interact within the peer group but also their actual level of competence. Observing with CA methods interventions which recruit children's natural speech partners could provide a better understanding of the mechanisms through which the interventions work, and help improve them.

Interaction between children and professionals during assessment tasks

When caregivers suspect that a child is having difficulties, either in developmental pace or mental health, they are likely to seek specialist help to assess and diagnose the child's problem. A psychological assessment will be then performed either through observation and dialogue or through standardised scales measuring performances on different behavioural and cognitive tasks.

It is important to evaluate the process whereby the outcomes of psychological assessments are generated, in that diagnostic definitions and test scores tend to 'stay' with the children. Assessment results set adults' expectations on what the children can achieve and determine decisions about school placement or intervention. Moreover, test design and the utilisation of tests for creating or confirming diagnostic categories presuppose theories about what a condition is, what defines those who have it, and ultimately how the human psyche works.

The history of IQ (intelligence quotient) measures is a case in point. The IQ test was originally developed by Binet and Simon (1916) as a tool for teachers.

As the authors painstakingly repeat in their writings, it was not intended to measure a stable characteristic of the children but to capture their intellectual capabilities at a given moment in time, and independently from the contents of school learning. This was done so that teachers could devise the best strategies to *improve* children's intelligence, which Binet also did not see as limited by innate dispositions. Famously, the IQ quickly came to be conceived of as a property of the individual ('X has an IQ of...') and to contribute to a folk understanding of intelligence as a fixed genetic endowment.

Binet also saw his test as just one method among others to build a picture of a child's intellectual level. He and his collaborators warned practitioners that test scores were not sufficient to understand a child's capabilities and resources, and, writing about the errors an inexperienced assessor could commit, they stated:

> [T]he first consists in recording the gross results [of the test] without making psychological observations, without noticing such little facts as permit one to give to the gross results their true value.
>
> (Binet, Simon, & Town, 1913, p. 57)

Today, psychologists are trained to consider psychological assessments' results as exhaustive, and not in need of additional 'little facts'. Standardisation is the procedure that is supposed to guarantee the objectivity of results; however, as CA has shown and as we shall see later in the chapter, standardisation in testing and surveying is chimeric: even standardised systems depend on mutual adjustments and interactional negotiations to be completed (Antaki, 1999; Maynard, Houtkoop-Steenstra, Schaeffer, & van der Zouwen, 2002). What regularly happens, though, is that the ways in which interactional adjustments contribute to the assessment outcomes are ignored, and omitted from the final report or evaluation based on the assessments' results.

The second aspect that makes tests and assessments worth studying is that they divulge scientific theories about mental health and disability, theories that will directly impact the way individuals – including the very people affected by these conditions – understand both the problem and themselves. The famous 'false belief' test, for example, is based on the psychological postulate of the existence of a 'Theory of Mind' behind the human capacity for intersubjectivity. In this test, children have to recognise the content of the mind of a character in an illustrated story (i.e. the character's belief about the location of a toy) and correctly guess her action (she will look in the wrong place as she was not present when the toy was moved). Children with autism fail the tests significantly more often than children of comparable age. The consequent notion that autistic people lack a Theory of Mind (Baron-Cohen, 1996), and are therefore unable to understand how fellow humans think, has now become common knowledge, influencing the social relations and identity of people with autism.

The 'false belief' test as such has not been analysed by conversation analysts, but critics of the paradigm have created variants of the experiment which show that children with autism can perform adequately, provided that they can use familiar objects, are given more time, or are made to do exercises for concentration before the test (Bara, Bucciarelli, & Colle, 2001). These findings raise questions about whether the performance in this test reflects autistic children's deficit in understanding other minds or rather their problems with social anxiety, speed of verbal production, and concentration. Anyhow, it is clear that assessment environments are not transparent with respect to which abilities are effectively tested and that investigating the experimental setting they are administered in within an interactional framework could shed light on how their results come about.

Maynard and collaborators (Maynard, 2005; Maynard and Marlaire, 1992) have analysed two different assessments regularly administered to children with autism, with the aim of investigating 'how participants make sense of the questions being asked' (Maynard, 2005, p. 500). Maynard and Marlaire (1992) document the 'interactional substrate' upon which the assessments rest, showing the difficulties in maintaining standardised procedures and how interactional incidents might cause erroneous answers. For example, the professionals delivering the test (in this case a subtest of the Woodcock Johnson Psychoeducational Battery) were not supposed to give feedback but, across different clinicians, cues about the correctness of the answers were common. In an assessment situation, the authors argue, such involuntary feedback, and the fact that children generally monitor practitioners' reaction after producing an answer, can create carry-over effects across different test trials, and also induce discouragement after failure, thus compromising the test's validity. Carry-over effects are discussed by Maynard and Marlaire (1992, p. 194) in terms of 'learning within the task'. While tests are supposed to measure pre-existing abilities, a good deal of learning seems to happen at the time of the examination itself:

> From within the interior of the exam experience, children and clinicians learn what they should do to give, receive, and answer test items properly and correctly.
>
> (1992, p. 194)

In a different study, Maynard (2005) analysed the subtest 'What do you when' of the Brigance Diagnostic Inventory of Early Development and compared the required question–answer sequences to similar sequences in ordinary conversation. The questions in this subtest require giving solutions to hypothetical problematic scenarios. Here is a sequence involving Tony (TO), a

five-year-old autistic boy (shortened from Maynard, 2005, pp. 515–516) and
a clinician (CL):

```
1    CL:    ↑What do you do when you're hungry?
2           (1.4)
3    TO:    You eat.
4    CL:    Okay. What do you do when you're sleepy.
5           (1.0)
6    TO:    Reh- (0.8) you res::::t:s.
7    CL:    What do you do when you're col:d.
8           (1.5)
9    TO:    And den you .hh and den you (1.4) and den you gets::
10          fwo::zen.
11          (0.2)
12   CL:    Y:e::::ah.
```

The child's first and second answers are correct, but he gets the third wrong:
instead of providing a solution Tony depicts a possible consequence of being
cold. This type of wrong response happens frequently in children with autism
and can be interpreted in the light of the theory of 'weak global coherence'
(Happé & Frith, 2006). Looking into the response as a conversational turn,
though, it is also possible to see what kind of operations are performed by the
child on the question. A different pattern is noticeable for the wrong answer
compared to the correct ones, namely the 'and then' preface (line 9). This,
Maynard notes, is a common way of introducing a new event in storytelling.
Tony, it seems, responds in this occurrence by creating a 'narrative tie' between
the question and his answer. Looking at this type of question in ordinary conversation, Maynard finds that the 'when' in 'what do you do when' is a less
straightforward marker of a request for hypothetical reasoning than 'if' would
be in the same question form and is more commonly used to request information about events already experienced by the recipient (although the abstract,
hypothetical use also occurs). These children, thus, are only a small step away
from ordinary use in their test responses. Maynard discusses this in terms of a
different gestalt governing autistic children's interpretation and responses: they
sometimes fail to assemble all the elements concurring to an 'academic' interpretation of the question (i.e. the more abstract option) and go for the 'local'
association with a next event tied to the first through narrative logic.

The analysis does not deny a peculiarity in the way the child responds (this
is not the aim of the exercise); however, it looks at his strategies and preferences with the aim to reach a constructive description of his language use.
The method of comparing instances of language use across atypical and typical
interaction also emerges here as helpful in identifying subtle points of continuity and discordance in the practices of children with difficulties compared with
those of their speech community.

Analysis of consultations and intervention settings

Another way of studying children's mental health is to look at counselling and psychotherapy as the settings in which understandings of children's problems are displayed and acted upon.

CA research on psychological interventions with adults is a relatively established field (Bysouth, 2012; Peräkylä, 2012), despite scholars lamenting too little cumulative findings due to the variety of therapy situations and perspectives. CA methods appear naturally suited to investigate a form of activity accomplished in and through talk, and in which participants attend intently to each other's turns (Peräkylä, 2008). CA research on psychological intervention on children is also growing, so far including counselling (Hutchby, 2007), family therapy (O'Reilly, 2008), and children's telephone helplines (Danby, Butler, & Emmison, 2009; Hepburn & Potter, 2010).

Analysing communication with children in a psychotherapeutic setting implies dealing with the fact that children rarely go into counselling or therapy out of their own choice. Differently from studies on psychotherapy with adults, therefore, it cannot be assumed that children are collaborating to therapeutic work as such when talking (or not talking) to professionals and parents. In a way, though, this makes the investigation of such settings all the more relevant, as researchers are exploring how children's status in therapy comes about and how professionals implicate the child in their domain of practice.

In family therapy settings research has showed that children's position is doubly weakened by their being a child and being imputed a learning or mental health difficulty (Aronsson & Cederborg, 1996).[3] O'Reilly (2008) shows how, in a family therapy situation, therapists do not orient to a child's speaking rights in the same way they do for adults, as exemplified in the lack of acknowledgements or apologies when they interrupt children. Here is an example from a recorded session; Steve is the child, Mr Niles is his father, and FT is the therapist (from O'Reilly, 2008, p. 518):

```
1   Mr Niles:   ↑Yeah (.) <but> tell Joe why >I won't give ↑you your
2               'phone< back
3   Steve:      'cause I swear [too much I sw-
4   FT:                        [But >but< I I >think< (.) that this
5               ne:eds to be re:ally cle:ar
```

In line 3 Steve, a child with suspected attention deficit hyperactivity disorder (ADHD), is explaining something to the therapist, following the father's prompt, but his turn gets overlapped by the therapist before completion. Throughout the therapy sessions, therapists are seen to apologise or otherwise mark their action when they cut off adults' talk, but interruptions to children, as in the case of Steve above, are left unmarked; Butler and Wilkinson (2013), drawing on data of a five-year-old child and his family, have observed that children have restricted rights to engage adults in a sustained interaction, and O'Reilly's findings above indicate that this may be also true for older children in therapy.

Following Shakespeare (1998), the peculiar status of children vis-à-vis adults is often referred to as 'half-membership', but what this status exactly entails varies within different situations and conversational activities. Hutchby (2010) shows that in counselling for children whose parents are separating (carried out with the child(ren) alone), counsellors attribute children a high degree of competence, for example asking them to discuss their parents' motives or to explain how they see their family in the future, but if children do not provide responses along these lines, counsellors often offer their views on the matter overwriting the child's perspective. Sometimes, it may be a burden for a child to be bestowed the identity of the 'primary knower' and to be solicited to explain themselves to a professional (as Clemente, Lee, & Heritage, 2008, find in paediatric visits to children with chronic pain). Certainly, the way children are engaged by practitioners – when and how they are accorded participation rights – is consequential for their position and involvement throughout the whole course of the encounter (Stivers, 2012).

In the analysis of psychological interventions there is also space for pursuing the agenda of discursive psychology, namely looking at the way psychological terms and labels are employed and made sense of by participants (Edwards & Potter, 2005). Such studies help identify the psychological theories and understandings of both professionals and clients, and, more generally, show how terms indicating mental health issues or disability travel through society (O'Reilly, 2005; O'Reilly, Taylor, & Vostanis, 2009).

Many psychological interventions are based on ideas about communication which are not grounded in empirical observations. Peräkylä and Vehviläinen (2003) introduced the notion of 'stock of interactional knowledge' (SIK) for those models and theories of interaction that have a direct bearing upon professional practice and that may be adopted and transmitted by professionals without ever verifying what exactly happens when they get translated into therapeutic practices.

The effect of such unexamined views has been observed, for example, in the delivery of a training in communication skills to autistic pre-adolescent children. In the following extract, Marco, a 13-year-old boy with high-functioning autism, is trying to tell the therapist a story about some sharks he has seen at an exhibition; the goal of the therapist (Anna) is to make him produce numerous turns in quick succession (Fasulo & Fiore, 2007, p. 243):

```
1    Anna:   →   >Quanti erano?<
                 >How many were they?<
2    Marco:      ME- eh, squali sono due.
                 U- uh, sharks are two.
3    Anna:   →   Due. ma quanto erano grandi?
                 Two. but how big were they?
4                (1.2) ((Marco looks down, then to Anna))
5    Anna:   →   Eh::? così?
                 Uh::? this big? ((extending arms))
6                (0.8) ((Anna stays in the position))
7    Marco:      .H non somiglia allo squalo martello ohh:::
                 .H it did not look like hammer shark uh:::
8                squalo di: ah spada (non era)
                 shark of: it (was not) swordfish
9    Anna:       No:: ascolta Marco. era grande coSI':? questo squalo
                 No:: listen Marco. was it THis: big? this shark
10               o era più grande?
                 or was it bigger? ((shows size with hands))
11   Marco:      Lo squalo? e⁴ e non lo
                 The shark? but I don't know
12               (0.8)
```

Marco's irritated answer in line 11 and the gap following it show that the boy is resisting the therapist's line of inquiry (see Hutchby, 2002, for children answering 'I don't know' as resistance). Up to that point Anna has asked very basic questions about the shark ('How many were they' 'how big were they'), and Marco had already given signs of low engagement by looking down and not answering (line 4). When Marco tries to get into a description of the particular species of shark by excluding other types of big fishes, he gets corrected ('no listen Marco') and then again addressed a simple yes/no question (about whether the shark was 'this big', line 9). The story ends at Marco's reaction in line 11.

Conversations with the therapists in the centre informed us that training in turn-taking was one of the main goals of the programme, implemented also through card and board games; the problem is that in the context of storytelling asking many questions can halt the development of the narrative and be perceived as disruptive by the narrator.[5] Furthermore, in the extract above, we see that the questions are set too low in terms of 'granularity', that is, the level of expertise or detail participants speak at (Schegloff, 2000): the questions are too gross and simple given Marco's competence on sea life.

A fortuitous comparison is offered for analysis when, at the end of the activity, Marco raises his hand and asks to tell his story again. Having finished training, Anna lets him do it and takes on this time the natural behaviour of a story-telling recipient (Sacks, 1992): she remains silent while Marco tells her about how sharks can break and digest turtles' shells, utters 'continuers' in predictable places and collaborates to the story conclusion about the power of sharks against all enemies. This second, more satisfactory interaction demonstrates that the therapist was available for the child, and that he valued her as an interlocutor, but also that the unrefined ideas about turn-taking contained in the local 'stock of interactional knowledge' had made her behave unnaturally and almost miss a good opportunity for a meaningful exchange.

To conclude, CA studies of psychological consultations and interventions could have several positive effects: they could foster mutually beneficial collaboration between researchers and practitioners (Richards & Seedhouse, 2007), lend practitioners more control over their practice, and be used in educational and training programmes.

When parents are the sufferers

It is difficult to study the impact on children of parents' mental health within a CA framework, because any claim on this regard needs either longitudinal observations or large samples, and both are not typical of CA methods. This is, however, a substantive domain in psychology, showing correlations between mental health conditions in parents, especially mothers, and later difficulties in children, so it would be worth shedding light on the interactional dynamics by which the documented effects occur.

Clinical psychologist Capps and linguist anthropologist Ochs analysed narratives told by a woman suffering from agoraphobia at the presence of their children, and found that the mother portrayed the outside world as dangerous, describing emphatically both the perceived threats and her fear of them. The authors suggest that repeated exposition to this kind of narratives could contribute to the higher levels of anxiety observed in children of mothers with agoraphobia (Capps, Sigman, Sena, Henker, & Whalen, 1996), gradually wearing out the sense of invulnerability typical of young age (Capps & Ochs, 1995). Even though establishing causal relationship is difficult, it would be useful to extend this kind of research to explore whether conversations between children and parents with different mental health issues are characterised by specific interactional patterns.

An area in which CA might offer a valuable contribution is the interaction between mothers diagnosed with depression and their infants. Studies in developmental psychology report that children of mothers who suffer from depression may develop psychiatric problems themselves or encounter difficulties in psychosocial adjustment (for a review, see Lovejoy, Graczyk, O'Hare, & Neuman, 2000). Observations of early interactions points to misalignments between mother and infants' behaviour (Beebe et al., 2008, Murray & Cooper, 1997; Reck et al., 2011), but behavioural coding categories are not always comparable across studies and the sequences of mothers' and infants' actions cannot always be clearly reconstructed (Fantasia, 2015). Research adopting CA sequential analysis for comparing mothers with and without a diagnosis of depression could contribute to understanding what might influence later difficulties in the children of the former, and also further our general knowledge about very early interactional exchanges.

Clinical relevance summary

Our review shows that many types of conditions and settings are still to be explored, and that, given that CA research is key to find alternatives to deficit approaches, scholars in the field could further extend their collaboration with other researchers and with practitioners to identify the important questions.[6] Conversation analytical work can give back to professionals pictures of their habitual practices thus, encouraging reflexive processes, as demonstrated in research on adults with mental health issues (McCabe, Khanom, Bailey, & Priebe, 2013; McCabe, Leudar, & Antaki, 2004). Finally, studying interactions around children and mental health sharpens our understanding of notions of mental illness, healthy psychological development and effective communication.

For a simple summary of the clinical implications, see Table 1.1.

Chapter summary and conclusions

I have examined conversation analytic research in a few domains concerned with children's mental health, exploring 'disorder' through the interaction order (Wootton, 2002) in children's communication with family members, peers, and professionals.

The review demonstrates how detailed analysis of interactions can change or refine existing theories on mental health, and on intellectual and communicative impairments. Despite CA mainly working on snapshots of the interaction, its focus on the dynamics of verbal exchanges can offer insights into how interactional mechanisms can play a role in the way a condition develops its observable manifestations in the course of a child's life.

Comparing examples of typical and atypical conversational sequences, furthermore, has been instrumental in identifying the strategies children adopt to

Table 1.1   Clinical practice highlights

---

1.   CA can offer up suggestions and recommendations for practice in a range of settings, including institutional settings such as counselling and therapy.
2.   Through the use of CA, we are able to identify ways of supporting children's abilities.
3.   CA can identify and describe how different psychological interventions are administered and can promote explore the interaction rights of children.

---

get as close as possible to adequate communication, despite the end product appearing sometimes odd or dysfunctional.

When considering institutional work, as in psychological assessments or interventions, CA shows the functioning of its apparatuses and – rather than collaborating to diagnosis or intervention – unpacks the activities that produce definitions of children and their mental health conditions in allegedly objective ways.

Finally, it has been suggested that there is potential for CA research to extend the use of its sensitive methodological instrumentation to conversations between children and parents with mental health conditions, going as far as the early interactions between infant and their caregivers.

Notes

1. See Sidnell (2010) for an account of how children's repairs differ at four and five years of age due to their different concerns at these ages. Linguistic socialisation studies had already pointed out how, in the acquisition of language, progression is dictated by the socio-cultural environment and not by cognitive development (Ochs & Schieffelin, 1983). For a discussion about the degree to which various authors in CA engage with the notion of development, see Kidwell (2012).
2. Autism has been studied extensively also in linguistic anthropology as a way to distinguish different components within linguistic practices and explore their role in the constitution of different forms of sociality (Bagatell & Solomon, 2008; Ochs, Kremer-Sadlik, Sirota, & Solomon, 2004; Ochs & Solomon, 2004).
3. Before CA, detailed analysis of filmed psychotherapy sessions involving children had been carried out in the 50s by the interdisciplinary group of the Palo Alto School, both to understand communication around psychopathological conditions and as training for the therapists. The method led to a revolutionary theory about the genesis of psychoses as adaptations to a dysfunctional communicative environment. Another important concept with regard to children's mental health was the idea of the child as 'symptom-bearer', namely a carrier of the problems of the family that the child 'pathology' would hide from sight (Bateson, 1972; Bateson, Jackson, Haley, & Weakland, 1956).
4. The conjunction 'e' ('and') in Italian has an oppositional value at turn-beginning.
5. The Centre is inspired by the theories and procedures developed by Theo Peeters (1997); however, the conversational techniques analysed here, which were also explained to us by the staff, could not be retrieved in his work.
6. For an example of profitable collaboration between clinical psychology and interactional linguistics, see DuBois, Hobson, and Hobson (2014).

References

Antaki, C. (1999). Assessing quality of life of persons with a learning disability: How setting lower standards may inflate well-being scores. *Qualitative Health Research, 9,* 437–454.

Antaki, C., Vehviläinen S., & Leudar, I. (Eds.) (2008). *Conversation analysis and psychotherapy.* Cambridge: Cambridge University Press.

Aronsson, K., & Cederborg A. C. (1996). Coming of age in family therapy talk: Perspective setting in multiparty problem formulations. *Discourse Processes, 21*(2), 191–212.

Atkinson, M. J., & Heritage, J. (Eds.) (1984). *Structures of social action: Studies in conversation analysis.* Cambridge: Cambridge University Press.

Bagatell, N., & Solomon, O. (Eds.) (2008). Rethinking autism, rethinking anthropology. *Ethos, 38*(1), i–ii (special issue).

Bara, B. G., Bucciarelli, M., & Colle, L. (2001). Communicative abilities in autism: Evidence for attentional deficits. *Brain and Language, 77*(2), 216–240.

Baron-Cohen, S. (1996). *Mindblindness: An essay on autism and theory of mind.* Cambridge, MA: MIT Press.

Bateson, G. (1972). *Steps to an ecology of mind.* Chicago, IL: University of Chicago Press.

Bateson G., Jackson D., Haley J., & Weakland J. (1956). Toward a theory of Schizophrenia. *Behavioral Science, 1*(4), 251–254.

Beebe, B., Jaffe, J., Buck, K., Cohen, P., Feldstein, S., & Andrews, H. (2008). Six-week postpartum maternal depressive symptoms and 4-month mother-infant self – and interactive contingency. *Infant Mental Health Journal, 29*(5), 442–471.

Berducci, D. (2010). From infants' reacting to understanding: Grounding mature communication and sociality through turn-taking and sequencing. *Psychology of Language and Communication, 14*(1), 3–27.

Binet, A., & Simon, T. (1916). *The development of intelligence in children: The Binet-Simon scale* (No. 11). New Jersey: Williams & Wilkins Company.

Binet A., Simon, T., & Town, C. H. (1913). *A method of measuring the development of the intelligence of young children.* Lincoln, IL: Courier.

Brouwer, C. E., Dennis, D., Ferm, U., Hougaard, A. R., Rasmussen, G., & Thunberg, G. (2011). Treating the actions of children as sensible: Investigating structures in interactions between children with disabilities and their parents. *Journal of Interactional Research in Communication Disorder, 1*(2), 153–182.

Butler, C. W., & Wilkinson, R. (2013). Mobilising recipiency: Child participation and 'rights to speak' in multi-party family interaction. *Journal of Pragmatics, 50*(1), 37–51.

Bysouth, D. (2012). Conversation analysis and therapy. In C. A. Chapelle (Ed.), *The encyclopedia of applied linguistics.* Blackwell Publishing: Wiley online.

Capps, L., & Ochs, E. (1995). *Constructing panic: The discourse of agoraphobia.* Cambridge, MA: Harvard University Press.

Capps, L., Sigman, M., Sena, R., Henker, B., & Whalen, C. (1996). Fear, anxiety, and perceived control in children of agoraphobic parents. *Journal of Child Psychology and Psychiatry, 37*(4), 445–452.

Clemente, I., Lee, S. H., & Heritage, J. (2008). Children in chronic pain: Promoting pediatric patients' symptom accounts in tertiary care. *Social Science & Medicine, 66*(6), 1418–1428.

Danby, S. J., Butler C., & Emmison, M. (2009). When 'listeners can't talk': Comparing active listening in opening sequences of telephone and online counselling. *Australian Journal of Communication, 36*(3), 91–114.

Drew, P. (2003). Comparative analysis of talk-in-interaction in different institutional settings: A sketch. In P. Glenn, C. D. LeBaron, & J. Mandelbaum (Eds.), *Studies in language and social interaction: In honor of Robert Hopper* (pp. 249–263). New Jersey: Lawrence Earlbaum Associates.

Du Bois, J. W., Hobson, R. P., & Hobson, J. A. (2014). Dialogic resonance and intersubjective engagement in autism. *Cognitive Linguistics, 25*(3), 411–441.

Edwards D., & Potter, J. (2005). Discursive psychology, mental states and descriptions. In H. te Molder & J. Potter (Eds.), *Conversation and cognition* (pp. 241–259). Cambridge: Cambridge University Press.

Esposito, G., Nakazawa J., Venuti P., & Bornstein, M. H. (2013). Componential deconstruction of infant distress vocalizations via tree-based models: A study of cry in autism spectrum disorder and typical development. *Research in Developmental Disabilities, 34*(9), 2717–2724.

Fantasia V. (2015). *Exploring infants' cooperative participation in early social routines.* Unpublished PhD dissertation, University of Portsmouth UK.

Fasulo, A., & Fiore F. (2007). A valid person: Non-competence as a conversational outcome. In A. Hepburn & S. Wiggins (Eds.), *Discursive research in practice: New approaches to psychology and interaction* (pp. 224–247). Cambridge: Cambridge University Press.

Forrester, M. (2010). Ethnomethodology and adult–child conversation: Whose development? In H. Gardner & M. Forrester (Eds.), *Analysing interactions in childhood: Insights from conversation analysis* (pp. 42–58). Oxford: Wiley-Blackwell.

Heath, C., & Luff, P. (2000). *Technology in action.* Cambridge: Cambridge University Press.

Hepburn, A., & Potter, J. (2010). Interrogating tears: Some uses of 'tag questions' in a child protection helpline. In A. F. Freed & S. Ehrlich (Eds.), *'Why do you ask?': The function of questions in institutional discourse* (pp. 69–86). Oxford: Oxford University Press.

Heritage J., & Clayman, S. (2011). *Talk in action: Interactions, identities, and institutions.* Oxford: Wiley-Blackwell.

Hutchby, I. (2002). Resisting the incitement to talk in child counseling: Aspects of the utterance 'I don't know'. *Discourse Studies, 4,* 147–168.

Hutchby, I. (2007). *The discourse of child counselling.* Amsterdam: John Benjamins.

Hutchby, I. (2010). Feelings-talk and therapeutic vision in child-counsellor interaction. In H. Gardner & M. A. Forrester (Eds.), *Analysing interactions in childhood: Insights from conversation analysis* (pp. 146–162). Oxford: Wiley-Blackwell.

Jefferson, G. (2004). Glossary of transcript symbols with an introduction. In G. H. Lerner (Ed.), *Conversation analysis: Studies from the first generation* (pp. 13–23). Philadelphia: John Benjamins.

Kaye, K., & Brazelton, T. B. (1971). Mother-infant interaction in the organization of sucking. *Paper presented at the Meeting of the Society for Research on Child Development,* Minneapolis, MN, April 1971.

Kidwell, M. (2012). Interaction among children. In J. Sidnell & T. Stivers (Eds.), *The handbook of conversation analysis.* Oxford: Wiley-Blackwell.

Happé, F., & Frith, U. (2006). The weak coherence account: Detail-focused cognitive style in autism spectrum disorders. *Journal of Autism and Developmental Disorders, 36*(1), 5–25.

Levinson, S. C. (1983). *Pragmatics.* Cambridge: Cambridge University Press.

Llewellyn, N., & Hindmarsh, J. (Eds.) (2010). *Organisation, interaction and practice: Studies of ethnomethodology and conversation analysis.* Cambridge: Cambridge University Press.

Local, J., & Wootton, A. (1995). Interactional and phonetics aspects of immediate echolalia in autism: A case study. *Clinical Linguistics and Phonetics, 9,* 155–184.

Lovejoy, M. C., Graczyk P. A., O'Hare, E., & Neuman, G. (2000). Maternal depression and parenting behavior: A meta-analytic review. *Clinical Psychology Review, 20*(5), 561–592.

Maynard, D. W. (2005). Social actions, gestalt coherence, and designations of disability: Lessons from and about autism. *Social Problems, 52*(4), 499–524.

Maynard, D. W. (2012). Everyone and no one to turn to: Intellectual roots and contexts for conversation analysis. In J. Sidnell & T. Stivers (Eds.), *The handbook of conversation analysis* (pp. 11–31). Oxford: Wiley-Blackwell.

Maynard, D. W., Houtkoop-Steenstra, H., Schaeffer, N. C., & van der Zouwen, H. (Eds.) (2002). *Standardization and tacit knowledge: Interaction and practice in the survey interview.* New York: John Wiley.

Maynard D. W., & Marlaire, C. L. (1992). Good reasons for bad testing performance: The interactional substrate of educational exams. *Qualitative Sociology, 15*(2), 177–202.

McCabe, R., Khanom, H., Bailey, P., & Priebe, S. (2013). Shared decision-making in ongoing outpatient psychiatric treatment. *Patient Education and Counseling, 91*(3), 326–328.

McCabe, R., Leudar I., & Antaki, C. (2004). Do people with schizophrenia display theory of mind deficits in clinical interactions? *Psychological Medicine, 34*(3), 401–412.

Murray, L., & Cooper, J. (Eds.) (1997). *Postpartum Depression and Child Development.* New York & London: The Guilford Press.

Ochs, E., Kremer-Sadlik, T., Sirota, K. G., & Solomon, O. (2004). Autism and the social world: An anthropological perspective. *Discourse Studies, 6*(2), 147–183.

Ochs, E., Kremer-Sadlik, T., Solomon, O., & Sirota, K. G. (2001). Inclusion as social practice: Views of children with autism. *Social Development, 10*(3), 399–419.

Ochs, E., & Schieffelin, B. B. (Eds.) (1983). *Acquiring conversational competence.* Boston: Routledge & Kegan Paul.

Ochs, E., & Solomon, O. (2004). Practical logic and autism. In R. Edgerton & C. Casey (Eds.), *A Companion to psychological anthropology: Modernity and psychocultural change* (pp. 140–167). Oxford: Blackwell.

O'Reilly, M. (2005). 'What seems to be the problem?' A myriad of terms for mental health and behavioural concerns. *Disability Studies Quarterly, 25*(4) (online journal article). Retrieved July 12, 2014 from www.dsq-sds.org.

O'Reilly, M. (2008). What value is there in children's talk? Investigating family therapists' interruptions of parents and children during the therapeutic process. *Journal of Pragmatics, 40*(3), 507–524.

O'Reilly M., Taylor, H., & Vostanis, P. (2009). 'Nuts, schiz, psycho': An exploration of young homeless people's perceptions and dilemmas of defining mental health. *Social Science and Medicine, 68*(9), 1737–1744.

Peeters, T. (1997). *Autism: From theoretical understanding to educational intervention.* New York: Wiley.

Peräkylä, A. (2008). Psychoanalysis and conversation analysis: Interpretation, affect and intersubjectivity. In A. Peräkylä, C. Antaki, S. Vehviläinen, & I. Leudar (Eds.), *Conversation analysis and psychotherapy* (pp. 100–119). Cambridge: Cambridge University Press.

Peräkylä, A. (2012). Conversation analysis in psychotherapy. In J. Sidnell & T. Stivers (Eds.), *The handbook of conversation analysis* (pp. 551–574). West Sussex: Wiley-Blackwell.

Peräkylä, A., & Vehviläinen, S. (2003). Conversation analysis and the professional stocks of interactional knowledge. *Discourse & Society, 14*(6), 727–750.

Peskett, R., & Wootton, A. J. (1985). Turn-taking and overlap in the speech of young Down's syndrome children. *Journal of Intellectual Disability Research, 29*(3), 263–273.

Rączaszek-Leonardi, J., Nomikou, I., & Rohlfing, K. J. (2013). Young children's dialogical actions: The beginnings of purposeful intersubjectivity. *IEEE Transactions on Autonomous Mental Development, 5*(3), 210–221.

Reck, C., Noe, D., Stefenelli, U., Fuchs, T., Cenciotti, F., Stehle, E., Mundt, C., Downing, G., & Tronick, E. Z. (2011). Interactive coordination of currently depressed inpatient mothers and their infants during the postpartum period. *Infant Mental Health Journal, 32*(5), 542–562.

Reddy, V. (2008). *How infants know minds.* Cambridge, MA: Harvard University Press.

Richards, K., & Seedhouse, P. (2007). *Applying conversation analysis.* Hampshire: Palgrave Macmillan.

Sacks, H. (1992). *Lectures on conversation,* edited by G. Jefferson. Oxford: Blackwell.

Sacks, H., Schegloff, E. A., & Jefferson G. (1974). A simplest systematics for the organization of turn-taking for conversation. *Language, 50,* 696–735.

Schegloff, E. A. (2000). On granularity. *Annual Review of Sociology, 26,* 715–720.

Schegloff, E. A. (2007). *Sequence organization in interaction, vol. 1: A primer in conversation analysis.* Cambridge: Cambridge University Press.

Schuler, A. L. (2003). Beyond echoplaylia. Promoting language in children with autism. *Autism, 7*(4), 455–469.

Shakespeare, P. (1998). *Aspects of confused speech: A study of verbal interaction between confused and normal speakers.* Mahwah, NJ: Lawrence Erlbaum Associates.

Sidnell, J. (2010). Questioning repeats in the talk of four-year-old children. In H. Garder & M. Forrester (Eds.), *Analysing interactions in childhood: Insights from conversation analysis* (pp. 103–127). Oxford: Wiley-Blackwell.

Sidnell J., & Stivers T. (Eds.) (2012). *The handbook of conversation analysis.* Oxford: Wiley-Blackwell.

Sterponi, L., & Fasulo, A. (2010). How to go on: Intersubjectivity and progressivity in the communication of a child with autism. *Ethos, 38*(1), 116–142.

Sterponi L., & Shankey, J. (2014). Rethinking echolalia: Repetition as interactional resource in the communication of a child with autism. *Journal of Child Language, 41*(2), 275–304.

Stivers, T. (2012). Physician–child interaction: When children answer physicians' questions in routine medical encounters. *Patient Education and Counseling, 87*(1), 3–9.

Tarlin, K., Perkins, M. R., & Stojanovik, V. (2006). Conversational success in Williams syndrome: Communication in the face of cognitive and linguistic limitations. *Clinical Linguistics and Phonetics, 20*(7–8), 583–590.

Tarplee, C., & Barrow, E. (1999). Delayed echoing as an interactional resource: A case study of a three-year-old child on the autistic spectrum. *Clinical Linguistics and Phonetics 6,* 449–82.

ten Have, P. (1999). *Doing conversation analysis: A practical guide.* London: Sage.

Trevarthen, C. (1979). Communication and cooperation in early infancy: A description of primary intersubjectivity. In M. Bullowa (Ed.), *Before speech: The beginning of interpersonal communication* (pp. 321–348). Cambridge: Cambridge University Press.

Trevarthen, C., Aitken, K., Papoudi, D., & Robarts, J. (1998). *Children with autism: Diagnosis and intervention to meet their needs.* London: Jessica Kingsley.

Wells, B., Corrin J., & Local, J. (2008). Prosody and interaction in atypical and typical language development. *Travaux Neuchâtelois de Linguistique, 49,* 135–151.

Wootton, A. J. (1997). *Interaction and the development of mind.* Cambridge: Cambridge University Press.

Wootton, A. J. (1999). An investigation of delayed echoing in a child with autism. *First Language, 19,* 359–381.

Wootton, A. J. (2002). Interactional contrasts between typically developing children and those with autism, Asperger's syndrome, and pragmatic impairment. *Issues in Applied Linguistics, 13*(2), 133–159.

Recommended reading

- Fasulo, A., & Fiore F. (2007). A valid person: Non-competence as a conversational outcome. In A. Hepburn & S. Wiggins (Eds.), *Discursive research in practice: New approaches to psychology and interaction* (pp. 224–247). Cambridge: Cambridge University Press.
- Peräkylä, A. (2008). Psychoanalysis and conversation analysis: Interpretation, affect and intersubjectivity. In A. Peräkylä, C. Antaki, S. Vehviläinen, & I. Leudar (Eds.), *Conversation analysis and psychotherapy* (pp. 100–119). Cambridge: Cambridge University Press.
- Sidnell, J. (2010). Questioning repeats in the talk of four-year-old children. In H. Garder & M. Forrester (Eds.). *Analysing interactions in childhood: Insights from conversation analysis* (pp. 103–127). Oxford: Wiley-Blackwell.
- Wootton, A. J. (1997). *Interaction and the development of mind.* Cambridge: Cambridge University Press.

2
The Value of Conversation Analysis: A Child Psychiatrist's Perspective

Khalid Karim

Introduction

The exponential growth of information available on the Internet, together with the increased volume of published literature produced daily, has led to many great strides in both academic and clinical progress. While a proportion of this literature has had significant impact on the research arena, on health matters only a limited amount gets translated into clinical practice. Nonetheless, this translation is essential for subsequent improvements in care. However, with this volume of research available, it is increasingly difficult to adequately appreciate all of the research which is pertinent to a particular field. As a clinician, there are multiple sources of information presented in a variety of settings, but the pressures from other aspects of work create problems in the ability to adequately incorporate research into practice. Consequently, much research goes unnoticed and there is a tendency for individuals to remain quite narrowly focused, exploring only the classical approaches to which they are familiar. In addition, the drive to utilise what is described as evidence-based approaches, which are often a synthesis of randomised controlled trials, further exacerbates this issue. This obviously leads to a situation where certain methodologies can seem to be much less relevant and therefore any conclusions drawn from them have the potential to be ignored in the clinical world.

The question that will be addressed in this chapter is whether conversation analysis (CA) should have a place in the research and clinical mindset of the child psychiatrist. Child psychiatrists work with children and adolescents who have a range of mental health problems. In the assessment and ongoing therapy with young people, there is a range of therapeutic options available to the clinician, and while some of these are pharmacological in nature, the predominant therapeutic medium is 'talk'. The therapeutic relationship with the child and their family is seen as central to the practice within this speciality and many of the treatments rely upon good communication practices. Thus in developing a fuller understanding of child psychiatry it becomes essential to have a greater appreciation of the nuances of language use and the social actions that talk achieves. Hence, CA has the potential to become a powerful tool in the micro-analysis of psychiatric conversation. The issue that really needs to be explored is whether the in-depth analysis of these clinical encounters using CA has any significant clinical place in the current healthcare culture.

Most practising clinicians, including child psychiatrists, want research that can be translated into something that will improve the care of their patients on a daily basis, rather than being seen as research for research sake. It is important to recognise that although the focus for this chapter is on conversation analysis and its position within the arena of applied research, many of the issues are equally relevant to other areas of academia.

As a way of exploring the role of CA in child psychiatry, it is essential to view how the context of contemporary child psychiatry practice and the prevailing medical discourse influence clinical care. Although it is useful to describe this broader context, the addition of a personal perspective hopefully provides insights into how conversation analysis may develop a sustainable impact.

Child mental health

Child mental health services consist of different professional groups who deliver care and provide a multidisciplinary approach to a child's needs. Child psychiatry is one such profession, but one where the professional is trained in the discipline of medicine. Child psychiatrists are therefore medically trained doctors who pursued a career in psychiatry once they have graduated. In addition to the initial medical degree, they then complete further qualifications initially in psychiatry and then complete training in child psychiatry. Although there are individual differences between child psychiatrists, this training has a significant influence on their practice and the mindset with which they approach any given situation with families. Child mental health is a fairly recent area of clinical practice and one that continues to undergo development. The explosion in the recognition and diagnosis of child mental health problems has been substantial and has placed extensive pressure on health services across many countries, and the development of new treatments has also led to significant changes in how this health need has been approached.

The field of child mental health needs to be contextualised within the overall social, cultural, and organisational rhetoric of mental illness. There have been many influences on the thinking and conceptualisation of mental health and mental distress, particularly during the last century. These have affected practice, particularly in the Western world, and include the influence of psychoanalysis, the biological movement, including the promotion of psychotropic medication, and the development of new talking therapies. In considering the discipline of child mental health there have been other considerable influences which are often defined by the contemporary position of children in society and the prevailing concept of the family. For example, societal attitudes towards children have changed, moving away from seeing them as adults in the making and as possessions of their parents, to more autonomous agents, with rights to participate in decisions that could affect their lives (Corsaro, 2011). This changing attitude towards children is now an international issue, with more global policies influencing the way in which we work with children, such as the United Nations Convention on the Rights of the Child (1989). Furthermore, the concept of the nuclear family and what constitutes the family is continually challenged.

Historically, for child mental health in the Western world, the most significant development was probably the introduction of universal education for children, which occurred in the United Kingdom during the late 19th century. Subsequently, this provided a mechanism to see children developing, and during this time there was increasing recognition that children were suffering from a multitude of physical and mental health problems which needed addressing, whereas previously this was often practically invisible. This increasing appreciation of child problems was thus paralleled by the development of areas such as educational psychology, social care, and the formation of the first child guidance units as a way of addressing these difficulties.

While it is necessary to recognise that there were some attempts to address child mental health difficulties prior to the 20th century, it was during this time period that many of the developments started. The influence of Sigmund Freud on the early development of psychoanalysis led to further developments by Anna Freud and Melanie Klein, who looked further at issues with children. There were significant debates between these individuals in the 1920s and 1930s, and it is Anna Freud who is credited with the development of play therapy, which emphasised the importance of play as a therapeutic medium (see, e.g. Freud, 1946). It is interesting to note that the first child psychiatry textbook was first published only in 1935 by Leo Kanner, and the term 'child psychiatry' originated in 1933 at a meeting of the Swiss Psychiatric Society. Paralleling these debates was the development of the Boston Habit Clinic (described as a specialised mental hygiene institution) in 1921 under psychiatrists such as Douglas Thom (Jones, 1999). This clinic promoted child guidance as a way of helping parents to manage the everyday problematic behaviours of their children. This child guidance model then proliferated, particularly in the first half of the 20th century, with considerable expansion after the Second World War (Jones, 1999). This was due to society's growing recognition that families and their children required support, and hence the aforementioned shift in the constitution of families. Additionally, societal perspectives regarding the care of young children have been significantly influenced by the work of John Bowlby, particularly his work on maternal care and mental health in 1951, and also the publication of *The Child, the Family and the Outside World* in 1964 by Donald Winnicot.

More recently, increasing recognition of child mental health within the diagnostic manuals such as the *International Classification of Diseases* (ICD) and the *Diagnostic and Statistical Manual* (DSM) has enabled problems to be diagnosed using more consistent and reliable criteria. However, as a consequence of increasing the number of diagnostic categories, which are often based on descriptive criteria, there have been concerns that this growth has increasingly medicalised the behaviour of children and reduced what is perceived as 'normal' (Wykes & Callard, 2010).

Another issue which has considerably impacted this field is the development of treatments which are specifically tailored for children. Particularly influential was the development of family therapy in the 1960s and 1970s which views children's problems as occurring within complex systems, most often the family (see Dallos & Draper, 2010). The appreciation that there are multiple influences on a child's mental health has been a considerable driver in developing interventions that are more suitable for children and consequently based on the action of talk. A further example has been the recent expansion of cognitive behavioural therapy for the treatment of child mental health problems (see Fuggle, Dunsmuir, & Curry, 2013). Alongside these talking therapies, there has been a rapid rise of pharmacological treatments, for example, methylphenidate for attention deficit hyperactivity disorder (ADHD), and also increased use of anti-depressants.

The history of child mental health thus demonstrates a diverse range of influences which still have considerable impact on the practice of clinicians in this field. Adding to this milieu is the influence on the field from a diverse range of agencies and professional disciplines that are involved with children. The conceptualisation of child mental health is often open to interpretation, and therefore professionals may approach the same difficulty from a different theoretical framework. While this approach can be beneficial, there can also be differences of opinion which can be challenging. Therefore, any research in this field can become valuable in further shaping a consistent approach to children.

While there has been increased recognition of mental health problems in children the changing criteria in this field have led to difficulties in understanding and researching the extent of these problems. However, it is now appreciated, that at any one time approximately 10% of children are affected (Ford, Goodman, & Meltzer, 2003). These problems can take many forms and present differently at different ages. In younger children for example, presentations tend to have a more behavioural element and there are certain diagnosable disorders which are more prevalent. Mental health difficulties tend to increase with chronological age and become much more obvious within the adolescent years. Although there can be many presentations of adult-like difficulties such as depression, anxiety, eating disorders, psychosis, and obsessive-compulsive disorder (OCD), it is sometimes better to consider child mental health problems as being in three broad areas: behavioural, emotional, and neurodevelopmental. Although there can be considerable crossover between these elements, this arrangement can be useful. The emotional difficulties often take the form of mood and anxiety problems and form a considerable part of the workload within a child mental service. In addition, other problems are classified as having a clear neurological basis, and these include disorders such as autism spectrum disorder (ASD), ADHD, tic disorder, and learning disabilities. These conditions present with a range of difficulties which are typically present from very early childhood and affect a child's development in specific ways. Other problems are classified as behavioural in nature, such as oppositional defiant disorder or conduct disorder, but there is less clarity regarding the parameters which constitute the diagnosis. This can be highly dependent upon what society expects from their children.

Medical training

To understand whether CA would be seen as important to child and adolescent psychiatrists it is necessary to explore the factors which influence medical training. Although there have been a number of changes in the field of medical education in recent years, particularly with changes to the curriculum led by organisations such as the General Medical Council (GMC), there are still many aspects of medical training that remain unchanged. Consider the criticisms that medical undergraduate selection does not adequately reflect the diversity within society, particularly with respect to social background or financial status, and that the selection process has tended to favour students who possess educational qualifications in science. This has led to concerns that only potential students with a particular mindset are being admitted into medical schools While there have been some changes to address this to some extent, medicine is still dominated by a scientific way of thinking and it is not inconceivable that those with a science background will favour research methods that ostensibly are grounded in objective quantifiable measures. Although this preference is not universal, and there are many doctors who are interested in the social aspect of medicine, the prevailing perspective in medicine (which spans practice, journals, funding, and educational curricula) could be considered a more reductionist perspective and one that values quantitative methods. Thus it can be more challenging for qualitative approaches and qualitative evidence to be influential in this field.

Classically, the undergraduate degree is divided into distinct stages, with the first two years dominated by the medical sciences. There is a substantial quantity of learning to digest during this phase, with diverse subjects such as anatomy, physiology, biochemistry, and genetics. The social sciences are often studied in the context of health, and this includes disciplines such as medical sociology and psychology. It is during this phase of the training that medical students are introduced to research methods, but predominantly they tend to study quantitative methods. The importance of epidemiology in the study of illness is very clearly emphasised and there is an introduction to evidence-based medicine. The latter years in the medical degree are spent in clinical specialities, understanding disease processes and the treatments available, before a final exam. In recent years, there has been a drive to integrate both these phases of study with the introduction of clinical subjects much earlier, and there has also been a greater emphasis on developing skills such as communication, inter-professional education, and team-working. Although the communication skills are taught in a variety of formats and are assessed throughout training, interestingly a core complaint still made by patients relates to poor doctor–patient communication (Beckman, Markakis, Suchman, & Frankel, 1994). Interestingly, despite psychiatry relying on a good therapeutic relationship and obtaining information, often of a sensitive nature, there tends to be only limited training in communication skills during the specialist postgraduate experience.

Following graduation and a period of universal postgraduate training, the medical graduate can specialise in a particular area of medical practice. In the United Kingdom, this involves a period of time as a junior doctor in different areas of psychiatry. Following the completion of postgraduate exams, the trainee can then undertake higher training in a particular specialist area, such as child and adolescent psychiatry. While psychiatrists are given some training in research methods, like in other areas of medicine, this is predominantly in quantitative methods to enable critical evaluation of research papers. For example, during my own postgraduate studies in the specialty area of child psychiatry, there remained an emphasis on epidemiology, randomised controlled studies, and the meta-synthesis of quantitative papers. Indeed, many of my colleagues still remain dismissive of qualitative methods and the value of adopting this approach.

Prevailing research perceptions

Now when considering how a child psychiatrist may view CA and if it could be considered a useful addition to the clinical cornucopia, it is important to consider the research influences which affect contemporary practice, both as a medical graduate and as a specialist in child psychiatry. Considering the drive for improved communication within the medical setting, it is interesting to see how this is being achieved and how CA has the potential to help with this.

There are certain seminal research studies which are often quoted with respect to child and adolescent psychiatry. Virtually all of these are quantitative in nature and are often epidemiological studies. An important series of research studies which are well known to all child psychiatrists is the *Isle of Wight* research by Rutter and his colleagues produced in the 1970s (see Rutter, Graham, & Yule, 1970; Rutter, Tizard, Yule, Graham, & Whitmore, 1976). This influential piece of research was a series of studies designed to clearly identify the rates of child mental health problems within a general population. These studies utilised classic quantitative methodological approaches to elicit a wide range of child mental health, physical health, and social problems. The project employed a two-phase longitudinal design together with multiple informants to obtain the essential quantitative data. The research team was clearly aware of the limitations of using standardised questionnaires in their project and made enormous efforts to account for this; however, the use of such devices is difficult when so much of the assessment of children has a subjective component. The authors utilised a range of standardised questionnaires and scales to define the boundaries of normality in terms of behavioural, psychiatric, and parental interview measures, and this approach continues to form the backbone of quantitative research.

The development of schedules of assessment is thus dependent on how well the instrument has been defined and validated, which may be difficult when accounting for factors as diverse as culture, age, or emotional maturity. This determination of 'caseness' (i.e. the decision of who falls outside of the boundaries of normality) therefore also needs clarity in understanding of what constitutes normality (see Part 3 of this volume). In mental health there are many factors which can be used to determine such caseness, but given the socially constructed nature of normality, psychiatry as a discipline will always have an element of subjective judgement. Adding to this is the extensive variability of children in factors as diverse as age, developmental ability, and capacity. Nonetheless, despite these difficulties and criticisms, epidemiology remains an essential part of child psychiatry and is important for identifying the prevalence of problems within defined communities.

When considering the treatments used in health, there has been a significant growth and support for the evidence-based medicine approach, a system which in research terms is dominated by the randomised control trial study. The randomised controlled trial study is a mechanism used to test the efficiency of a particular treatment. These studies are characterised by well-known structural dimensions, including a randomisation process, an intervention applied to the subjects, and a group that may have a different (or no) intervention – the control group. A carefully designed randomised controlled study is considered the most reliable evidence in the hierarchy of evidence (Marks, 2002) by many doctors, and the strength is further enhanced if combined in a systematic review and meta-analysis. The hypothesis addressed by randomised controlled study should then be definitively answered. However, there are concerns that randomised controlled studies may lack validity when applied to normal clinical situations when their strict criteria for selection are no longer applicable (see Hesse-Biber, 2012, for discussion). These exclusion criteria may be significant and unreflective of a naturalistic environment. The studies are also costly and the research questions may be very narrow. More recently, there have also been concerns that many of the published studies, particularly those involving the pharmaceutical industry, have been the subject of selective publication of data and their influence overemphasised through publication bias (Melander, Ahlqvist-Rastad, Meijer, & Beermann, 2003).

Evidence-based medicine: A more detailed examination

The term of evidence-based practice was coined in the 1980s by David M. Eddy and has become a guiding principle in the development of treatment pathways and organisational and governmental policies (Eddy, 2011). This process adopts a rigorous strategy in assessing which evidence is appropriate for answering a particular question; for example, whether one method is better than another. Evidence-based medicine more specifically has been argued to enable practitioners to make use of the available evidence when making decisions about the treatment and care of patients. This approach has evolved since its inception and originally included the integration of clinical expertise (Sackett, Strauss, & Richardson, 2000), and debates still exist regarding the tension between the application of evidence and the clinical judgement on an individual basis. These definitions and the direction of evidence-based practice have seen the development of an evidence-based hierarchy, and although there are slight variations in this, meta-analysis and synthesis of randomised controlled trial studies are always placed at the top. As a consequence, these hierarchies of evidence thus privileged research that utilised this quantitative method (Kovarsky, 2008).

The perception that there is a clear hierarchy of evidence thus has a considerable effect on the clinical practitioner and their mindset. The child psychiatrist is no exception and there has been a significant rise of evidence-based practice in the field of mental health (Tanenbaum, 2003). Thus, those who work in the area of mental health are working in an environment that stresses the use of current evidence in the direction of their treatment decisions and the use of interventions (Rice, 2008).

However, there are concerns that this practice does have significant limitations. It is utterly dependent on the evidence available and therefore is not suitable for rarer conditions, and hence many areas of medical practice, including those in child mental health, tend to be under-researched or unreported. Criticisms of evidence-based medicine tend to be that it adopts a population-based approach to an illness and that clinicians should still be able to display discretion on an individual basis, that is, clinical judgement is seen as important (French, 1999; Strong, Busch, & Couture, 2008). Perhaps most notably, as discussed above, the earlier definitions placed an emphasis on the integration of the use of evidence with clinical experience (Sackett, Rosenberg, Gray, Hayes, & Scott Richardson, 1996), but the contemporary development of this seems to have waylaid the importance of clinical judgement or patient perspectives (Hesse-Biber, 2012). Thus arguably there is scope for qualitative evidence to make an impact, despite its current position at the bottom of the evidence pyramid (Lester & O'Reilly, in press). The different approach taken by qualitative research should however be seen as complementary to quantitative methods, as it may demonstrate a different perspective. There is increasing interest in mixing the methods and this probably needs further exploration.

The qualitative position

As aforementioned, the current environment provides a predominantly quantitative research perspective and it can take some time for clinicians to see the value of the qualitative research approach and appreciate what role this may have in their practice. It is important to recognise that both areas are not mutually exclusive; however, it may not be immediately clear on the relationship. The recognition that qualitative methods can be useful in looking at how and why things happened, particularly in the decision-making process, is a useful step forward in this understanding. Thus, qualitative research as a method for looking at processes is a useful way of promoting its applicability.

However, to the uninitiated there appears to be a bewildering array of qualitative methods looking at the different aspects of these processes that can range from more descriptive approaches, such as thematic analysis, to more in-depth methodologies, such as grounded theory, narrative analysis, ethnography, phenomenological analysis, or discourse studies. Although qualitative research is increasingly recognised as producing a significant body of evidence on many aspects of health research in both the physical and mental health arena, at times it becomes a difficult approach to promote. While certain methodologies such as thematic or qualitative content analysis or grounded theory have gained particular credibility, there continues to be difficulties in ensuring apparent methodology consistency as compared to quantitative studies. Consistency between studies tends to be much easier to demonstrate using the questionnaires and the statistical methods inherent to quantitative work. For example, a common problem for qualitative research, due to its heterogeneity, is the lack of standardised quality markers for assessing the value of the work. There is considerable tension within the qualitative community regarding the need for universality of quality indicators, with some arguing for a checklist for assessment and others arguing that each methodology needs to be judged against its own quality standards (see O'Reilly & Kiyimba, 2015, for an overview of these). Problematically for this research method, medical practitioners view standardisations and measures of quality as essential as markers for their practice, are more likely to seek it out when judging research studies. While there is increasing interest in producing meta-syntheses and systematic reviews of qualitative work (Dixon-Woods & Fitzpatrick, 2001), this has yet to achieve the respect that is given to meta-analysis and systematic reviews of randomised controlled trials, which reflects the hierarchy of evidence perspective mentioned earlier in the chapter.

Conversation analysis at first glance

While CA is part of the qualitative tradition, the qualitative approach is a heterogeneous rubric, encompassing a range of methodologies and epistemological positions (O'Reilly & Kiyimba, 2015). This can result in CA seeming a very different approach to that of other methods which people may be used to.

Conversation analysis is described as a method for studying social interaction utilising talk. This methodological approach was developed by Harvey Sacks together with Emmanuel Schegloff and Gail Jefferson in the 1960s and 1970s, and it emerged from Harold Garfinkel's work in ethnomethodology. This is an approach that has continued to evolve over time, with various developments within the field. It is described as a data-driven analysis of naturally occurring talk and looks at the recurring patterns in any given interaction between individuals or groups. As a method it is particularly interested in aspects of social dialogue, such as turn-taking in the conversations and sequence organisation of certain aspects of the talk (Hutchby & Wooffitt, 2008). By analysing these interactions at a micro level certain patterns can emerge from the data. It is a method that has been used in a variety of situations and is increasingly used to look at the interactions in a health environment. When considering CA as an approach, it is important to recognise that it is based upon certain assumptions (Peräkylä, 2004). First is that talk is action and is a vehicle for human action, such as opening or closing a conversation. Second, this action is structurally organised, with individuals orientating themselves to rules and structures that make the actions possible. Third, this talk in turn creates and maintains an intersubjective reality and the understanding of these meanings is publicly displayed. Fourth, the approach is data driven and is strongly biased against the motives and orientation of the speaker taking the meaning from the actual actions within the text, which is an issue which has generated intense debate (Billig, 1999; Schegloff, 1997). This is a different position from most other qualitative research, and most conversation analysts argue that the context is derived in and through the interaction and this in itself should identify the relevance of the context. For example, in research examining mental health clinical situations, there is debate regarding how much contextual detail the analyst requires of the patient's history, demographics, and problems to perform a detailed analysis. The general orientation of CA is that if and when relevant this contextual detail will be created through the interaction and is only analysable if and when it occurs. This lack of context may at first seem an unusual position to those first exposed to the approach, and while there is no clear conclusion to the argument, this approach does enable data to be viewed in an alternative way.

Discovering CA

My own exposure to CA was entirely by chance and this is probably the way clinicians find their way into this area of research. I am presently a clinical teaching fellow and honorary consultant in child and adolescent psychiatry. Following graduation I had no intention to pursue a career in psychiatry and looked to paediatrics. However, a chance encounter with a child psychiatrist during one of my rotations changed this view and I subsequently changed career. I decided to pursue an academic career which would allow a wider exposure to both research and teaching opportunities and enable a further development of the skills. It was fairly apparent from an early point that my understanding of research was fairly limited and I had adopted a clearly quantitative perspective to research questions.

Through interactions with non-clinical academic colleagues an increasing awareness grew regarding the role of qualitative methods, particularly their use in showing how a change may have occurred in a given situation rather than just the outcome. This exposure to other research methodologies, together with a further recognition that the therapeutic interaction is so essential in child mental health, further increased my interest in other ways of exploring healthcare matters. Over time, there was a growing realisation that much of the clinical evidence presented had significant limitations when utilised in real-world environments and that there was actually a paucity of naturally occurring data. This led to further exploration of discourse studies and following discussions with a particular colleague (O'Reilly) to CA. Unfortunately, my first exposure to this particular method was not favourable due to a struggle with the academic language and more problematically the length of some published papers. Anyone who has published in a medical journal would soon realise that there is a tight word count and the style expected is quite punchy with an emphasis often on what benefits these findings will have on clinical work. I am not sure if this is a reflection of editors trying to put as much in the journal as possible or a readership which has often little time to assimilate the research.

The value of CA to child psychiatry

While it is clear to an individual such as myself that CA does have a place within child psychiatry, the question is whether many others could be convinced of its value. As an approach, there are obviously the positive aspects such as its use of naturally occurring situations and that it looks at talk which is a powerful therapeutic medium. However, overcoming the first impressions of any new method is essential to those who wish to integrate CA within the mainstream healthcare arena. What will actually surprise many within clinical communities such as child psychiatrists is that there is a large range of published books and journal articles already using CA that examine clinical settings, in both physical and mental health. Some of this is reaching clinical audiences through publication in clinically focused journals, but the level of impact is questionable and probably remains quite low. This is problematic, given that some of the studies have great potential to inform healthcare practice, but clinicians sometimes have to actively seek it out. This is challenging if clinicians are unfamiliar with the approach or its potential to inform their work. Unfortunately, in postgraduate education there is also a reduced emphasis on the importance of clinical interactions, so any published literature in CA has to be easily accessible to promote any interest.

On a personal level there are a few studies which stand out from the literature as useful for informing my own practice. One study which is especially pertinent to child mental health and therapeutics was written by Ian Hutchby, who explores the discourses of child counselling using CA (Hutchby, 2005; see also Hutchby, Chapter 29, this volume). In the 2005 paper, Hutchby addresses an important mental health communication issue of 'active listening'. This process is described within the National Institute of Clinical Excellence (NICE, 2009) as part of the management of individuals with depressive illness. Although there is a brief description of this process within the NICE guidelines, the reality of the process is explicated through the analysis in Hutchby's publication. It becomes clear that there is no true definition of what constitutes active listening, and that only by using CA are we able to see how and if it is achieved in practice. Indeed, Hutchby highlights that in some cases it was not fully achieved.

Second, in an ongoing study of primary care (Heritage & Robinson, 2011), CA revealed the intricate and specific language use in doctor–patient communication. When reviewing the literature on CA, it became apparent that the use of words 'any' versus 'some' achieved different outcomes of different interactions, with 'any' being used as a way of closing down the conversations. In health situations this was constructed as a negative marker in the interaction, effectively reducing or stopping the presentation of additional patient concerns. Conversely, the recognition of the simple substitution of 'some' had a positive influence encouraging further disclosure. It was only as a consequence of the methodological approach utilised that this simple but valuable interactional process was identified. This therefore formed the basis for a randomised controlled trial application to test the different outcomes when doctors use either of these words when addressing additional patient concerns towards the end of the primary care appointment (Heritage, Robinson, Elliot, Beckett, & Wilkes, 2007). On a personal level I have also tried this in lectures, and it appears to work.

Although I have picked out these two excellent examples of how CA can be applied to medical practice, there are many studies that have contributed to our understanding of issues related to mental health or healthcare communication more generally. In the psychiatric literature, this approach has been utilised to explore some very difficult areas of mental health practice. There are a range of different areas explored, such as in explaining non-epileptic seizures (Plug, Sharrack, & Reuber, 2009), medically unexplained symptoms (Burbaum et al., 2010), and engaging those patients with psychosis (McCabe, Heath, Burns, & Priebe, 2002). For example, differentiating epileptic and non-epileptic seizure disorders is notoriously problematic even for experienced consultants, and any way which can help illuminate the diagnosis is important for clinical practice. CA studies of this kind thus help to explicate the ways in which these situations are realised in the clinical encounter. Only by using this approach can the nuances of the interaction become realised (Plug et al., 2009). The clinical situations described create significant difficulties for which there is no easy answer in clinical practice and there has been little research guidance on how to progress. What is significant is that the in-depth analysis may actually provide a way forward and this method could contribute to better patient care.

While there is a smaller quantity of conversation analytic literature on children, and particularly child mental health, there is a body of research exploring facets of this field. Some studies have focused on specific disorders such as autism (Sterponi, De Kirby, & Shankey, Chapter 15, this volume; Stribling, Rae, & Dickerson, 2009), but there have been many studies that have explored the actual practices of therapy and counselling with children. For example, there is analysis of children attending child psychiatry appointments (O'Reilly, Karim, Stafford, & Hutchby, 2014), children in family therapy (O'Reilly & Parker, 2013; Parker & O'Reilly, 2012), child counselling (Hutchby, 2007), and counselling helpline interactions (Danby et al., Chapter 15, this volume). Considering the importance of these encounters and the increasing emphasis on ensuring children enjoy better mental health, any evidence that can enhance the process should not be ignored.

In our recent work on child psychiatric assessments (O'Reilly et al., 2014; Stafford, Hutchby, Karim, & O'Reilly, 2014), the employment of the conversation analytic methodology allowed us to explicate the nuances of interaction and the relevant social actions that occurred within that setting. Surprisingly, for a speciality that prides itself on the need for therapeutic engagement and focuses on communicative practices, this had not been explored in much detail before, and especially in such a detailed way using an approach like CA.

Clinical relevance summary

Conversation analysis provides a unique perspective on clinically related matters. It enables nuances of clinical interactions to be explored in substantial detail. Although the doctor–patient relationship is being espoused as an essential component of modern medical practice, in many ways this is not utilised to its full potential. There are clinical examples in practice, where the use of words achieves a better outcome for patients, in the area of diagnosis and treatment, and it is important to recognise how the use of language can be a powerful therapeutic modality.

If conversation analysts want to break through into mainstream medicine, and want a greater impact in the field of child psychiatry, and child mental health more broadly, then some adaptation may be needed. The vocabulary and analytical design will need some translation for those who are less familiar with the general concepts. The length of the analysis will need to be more concise and clear points and practical relevance will need to be more obviously explicated. This is an essential way of working in the current competitive environment for research and clinical resources. Thus, it becomes more important to demonstrate that a methodology can deliver a cost-effective and usable interface. See Table 2.1 for a general overview of the practice implications.

Table 2.1   Clinical practice highlights

---

1. Psychiatrists and other medical professionals tend to affiliate with quantitative ways of thinking due to their training and professional backgrounds and adopt an evidence-based practice approach.
2. Qualitative evidence is however useful for exploring the patient and clinician perspective, and naturally occurring data particularly are able to identify and specify the nuances of interactions for showing why something may or may not work in practice.
3. In psychiatry, clinical interaction, particularly with children in child psychiatry, has an important therapeutic role in all aspects of treatment, but it remains a fairly under-researched domain.
4. Conversation analysis particularly is an invaluable methodology for illuminating the process of clinical interaction and is especially useful for exploring many areas of child mental health.
5. There are some limitations of the conversation analytic evidence that the community needs to address to make their research more mainstream in psychiatry.

Summary

In this chapter, I have presented a personal perspective on the use of CA for child psychiatry and the value of this approach for exploring child mental health. It is important to appreciate when working with medical professionals the inherent organisational psychology and ways of thinking in which these individuals are culturally embedded and how this affects their appreciation of research. Child psychiatry is a fairly new speciality but has only limited therapeutic options, and the interactions with families are seen as paramount to the psychiatric process. However, there has only been limited research within this area on this interaction and the field of CA could be more central to this issue. There are nonetheless some barriers which would need to be circumvented if this is to be achieved, particularly in terms of the need to show direct clinical benefits.

References

Beckman, H. B., Markakis, K. M., Suchman, A. L., & Frankel, R. M. (1994). The doctor-patient relationship and malpractice: Lessons from plaintiff depositions. *Archives of Internal Medicine, 154*(12), 1365–1370.

Billig, M. (1999). Whose terms? Whose ordinariness? Rhetoric and ideology in conversation analysis. *Discourse and Society, 10*(4), 453–582.

Bowlby, J. (1951). Maternal care and mental health. *World Health Organization Monograph* (Serial No. 2).

Burbaum, C., Stresing, A-M., Fritzsche, K., Auer, P., Wirsching, M., & Lucius-Hoene, G. (2010). Medically unexplained symptoms as a threat to patient's identity? A conversation analysis of patient's reactions to psychosomatic attributions. *Patient Education and Counseling, 79*(2), 207–217.

Corsaro, W. (2011). *The sociology of childhood* (3rd edition). California: Pine Forge Press.

Dallos, R., & Draper, R. (2010). *An introduction to family therapy: Systemic theory and practice* (3rd edition). Berkshire: Open University Press.

Danby, S., Cromdal, J., Rendle-Short, J., Butler, C., Osvaldsson, K., & Emmison, M. (this volume). Parentification: Client and counsellor talk on a helpline for children and young people. In M. O'Reilly & J. N. Lester (Eds.), *The Palgrave handbook of child mental health*. Hampshire: Palgrave MacMillan.

Dixon-Woods, M., & Fitzpatrick, R. (2001). Qualitative research in systematic reviews. *British Medical Journal, 323*, 765–766.

Eddy, D. (2011). History of medicine. *American Medical Association Journal of Ethics, 13*(1), 55–60.

Ford, T., Goodman, R., & Meltzer, H. (2003). The British child and adolescent mental health survey 1999: The prevalence of *DSM-IV* disorders. *Journal of the American Academy of Child & Adolescent Psychiatry, 42*(10), 1203–1211.

French, P. (1999). The development of evidence-based nursing. *Journal of Advanced Nursing, 29*(1), 72–78.

Freud, A. (1946). *The psycho-analytic treatment of children*. London: Imago.

Fuggle, P., Dunsmuir, V., & Curry, V. (2013). *CBT with children, young people and their families*. London: Sage.

Heritage, J., & Robinson, J. (2011). 'Some' versus 'any' medical issues: Encouraging patients to reveal their unmet concerns. In C. Antaki (Ed.), *Applied conversation analysis: Intervention and change in institutional talk* (pp. 15–31). Hampshire: Palgrave MacMillan.

Heritage, J., Robinson, J. D., Elliott, M. N., Beckett, M., & Wilkes, M. (2007). Reducing patients' unmet concerns in primary care: The difference one word can make. *Journal of General Internal Medicine, 22*(10), 1429–1433.

Hesse-Biber, S (2012). Weaving a mutlimethodology and mixed methods praxis into randomised control trials to enhance credibility. *Qualitative Inquiry, 18*(10), 876–889.

Hutchby, I. (2005). 'Active listening': Formulation and the elicitation of feelings-talk in child counselling. *Research on Language and Social Interaction, 38*(3), 303–329.

Hutchby, I. (2007). *The discourse of child counselling*. Amsterdam: John Benjamins.

Hutchby, I., & Wooffitt, R. (2008). *Conversation analysis* (2nd edition). Cambridge: Polity.

Jones, K. (1999). *Taming the troublesome child*. Harvard: Harvard University Press.

Kanner, L. (1935). *Child psychiatry*. Springfield, IL: C.C. Thomas Publishing.

Kovarsky, D. (2008). Representing voices from the life-world in evidence-based practice. *International Journal of Language & Communication Disorders, 43*(S1), 47–57.

Lester, J. N., & O'Reilly, M. (in press). Is evidence-based practice a threat to the progress of the qualitative community? Arguments from the bottom of the pyramid. *Qualitative Inquiry* (special issue).

Marks, D. (2002). Perspectives on evidence-based practice. *Health development agency: Public health evidence steering group*. Retrieved April 29, 2013 from http://admin.nice.org.uk/aboutnice/whoweare/aboutthehda/evidencebase/publichealthevidencesteeringgroupproceedings/perspectives_on_evidence_based_practice.jsp.

McCabe, R., Heath, C., Burns, T., & Priebe, S. (2002). Engagement of patients with psychosis in the consultation: Conversation analytic study. *British Medical Journal, 325*, 1148–1151.

Melander, H., Ahlqvist-Rastad, J., Meijer, G., & Beermann, B. (2003). Evidence b(i)ased medicine – selective reporting from studies sponsored by pharmaceutical industry: Review of studies in new drug applications. *British Medical Journal, 326*(7400), 1171.

National Institution of Clinical Excellence (NICE) (2009). Retrieved November 17, 2014 from http://www.nice.org.uk/guidance/cg90.

O'Reilly, M., Karim, K., Stafford, V., & Hutchby, V. (2014). Identifying the interactional processes in the first assessments in child mental health. *Child and Adolescent Mental Health,* doi: 10.1111/camh.12077.

O'Reilly, M., & Kiyimba, N. (2015). *Advanced qualitative research: A guide to contemporary theoretical debates*. London: Sage.

O'Reilly, M., & Parker, N. (2013). 'You can take a horse to water but you can't make it drink': Exploring children's engagement and resistance in family therapy. *Contemporary Family Therapy, 35*(3), 491–507.

Parker, N., & O'Reilly, M. (2012). 'Gossiping' as a social action in family therapy: The pseudo-absence and pseudo-presence of children. *Discourse Studies, 14*(4), 1–19.

Plug, L., Sharrack, B., & Reuber, M. (2009). Conversation analysis can help to distinguish between epilepsy and non-epileptic seizure disorders: A comparison. *Seizure, 18*, 43–50.

Peräkylä, A. (2004). Conversation analysis. In C. Seale (Ed.), *Qualitative research practice* (pp. 153–167). London: Sage.

Rice, M. (2008). Evidence-based practice in psychiatric care: Defining levels of evidence. *Journal of American Psychiatric Nurses Association, 14*(3), 181–187.

Rutter, M., Graham, P., & Yule, W. (1970). *A neuropsychiatric study in childhood*. London: Heinemann/SIMPP.

Rutter, M., Tizard, J., Yule, W., Graham, P., & Whitmore, K. (1976). Isle of Wight studies, 1964–1974. *Psychological Medicine, 6*, 313–332.

Sackett, D., Rosenberg, W., Gray, M., Hayes, B., & Scott Richardson, W. (1996). Evidence-based medicine: what it is and what it isn't. *British Medical Journal, 312*, 71–72.

Sackett, D., Strauss, S., & Richardson, W. (2000). *Evidence-based medicine: How to practice and teach EBM*. Edinburgh, UK: Churchill Livingstone.

Schegloff, E. (1997). Whose text? Whose context? *Discourse and Society, 8*(2), 165–185.

Stafford, V., Hutchby, I., Karim, K., & O'Reilly, M. (2014). 'Why are you here?' Seeking children's accounts of their presentation to CAMHS. *Clinical Child Psychology and Psychiatry*, doi: 10.1177/135910451454395.

Stribling, P., Rae, J., & Dickerson, P. (2009). Using conversation analysis to explore the recurrence of a topic in the talk of a boy with autism spectrum disorder. *Clinical Linguistics and Phonetics, 23*(8), 555–582.

Strong, T., Busch, R., & Couture, S. (2008). Conversational evidence in therapeutic dialogue. *Journal of Marital and Family Therapy, 34*(3), 388–405.

Tanenbaum, S. (2003). Evidence-based practice in mental health: Practical weaknesses meet political strengths. *Journal of Evaluation in Clinical Practice, 9*, 287–301.

United Nations Convention on the Rights of the Child (1989). As retrieved from UNICEF on December 10, 2014 from http://www.unicef.org.uk/Documents/Publication-pdfs/UNCRC_PRESS200910web.pdf.

Winnicott, D. W. (1964 [1987]). *The child, the family and the outside world*. Harmondsorth: Penguin; Reading, MA: Addison-Wesley.

Wykes, T., & Callard, F. (2010). Diagnosis, diagnosis, diagnosis: Towards DSM-5. *Journal of Mental Health, 19*(4), 301–304.

Recommended reading

- Heritage, J., & Robinson, J. (2011). 'Some' versus 'any' medical issues: Encouraging patients to reveal their unmet concerns. In C. Antaki (Ed.), *Applied conversation analysis: Intervention and change in institutional talk* (pp. 15–31). Hampshire: Palgrave MacMillan.
- Hutchby, I. (2005). 'Active listening': Formulation and the elicitation of feelings-talk in child counselling. *Research on Language and Social Interaction, 38*(3), 303–329.
- Hutchby, I. (2007). *The discourse of child counselling*. Amsterdam: John Benjamins.
- O'Reilly, M., Karim, K., Stafford, V., & Hutchby, V. (2014). Identifying the interactional processes in the first assessments in child mental health. *Child and Adolescent Mental Health*, doi: 10.1111/camh.12077.
- O'Reilly, M., & Kiyimba, N. (2015). *Advanced qualitative research: A guide to contemporary theoretical debates*. London: Sage.

3
The Value of Discourse Analysis: A Clinical Psychologist's View

Nikki Kiyimba

Introduction

As a discipline, clinical psychology has historically favoured a positivist approach to understanding human behaviour, and clinical psychological practice has been largely informed by research based on quantitative methods. Psychologists have tended to exploit the methodologies of the natural sciences by measuring phenomena (Peters, 2010). This bias towards quantitative research may at least in part be a function of how the discipline of psychology has from its genesis been careful to define itself as a science. It seems that the development of a broader engagement with alternative methods has largely grown from challenges to psychology's conception of what constitutes science and debates regarding its merits developed in relation to our thinking about science (Biggerstaff, 2012). The growth of qualitative methods in psychology is therefore relatively new despite the rich history of the approach (Howitt, 2010). However, now that psychology is more securely established, this has begun to lead to a refreshing openness to embrace qualitative process methodology as well as outcome research.

This evolution is in part due to more recent recommendations for greater methodological diversity in research and a growing recognition of the critical role that process research can play in supporting the integration of science and clinical practice (Rhodes, 2011). Acknowledgement of the value of examining the implementation of procedures involved in clinical practice has grown in part from recognition that quantitative approaches have not been able to adequately consider the role of the clinician in guiding the client's recovery (Duncan & Miller, 2005). It is also becoming more widely appreciated that a straightforward reliance on manualised treatments may fail to account for the critical role of the therapeutic relationship in determining treatment efficacy (Wampold et al., 1997).

The limitation of outcome-focused research has been that information about *how* the application of certain interventions is effective has not been fully understood, and practitioners and researchers alike have been left with questions regarding the process of change (Burck, 2005). Arguably however, although qualitative research has begun to be more widely used in many healthcare settings, qualitative research in mental health has not been as fully utilised (Peters, 2010). It is investigation of the dynamic aspects of the therapeutic process and appreciation of its situational and relationship variables as expressed through discourse that can illuminate the answers to the question of how therapy works. This redirection of attention from content to process may demonstrate what the active ingredients are within the dynamic process of change (Rhodes, 2011). A useful way, therefore, of understanding the clinical relationship is to discursively interrogate how the therapeutic relationship is talked into being and maintained (Roy-Chowdhury, 2006). This chapter explores the utility of discursive approaches as mechanisms for explaining the social processes involved in clinical therapeutic interactions and in particular the activity of psychological therapy with children. In doing so, I seek to illustrate how adopting a discourse approach can help us better understand the particular aspects of the therapeutic process which aid or inhibit attainment of the goal of improvements in child mental health. The aim is to examine how insights gained from these processes can inform clinical practice. As a practicing clinical psychologist engaged in empirical discursive research, I argue that the use of discourse analysis (DA) can be extremely useful in gaining better understanding of psychological processes and phenomena and has the potential to advance our knowledge in this field. Rhodes (2011) argued that clinical professionals need to be more informed on the different process methods, and I present an argument as to how a discursive understanding of child mental health can be useful to clinical practitioners and encourage future work of this nature.

DA and social constructionism

DA has become an umbrella term for a number of different approaches to the study of language and encompasses a wide range of theories, topics, and analytic approaches for explaining language in use (Shaw & Bailey, 2009). It is an approach that spans several disciplines wherein there is significant variation in how discourse is conceptualised and analysed (Georgaca, 2012). Although there are different origins and definitions, they do share several conceptions about social life: that context is relevant to language and interaction, that social reality is socially constructed, and that it is important to look beyond the literal meaning of language (Shaw & Bailey, 2009). What is crucial to appreciate is that the social constructionist epistemological framework of DA is based on challenging the taken-for-granted assumptions about the role of research as a mechanism for accessing internal cognitive states. Rather DA reflexively demonstrates that all data collected are situated artefacts of a particular context. As such, all forms of DA share the social constructionist perspective that language is context-bound, functional, and constructive (Wetherell, Taylor, & Yates, 2001). In the case of therapeutic process research, data are produced within the particular institutional setting, in response to particular interactional stimuli, and must therefore be analysed with a sensitive appreciation of the specificity of these variables of the social, institutional, and relational features of data production. DA also respects that all knowledge gathered is historically, socially, and culturally specific (Harper, 2006).

Language is not a neutral activity, nor is it an unbiased portal to cognitive processes. Rather, language is always functional in that it performs social actions, with speakers constructing what they say in order to accomplish certain activities that serve particular interests (Potter & Wetherell, 1987). When people talk, they are constantly in the process of performing social actions such as justifying, criticising, complementing, arguing, blaming, thanking, asking, and the like. It is in this sense that all language is constructed within a particular social context, and what is said is designed for that particular encounter with that particular person or persons. Even when talk is designed to appear as merely descriptive, the selection of which words to use and which aspects to highlight can vary immensely. For example, two people may be describing a tree. In the situation of a nature documentary, the tree may be described in terms of its evolutionary history, its adaptive qualities, and its botanical features. Being described by a poet to his lover it may be described in romantic metaphorical terms, emphasising its aesthetic qualities. Thus even description can never really be considered to be activity-free. Rather, the individual features of the speaker, the audience, and the social action that the words are vehicles are always factors that influence the way descriptions are presented.

Furthermore, the way that people talk about things goes beyond merely describing, but it actually creates and constitutes the reality of the world through our representation of it (Spong, 2010). The analysis of discourse therefore is a means to examine in detail the meaning and function of language within the context that the interaction takes place (Shaw & Bailey, 2009). As such the epistemological foundation of DA is social constructionism that holds to a number of assumptions. These are to doubt a 'taken-for-granted' world; to appreciate that knowledge is culturally, socially, and historically specific; that what is accepted knowledge persists through social processes; and that explanations and descriptions are not neutral but constitute social actions (Gergen, 1985). The premise of social constructionism is that all human understanding and experiences are fashioned by discourse, which is the predominant socially available means of conceptualisation (Georgaca, 2012). It is important therefore that therapists consider how their clients' problems are constituted through their discourses (Spong, 2010) and how those clients, particularly children, construct those problems (see e.g. Stafford & Karim, Chapter 17, this volume). The goal of DA is not to uncover 'truth' per se but to understand how meaning is constructed through accounts and the conditions that produce them (Hollway, 1989).

The social construction of mental health problems

Social constructionism as an epistemological position recognises that psychological phenomena are constituted through social and interpersonal processes (Georgaca, 2012). Davies and Harré (1990) argued that speakers both position themselves through language and are positioned by others by language. This viewpoint when directed towards the constitution of categories that organise mental health difficulties proposes that these categories are themselves produced through language that may construct identities for clients in such a way as to diminish their status (Harper, 1995). This message is at the heart of how diagnostic categorisations are used by medical professionals to construct the identities of the patients and clients that they work with. While diagnostic categories serve a purpose in identifying criteria for the justifiable prescribing of appropriate medication, categorisation by diagnosis can often impose an inflexible rigidity to how clients (re)conceptualise their identity within the construct of that diagnostic label. Understanding diagnostic categories as constructs does not mean that they should be dispensed with, but it does help us to remember that they are simply a means to an end and not an end in themselves. They are perhaps better understood as convenient medical shorthand ways of indicating to other professionals that a client has presented through their discourse a narrative that encapsulates a familiar pattern of characteristics. Additionally, they are based on an assumption not shared by all professions or researchers – that the problem is located within the individual (Harper, 1995). This epistemological position is contrasted with the social constructionist perspective that conceptualises mental health difficulties as constructed through discourse processes within a social, political, historical, and cultural context.

Within their professional role, clinical psychologists are often in a position where they are obliged to engage in the dominant discourses of categorisation. However, difficulties can arise between psychologists and medical colleagues when diagnostic categories are used as the basis for referral for a specific 'prescribed' therapeutic treatment. This is a potential arena for misunderstanding, as psychologists who are trained in multiple modalities of therapy work on the basis of assessment and formulation that will by clinical judgement indicate a particular treatment approach or combination of treatments. Formulation is a crucial element in the process of developing a treatment plan, and it seeks to incorporate the complexity of multiple elements of clients' life experience, including social, biological, psychological, and dynamic aspects. A move away from viewing mental illness as objective and towards engaging with the importance of language as constitutive means assumptions about mental health can be re-considered (Fee, 2000). In particular, language can be studied in its own right as something which is constructive of human experience rather than as a means to access a presupposed underlying psychological phenomenon (Georgaca & Avdi, 2009). In this way, those currently defined by the dominant ideology as 'mentally ill' may be more empowered to refocus their attribution discourses from internal to external, from 'being' to 'having' (Harper, 1995). Discourse analytic research therefore provides a valuable approach that demonstrates that the labels of psychopathology do not necessarily point to a pre-existing entity, but they are constructs which are produced within specific conditions (Fee, 2000). Thus, an approach such as DA, which is underpinned by social constructionism, is a useful way to study mental health (Harper, 1995).

Using DA to study mental health

As has already been outlined, discourse, whoever it is produced by, does not occur in a vacuum, and it is specifically designed for that particular interaction to perform a social action. For example, DA can demonstrate the way in which individuals use discursive strategies to justify their versions and construct a view of themselves as credible (Willig, 2001). A central function of DA studies in mental health has been to illuminate the historically contingent and socially constructed character of professional knowledge and practice (Georgaca, 2012). Crucially therefore, DA research can be a great aid to psychologists in the process of formulation, as with ongoing therapy, as it offers a way of extracting meaning within the context that the discourse occurred. This is particularly pertinent when working with client diversity, as clients from minority or under-represented groups are likely to use discourse frames that are less familiar and may not be the same as the cultural norms of the psychologist. Additionally, clients unfamiliar with the institutional context and relationship within which therapy occurs typically present their experience of difficulties in the same 'mundane' or familiar language that they use in their everyday lives. Often these discourses are non-medical, and therapeutic effectiveness can be enhanced by engaging with clients in the style of discourse and the constructs that are representative and meaningful to them.

Apart from the matter of 'mundane' versus 'institutional' styles of talk and the differences that lay clients and professional therapists may have in making sense of experiences and expressing them to others in therapy involving children, developmental aspects of communication and power are also relevant. In the case of working with children, there is the added complexity that the adult therapist may have different ways of communicating than the child.

The art of good therapy is to be able to subsume the client's own sense-making framework into one's own, to interpret it in terms of one's own therapeutic theoretical perspective, and to translate that back into language which the client can readily access. When working with children, it is not always possible in the moment for therapists to be so acutely attuned as to notice the nuances of children's modes of expressing their sense-making practices, and hence important information may be missed. DA therefore is an incredibly helpful tool that can enable therapists the luxury of opportunity to analyse recorded data in a way that can help illuminate these communication practices more clearly. Thus discourse becomes a direct resource for practice by enabling the opportunity to look at how meanings are constructed in conversation (Spong, 2010), which is especially important when working with children. There are a number of ways that Spong suggested that DA can facilitate our understanding of therapy, which are outlined in Table 3.1.

Table 3.1 How discourse analysis can facilitate understanding in therapy (*Spong, 2010*)

Facilitators

1. It allows the examination of language use in therapy.
2. Concepts from discourse are available for use as direct resources for practice.
3. It allows the exploration of client issues, populations and diagnoses.
4. It allows the analysis of discourses of particular models of therapy.
5. It allows analyses which focuses on broader social structures, power relations and meanings.

By investigating therapeutic conversations closely, discourse analytic research, including conversation analysis (CA), can demonstrate how therapeutic techniques are applied within the dynamic interactions within therapy (Rhodes, 2011). The current field of psychotherapy is a heterogeneous one that comprises of a number of different positions and opinions that are often grouped together and referred to as 'psychoanalysis' (Streeck, 2010). To date, all of the different modalities of therapy have not been fully investigated through the use of DA, and so the literature is limited. However, the following sections serve as examples from family therapy and dynamic psychotherapy as to how DA can be a beneficial analytic approach for two different clinical approaches.

DA in family therapy

Discourse analysis is becoming more established in family therapy, as it can offer a sophisticated insight into the complexity of family practice (Shaw & Bailey, 2009). Additionally, as Harper (1995) argued, DA can bridge the individual–social divide by demonstrating how the two are inseparable and intricately intertwined. The debate relating to the individual versus the social is at the heart of what family therapy is about. Essentially, the premise of family therapy is that it is not the individual child that is or has the problem, but that the difficulties that are being displayed by the individual child are symptomatic of a breakdown in the family system. This is an excellent example of the recognition that mental health difficulties are not solely a biological idiosyncrasy of the individual patient but in part constituted by and in turn constitutive of a wider social system. It makes a lot of sense therefore that discourse analytic researchers would find an affinity with the socially constructed rational that is at the basis of the concept of family therapy. Furthermore, some analysts of family therapy working within the discursive arena argue that it is important to consider the achievement of the therapeutic interaction as something that is a collaborative act between the therapist and clients who are all involved in 'performing' therapy, as well as co-managing both the content and process of talk (Sutherland & Strong, 2011). Thus, there have been a limited but growing number of studies that have used DA or CA to study the interactional processes within family therapy, and it has been argued that DA holds some promise for the study of systemic family therapy and for psychotherapy (Tseliou, 2013). One of the arguments for this is that DA provides an additional dimension to family practice as it draws upon theories and approaches from a range of disciplines. It also brings a different 'lens' that can usefully add to and deepen our understanding of family therapy interactions (Shaw & Bailey, 2009). In other words, by taking a more reflective stance, and considering the construction of reality as something that occurs between interactants, this can facilitate a greater understanding and appreciation of the systemic nature of difficulties that occur within families.

However, despite these promising recommendations, a review of family therapy research by Tseliou (2013) found that although DA as a methodology was very well suited to addressing the family therapy research questions, there were methodological shortcomings in the practical application of the approach. Avdi and Georgaca (2007) also noted that many discursive research studies they examined used a mixed analysis which comprised elements of discourse analysis, conversation analysis, and discursive psychology, and some were not clear in systematically defining the research question and the consistency between the method and epistemology (Avdi & Georgaca, 2007). Thus, in advocating the value of using DA as the research approach of choice as a clinical psychologist, I do so with the clear caveat that it must be conducted in a rigorous and methodologically congruent way, ensuring that the quality criteria for the particular methodology are carefully adhered to and that epistemological congruence is transparent. Avdi and Georgaca cite O'Reilly (2005) as a good practice example of DA research with family therapy data that follows this model.

DA in dynamic psychotherapy

Dynamic psychotherapy is based on a premise that the words of the client reveal underlying conscious and unconscious processes, and the analyst should be able to verbalise what is understood of the unconscious mental processes of the client (Streeck, 2010). Different therapeutic approaches place differential emphasis on the sequential aspects of how therapy talk unfolds. At a foundational level, there is a universal attentiveness to pairs of question/answer sequences; however, in dynamic psychotherapy, for example, there is a particular attentiveness to the specific sequential placement of statements and responses from clients in relation to the therapists' turns. Noticing these turn-by-turn details is intrinsic to the understanding and interpretation of the client's discourse. CA is a particularly detailed form of DA that attends very carefully to the minutiae of turn-taking placement and content to explicate interactional processes. Similarly, dynamic psychoanalytic therapy also acknowledges that the sequential placement of an utterance within a therapeutic conversation is not arbitrary and has clinical relevance. In general, most dynamic psychoanalysts are trained to monitor these fluctuations and notice links between utterances in this patterned way in order to reflectively analyse their importance within the trajectory of the conversation. This is usually done simultaneously to maintaining the interaction with the client in the present moment. By monitoring this trajectory on the basis of unfolding turns of interaction, psychoanalysts position themselves to offer an interpretation and then to work with the client's response after the interpretation (Peräkylä, 2011).

This is a highly skilled process to master well and is reflected in the length and rigour of training for dynamic psychoanalysts. In this approach, the therapist does not take notes during the session in order to attend fully to the client's discourse. The expectation of this model is that case notes written after the session will be very detailed, noting turn-by-turn interjections and responses. It is often during this reflective space of clinical note writing that therapists are able to formulate analytic links between utterances as they are recalled in sequential context. This combination of in situ sequential analysis and post-session analysis lends CA as an eminently compatible methodology for analysing data collected from therapy sessions conducted within this psychoanalytic frame. This is because CA also emphasises the importance of analytically taking into account the characteristics and sequential placement of utterances within the unfolding interaction (Peräkylä, 2011).

Clinical psychology outcomes and treatment efficacy

Therapy is about healing through verbal communication, and therefore at the very crux of everything that happens in therapy is the crucial interplay in the verbal realm between therapist and client. Different therapeutic approaches differ in regard to the emphasis placed on the relationship between therapists and their clients, and these approaches can be broadly categorised as following either the 'medical model' or the 'interactional model'. In the medical model, the therapist is viewed as an expert using a therapeutic method for the treatment of a diagnosed pre-existing illness, whereas in the interactional model, the therapist is an active participant whereby those factors ascribed to the mental disorder are shaped both by the client and by the influence of the therapist (Streeck, 2010). It is common practice now that providers of psychological therapy are required to justify the therapeutic approaches that they rely on by ensuring that the approaches are empirically grounded. One of the ways that has been put forward (and to a large extent taken up) has been the matching of specific treatments to specific conditions in a fairly protocol-driven or 'manualised' fashion. However, although this approach has had some success, many clients present with a complex array of co-morbid presentations and idiosyncratic symptoms that defy categorisation in this way.

Additionally, there is a growing body of research evidence that indicates that there is very little if any marked difference in the efficacy of any particular therapeutic model over another (Imel & Wampold, 2008; Wampold, 2001). A large-scale study carried out by Human Affairs International involving more than 2,000 therapists and 20,000 clients across 13 different approaches, including medication and family therapy, indicated that there were no significant differences in outcome (Brown, Dreis, & Nace, 1999). This area of research suggests that different types of therapy are equally effective and that improvement may be due to what is generally referred to as the 'therapeutic alliance' that transcends method. This alliance is more than an empathic and caring relationship; it is also an agreement about the goals and tasks of psychotherapy. Considering the importance of relationship when working with children in particular, it seems very sensible and potentially profitable for researchers to place more emphasis on seeking to understand the interactional dynamics that lead to effective treatment. It is always important to remember that psychotherapy requires clients to bravely face some of their deepest fears and also to risk the implications of changing core aspects about their beliefs in themselves and others. Clients are willing to engage in this process only if they believe the psychotherapist understands them and that the treatment offered will benefit them (Wampold, 2007). For children, I would argue that the matter of trust and of empathic rapport are even more essential than working with adults, as this relationship is central to effectively engaging the child in the treatment and enabling them to feel safe in an unfamiliar environment.

It has also been shown that in clinical practice, some therapists working within the same model as their colleagues consistently attain better outcomes than their counterparts regardless of the client's presenting problems, age, developmental stage, or medication status (Wampold & Brown, 2005). Therefore, if the therapeutic modality is not the primary determinant of positive clinical change, then there must be other factors involved in the interaction that account for outcome variability between therapists (for an interesting discussion of therapy, outcomes, and the role of discourse, see Roy-Chowdhury, Chapter 11, this volume). Thus, there is a need to explore the interplay of client and therapist variables in order to understand the means by which outcomes are established. Studies suggest that there are a number of common factors inherent within therapeutic interventions which have more successful outcomes, which has led to what is known as 'common factors theory' (Imel & Wampold, 2008). Common factors theory claims that there are interactional meta-factors that supersede specific treatment protocols in treatment efficacy. With this in mind, some researchers believe that investing resources in the development of treatment manuals is misguided and that instead of trying to fit clients into manualised treatments via evidence-based practice therapists should instead tailor their work to individual clients' needs by using 'practice-based evidence' (Duncan & Miller, 2005). The utilisation of practice-based evidence is proposed as a practical solution to the management of the complexity and unpredictability involved in the process of psychotherapy and involves taking a client-led approach to monitoring clients' progress and their view of the alliance and adjusting the intervention accordingly (Duncan, Miller, & Sparks, 2004).

Clients engage best in psychotherapy when they receive a treatment that is consistent with their expectations, have positive expectations for success, and feel understood by the psychotherapist (Wampold, 2007). This means that psychologists need to be attuned to patients' attitudes, values, culture, and expectations and of their motivation for change, coping styles, and their tendency to resist and select or adapt treatments accordingly (see Norcross, 2002). Awareness of all of these factors can certainly help to facilitate a containing therapeutic alliance within which clients can feel safe to explore difficult aspects of themselves and their interactions with others. It is also very common in mental health services to find that clients have a poor attachment history, which impacts on their difficulty in forming a therapeutic alliance. When working with children and young people in child and adolescent mental health services, it is always important to bear in mind that these children more than most are likely to struggle to make secure attachment relationships with others, including the therapist. Thus, it is the responsibility of the therapist to skilfully manage this challenge, as it is arguably the psychotherapist's contribution, not the client's contribution to the alliance that makes a difference (Baldwin, Wampold, & Imel, 2007).

Clinical psychology and DA

The medical model approach tends to hold a position that efficacy in psychotherapy is related to the application of specific critical ingredients to specific identified clinical problems (Imel & Wampold, 2008). Within the profession of clinical psychology, the medical model is generally ill received due to the perception that it is unhelpful to try to operate within a framework that dictates specific therapeutic interventions for specific diagnoses. Rather, the process of treatment followed by clinical psychologists is to develop a formulation based on clinical and psychometric assessment that will then inform the implementation of psychological therapy appropriate to the presenting problem and to the psychological and social circumstances of the client (Health Care Professions Council, 2012).

In relation to the specifics of using DA as a viable and preferable mode of analysis for a range of research topics pertinent to the field of clinical psychology, there are a number of important considerations to bear in mind. One issue is that there is a need for adequate supervision of trainee clinical psychologists working at a doctoral level to gain their professional qualification status. My personal experience of this both as a trainee and as a qualified psychologist teaching and supervising trainees has been that there is generally a shortage of confident, qualified qualitative researchers in institutions where clinical psychology training courses are run. This limits the possibilities that trainees have in terms of the research projects that they are able to undertake as students and consequently the likelihood of them having the necessary skills to engage in further research post-qualification. It is important therefore that academic institutions offering clinical psychology courses ensure not just an adequate percentage of staff are qualified and experienced in conducting and supervising some of the more thematic-based qualitative approaches such as interpretative phenomenological analysis, thematic analysis, and template analysis but that there are also staff who are competently able to supervise good quality Discourse Analytic student projects.

Rhodes (2011) argued that CA in particular aligns closely with the ethos of clinical psychology as it is concerned with understanding linguistic repertoires and that it has been already usefully applied in exploring interactions between professionals and clients. In other areas of healthcare research, these micro-level studies that involve the detailed study of language in use tend to be concerned with the techniques and competencies involved in conversation (Heritage & Manyard, 2006). It is precisely these competencies that are exhibited in psychotherapy interactions that can be fully explicated through forms of DA including conversation analysis. This in turn allows for empirically grounded and concrete specifications of what is done during the interaction and how it is achieved (Halkowski & Gill, 2010). Thus, the knowledge produced through discursive approaches such as CA has the advantage of revealing the subtleties of effective interactions with clients by providing 'a means by which students and clinicians can discover what therapy really looks like, not a didactic phenomenon, but one that requires significant interpersonal skill' (Rhodes, 2011, p. 5).

Clinical relevance summary

With a move towards the manualisation of treatments, this has fostered a more reductionist approach to both clinical practice and clinical research. In other words, in a desire for greater efficiency within mental healthcare provision, there has been an emphasis placed on categorisation of mental health difficulties, not just so that appropriate pharmacological interventions can be prescribed, but also so that psychotherapeutic interventions can be prescribed in therapy-specific and session-limited 'dosages'. While it makes good economic sense to have a way of being able to sensibly predict the likely cost of a treatment package for a particular client with mental health difficulties, often these attempts to sort clients into diagnostic categories and apply recommended treatment options fail to take into account the complexities of many individuals' co-morbid and interconnected difficulties. Manuals furthermore do not adequately consider the role of the clinician in guiding clients' recovery (Duncan & Miller, 2005).

The assignment of particular treatment options also fails to adequately account for the literature that indicates quite clearly that of particular significance in the efficacy of any treatment is the therapeutic alliance. It may therefore be very beneficial to those both training and practicing in clinical psychology as a profession to have access to good quality research which is able to explicate the processes that occur within therapy that demonstrate how a good therapeutic alliance can be attained. This is a slightly different way of approaching and understanding treatment efficacy from the choice of model approach, which can be largely assessed through outcome measures. If we are truly to advance our knowledge and understanding of what works for clients, a more in-depth analysis of how it works is required.

The answers to how successful therapy is accomplished, I suggest, are most readily discovered through the detailed analysis of the processes at work within therapeutic exchanges by recording and analysing actual naturally occurring therapy data. However, as Harper rightly pointed out, 'a discourse analysis that does not have any implications for the practical organisation of mental health services is an impotent one' (Harper, 1995, pp. 353–354). The necessity for close working partnerships between researchers and psychologists is essential in order that information learned from clinical practice is used effectively to inform the growing evidence base about not just treatment efficacy but process and therapeutic relational efficacy. This growth of practice-based evidence will be invaluable in developing trainee clinical psychologists' core clinical competencies as well as adding quality and depth to the clinical practice of their already qualified colleagues. In turn, the availability of good quality qualitative process research will be of great benefit in informing clinical practice and influencing decisions made about how treatment options are decided.

Additionally, the work of psychotherapy is a uniquely reflective act which requires the therapist to give as much focus on themselves as the instrument of therapeutic change, as it does on reflection and formulation of the client's world to make meaning of their struggles. What is unusual about psychotherapy is that therapists must simultaneously be aware of their own reactions, biases, world-views, beliefs, and perceptions while listening to clients' narratives. Inevitably, all of these inner attitudes, beliefs, and presuppositions that the therapist brings to the interaction will have some kind of impact on how the interaction plays out. DA offers an accessible tool for aiding the process of reflective practice that all therapists must engage with. At a theoretical level DA offers a framework for consideration of how each therapist's world-view affects their priorities, ideas about pathology, and treatment processes. I would suggest that the active effort to engage with and consider one's own ways of constructing the social world will enhance the status of reflectivity within clinical practice to the benefit of clients. At a practical level, engaging in discursive analysis of therapy sessions is also a valuable way to illuminate these social actions and processes more fully and thus to demonstrate the real-life ways that realities are created by and between therapists and their clients. For a simple summary of the implications for practice, see Table 3.2.

Table 3.2   Clinical practice highlights

1. Discourse analysis is an excellent tool to investigate the process of therapy.
2. Discourse analysis can help to explicate the nuances of therapeutic alliance.
3. There is a need for epistemological congruence in discourse analytic research in therapeutic interactions.
4. Clinical psychology doctoral training courses need to ensure that staff are experienced and skilled in discourse analysis in order to support trainee research.
5. There are significant advantages in developing collaborative partnerships between clinicians and academics to produce good quality applied discursive research.

Summary

In this chapter, I have sought to offer my own fairly unique perspective as one of the few practicing clinical psychologists who is also active as a discourse analytic researcher. Having previously once been a purely DA researcher investigating the interactions of adults and children in family therapy encounters, my concerns were not about what the therapist was 'trying to do' but simply on taking an observer stance to follow the unfolding process of how one turn of talk would lead to another turn of talk from the recipient. The advantage of not being a trained therapist at the time was that I was not 'cluttered' by meta-cognitions about potential intentionality on either part, but I was solely interested in how utterances were treated in those interactions. I still strongly believe that there is a very important role for academics working in applied psychology to retain this singularity of focus. However, having subsequently qualified as a clinical psychologist and spent several years in clinical practice, that attentiveness to the detail of talk, I believe, has served me well as a clinician and I now have the very good fortune of being able to combine the two things that I love to do – therapeutic work and discourse analytic research.

Having worked with both adults and children, attachment difficulties are clearly present in a large percentage of clients and are always played out in the therapeutic relationship. The way in which these relationship patterns are demonstrated varies enormously however and can be particularly challenging when working with children and young people. The nature of psychological therapy is often to find ways to gently approach areas that clients have tried very hard to protect themselves from through a variety of avoidance strategies. Skilfully working with clients in exposure work requires a level of trust in the relationship for the client to be willing to take the risk. For those clients with greater incidents of trauma in their lives, there is a necessity for a longer period of time initially in order to build this safe space. It may seem that during this stabilisation phase no obvious clinical change in symptoms can be seen, but nevertheless it is crucial to the process in order to achieve a stable therapeutic relationship that the client can trust. I suggest that DA is an excellent analytic approach which can help us to investigate the subtle nuances of the verbal interactions in these phases of treatment in order to help us have a fuller understanding of the essential ingredients within the therapeutic alliance which act as a vehicle for the delivery of therapeutic 'content'.

For those readers who may be practicing clinicians and who would love to be more involved in research, but perhaps have not had the opportunity to study this methodology in detail, I would recommend seeking out mutually advantageous relationships with academics in the field. It is this collaboration of working between academics and clinicians that I feel will give the best results in terms of producing really good quality discursive research that is robustly epistemologically grounded and also sensibly applied to informing good clinical practice. Economically and ethically, more than ever, there is an onus on researchers to be demonstrating that their research has a useful real-world application that justifies its 'indulgence'. In turn there are also clinicians who can see the need for research in particular areas and have ready access to clients who may be willing to allow their therapy sessions to be recorded for the benefit of knowledge growth. For the sake of the credibility of qualitative research as a discipline, and of DA in particular, there needs to be really good quality work produced which is of a credible quality standard. I believe this is best achieved by the promotion of effective clinical/academic partnerships that can maintain the academic rigour of discourse analytic research.

References

Avdi, E., & Georgaca, E. (2007). Discourse analysis and psychotherapy: A critical review. *European Journal of Psychotherapy and Counselling, 9*(2), 157–176.

Baldwin, S. A., Wampold, B. E., & Imel, Z. E. (2007). Untangling the alliance-outcome correlation: Exploring the relative importance of therapist and patient variability in the alliance. *Journal of Consulting and Clinical Psychology, 75*, 842–852.

Biggerstaff, D. (2012). Qualitative research methods in psychology. In G. Rossi (Ed.), *Psychology – selected papers* (pp. 175–206). Croatia: In Tech Open Science.

Brown, J., Dreis, S., & Nace, D. K. (1999). What really makes a difference in psychotherapy outcome? Why does managed care want to know? In M. A. Hubble, B. L. Duncan, & S. D. Miller (Eds.), *The heart and soul of change: What works in therapy* (pp. 389–406). Washington, DC: American Psychological Association Press.

Burck, C. (2005). Comparing qualitative research methodologies for systemic research: The use of grounded theory, discourse analysis and narrative analysis. *Journal of Family Therapy, 27*, 237–262.

Davies, B., & Harre, R. (1990). Positioning: The discursive production of selves. *Journal for the Theory of Social Behaviour, 20*(1), 43–63.

Duncan, B. L., & Miller, S. D. (2005). Treatment manuals do not improve outcomes. In J. Norcross, L. Beutler, & R. Levant (Eds.), *Evidence-based practices in mental health: Debate and dialogue on the fundamental questions* (pp. 140–148). Washington, DC: American Psychological Association Press.

Duncan, B. L., Miller, S. D., & Sparks, J. (2004). *The heroic client: A revolutionary way to improve effectiveness through client directed outcome informed therapy* (revised edition). San Francisco: Jossey-Bass.

Fee, D. (2000). The broken dialogue: Mental illness as discourse and experience. In D. Fee (Ed.), *Pathology and the postmodern: Mental illness as discourse and experience* (pp. 1–17). London: Sage.

Georgaca, E. (2012). Discourse analytic research on mental distress: A critical overview. *Journal of Mental Health*, doi: 10.3109/09638237.2012.734648.

Georgaca, E., & Avdi, E. (2009). Evaluating the talking cure: The contribution of narrative, discourse, and conversation analysis to psychotherapy assessment. *Qualitative research in Psychology, 6*, 233–247.

Gergen, K. J. (1985). The social constructionist movement in modern psychology. *American Psychologist, 40*(3), 266.

Halkowski, T., & Gill, V. (2010). Conversation analysis and ethnomethodology: The centrality of interaction. In I. Bourgeault, R. DeVries, & R. Dingwall (Eds.), *Handbook of qualitative health research* (pp. 212–228). London: Sage.

Harper, D. (1995). Discourse analysis and 'mental health'. *Journal of Mental Health, 4*, 347–357.

Harper, D. (2006). Discourse analysis. In M. Slade & S. Priebe (Eds.), *Choosing methods in mental health research: Mental health research from theory to practice* (pp. 47–67). Hove: Routledge.

Health and Care Professions Council (2012). *Standards of proficiency: Practitioner psychologists*. London: HCPC.

Heritage, J., & Maynard, D. (2006). *Communication in medical Care: Interaction between Primary care physicians and patients.* Cambridge: Cambridge University Press.

Hollway, W. (1989). *Subjectivity and method in psychology: Gender, meaning and science.* London: Sage.

Howitt, D. (2010). *Introduction to qualitative methods in psychology* (2nd edition). Essex: Pearson Education Limited.

Imel, Z., & Wampold, B. (2008). The importance of treatment and the science of common factors in psychotherapy. In S. Brown & R. Lent (Eds.), *Handbook of counseling psychology* (4th edition) (pp. 249–262). London: John Wiley & Sons Inc.

Norcross, J. C. (Ed.) (2002). *Psychotherapy relationships that work: Therapist contributions and responsiveness to patients.* New York: Oxford University Press.

O'Reilly, M. (2005). The complaining client and the troubled therapist: A Discursive investigation of family therapy. *Journal of Family Therapy, 27,* 371–393.

Peräkylä, A. (2011). A psychoanalyst's reflection on conversation analysis's contribution to his own therapeutic talk. In C. Antaki (Ed.), *Applied conversation analysis: Intervention and change in institutional talk* (pp. 222–242). Hampshire: Palgrave MacMillan.

Peters, S. (2010). Qualitative research methods in mental health. *Evidence Based Mental Health, 13*(2), 35–40.

Potter, J., & Wetherell, M. (1987). *Discourse and social psychology: Beyond attitudes and behaviour.* London: Sage.

Rhodes, P. (2011). Why clinical psychology needs process research: An examination of four methodologies. *Clinical Child Psychology, 17*(4), 495–504.

Roy-Chowdhury, S. (2006). How is the therapeutic relationship talked into being? *Journal of Family Therapy, 28,* 153–174.

Shaw, S., & Bailey, J. (2009). Discourse analysis: What is it and why is it relevant to family practice? *Family Practice Advance Access, 29,* 413–419.

Spong, S. (2010). Discourse analysis: Rich pickings for counsellors and therapists. *Counselling and Psychotherapy Research, 10*(1), 67–74.

Streeck, U. (2010). A psychotherapist's view of conversation analysis. In A. Peräkylä, C. Antaki, S. Vehvilainen, & I. Leudar (Eds.), *Conversation analysis and psychotherapy* (pp. 173–187). Cambridge: Cambridge University Press.

Sutherland, O., & Strong, T. (2011). Therapeutic collaboration: A conversation analysis of constructionist therapy. *Journal of Family Therapy, 33*(3), 256–278.

Tseliou, E. (2013). A critical methodological review of discourse and conversation analysis studies of family therapy. *Family Process, 52*(4), 653–672.

Wampold, B. E. (2001). *The great psychotherapy debate: Models, methods, and findings.* Mahwah, NJ: Lawrence Erlbaum Associates Publishers.

Wampold, B. E. (2007). Psychotherapy: The humanistic (and effective) treatment. *American Psychologist, 62*(8), 857–873.

Wampold, B. E., & Brown, G. S. (2005). Estimating therapist variability: A naturalistic study of outcomes in managed care. *Journal of Consulting and Clinical Psychology, 73,* 914–923.

Wampold, B., Mondin, G., Moody, M., Stich, F., Benson, K., & Ahn, H. (1997). A meta-analysis of outcome studies comparing bona fide psychotherapies: Empirically 'all must have prizes'. *Psychological Bulletin, 122,* 203–215.

Wetherell, M., Taylor, S., & Yates, S. (Eds.) (2001). *Discourse theory and practice: A reader.* London: Sage.

Willig, C. (2001). *Introducing qualitative research in psychology: Adventures in theory and method.* Maidenhead: Open University Press.

Recommended reading

- Gergen, K. J. (1985). The social constructionist movement in modern psychology. *American Psychologist, 40*(3), 266.
- O'Reilly, M. (2005). The complaining client and the troubled therapist: A discursive investigation of family therapy. *Journal of Family Therapy, 27,* 371–393.
- Peräkylä, A. (2011). A psychoanalyst's reflection on conversation analysis's contribution to his own therapeutic talk. In C. Antaki (Ed.), *Applied conversation analysis: Intervention and change in institutional talk* (pp. 222–242). Hampshire: Palgrave MacMillan.
- Roy-Chowdhury, S. (2006). How is the therapeutic relationship talked into being? *Journal of Family Therapy, 28,* 153–174.

4

Looking or Spotting: A Conversation Analytic Perspective on Interaction between a Humanoid Robot, a Co-present Adult, and a Child with an ASC

Paul Dickerson and Ben Robins

Introduction

There is a danger that when we look at the behaviour of children with an Autistic Spectrum Condition (ASC) we can wind up behaving like the character searching for his key in a well-known Mulla Nasruddin Sufi tale. In *The Lamp and Key* the protagonist is described as searching in his garden for a key that he had lost inside his house. When a neighbour asked why he searched in the garden, when the key had been lost indoors, he replied, 'Because there is much more light here than in my house'. This chapter draws on conversation analytic ideas to suggest that checklists and coding schedules can be a bit like Mulla Nasruddin's lamp – perhaps highlighting some interesting behaviours at the expense of important others that lie elsewhere, outside of that which the list illuminates. In examining these issues, the chapter will consider some of the potential limitations of using checklists to either make diagnoses of ASCs or to measure the manifestation of observable behaviours that are associated with ASCs. This is particularly important in light of contemporary debates regarding the classification of autism in terms of the language of mental health and disability (see O'Reilly, Karim, & Lester, Chapter 14, this volume).

The argument developed is that using checklists to identify and record pre-specified behaviours may risk creating two blind spots. First, the checklist runs the risk of focusing more on the target behaviour than the nuanced detail of the sequential context in which that behaviour is found. Second, by identifying only certain pre-specified behaviours as those to be observed and recorded, a checklist inevitably results in other, potentially relevant, behaviours being omitted. This introduction will briefly consider two highly influential forms of checklist, the E-2 and the Autism Diagnostic Observation Schedule, before outlining how conversation analysis can raise a challenge for checklist-led observations, but a challenge that may perhaps give far more than it takes away.

Checklists have played a crucial part in clinical assessments, interventions, and empirical work involving children with an ASC at least since Rimland's (1962, discussed in Rimland, 1971) E-2 Checklist. Published 19 years after Kanner's (1943) influential paper, the E-2 was aimed to diagnose children with what was then known as 'Kanner's Syndrome' or 'Classical Autism'. This early checklist, in contrast to some that followed, was to be completed by parents, rather than professionals, or those with special assessment training. Furthermore, the questions asked did not relate to behaviour within specific assessment or observation sessions but instead were more concerned with observations gained as a result of parental contact across the lifespan of the child. The E-2 covers an enormous range of issues (e.g. medical conditions, physiological characteristics, the child's preferences, behavioural patterns, and competencies); while some of its items are rarely used in current checklists, some remain important. Items such as Part One item 51 – 'Is it possible to direct the child's attention to an object some distance away or out of a window' – and Part One item 53 – 'Does the child look up at people (meet their eyes) when they are talking to him?' – relate to issues that are of importance in many contemporary checklists, including what is often seen as the 'gold-standard' observation-based diagnostic tool, the Autism Diagnostic Observation Schedule (ADOS).

ADOS (ADOS-G and ADOS-2), discussed in Lord et al. (2000), is quite a different entity to the E-2. While the E-2 relies on parental judgements of children's behaviour over their lifespan, ADOS relies on specific assessment sessions conducted by a professional trained in the ADOS technique. Furthermore, while E-2 relies on characterising the child's typical behaviour in everyday life, ADOS places the emphasis very much on recording the success or failure of a child within structured interaction, with criteria that define what constitutes a success. Typically, though not exclusively, the interactions that ADOS examines are between the trained assessor and the child. These contrasts between E-2 and ADOS reflect the deep concern with standardisation at the heart of ADOS – by having assessors who have completed the same training, observing clearly defined behaviours within specific structured interactions the hope is that the ADOS checklist is – as far as possible – an unchanging metric with which data can be consistently measured and meaningfully compared.

On the one hand, it would be wrong to build a 'straw person' of ADOS or any other assessment tool that has a checklist component. The accusation of rigid inflexibility, for example, would not do justice to the way in which tools such as ADOS have endeavoured to capture the range of ages and abilities of those who may be assessed with different modules. ADOS-2 shows still more flexibility in suggesting, for example, that if the assessor finds that one component (or module) of the assessment seems inappropriate to the interests or abilities of the child, the assessor may switch to a module that is more appropriate. However, on the other hand, the checklist approach of spotting pre-defined 'successful' and 'unsuccessful' behaviours in interaction scenarios that are defined, initiated, and structured by the assessor can be seen as losing something at the shrine of standardisation.

In contrast to coding that is informed by a pre-specified checklist of target behaviours, Sacks (1984) stressed the importance of 'unmotivated looking', which entails 'giving some consideration to whatever can be found in a particular conversation you happen to have your hands on' and 'subjecting it to investigation in any direction' (Sacks, 1984, p. 27). Sacks further notes that 'from close looking at the world we can find things that we could not, by imagination, assert were there' (Sacks, 1984, p. 25). From this perspective then, checklists of pre-specified behaviours and pre-defined notions of what constitutes task 'success' may – like any guiding set of expectations – be limited by our thoughts about what we might find. This, in turn, may unintentionally hinder us from noticing what may be unanticipated and yet particularly important features of the interaction we are examining. Furthermore, any *a priori* list of target behaviours may encourage us to focus on the specified behaviour in isolation, rather than examining interaction within the sequential context where it occurs. Finally, any attempt to specify 'success' or 'failure' in advance of examining the detail of the sequence of interaction as it actually occurred may rely on our, albeit well informed, *ideas* rather than the often counter-intuitive interactional *reality*.

This conversation analytic emphasis on the careful examination of sequences of interaction and sensitivity to the ways in which behaviours may orientate to that sequence has increasingly informed analyses of interactions involving one or more children with an ASC. In a number of cases this has led to rethinking behaviour that may, out of context, appear to be a mere manifestation of individual pathology in terms of *what it is doing interactionally* (Barrow & Tarplee, 1999; Stribling, Rae, & Dickerson, 2005, 2007, 2009). To take one example, Dickerson, Stribling, and Rae (2007) argued that what may appear to be mere repetitive tapping by a child with an ASC when looked at in isolation can be seen as interactionally meaningful by virtue of *where it occurs in the sequence of interaction*. This approach has brought to analysis of interaction involving children with an ASC and robotic platforms, a specific interest in examining *sequences of interaction* (Dickerson, Rae, Stribling, Dautenhahn, & Werry, 2005; Robins, Dautenhahn, & Dickerson, 2009; Robins, Dickerson, Stribling, & Dautenhahn, 2004).

This chapter adopts the conversation analytic maxims of treating all interactions as potentially relevant data, of unmotivated looking (rather than heavily pre-specified spotting), and of detailed attention to sequential context. These guiding thoughts are brought to bear on data derived from a series of encounters between a humanoid robot, one or two co-present adults, and a child with an ASC that are described below. The data are investigated so as to illustrate the importance of treating all interactions as potentially relevant and of understanding sequential context as crucial for making sense of any specific instance of behaviour that is being examined. Thus, conversation analysis enables consideration not just of *how much* of a target behaviour (such as gaze) is present but crucially *where it is positioned* in the interactional sequence. The analysis presented here highlights some of the potentially relevant behaviours that might have been unnoticed, or differently construed, had an alternative guiding framework – such as a checklist of target behaviours, or an a priori definition of task success and failure – been used.

Project overview

The data analysed below are taken from a series of studies in which children who had been diagnosed as having an ASC were filmed across approximately eight sessions interacting with KASPAR, a minimally expressive humanoid robot who is approximately the size of a one-year-old child. These data were chosen for this study simply because they comprised video recordings of extensive child–adult interactions. In this sense, the robot provided a shared object of attention for much of the interaction. The two extracts presented here are drawn from a bank of data which was in part funded by the ROBOSKIN project (ROBOSKIN, 2012). The ages of the children reported in this chapter were 8 (Extract 1) and 9 (Extract 2) at the time of the recording being made.

The robotic platform KASPAR

KASPAR is a small robot developed by the Adaptive System Research Group at the University of Hertfordshire, UK. It is a small child-sized minimally expressive humanoid robot which acts as a platform for human robot interaction (HRI) studies using bodily expressions (movements of the hand, arms, and facial expressions), gestures, and voice to interact with a human (Figure 4.1).

The robot has a moveable head, arms, and torso, as well as eyes that can open and close and a mouth also capable of opening and closing and of displaying a smiling and as well as a sad expression. In addition, the robot has a pre-recorded bank of utterances, such as 'Hello my name is KASPAR' and 'Ouch, that hurts'. In the data presented below, all gestural and vocal actions on the part of KASPAR were triggered by means of a remote control keypad; this was sometimes operated by the adult and sometimes by the child. Details of the robot's design rationale, hardware, and application examples are covered in Dautenhahn et al. (2009).

Figure 4.1   The robotic platform KASPAR
The figure on the left shows some of the tactile skin patches mounted on the robot which enable responses to being touched by the child.

The interaction setting

The data presented here are derived from interactions that took place in a small room within the school that each child attended. The children were brought into the room one at a time, with sessions lasting approximately 10 minutes per child. The adult who designed KASPAR ('B' in the transcripts below) typically suggested actions that the child might try, interacted with the children, and provided the opportunity for the children to initiate actions, to check KASPAR's response to different forms of touch and to control KASPAR via the control pad. As the analysis below suggests, the relatively unscripted nature of the adult's interaction with the child enabled behaviours that had not been envisaged to form part of the analysis.

Analysis of data

Extract 1

In Extract 1 the child ('C') is introduced to the buttons on the control pad, enabling them to explore the way in which the 'cause' of pressing a button has a specific consequence or 'effect' in terms of KASPAR's actions; for example, seeing how KASPAR blinks once the blinking button on the control pad has been pressed. The child's support worker ('A') is present along with KASPAR ('K') and the researcher who designed KASPAR ('B'). In the following data extract the analysis seeks to make sense of a puzzle; the child does gaze at KASPAR a great deal but does not appear to readily comply with adult instructions to do so. In addressing this issue, an understanding of what is taking place in the sequence of interactions is drawn on to make sense of this apparent discrepancy.

It should be noted that these data, and this analysis, do not seek to deny what may be genuine differences between the behaviour of the child with an ASC here and the possible behaviour of 'neuro-typical' children – it might well be the case that more obvious obedience to instructions could be found in populations of children who are not diagnosed as having an ASC. However, the analysis seeks to come to a more detailed understanding than a checklist look the data might allow. With our checklist in hand, we might simply spot a child ignoring an adult instruction to look at a specific target. We might further note that the instruction is ignored even when it is accompanied with adult gaze and pointing at the target object. Our knowledge of ASCs might then be turned to in order to make sense of these findings. By following this alternative course of examining features of the interaction that we might not have anticipated in advance, we have the opportunity to consider the ways in which features of the interaction as it occurs, in all of its detail, may shed light on the behaviour of the child with an ASC (Figure 4.2).

In the extract below, C is gazing at KASPAR's control pad throughout, except where indicated. In addition to conventional CA transcription symbols (see the transcription conventions at the end of this chapter), + is used to indicate C's independent press of a button on KASPAR's control pad and X to indicate C's gaze at specific locations, apart from the control pad, with a suffix indicating

Figure 4.2   C gazing at KASPAR

the target of that gaze; for example, XK indicates that C is gazing at KASPAR and XCP indicates that C is gazing at KASPAR's control pad. Arrows pointing to the numbered lines of transcript indicate exactly where descriptions of actions, given in double parenthesis, and numbered photographic images are located within the transcript. To protect the anonymity of the adult 'A' and the child 'C' their faces have been distorted and arrows have been used to indicate the direction of their gaze. Each image is numbered (e.g. 1.1) and those numbers are positioned next to downward arrows to indicate precisely where these images are positioned in the sequence of interaction.

```
                      +           ↓((C gazes at K))
1   B    You-you You'll decide what you want you want him happy or sad
2   B    Look at (it)
         ↓((All gaze at control pad, CP))
3        - - - - - - - - (0.8)
```

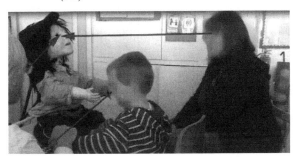

(1.1)

```
         ((Ben pushes button with C's finger))↓      (1.1)↓((A & B look at K))
4.  B    Ahh the blinking (0.5) this is the blinking (0.3) look at his eyes
         ↓(A & B gaze at K's eyes, C gazes at CP)
5        - - - - - - - - -1 - - - - (1.4)
```

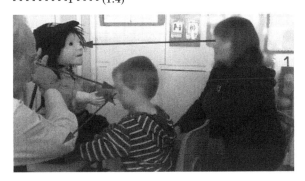

(1.2)

```
              (1.2)↓ +      ↓XK((C glances up then back to CP))
6   B    Do you look at his eyes? =
         ↓((A & B gaze at K's eyes, A points at K, C moves gaze to CP))
7   A    = °look°
         ↓((A & B gaze at K's eyes, A points at K, C gazes at CP))
8        - - - - - - - -(0.9)
```

(1.3)

((B moves C's hand from CP))↓ (1.3)

9   B   No you [don't just press the] button you have to look at the eyes =
    A   [KASPAR's eyes   ]
                        ↓ +
10 B   = if you press the buttons::
                ↓ +((C's gaze into K's eyes begins at 0.8, all gaze at K's eyes))
11   - - - - - - - - -1-(1.1)

(1.4)

                    ↓(1.4)
12 A   look at KASPAR's eyes =
        ↓((C reaches out and touches K's hands))
13 B   = (what is this) () ˙hhhh ((audible inbreath))
                ↓((A moves her hand towards C's arm))
                ↓((C removes his hands from K))
14 B   >No no [nono no<

(1.5)

        ↓((C gazes to CP then K))↓(1.5)
    A   [don't touch (.) just look

```
              ↓((C gaze at CP))              ↓((glance up to K and back to CP))
15            --------1---------2---------3---(3.3)
              ↓((gaze at CP))                +
16    A       Look ---------1---- (1.4) KASPAR's eyes
                                     ↓((C gazes at K)) ((B stands))↓
17            (- - - - - - - - 1 - - - - - - - - 2 - - - - - - - - 3 - - - - - - - - 4- - - -)
```

Extract 1 is a 45-second segment, taken from a two-minute 30-second sequence of interaction, involving an eight-year-old child who received a diagnosis of having an ASC, a researcher, and support worker. In the sequence below we find six verbal instructions, of varying degrees of explicitness by the co-present adults, for C to look at KASPAR's eyes (lines 4, 6, 9, 12, 14, and 16). Two of these instructions are accompanied by one or both of the adults pointing to KASPAR (lines 9, 12) and each are accompanied by one or both of the adults gazing at KASPAR's eyes. The initial impression is that C is reluctant to follow the instruction to look at KASPAR's eyes – only once following the explicit adult instruction in the segment of data in Extract 1 (line 15).

Examining a sequence like this, with checklist in hand, we might readily record some failure to follow an instruction to look, though it must be acknowledged that some observation schedules (such as ADOS-2) have a fairly detailed sequence of verbal and gestural prompts to be used before a failure to look is recorded. However, closer inspection of the sequence reveals a more complex picture. If we first broaden our gaze to consider the entire two-minute 30-second sequence, we find that C does look at KASPAR's eyes 46 times. Even within the data shown in Extract 1 we find there are five occasions when C looks at KASPAR. If we approach the behaviour of looking to KASPAR's eyes, or the subset of looking to KASPAR's eyes when instructed to do so, as isolatable behaviours that we can simply spot, then we may run into the problem of how to make sense of the child looking at the required target (KASPAR's eyes) a great deal, but not necessarily when instructed to do so.

If we were following a checklist approach we might well restrict our examination of the data to the behaviour of the child when requested to look by an adult and may explain what we spot in terms of the child's underlying pathology. If, by contrast, we followed Sack's (1984) injunction to engage in *unmotivated looking* however, we might notice other things that could provide a more nuanced understanding of the interaction. In Extract 1, we can first note that C looks at KASPAR's eyes following each of the four times that he presses the button. We can see these button presses in lines 1, 6, 10, and 16 (marked with a '+') and the corresponding gazes to KASPAR in lines 1, 6, 11, and 17. Indeed, across the entire 2.5-minute sequence, C gazes at KASPAR 39 times following his press of a button. Thus, it seems that C is keen to gaze at

KASPAR's eyes in order to examine the consequence of having pressed a button on KASPAR's control pad. Examining behaviour that might not be our immediate focus – in this case C's pressing of the button – can therefore help to build a better understanding of the target behaviour (gazing, and not gazing, at KASPAR) that we are trying to understand.

This however does not entirely explain why C appears reluctant to follow both A's and B's verbal instructions and accompanying gestures to look at KASPAR's eyes. In order to address this issue, it is again worth guarding against the 'tunnel vision' that any pre-specified checklist may encourage. If we examine what C is doing as instructions to look are given, we see that he is in a sense 'on task', in that he is concerned with pressing buttons on the control pad and examining the effects of doing so. However, C appears to be concerned with investigating the effects of his *independent* pressing of the button on the control pad.

If we look at the series of seemingly ignored instructions for C to look at KASPAR's eyes, we find that the first of the instructions from B in line 4, 'look at his eyes', follows B pushing the button with C's finger. As the instruction is given, C is gazing at the control pad (see Figure 1.1). In line 6, as B asks, 'Do you look at his eyes', C is gazing at the control pad and independently presses a button after which he glances up at KASPAR's eyes. A gives the instruction to 'look' in line 7, whereas C appears to do the opposite, moving his gaze, which was at KASPAR briefly in line 6, back to the control pad. C continues to gaze at the control pad while A and B gaze at KASPAR's eyes and A points at KASPAR. In lines 9 and 10, B removes C's hand from the control pad, telling him that 'No you don't just press the button you have to look at the eyes if you press the button' – at the end of line 10, C presses the button and then gazes, for a prolonged period, at KASPAR's eyes (lines 11–14). In line 16, as the instruction 'Look (1.4) KASPAR's eyes' is uttered, C first gazes at the control pad and then presses the button, gazing to KASPAR in line 17, after the button has been pressed.

In each of these cases when the adult instructions to look at KASPAR were not apparently obeyed, C could be seen to be gazing at the control pad and gazing at KASPAR only *after he had independently pressed a button*. In an interesting sense then, C was complying with the cause-and-effect scenario that the adults were attempting to structure, but in a way that enabled him to investigate the consequence of his agency in controlling KASPAR – he looked after *he* had pressed the button.

The one case where C appeared to unproblematically follow the adult instruction to look is, ironically, perhaps the least explicit instruction of the six, where A says, 'don't touch (.) just look' (line 14). It is noteworthy here that C, having touched KASPAR and had his hands removed (line 14), has received a mild rebuke for touching, and as Figure 1.5 reveals, his hands which had touched, or attempted to touch, the control pad throughout are removed from the table altogether. Here then is a place in the interactional sequence where touching the keypad by C has been explicitly prohibited by one of the adults. In this specific segment, where C is not able to investigate the effect of his independent pressing of buttons on KASPAR's control pad, he is indeed found to 'look' at KASPAR as instructed.

If the child's behaviour in Extract 1 was examined using a checklist, then, it could quite easily be coded as revealing a reluctance on the part of the child to follow adults' instructions to look at KASPAR's eyes. It would not take us long to draw on ready-to-hand understandings of ASCs in order to make sense of this 'poor performance'. However, freed from the constricting gaze of the checklist, the above analysis provides a different understanding that suggests the child may not be avoiding looking, or failing to follow instructions, so much as attempting to investigate something of their own efficacy. In this sequence then, the child appears to investigate what happens when they press a button – gazing at KASPAR after each time that they independently press a button. Once this possibility appears to be 'off-limits', having been told 'don't touch (.) just look', then they do follow the adult instruction to look at KASPAR. In this way, examining the child's behaviour within its sequential context provides a sense of interactional competencies rather than deficits on the part of the child – their behaviour appears better fitted to the details of the interactional context than may have first been apparent.

Extract 2

In Extract 2 the child is introduced to a button on KASPAR's control pad which has a specific consequence, while the child displays an interest in pressing a different button on the control pad, one that activates KASPAR's greeting response.

(2.1)

↓(2.1, B & C mutual gaze)
1.   K   **let's play   together**
                ↓((B gazes at KASPAR's control))
                        ↓((C's gaze from B to K's control pad))
2   (- - - - - - - - - 1- - - - -)
                            ↓((B gazes into C and back into K's control pad))
3   B   **Come (.) what is this**
                ↓((C leans in closer to KASPAR's control pad))
4   (- - - - - - - -)
                ((C moves hand towards control pad))

                                                                                (2.2)

                ↓(2.2)
5   B   **You see this ↑picture**

                                                                                (2.3)

                ↓((2.3, C presses control then C and B gaze up at KASPAR))
6   (- - - - - - - - -1- -)
                ↓((B mimics KASPAR's movement while gazing at C))
7   B   **Yeeeees** =
                ↓((C shifts gaze to B then down into KASPAR's control pad))
8   C   = **an:::** [**Nee yaa hhh eeeeehh**   ((in a delighted tone))
                                        ↓((C presses KASPAR's control pad))
9   B   **[huh huh huh huh huh huh huh** ((laughter))
                ↓((B briefly gazes down at KASPAR then away))
10   K   **hello[(.)   [I am KASPAR**
                        ↓((C gazes at B))
                            ↓((B gazes at C – B and C share mutual gaze))

11	B	[HELLO! Huh[huh [huh huh (.) huh ((laughter))
12	C	[heh heh! huh huh ((laughter))

↑(2.4)

(2.4)

Extract 2, examined by reference to a checklist, could present a problem. It is not that a checklist would not be able to make sense of these data, but rather that its seemingly unproblematic coding of behaviour may present a markedly different picture of the interaction to that which appears when the sequence itself is examined. One illustration of this is provided by imagining that we were coding for the child's gaze behaviour in this extract. On the one hand, the sequence suggests somewhat avoidant gaze activity, apart from a brief glance in line 1 and a longer look at the adult in lines 11 and 12; C appears to gaze very little at B. However, this does not do justice to the possibility investigated below that the gaze behaviour of C in this extract – even where it appears 'avoidant' – can be understood in terms of its sequential context. The point argued in the following analysis is that attempts to code the behaviour as isolatable fragmented parts miss something of their embedded meaning and the organisation of interaction across the sequence as a whole.

C's switch from gazing at a co-present adult (line 1) to gazing at KASPAR's control pad (line 2) is revealed through a detailed examination of the sequence as occurring after the adult, B, does so. Thus, C's shift in gaze (from B to KASPAR's remote control pad) can be seen as treating B's gaze as identifying the relevant focus for visual attention – C moves his gaze to the object that B's gaze identifies as relevant to be looked at. B's pointing and his words, 'Come (.) what is this' (line 3), further underscore the relevance of gazing at KASPAR's control pad – and C demonstrably orientates to this in line 3 by not only gazing but also leaning in towards the control pad. In this way C not only follows B's gaze, pointing, and verbal reference to the control pad, but he *demonstrably* does so – that is, he does so in a way which makes his doing explicitly available for B to see.

C's next gaze shift (line 6) is from KASPAR's remote control pad to the part of KASPAR's body affected by the button that he has just pressed. Taken in isolation, this might again be seen as some form of avoidance of eye contact, as C is not at this point gazing at the co-present adult B. However, the relevance of C's gaze shift is revealed by considering that it follows immediately from his pressing of a button on KASPAR's remote control pad – the gaze then moving to KASPAR's upper body and head region (the part affected by the button that was pressed). The relevance of gazing here is further underscored by the fact that B also gazes at the same part of KASPAR at the same time as C. In this light, what might be coded as 'avoidant gaze' or 'unusual eye contact' when taken in isolation becomes highly relevant, appropriate, and skilled use of gaze when viewed in its sequential context.

In line 8, C shifts his gaze fleetingly to B before moving it down to KASPAR's control pad. This brief period of gaze abstracted from its sequential context could again, potentially at least, be noted as unusual – perhaps unusually brief. But when considered in terms of the projected course of actions, the brief gaze appears interactionally skilful. C's gaze shifts from B to KASPAR's remote control pad; he then presses his preferred remote control button, then gazes back at B (where a period of sustained mutual gaze emerges). However, C's shift of gaze from B to KASPAR's remote control not only facilitates the action of C (pressing his preferred button on the remote control); it can also be seen as enabling the establishment of prolonged mutual gaze and shared laughter *after* C's action. It is here, once C has gazed down at KASPAR's remote control and operated his preferred button, that we find C's sustained gaze at B (and the subsequent establishment of mutual gaze). In line 11, C not only gazes at B and establishes and maintains mutual gaze, but he does so with a smiling face.

In Extract 2 an examination of C's behaviour, in context rather than in isolation, reveals that it orientates to the sequential environment; he moves his gaze to a target that the adult is gazing and pointing at, he makes his attentional focus available to the adult, and then he gazes at the adult as shared laughter emerges. However, beyond these moment-by-moment orientations to the sequential environment, C can be understood as orchestrating his actions to project a longer course of action which again orientates to the sequential context. Because B and C have had a battle over the button to be pressed – with C showing a preference for the 'Hello my name is KASPAR, let's play together' button – C is able to perform the pressing of his preferred button as a source of shared humour. This entails gazing at B briefly, initiating laughter (which B joins), pressing the button, then gazing at B as both laugh. Doing this pressing as humour takes a considerable amount of skill, orchestrated across several turns of interaction, and depends on being positioned immediately after having briefly followed B's instructions to press a different button (without which it would be a form of non-compliance).

Examining C's gaze activity in its sequential context, we see that characterisations of C's gaze behaviour have little 'standalone' or context-free meaning – they can be best understood in terms of a detailed examination of sequential context. The point for the current chapter then is that coding the behaviour in isolation does not do justice to the interactional nuances that are at work. Coding an isolatable behaviour whether it denotes competence or impairment runs the risk of a blind spot when it comes to complex actions built over a number of turns or involving the orchestration of different actions. The complexity of these courses of action is further revealed when we consider how they can be sensitive to the immediate interactional context as well as orientating to both the immediate interactional history between the participants and the projected 'post-course of action' context. As with Extract 1 the alternative perspective that this analysis provides on the child's behaviour is especially consequential as it speaks to the issue of competency and deficit. By paying attention to behaviour *within its sequential context* the target behaviours investigated here appear to orientate to sequential context. In this way the behaviours investigated can be understood as indicative of some level of interactional competency on the part of the child, rather than being merely symptomatic of their underlying pathology.

Clinical relevance summary

In order to better understand how conversation analysis of these data is relevant for clinical, as well as therapeutic and educational practice, it is worth briefly considering Schegloff's (2003) examination of interactions within a clinical assessment of 'Alvin', an adult with Aphasia. Schegloff (2003) argues that much of Alvin's seemingly incidental behaviour, which was not formally viewed as part of the test, such as a sequence in which Alvin adjusts his chair, is rich in the interactional competencies that it reveals. What is of particular relevance for clinicians, therapists, and educationalists is that these behaviours, which revealed precisely the very competencies that the assessment was concerned with, were not spotted because the clinical gaze was directed towards a narrowly pre-specified range of 'correct' responses to clinical prompts.

The current chapter emphasises that the pre-specification of relevant behaviours found in various forms of checklists can create a twofold problem. First, similarly to Schegloff (2003), the checklist can result in certain behaviours being unnoticed. By directing attention towards certain behaviours, often occurring within particular observation parameters, other behaviours – while highly relevant – may not be noticed. Second, even where a relevant behaviour does come into clinical, educational, or therapeutic view, the understanding of it will be severely curtailed if it is seen as understandable *in isolation from the sequential context in which it was positioned.* Examining behaviours independently of their sequential context means that the extent to which they might be orientated to that context cannot be seen and may thus result in an underestimation of the interactional competence – specifically the fittedness to the interactional context – that those behaviours reveal. For a simple summary of the implications for practice, see Table 4.1.

Summary

This chapter drew on semi-structured interactions involving one or two adults, a humanoid robot, and a child with an ASC in order to contrast two ways of examining observational data involving children with an ASC. On the one hand, observations of such interactions could be coded using some form of checklist or coding scheme, in which pre-specified behaviours are identified and successful or unsuccessful responses to any (assessor initiated) scenarios are pre-defined. On the other hand, these data could be examined by looking at any target behaviour or courses of action in terms of the detailed sequential context in which they are positioned. Overwhelmingly, clinical, educational, and academic approaches to making sense of observational data involving a child with an ASC are informed by the checklist approach of spotting pre-specified behaviours and employing predefined measures of competencies.

Behind this dichotomy, between spotting pre-specified behaviour and looking at sequences of interaction, lies a different understanding of behaviour – from one perspective behaviour can be understood as an isolatable phenomenon that can be defined separately from, and in advance of, any specific interaction. From the other perspective any behaviour can only be meaningfully understood in terms of its position within the precise sequence, *as it actually occurred,* and we cannot envisage in advance what behaviours may emerge as important within a sequence of interaction. This idea about examining the interaction that actually occurred is worth further clarification: while some checklists and coding schemes do attempt to treat the context as important, it is *context as pre-envisaged*, rather than context *as discoverable in the data.* That is, checklists and coding schemes treat context as something that the person administering them can create by, for example, enacting some prespecified, scripted behaviour. This stands in sharp contrast to the practice, exemplified in this chapter, of examining, *post hoc,* the actual interactional context than can be discovered in the detail of the data.

What conversation analysis repeatedly shows is that actual interaction differs to what we might envisage in significant ways. Our attempt to script a 'natural' prompt to get the child's gaze may be anything but natural because of its pre-scripted, non-spontaneous form. Our hope that the scenario is repeatable without variation fails to recognise that any adult behaviour is itself shaped by context. While the adult's behaviour may endeavour to be standardised, the

Table 4.1   Clinical practice highlights

1. This chapter has suggested that coding schemes and checklists run the risk of conceptualising target behaviour as isolatable from the interactional context in which it occurs and ignoring the potential relevance of certain behaviours because they are not part of a pre-specified list. In the light of these considerations, the following practices are suggested:

    a. Assessment tools could be developed and modified such that they approach *action in interaction* as the basic unit of interest, rather than *behaviour in isolation*. Taken seriously, this has quite radical implications that would entail a move to examining and articulating the specifics of interactional context and thereby sacrificing the pre-specification of standardised target behaviours for a nuanced understanding of individual sequences of interaction.

    b. Making video data far more prominent within clinical practice is a further implication of this chapter. This is not without potential practical and ethical challenges, but this chapter suggests that both pre-existing video, (including that which families themselves may have collected), and clinically instigated video recordings, can add an enormous amount of valuable information about the nuanced competencies that the child with an ASC may have and the specific interactional challenges that they may face. These video data are valuable not least for the very reason they are typically ignored – namely that they usually comprise children with an ASC interacting in *typical* environments with *known others*, rather than in highly pre-specified (and hence atypical) 'test' environments, with people that they would not otherwise interact with.

    c. The chapter further suggests that clinical interventions could be informed and monitored by information about the sequential nature of the child's behaviour. Getting more eye contact might be a generally desirable aim, but where and when eye gaze is present or absent provides a more focused starting point, and a more specific way of framing an intervention. Similarly, if an intervention creates interactional side effects, for example if efforts to increase eye contact result in unrelenting staring, then data about eye gaze within sequences of interaction may provide a means of monitoring such unintended effects (as well, of course, as unintended benefits).

    d. Finally, this chapter points to the idea that issues of clinical concern need to be articulated in ways that recognise their sequential location. Thus rather than de-contextualised formulations such as; 'repetitive tapping' or 'avoidant gaze behaviour', more interactionally informed alternatives could be developed such as; 'repetitive tapping in specific post question environments' or, 'avoidance of gaze when visually occupied with examining the consequence of an action they have initiated'. This contextually enriched language for articulating clinical issues brings not only more precision but also an awareness of the importance of the interactional environment as being at the very heart of issues faced by children with an ASC.

context into which that behaviour is positioned, being, in part, shaped by the child, may not be. The context of where the child is looking and what the child is saying or doing – prior to and alongside any stimulus word or action on the part of the adult – actually shapes the context within which the adult's words, gaze, and body movement are positioned, making them different types of *action in interaction*, despite any appearance (on paper) of being standardised. For example, an adult's injunction to 'look' at a specific target takes on a different meaning if the child is looking at the adult, looking at the target, or looking elsewhere – when the instruction is given – and varies still more if other aspects of the child's talk and behaviour are considered.

In this chapter we have argued that our quest to achieve standardisation through the uniformity of coding schemes and checklists is not only impossible to obtain in actual practice but it also limits what we see. As Sacks (1984) noted, in a criticism of theoretically informed sociological analyses, thoughts and expectations designed to help analysis may, inadvertently, constrain it as we cannot know in advance the detailed ways in which any interaction will be organised, nor even which behaviours may be especially important within that interaction. This chapter, along with Sacks, calls for clinicians, educationalists, and academics to genuinely look at interaction involving children with an ASC. In doing so we might sacrifice something of the security of an apparently standardised tool, but we have the possibility of moving beyond what we expect to find and coming closer to seeing what is actually there in the interaction.

References

Barrow, C., & Tarplee, E. (1999). Delayed echoing as an interactional resource: A case study of a 3-year-old child on the autistic spectrum. *Clinical Linguistics & Phonetics, 13*(6), 449–482.

Dautenhahn, K., Nehaniv, C. L., Walters, M. L., Robins, B., Kose-Bagci, H., Mirza, N. A., & Blow, M. (2009). KASPAR – A minimally expressive humanoid robot for human-robot interaction research. Special Issue on «Humanoid Robots», *Applied Bionics and Biomechanics, 6*(3), 369–397.

Dickerson, P., Rae, J., Stribling, P., Dautenhahn, K., & Werry, I. (2005). Autistic children's co-ordination of gaze and talk: Re-examining the 'asocial autist'. In K. Richards & P. Seedhouse (Eds.), *Applying conversation analysis* (pp. 19–37). Basingstoke: Palgrave MacMillan.

Dickerson, P., Stribling, P., & Rae, J. (2007). Tapping into interaction: How children with autistic spectrum disorders design and place tapping in relation to activities in progress. *Gesture, 7*(3), 271–303.

Jefferson, G. (2004). Glossary of transcript symbols with an introduction. In G. H. Lerner (Ed.), *Conversation analysis: Studies from the first generation* (pp. 13–23). Amsterdam: John Benjamins.

Kanner, L. (1943). Autistic disturbances of affective contact. *Nervous Child, 2*, 217–253.

Lord, C., Risi, S., Lambrecht, L., Cook, E. H., Leventhal, B. L., DiLavore, P. C., & Rutter, M. (2000). The autism diagnostic observation schedule – generic: A standard measure of social and communication deficits associated with the spectrum of autism. *Journal of Autism and Developmental Disorders, 30*(3), 205–223.

Rimland, B. (1971). The differentiation of childhood psychoses: An analysis of checklists for 2,218 psychotic children. *Journal of Autism and Childhood Schizophrenia, 1*, 161–174.

Robins, B., Dickerson, P., Stribling, S., & Dautenhahn, K. (2004). Robot-mediated joint attention in children with autism: A case study in robot-human interaction. *Interaction Studies, 5*(2), 161–198.

Robins, B., Dautenhahn, K., & Dickerson, P. (February 1–7, 2009). *From isolation to communication: A case study evaluation of robot assisted play for children with autism with a minimally expressive humanoid robot.* Proceedings of the Second International Conferences on Advances in Computer-Human Interactions, ACHI 09. Cancun, Mexico: IEEE Computer Society Press.

ROBOSKIN (2012). Retrieved May 26, 2012 from http://www.roboskin.eu.

Sacks, H. (1984). Notes on methodology. In J. M. Atkinson & J. Heritage (Eds.), *Structures of social action: Studies in conversation analysis* (pp. 21–27). Cambridge: Cambridge University Press.

Schegloff, E. A. (2003). Conversation analysis and 'communication disorders'. In C. Goodwin (Ed.), *Conversation and brain damage* (pp. 21–55). Oxford: Oxford University Press.

Stribling, P., Rae, J., & Dickerson, P. (2005). 'Spelling it out': The design, delivery, and placement of 'echolalic' utterances by a child with an autism spectrum disorder. *Issues in Applied Linguistics, 15*, 3–32.

Stribling, P., Rae, J., & Dickerson, P. (2007). Two forms of spoken repetition in a girl with autism. *International Journal of Language and Communication Disorders, 42*, 427–444.

Stribling, P., Rae, J., & Dickerson, P. (2009). Using conversation analysis to explore the recurrence of a topic in the talk of a boy with an autism spectrum disorder. *Clinical Linguistics and Phonetics, 23*, 555–582.

Recommended reading

- Dickerson, P., Stribling, P., & Rae, J. (2007). Tapping into interaction: How children with Autistic Spectrum Conditions design and place tapping in relation to activities in progress. *Gesture, 7*(3), 271–303.
- Goodwin, C. (Ed.) (2003). *Conversation and brain damage.* Oxford: Oxford University Press.
- Richards, K., & Seedhouse, P. (Eds.) (2005). *Applying conversation analysis.* Basingstoke: Palgrave MacMillan.
- Robins, B., Dickerson, P., Stribling, S., & Dautenhahn, K. (2004). Robot-mediated joint attention in children with autism: A case study in robot-human interaction. *Interaction Studies, 5*(2), 161–198.
- Sacks, H. (1984). Notes on methodology. In J. M. Atkinson & J. Heritage (Eds.), *Structures of social action: Studies in conversation analysis* (pp. 21–27). Cambridge: Cambridge University Press.

Transcription conventions

The transcription conventions are a modified version of Jefferson's (2004) transcription conventions. For an overview of the Jefferson symbols, please see the preface of the book. The version used in this chapter is presented here.

[	Left square bracket	Onset of overlap
]	Right square bracket	Termination of overlap
= =	Equals signs	Latching of talk
(1)	Numerals in parenthesis	In the transcript, pause in seconds above transcript, image position
(- - -)	Hyphens in parenthesis	One hyphen indicates 0.1 seconds
(.)	Dot in parenthesis	Un-timed micro-interval
?	Question mark	Rising, or questioning, intonation
NO	Capital letters	Delivered with relative loudness
°no°	Degree signs	Enclosed talk delivered with relative quietness
↑	Upward arrow	Rising intonation
↓	Downward arrow	Within transcribed talk, falling intonation – above transcribed talk, indicates where an image or an action described in double parenthesis is positioned in the transcript
e:::	Colon	Indicates a sound stretch
(())	Double parenthesis	The enclosed text describes actions

Additional transcription symbols used

X	The capital letter 'X'	Used to indicate the child's gaze at a target, 'X' denotes the precise point at which gaze is directed at a specific target
+	Plus sign	Used to indicate the point at which the child had pressed a button on the robot's control pad

5
ADHD: Three Competing Discourses

Adam Rafalovich

Introduction

What follows in this chapter is a discourse analysis of the ADHD phenomenon from three major perspectives. To this aim, I have attempted to focus upon the most pre-eminent discussants in the conversation about ADHD, which include the medical, psychodynamic, and sociological viewpoints. On the medical front, I will focus upon the early conversations surrounding ADHD. As we will see, ADHD has an important presence in the medical record, stretching back to the late 19th century, which has focused upon the increasing interest of medicine in childhood deviance, academic performance, and life chances of children. The psychodynamic viewpoint was largely practised in clinical settings prior to the advent of widespread psychopharmacology in therapeutic circles. The psychodynamic viewpoint essentially argues that patterns of disruptive childhood behaviour emanate from interactive dynamics within the family and a lack of psychological well-being in children. From the sociological view, ADHD is a product of a variety of social forces that have created a type of perfect storm within which the ADHD diagnosis has gained momentum over time, and therefore an increasing legitimacy. The legitimacy of ADHD, or any social phenomenon for that matter, is a product of the conglomeration of enough agreement, or enough overlap of perspectives, rather than the discovery of a bona fide neurological or psychological cause.

Introducing ADHD

Depending upon perspective, Attention Deficit Hyperactivity Disorder (better known through its acronym ADHD; see American Psychiatric Association 1980, 1987, 1994, 2013) is interpreted in multiple ways. From the psychiatric viewpoint, ADHD is defined as a behaviour disorder that presents through a failure to pay close attention to details, having difficulty in organised activities, and an inability to remain calm in certain social contexts. Although widely recognised in clinical, journalistic, and popular culture circles, the way that ADHD is specifically recognised and regarded lends itself to protracted and intense debate. Such debate centres upon ADHD's epidemiology, legitimacy as an illness category, primary treatments, diagnostic protocols, and a litany of other concerns. These tensions thus have consequences for the moral accountability of those children and their families (see Davies & Horton-Salway, Chapter 9, this volume for a discussion).

Given the intense scrutiny that ADHD has received over the past several decades, this disorder lends itself to an analysis that focuses upon how the various conversations surrounding it 'socially construct' this disorder in contemporary society. This chapter provides a discourse analysis of the ADHD phenomenon from a variety of perspectives. I will focus upon the medical, psychodynamic, and sociological viewpoints.

It is not the intention to demonstrate the truth or falsity of one perspective over another. Indeed, this way of approaching ADHD would run counter to the logic of a discourse analysis. Instead, this chapter will focus upon the various conversations that have given rise to ADHD, and raised its salience in lay, clinical, and academic circles. As we will discuss, ADHD is a topic of debate for a variety of reasons, not the least of which concerns the dramatic rise in ADHD diagnoses in Western Europe and the United States, and the consequent skyrocketing number of prescriptions written for stimulant medications, which are the primary treatment for this disorder. ADHD is also a significant topic of debate because the diagnosis of the disorder itself provides a type of commentary about the nature of modern society, our expectations of children (and increasingly, adults), and whether or not social structures can be exonerated from responsibility in creating ADHD symptoms. For example, a key point of the debate surrounding ADHD concerns the extent to which parties identify the cause of ADHD inside the neurological structures of the brain, within family dynamics, or within the organisational structure of our social institutions.

These three viewpoints summarise the medical, psychodynamic, and sociological perspectives towards ADHD. On the medical front, we will focus upon the early conversations surrounding ADHD. As we will see, ADHD has an important presence in the medical record, stretching back to the late 19th century, which has focused upon the increasing interest of medicine in childhood deviance, academic performance, and life chances of children. The rationale for the medical inquiry into ADHD concerns the structures of the brain and the dopaminergic circuitry that enables impulse control and sustained attention. The medical argument has been bolstered by the apparent effective use of stimulant medications in changing children's behaviour. In short, a positive response to medication affirms the presence of some type of physiological abnormality in the brain.

With respect to the psychodynamic viewpoint towards ADHD, we witness a type of neo-Freudian viewpoint towards inattention and general child disruptiveness. Throughout the 1920s and into the 1940s, the psychodynamic viewpoint was largely practised in clinical settings, prior to the advent of widespread psychopharmacology in therapeutic circles. The psychodynamic viewpoint essentially argues that patterns of disruptive childhood behaviour emanate from interactive dynamics within the family and a lack of psychological well-being in children. ADHD, in short, is a developmental issue that has become a type of crystallised habit of conduct, more as a survival mechanism of a dysfunctional family than a bona fide neurological condition. Hence, the psychodynamic approach to ADHD emphasises the use of psychotherapy more so than the use of medications.

Finally, the sociological viewpoint places the cause of ADHD outside the individual and family institutional frameworks. Sociologists are quick to point out, for example, that the vast majority of ADHD cases are diagnosed within the United States and that 90% of the world's stimulant medications are consumed within North America. Further, sociologists point to the expanding province of medicine in a variety of human troubles – a process known as 'medicalization' – as a key concern in understanding the expansion of ADHD and other behaviour disorders. ADHD, then, is really a cultural construction, which is real in its consequences, but only because institutional representatives (physicians, school counsellors, clinical psychologists, and so on) believe it to be so. Before I begin a detailed expansion of these three perspectives, and show the different conversations among them, it is very important to highlight the framework of discourse analysis.

Discourse analyses: Key concepts

Discourse analyses of medical phenomena, especially those that fall under the rubrics of mental health, focus upon the points of agreement and divergence about a particular phenomenon over a considerable period of time. Because of this, discourse analyses often take on a historical tone which enables a broader perspective towards the evolution of a particular idea. This is not to be confused with a unilateral historical viewpoint which examines the discovery and development of one type of technology over time. Rather, a discourse analysis imbues neutrality as to the legitimacy of a category under question. For example, in looking at the phenomenon of ADHD, the goal of a discourse analysis is to raise awareness of the variety of viewpoints that converse about, and consequently give rise to, the ADHD phenomenon. Hence, it is important to note that this can be frustrating for readers who are looking for a simple, linear history of ADHD. Although a discourse analysis provides historical components, it does not assume that the conversation about ADHD is 'advancing' as we would if we were focusing upon a history of 'the effectiveness of stimulant medications', or 'advances in diagnostic technology', and so on.

Very importantly, discourse analyses focus upon how a variety of viewpoints towards a particular phenomenon 'create' that phenomenon. What we 'know' about ADHD, for example, is a product of ongoing conversations within a much larger field of discourse, rather than the achievement of indisputable scientific facts. Indeed, with respect to ADHD and other mental disorders, evidence that appears indisputable today becomes suspect or even discarded tomorrow. In some ways, this is part of the logical sequence of scientific inquiry. However, within the field of mental health, in which so much of the aetiology is based upon theories of brain physiology, we witness a continuous amount of fundamental reappraisal. For example, although there are ongoing and productive debates within somatic medicine about the treatments for diabetes, there remains little fundamental debate within the medical community about what defines diabetes, and its basic biomechanics. Although no diagnostic category is etched in stone, debates within somatic medicine, especially about the most well-known illness categories, retain a great degree of stability. This is not the case (at least, not yet) with psychiatric medicine, which has struggled for decades to find protocols that substantiate its diagnostic nomenclature. Reliability concerns, highlighted by David Rosenhan's (1973) famous study, continue to plague modern psychiatry and are, in part, the basis for widespread scepticism towards psychiatric illness categories.

This brings us to a very important conceptual framework developed by Michel Foucault (1978) in the mid-1970s. Throughout most of his career, Foucault wrestled with the Marxian notion of ideology, namely, that these large, overarching belief systems moved in a unidirectional pattern. From Karl Marx's perspective, ideology emanated from those focal points of power in society and moved uni-directionally to those who had less power. Hence, the ideas of less powerful were those given to them by the elites in society. For example, with respect to ADHD, the Marxian viewpoint would simply analyse ADHD as a phenomenon that has been spoon fed to the masses by those in the most powerful circles of academic and clinical medicine, the powerful interests in the pharmaceutical industry, and the various marketing campaigns that disseminate the viewpoint of such entities to a naïve and ill-informed public.

In his analysis of mental hospitals, prisons, and hospitals, Foucault utilises an important set of theoretical ideas to explain how these institutions, and their consequent practices, attain legitimacy within society. Instead of focusing upon the monolithic rise of these institutions in society, Foucault examines how such institutions and the discourses that prop them up are a product of a continuous and ongoing dynamic between different discursive frameworks. This is perhaps summarised best by Foucault's discussion of his three discursive interdependencies: (1) the inter-discursive; (2) the intra-discursive; and finally

(3) the extra-discursive. For the purposes of this chapter, it is important to provide a working definition of the inter-discursive and extra-discursive dependencies. Although an important part of his theoretical perspective, this chapter will not make any meaningful use of the intra-discursive dependency.

The inter-discursive refers to a conversation (or debate, if you like) between two distinct theoretical frameworks; for example, the difference between the neurological and psychological viewpoints towards ADHD. This conversation is often antagonistic in nature and commonly has points of disagreement about the fundamental nature of a particular topic of study. For example, for decades, medical practitioners and Freudian psychologists debated about the essential cause of schizophrenia, where the former argued that schizophrenia was a specific biological abnormality, and the latter argued that schizophrenia resulted from any one of a host of psychosexual developmental problems.

The extra-discursive dependency refers to the relationship between institutional practices and the way that people behave in everyday life. Stated simply, the extra-discursive is a way of describing how the types of knowledge, lines of reasoning, and modes of behaviour of institutions ultimately shape our perceptions of self, other, and infiltrate our daily interactions. The extra-discursive is a great way of describing why it is that people 'see' ADHD in themselves, even without a formal diagnosis. In this sense, ADHD 'infiltrates' public discourse and our popular culture in a way that far supersedes the supposedly manipulative powers of these powerful interests about which Karl Marx spoke so explicitly. We might look at the extra-discursive as an ultimate indicator about the power relationships that are associated with the development of a particular type of knowledge. On one side of this coin, we can explain the incredible rise in the amount of children diagnosed with ADHD, but on the other side, we may also explain the growing scepticism towards the ADHD diagnosis, parents who push back against suspicious teachers and clinicians.

The origins of the medical viewpoint towards ADHD

In order to posit a meaningful discourse analysis, it is important to establish the history of the medical conversation surrounding ADHD. These earliest discussions are firmly entrenched in clinical medicine beginning in the late 19th and early 20th centuries. There are four major components to this conversation, which will prove crucial in establishing the trajectory of the neurological viewpoint towards ADHD: First, the late-19th-century conceptualisation of the medical concepts of *imbecility* and *idiocy*; second, a series of lectures given by George Frederic Still in 1902 (commonly regarded by ADHD proponents and detractors alike as the origins of the medical conversation surrounding the disorder; see Barkley, 1990, 1998, 2005; Kessler, 1980; Schrag & Divoky, 1975); third, the discussion throughout the 1920s concerning the brain-behaviour manifestations of *encephalitis lethargica* in children; and finally, the earliest clinical discussions of the effects of stimulant medications in children, beginning in 1937 (Bradley, 1937).

Idiocy and imbecility

Although their meanings are completely enveloped in colloquial language today, idiocy and imbecility were considered to be bona fide medical concepts at the turn of the 20th century. Indeed, imbecility was a particularly useful concept for medicalising conditions that resided in that netherworld between severe mental incapacitation and relative mental normalcy. Contrary to idiocy which was a condition demarcating a specific, disabling, and acute incapacitation, imbecility was a much more flexible (and a few years later, much more useful) category. Utilising Foucault's (1965, 1973, 1977) viewpoint, imbecility enabled the early medical discourses surrounding ADHD to begin speaking about brain conditions and behaviours that were previously outside the conceptual reach of medicine. As William Ireland (1877) and Charles Mercier (1890) pointed out, the condition of idiocy should be regarded as extreme in its very essence, whereas the condition of imbecility denoted a less severe type of mental disability. In the early stages of its conceptualisation, imbecility was 'finding its way' as a category in medical nosology. It crystallised into terms that were more useful in expanding the province of medicine to become interested in, and advocate the diagnosis of, various forms of social deviance that were previously outside the medical purview.

William Ireland's work evolves the concept of imbecility to include a specific 'moral' component. This, in addition to the conceptual framework of imbecility, was ground-breaking for the discourse of medicine as it brought into view for the first time those behaviours that were not markedly physiological in nature, but rather characterised immoral behaviour, such as alcoholism, drug addiction, delinquency, and general patterns of criminal activity. Ireland's use of the term 'moral imbecile' described not only a medical condition in which individuals felt a type of compulsion to commit crime, drop out of school, or cease attending church, but also a condition exhibited by an individual who was not markedly mentally impaired. As Ireland's work described, a moral imbecile might be as intelligent and mentally adept as a person without the moral imbecility affliction. Alfred Tredgold's work in 1917 expanded the conversation surrounding moral imbecility to include its detection in children. As I have quoted in previous work (Rafalovich, 2001), Tredgold argued that moral imbecility characterised 'Persons who from an early age display some permanent mental defect coupled with strong vicious or criminal propensities on which punishment has had little or no deterrent effect' (Tredgold, 1917, p. 43, in Rafalovich, 2001). The key idea from this quote is the fact that moral imbecility is defined as a condition demonstrated from an early age. It is not surprising, then, that a significant amount of the conversation surrounding moral imbecility and other related medical conditions focuses attention upon children. As we will discuss later, this has the benefit of focusing the medical lens upon a population that not only can easily be studied but also frames medical practitioners as advocates for medical intervention into children's lives.

Of course, this poses a whole new set of problems for medicine. Chief among these: how to problematise a key distinction between behaviours that were 'purely' criminal and not representative of moral imbecility and those behaviours that were a direct result of the moral imbecility diagnosis. The implications of this are profound, and still a major part of the discussion surrounding ADHD today. For example, how do practitioners in clinical psychology, neurology, and psychiatry distinguish between those behaviours that are deviant, yet non-medical in nature, versus those behaviours that can be attributed to a bona fide medical condition. There is, of course, more to be said about this key distinction, and the continuous problem that it poses for modern psychiatry, which we will address later. This is highly relevant to the conversation surrounding ADHD, as it demonstrates the expansion of medical discourse into those areas that may be reserved for those in the emerging social sciences of that time period – sociology, anthropology, psychology, and so on. In essence, the concept of moral imbecility throws medicine's 'hat into the ring' with respect to social problems.

The work of George F. Still in 1902 is regarded by most who work in the ADHD field as the formal beginning of the medical interest in the moral control of children. Published in *The Lancet* as a series of lectures, Still's address was unquestionably influenced by the decades-long conversation about imbecility and moral imbecility. The key question, of course, is why this particular set of lectures substantiates the beginnings of the medical conversation about what we would call ADHD almost 80 years later. There are at least two significant reasons for this.

First, Still's address specifically linked problems of moral restraint to children in particular. Prior to this time, medical conversations certainly connected immorality to physiological problems, but for the first time, medicine now focused specifically upon an untainted group. Children were particularly vulnerable to this type of medical inquiry because, as a matter of course, they have not attained the life experience that would shape their behaviour, moral or otherwise. Hence, the display of repeated immoral behaviour from children could more easily be associated with an under-riding physiological condition, rather than one resulting from the social environment. This began a trend of exonerating societal conditions, including key institutions such as religion, family, and politics, from having a role in influencing children's behaviour. Such a medical conversation localises the cause of immorality within the brain structure of children.

Second, Still is the first to specifically list out the different symptoms of this 'new condition' in children. In addition to claiming the medical validity of children's immorality, his address may be regarded as the first sincere attempt to insert moral ineptitude into medical nosography. His list of symptoms for this condition included: '(1) passionateness; (2) spitefulness-cruelty; (3) jealousy; (4) lawlessness; (5) dishonesty; (6) wanton mischievousness-destructiveness' (Still, 1902, p. 1009).

Encephalitis lethargica

Although Still's address demonstrates a passionate plea to the medical community, it fails to substantiate any scientifically verifiable connection between childhood immorality and a specific disease process. Throughout the 1920s entered the medical conversation of *encephalitis lethargica* (EL) throughout the 1920s, which is regarded by many researchers as an important stage in the evolution of ideas that would ultimately become what we today call ADHD (Stewart, 1970; Kessler, 1980; Cantwell, 1981; Barkley, 1990). According to many clinicians writing at the time, EL reached epidemic proportions in the population after the First World War and was regarded as an illness with no specific cause or cure. It was an ailment that began with tremendous sluggishness and irritability, and if it did not lead to death, caused intractable brain damage. (For a detailed breakdown of the clinical discussion surrounding EL during this time, please see Abrahamson, 1920a, 1920b; Hohman, 1922; Paterson & Spence, 1921; Stryker, 1925.)

The after-effects of EL infection were of particular concern to physicians writing during this time. It was repeatedly noticed, for example, that children who had been diagnosed with EL infection would often have a permanent change in their personalities as a result. Some of these symptoms of the 'behaviour residuals' of EL included 'sleep reversals, emotional instability, irritability, obstinacy, lying, thieving, impaired memory and attention, personal untidiness, tics, depression, poor motor control, and general hyperactivity' (Kessler, 1980, p. 18). Again, similar to Still's 1902 address, the clinical discussion surrounding EL focused primarily upon the behaviour of children. Further, the discussion of EL demonstrated the next evolutionary stage in the collection of ideas that would later become ADHD. While the literature addressing imbecility and moral imbecility demonstrated medical interest in the social control over children, the medical category of *encephalitis lethargica* proposed the existence of a specific disease entity which, however inelegant, portended diagnosis and treatment. For the first time, children's social problems were attributed to a discernible (and infectious) illness.

Charles Bradley and Benzedrine

As *encephalitis lethargica* demonstrated medicine's quest to draw the connection between a specific disease and a collection of socially undesirable behaviours, it was only a matter of time before such behaviours were to receive some type of treatment. The first documentation of such treatment, and clearly one that charted the use of stimulant medication in ADHD children today, is the work of Charles Bradley in 1937. Experimenting with a small group of institutionalised boys, Bradley administered small dosages of the stimulant drug Benzedrine in order to see what impact it may have. Surprised at the results, Bradley wrote up his findings in a well-circulated paper that many regard as the single most significant breakthrough in ADHD treatment (Bradley, 1937). Among his findings, Bradley noticed that many of the boys who were administered Benzedrine focused more clearly upon their schoolwork, did their chores more compliantly, and had an overall happier disposition.

Perhaps more significant than these findings which demonstrated the acute effects of any stimulant medication was Bradley's assertion that among a few of these children, Benzedrine seem to have a type of 'paradoxical effect'. Such an effect was typified by children who appeared to be becalmed by the stimulant medications, rather than 'revved up' as one may expect from the typical brain behaviour response to stimulants. As Benzedrine increased focus, and on-task behaviour, much more markedly in some of these training school boys, Bradley hypothesised that children who exhibited such a paradoxical effect may have a more fundamental physiological condition in the brain that the drug was affecting.

The psychodynamic viewpoint

Up to this point, we have discussed the development of the physiological viewpoint towards the collection of childhood troubles that we today associate with ADHD. The several decades of research demonstrated in the medical literature illustrates the growing concern and substantial cultivation of legitimacy that medicine developed during the early years of theorising about the causes and treatments for ADHD. However, it is important to examine another significant viewpoint that addressed issues of childhood dishonesty, detachment, thievery, and other antisocial behaviours. This brings us to the psychodynamic viewpoint towards such behaviours, which in many ways opposes the more strict physiological perspective exhibited by medicine. The difference between the psychodynamic viewpoint and the medical one provides a good example of the inter-discursive dependency. Although both of these viewpoints perceive these collections of childhood troubles as problematic, the theories about the causes and treatments for such behaviours are major points of disagreement.

Two of the most significant proponents of the psychodynamic viewpoint were Anna Freud and Melanie Klein, both psychoanalysts who traced a lot of childhood anxiety and impulsive behaviour to psychosexual processes (Freud 1926; Klein 1932). As Klein and Freud both argued extensively, the inability to regulate impulses could be easily traced to family dynamics and the first few years of psychosexual development. Although we do not have space here to elaborate the psychoanalytic theory of psychosexual development, it is important to note that most psychoanalysts viewed the early years of childhood development as crucial in understanding problems in later childhood and adulthood. In other words, childhood therapy was really a way of 'righting the ship' of psychological development. Of particular significance for these and other psychoanalysts were family dynamics, which were invariably implicated in childhood misbehaviour.

The psychological impact of dysfunctional family dynamics upon children is described by Beata Rank (1954) as the 'fragmented ego'. A condition that can be traced to infancy, Rank argues that a fragmented ego results from an 'infant's unsuccessful struggle to obtain vital satisfaction from his parents' (Rank 1954, p. 495). Relying upon the Freudian orthodoxy of the time, such a viewpoint demonstrates the belief that social environment is not only an important factor in contributing to a well-adjusted childhood and adulthood but might be the single most important variable in determining mental well-being and behavioural compliance.

The sociological viewpoint

At the inter-discursive level, we can clearly see major disagreements between the psychodynamic viewpoint towards ADHD and the medical viewpoint. Although it can easily be argued that these two perspectives find points of confluence, they differ strongly in their approaches to the causes and treatments for ADHD. From the medical perspective, ADHD is a physiological ailment that responds to brain chemical changes brought about through stimulant medications. In the contemporary context, we see how the rampant use of stimulant drugs such as Ritalin demonstrates the application of such a perspective. From the psychodynamic viewpoint, ADHD is a product of a developmental problem and needs to be treated within the context of the family. In this sense, medication is regarded as a Band-Aid for a problem, rather than something which addresses the fundamental cause of childhood misbehaviour.

The debate between the medical and psychodynamic viewpoints brings up a very important question with respect to the way ADHD is viewed today: irrespective of their 'correctness', which of these perspectives seems to be the most predominant in contemporary society? This is where we examine the extra-discursive dependency, or the actual impact that a perspective has had upon everyday behaviour. The analysis of the extra-discursive dependency clearly indicates that a medical viewpoint is the most predominant way that ADHD is suspected, diagnosed, and treated. In order to better understand the extra-discursive dependency, we will now introduce our third viewpoint towards ADHD, the sociological one.

The sociological viewpoint towards ADHD is essentially a critique of the validity of this widely diagnosed behaviour disorder. There are several key concepts that sociology uses in order to level this critique. First, sociology relies upon the concept of medicalisation, which describes the expanding province of medicine into areas thought to be outside the medical purview (Conrad, 1975; Conrad & Schneider, 1980). As we have seen, medicine has a long history of attempting to problematise childhood difficulties. Medicalisation has also had a startling impact upon the lives of school-age children. As we have such a striking number of school-age boys currently taking stimulant medications for behavioural problems, it is clear that at the extra-discursive level the discourses of medicine which have realised themselves through medical practices are the most dominant in the ADHD conversation. That is, the everyday practices of individuals and clinicians in response to the discursive field of ADHD demonstrate a widespread subscription to the legitimacy of the neurological argument – ADHD is a bona fide brain impairment that requires long-term (if not permanent) medical intervention in the form of stimulant medications. How did we get here? On one side of this analysis, we could easily state that 'modern scientific practices' have honed their skills to such an extent that they are now incontrovertibly the way to go when it comes to the evaluation, diagnosis, and treatment for ADHD and a host of other childhood behaviour disorders. Through their diagnostic methods, demonstration of effective treatments, and considerable theoretical acumen, the neurological viewpoint has simply been the superior one and is therefore the most worthy in terms of directing how clinicians and laypeople are going to proceed.

As enticing as this argument seems, the drastic increase in cases of ADHD highlight other, less discussed concerns about the legitimacy of the ADHD diagnosis, and the effectiveness of ADHD treatment. As Lawrence Diller (1998) pointed out, the drastic increase in the amount of ADHD diagnoses should raise considerable alarm. After all, if we saw a 700% increase within a 10-year period of time (see Diller, 1998) in the diagnosis of diabetes, cancer, heart disease, or any other bona fide somatic medical condition, this would spark a litany of heated questions. Was there something in our environment that was contributing to these high rates of such illnesses? In our drinking water? Were the methods of diagnosis scientifically sound? Should we redefine what some of these illnesses are? These questions could continue ad infinitum. Not the case with ADHD. For the most part, voices of dissent or concern surrounding these incredibly high rates of diagnosis and treatment—especially for an illness in which the vast majority of positive diagnoses are made through self-reporting data like the various iterations of the Conners Scale—fall on relatively deaf ears. While the publication of a piece of journalism or a sociological article may underscore the concerns with the diagnosis and treatment process surrounding ADHD, those who work in the 'ADHD industry' continue to operate against very little headwind.

Second, the sociological viewpoint takes issue with the fact that modern medicine has still been unable to definitively distinguish between deviant behaviours that have medical origins and those that do not. In other words, what is a 'purely' deviant act, versus one that is prompted by brain pathology? Is it the frequency of the deviant behaviour? What about the inability of the deviant individual to respond to remedial measures? In order to address this distinction at greater length, let us bring in the example of the medically accepted somatic condition of diabetes. Physicians can clearly discern the distinction between a rise in blood sugar that is a result of eating a piece of cake or drinking a sugary soda and a chronic blood sugar regulation condition that is indicative of diabetes. Although both of these conditions can be evaluated according to medical procedures, the former illustrates a condition that generates no long-term medical interest, whereas the latter most certainly does. Indeed, we might argue that, depending upon the physician, the diagnostic tools utilised, and the reporting from the patient, the former condition might be diagnosed as a 'false positive'. and the latter, given some inadequacy in the diagnostic process, a 'false negative'. But how is this distinction resolved? The answer: through the tools of somatic medicine that have demonstrated both validity and reliability over time.

In looking at the condition of the 'false negative' diabetic, the patient may revisit the physician, complain about other symptoms, ask for a follow-up examination or a change of medical venue. Though the process of somatic medicine is far from perfect, and this patient may never receive the diagnosis and treatment she needs, chances are that the false negative will be detected and a new treatment protocol will be followed. In the case of ADHD and other behaviour disorders, there are little if any safeguards to help distinguish between a false positive and an actual case of the condition. Indeed, the clinical literature contains virtually nothing with respect to the detection and rectification of ADHD false positives. It is simply assumed that a favourable response to stimulant medication demonstrates the presence of the ADHD condition. This is the standard mode within modern psychiatry, which is a type of reverse engineering (see Rafalovich, 2004) of a diagnosis: begin with drug treatment, and based upon the impact of such treatments, either affirm or 'adjust' a diagnosis. This harkens back to Bradley's early studies on the impact of Benzedrine with boys housed at a training school. He assumed (as it turns out, incorrectly) that these stimulant medications had a type of paradoxical effect on certain children, which therefore affirmed the presence of some physiological lesion in the brain.

Third, sociology raises scepticism about the perspective from which a drug's impact is interpreted. Although the sociological viewpoint does not dispute (would anyone?) that taking stimulants like Ritalin, cocaine, methamphetamine, caffeine, and so on change behaviour, the evaluation of certain types of drugs is clearly mediated by the discursive viewpoint that seems to have the most legitimacy. As Ritalin is a drug seen almost exclusively as a treatment rather than a recreational drug, the interpretation of its impact upon individual behaviour is cast in medical terms. Because medicine has owned the discourse surrounding stimulant medications for so long, the evaluation of the impact of such medications is cast in detached, objective, 'clean' medical terms: Ritalin reduces the incidence of disruptive behaviour in class, increases academic performance, and enables children to maintain on-task behaviour.

In order to better understand this, let us look at the hypothetical case of the impact of marijuana upon behaviour. It is a matter of course that smoking pot changes behaviour, which can be measured through a variety of scientifically validated variables: ocular pressure, tests of short-term memory, blood pressure, and changes in appetite. However, we see very little, if any, conversation about whether the physiological response to marijuana, or the compulsive use of the drug, indicates some type of medical condition, other than, say, 'marijuana addiction'. Although the neurological evaluation of Ritalin which validates the existence of ADHD implies that an individual has a unique, favourable response to Ritalin, there is no clinical discussion about individuals having this same type of need for marijuana, cocaine, heroin, or any other recreational drug. People who use recreational drugs may be 'self-medicating', but as modern medical discourse mandates, they would need to cease the use of such recreational drugs, 'get clean', and then have an effective medical evaluation.

Clinical relevance summary: ADHD and the problem of inelegance

The sociological viewpoint is a good place to complete this analysis, as it highlights some of the validity and reliability issues that medicine faces when it comes to the diagnosis and treatment of ADHD. One need only to look at the alarming rise in the number of ADHD diagnoses. The Centers for Disease Control (CDC, 2014) affirms Diller's (1998) concerns, stating that the incidence of ADHD continues to rise dramatically (CDC, 2014). ADHD diagnoses jumped from 7.8% of the school-age population in 2003, to 9.5% in 2007, to 11.0% in 2011. Sociologists argue that this demonstrates both the danger of culturally interpretive medical categories and also the benefit such categories provide to modern medicine. The diagnostic criteria for ADHD as outlined by the American Psychiatric Association (1980, 1987, 1994, 2013) describe a whole host of symptoms that, when taken in their entirety, are easily interpreted through a cultural lens. For example, how can we look at 'fidgeting' as an illustration of a medical condition, when this seems to be a rather normal part of childhood behaviour? On a more general level, what criteria determine what constitutes a normal bout of fidgeting, versus one that is indicative of a pathological brain condition? Indeed the boundaries of normality are fluid and in themselves socially constructed (see Brownlow & Lamont-Mills, Chapter 13, this volume for a full discussion). In order to further unpack this distinction, we may look at the cases of homosexuality and post-traumatic stress disorder.

For the vast majority of its existence, psychiatry asserted that homosexuality was a bona fide mental illness (Freud initially called it sexual 'inversion') that could be treated through a host of methods – psychotherapy, medications, institutionalisation, shock therapy, and so on. In the early 1970s, several special interest groups lobbied the American Psychiatric Association to have homosexuality removed from its list of mental disorders. After minimal debate, the APA conceded, and in 1973, homosexuality was no longer listed as an official mental illness. The illness category itself was watered down significantly, and only labelled as pathological if one's homosexuality caused significant mental distress. During roughly the same period of time, veterans returning from the war in Vietnam lobbied the APA very strongly to include post-traumatic stress disorder (PTSD) among its list of bona fide mental illnesses. Again, the APA conceded, placing PTSD into the publication of *DSM-III* in 1980. Such responses from the APA were regarded as a victory for both groups: homosexuals won a significant battle against the stigma of sexual preference, and war veterans achieved a new nomenclature to describe, diagnose, and treat psychological injuries from the battlefield.

This type of response from modern psychiatry begs important questions that highlight the key distinction between psychiatric medicine and somatic medicine. For example, it would seem preposterous that people struggling with oesophageal cancer would lobby the American Medical Association to remove this type of cancer from its list of illnesses for fear that such nomenclature unfairly stigmatises them. Because the cases of homosexuality and PTSD have a profound social and cultural importance, and because modern psychiatry directs a large part of its attention to social and cultural matters, the entire psychiatric enterprise is vulnerable to changes in cultural sentiment and sensibility.

The connection between ADHD and the modern American classroom is another great case in point. As my own research demonstrates, the rise in the number of diagnosed cases of ADHD can be directly correlated with the advent of the 'open classroom' era in modern American education (Rafalovich, 2004, 2005). As public schools moved away from a typically Germanic model of education, and hence, a much less authoritarian one, teachers began to see more incidence of disruptive behaviour among their male students. Because it is responsive to changes in culture, and changes in institutional practices such as those that occurred in our public schools in the early 1970s, modern psychiatry responds by 'solving' these institutional problems. The crux of the sociological argument with respect to the incidence of ADHD concerns the disingenuousness of these medical practices. For example, instead

Table 5.1 Educational/clinical practice highlights

1. There has been a rise in the rates of ADHD being diagnosed and this has been thought of as quite alarming. The increase in prevalence has created important debates regarding diagnosis and the language of ADHD.
2. ADHD as a childhood condition has implications for the classroom as it links considerably with the advent of the 'open classroom' in American schools.
3. There is little from modern medicine to explain the significant gender differences that seem to exist in the rates of ADHD.

of acknowledging that their role in diagnosing ADHD is really a type of medical social control, psychiatrists 'double down' and passionately argue for the validity and reliability of their diagnostic and treatment practices.

Another significant point of contention for sociologists concerns the gender discrepancy in the diagnosis of ADHD. As of 2013 (the most recent year in which data were available), 13.2% of the ADHD cases were boys, whereas only 5.6% of girls were diagnosed with the disorder (CDC, 2014). There is little by way of response from the modern medicine camp to explain this. Although the rate of diagnosis in girls has increased, there is no 'disease theory' explaining why this condition afflicts boys almost three times as often as it does girls. For sociologists, this is another example of how the diagnostic procedures and practices of modern psychiatry are heavily influenced by culture – in this case, gender norms – rather than by standards of scientific inquiry. As any sociologist or developmental psychologist will point out, boys are generally more extroverted, and in institutional settings such as schools, boy behaviour, whether labelled pathological or not, is more visible and therefore disruptive. In this sense, the overrepresentation of boy cases of ADHD is more an extension of gender discrimination in the institutional context of schools than differences in the male and female brain. For a simple summary of the implications for practice, see Table 5.1.

Summary

As we look at these and other key social variables that contribute to the social phenomenon of ADHD, we cannot help but be cautious about the validity of the diagnosis and the high rates of prescriptions being written for stimulant medications for our children. As ADHD has become a bona fide industry, it is very important for academics, clinicians, journalists, and laypeople to interrogate the various interests that make up the discursive landscape of ADHD today. The above examples demonstrate how vulnerable the concept of ADHD is to criticism when cultural conditions and trends in medicine are more carefully examined.

References

Abrahamson, I. (1920a). The epidemic of Lethargic Encephalitis. *New York Medical Record,* December 11.

Abrahamson, I. (1920b). The chronicity of Lethargic Encephalitis. *Archives of Neurology and Psychiatry, 4,* 428–432.

American Psychiatric Association (1980). *Diagnostic and statistical manual of mental disorders* (3rd edition). Washington, DC: Author.

American Psychiatric Association (1987). *Diagnostic and statistical manual of mental disorders* (3rd edition revised). Washington, DC: Author.

American Psychiatric Association (1994). *Diagnostic and statistical manual of mental disorders* (4th edition). Washington, DC: Author.

American Psychiatric Association (2013). *Diagnostic and statistical manual of mental disorders* (5th edition). Washington, DC: Author.

Barkley, R. A. (1990). *Attention Deficit Hyperactivity Disorder: A handbook for diagnosis and treatment.* New York: Guilford Press.

Barkley, R. A. (1998). *Attention Deficit Hyperactivity Disorder: A clinical workbook* (2nd edition). New York: Guilford Press.

Barkley, R. A. (2005). *ADHD and the nature of self-control.* London: Guilford Press.

Bradley, C. (1937). The behavior of children receiving Benzedrine. *American Journal of Psychiatry, 94,* 577–585.

Cantwell, D. P. (1981). Foreward. In Barkley (Ed.), *Hyperactive children: A handbook for diagnosis and treatment* (pp. vii–x). New York: Guilford.

Centers for Disease Control and Prevention (2014). *Attention-Deficit/Hyperactivity disorder (ADHD): Data & statistics.* Retrieved October 15, 2014 from http://www.cdc.gov/ncbddd/adhd/data.html.

Conrad, P. (1975). The discovery of hyperkinesis: Notes on the medicalization of deviant behavior. *Social Problems, 23,* 12–21.

Conrad, P., & Schneider, J. (1980). *Deviance and medicalization: From badness to sickness.* St. Louis: Mosby.

Diller, L. H. (1998). *Running on Ritalin: A physician reflects on children, society, and performance in a pill.* New York: Bantam Books.

Foucault, M. (1965). *Madness and civilization: A history of insanity in the age of reason.* New York: Random House.

Foucault, M. (1973). *The birth of the clinic: An archaeology of medical perception.* New York: Vintage.

Foucault, M. (1977). *Discipline and punish: The birth of the prison.* New York: Pantheon.

Foucault, M. (1978). Politics and study of discourse. *Ideology and Consciousness, 3,* 7–26.

Freud, A. (1926). *Psycho-analytical treatment of children: Technical lectures and essays by Anna Freud.* New York: International Universities Press.

Hohman, L. B. (1922). Postencephalitic behavior disorders in children. *Johns Hopkins Hospital Bulletin, 33,* 372–375.

Ireland, W. W. (1877). *On idiocy and imbecility.* London: J. & A. Churchill.

Kessler, J. W. (1980). History of minimal brain dysfunctions. In H. Rie & E. Rie (Eds.), *Handbook of minimal brain dysfunction: A critical view* (pp. 18–42). New York: Wiley-Interscience.

Klein, M. (1963). *The psychoanalysis of children.* London: The Hogarth Press Ltd. (originally published as *Die Psychoanalyse des Kindes,* 1932).

Mercier, C. A. (1980). *Sanity and insanity.* London: Scott.

Paterson, D., & Spence, J. C. (1921). The after-effects of Epidemic Encephalitis in children. *The Lancet*, 491–493.

Rafalovich, A. (2001). Psychodynamic and neurological perspectives on ADHD: Exploring strategies for defining a phenomenon. *Journal for the Theory of Social Behaviour, 31*, 397–418.

Rafalovich, A. (2004). *Framing ADHD children: A critical examination of the history, discourse, and everyday experience of Attention Deficit/Hyperactivity Disorder.* Lanham, MD: Lexington Books.

Rafalovich, A. (2005). Relational troubles and semi-official suspicion: Educators and the medicalization of 'unruly' children. *Symbolic Interaction, 28*, 25–46.

Rank, B. (1954). *Intensive study and treatment of preschool children who show marked personality deviations or 'atypical development' and their parents.* Paper presented to International Institute of Child Psychiatry, Toronto.

Rosenhan, D. L. (1973). Being sane in insane places. *Science, 179*, 250–258.

Schrag, P., & Divoky, D. (1975). *The myth of the hyperactive child and other means of child control.* New York: Pantheon.

Stewart, M. A. (1970). Hyperactive children. *Scientific American, 222*, 94–98.

Still, G. F. (April 12, 19, 26, 1902). Some abnormal psychical conditions in children. *The Lancet*, 1008–1012, 1079–1082, 1163–1067.

Stryker, S. B. (1925). Encephalitis Lethargica: The behavior residuals. *The Training School Bulletin, 22*, 152–157.

Tredgold, A. F. (1917). Moral imbecility. *Practitioner*, July, 43–56.

Recommended reading

- Barkley, R. A. (2005). *ADHD and the nature of self-control.* London: Guilford Press.
- Rafalovich, A. (2004). *Framing ADHD children: A critical examination of the history, discourse, and everyday experience of Attention Deficit/Hyperactivity Disorder.* Lanham, MD: Lexington Books.

Part II

A Critical Approach to Child Mental Health

6
Discourses on Children's Mental Health: A Critical Review

Tom Strong and Monica Sesma-Vazquez

Introduction

A minor furore was set off back in the early 1960s when amateur historian Philips Ariés (1962) reported that childhood was *discovered* in the 17th century. The furore partly came from Ariés' claim that childhood was relatively a recent notion formerly indistinguishable from adulthood. It took until 1842, for example, for the British parliament to enact the Mines Act, forbidding children under 10 to work in coal mines. At about the same time, Charles Dickens' stories of children (e.g. *Oliver Twist*, 1838) on the streets of London sensitised readers to the plight of children in poverty. For Ariés, the bigger hermeneutic story was how our notions of childhood changed across historical and cultural contexts. Similar things could be said about children's mental health.

Mental health, as we currently know it, is a relatively recent construct though an enduring historical record, dating back to Assyrian tablets, mentions 'sickness of the head' (Ehrenwald, 1991). It was with the rise of psychiatry and psychology as professional disciplines in the early 20th century that mental health took on its modern thrust. That thrust was animated by an ideal that human science could produce knowledge for engineering mental health (Cushman, 1995; Gergen, 1982; Shotter, 1975), much as the applied knowledge of the natural sciences had enabled astounding technological innovations.

Equipped with psychological tests, laboratory observations, and statistical procedures, the disorders and abnormalities of children were categorised, explained, and modified (Burman, 1994; Rose, 1990). Accordingly, school curricula were revised to optimise children's mental health on such bases despite changing ideals for the self purportedly being educated (Martin & McLellan, 2013). More recently, 'epidemics' (see Frances, 2013, on 'diagnostic inflation', as well as Giles, Chapter 12, this volume) of children's mental illnesses such as attention deficit and hyperactivity disorder (ADHD) have coincided with a further 'epidemic' of medication use for these illnesses (Timimi & Timimi,

Chapter 8 this volume; Whitaker, 2010). By contrast, one finds a single paragraph on children's disorders in Karl Jasper's (1963) 900-page classic, *General Psychopathology*. No consensus on what is meant by children's mental health exists today.

Across the helping professions, one finds multiple discourses of understanding and practice, and discourses that in turn inform self-help resources accessible through online and other media (Illouz, 2008). Increasing preoccupation with mental illness has produced new forms of vulnerability and a zeitgeist Frank Furedi (2004) refers to as 'therapeutic culture'. The languages used to classify mental health and illness tend to draw from recycled moral sentiments (Danziger, 1997) associated with symptoms formerly unconsidered as medical (Conrad, 2007). For deVos (2012), too much psychologising of individual deficits and symptoms is occurring, while Western societies export ideas and practices regarding adults' and children's mental health/illness globally (Watters, 2010).

There is plenty of conflicting mental health advice, as researchers and practitioners take up quite varied discourses themselves (e.g. Freeman, Epston, & Lobovits, 1997; Rose & Abi-Rached, 2013). In the self-help sections of bookstores, or in Google or Youtube searches for advice on children's mental health, one finds many experts to read or view (e.g. Grohol, n.d.; Starker, 2002). How such self-help initiatives translate to everyday life can be both positive and negative, though most affirming (or disconcerting) are discourses that come to shape understandings of self and other (Prince, 2008; Thompson, 2012).

For children, the discourses for understanding and addressing mental health concerns that parents, educators, professionals, and children themselves turn to are seldom benign. Such discourses not only furnish descriptions for these concerns, the science associated with these discourses furnishes related *prescriptions* for the mental health of children. A discourse of children's mental health can seem bidirectional, if one considers the hopes and concerns associated with childhood (i.e. about a child's mental health or mental illness). However, health involves more than an absence of illness symptoms. Therefore, we trace the development of three discourses of childhood: well-being, mental health, and mental illness while inviting reader reflection on the descriptive and prescriptive elements of each discourse.

A recent focus on promoting children's 'mental health' and treating their 'mental illnesses' (e.g. Barry, 1998; Muzychka, 2007; Roberts & Grimes, 2011; Rogers & Pilgrim, 2010) seems to be narrowing society's efforts at enhancing children's 'well-being'. Our review of these literatures also suggests that these three discourses (well-being, mental health, and mental illness) can be unhelpfully intertwined or conflated. However, as our selective historical review will show, considerable differences developed within and across these three discourses over time.

Challenges in using these related yet dissimilar discourses become particularly evident when reading the recent *Diagnostic and Statistical Manual of Mental Disorders, Fifth Edition* (*DSM-5*, American Psychiatric Association, 2013). No definition for what is meant by a 'mental disorder' is provided to distinguish mental health from mental illness (Frances, 2013). Other challenges relate to cultural and other context-sensitive value judgements when assessing well-being, mental health, or mental illness (cf. Watters, 2010) though our focus here will be on common Euro-American discourses distinguishing well-being, mental health, and mental illness.

We trace these discourses of children's well-being, mental health, and mental illness through developments in their social construction and circulation, though modern efforts to optimise children's mental health began in the early 20th century. Along with others, we wonder if parents' and professionals' preoccupations with 'medicalising misery' (Rapley, Moncrieff, & Dillon, 2011) or 'psychologising' aspects of individual life, formerly considered social, cultural, or institutional (deVos, 2012), have been narrowing and diluting a focus on children's well-being. Seeking a 'history of the present' (Foucault, 1975), we trace developments across children's wellness, mental health and mental illness discourses and conclude by discussing two complementary concepts for a 'discursive sensibility' (Strong, 2011) towards practice: discursive capture and discursive resourcefulness.

Children's 'well-being'?

History shows how differently parents and adults have understood and promoted children's well-being over time. 'Well-being', as we are using the term, refers to the kind of wholeness and vigour associated with the Old English term 'hale' (as in hale and hearty). Discourse developments, particularly those evident in texts that reflect the dominant thinking, talking, and acting of particular eras – such as in books, new media, and artistic representations – show how notions such as children's well-being have been differently represented and acted upon. For example, until recently parents used corporal punishment to educate or discipline their children – 'for their own benefit' (Gershoff & Larzelere, 2002). In other words, what is now typically seen as a form of violence was an accepted and often encouraged way to raise a well-adapted child. However, children's corporal punishment at school is still debated (and legal) in at least 20 states in the United States (Human Rights Watch, 2010); whereas, in the United Kingdom, corporal punishment was abolished in public schools in 1987 and in private schools in 1999 (Gould, 2007; Society for Adolescent Medicine, 2003). Researchers now cite such physical abuse as negatively impacting children's 'mental health', or as a risk factor for becoming mentally ill adults (Choi, Reddy, & Spaulding, 2012; Spring, Sheridan, Kuoc, & Carnes,

2007; Sugaya et al., 2012). A 'well' child was well-disciplined or well-adapted, though needing occasional punishment.

Sometimes notions of children's well-being seemed suspiciously convenient for adults, by today's standards. In Victorian times, young girls were considered *lucky* to be married by the age of 13, or pushed to marry after 14, since otherwise they were considered a burden. In the European wars of recent centuries, male children were commonly sent as soldiers to fight battles (it still happens in some countries today). For Moran (2001), 'well-being', 'mental health', and 'mental illness' have been terms related to how lives were socially and economically organised. Children's work and contributions to their family's financial resources were expected, though their victimisation at work eventually required political intervention by the mid-19th century (Bagchi, 2010; Kirby, 2003; Schmitz, Collardey, & Larson, 2004). Such changing expectations of children seem harsh in contrast to that quintessentially healthy activity of childhood: play. While children's games and toys are found across time in all cultures, 'to play' during childhood has also been an evolving concept. Currently, unstructured playtime, games, and use of toys have been promoted as key elements helpful to sustaining and optimising children's well-being (Jacobson, 2008; Milteer & Ginsburg, 2012). This notion contrasts with an approach to childhood well-being where highly structured activities and expectations are seen as good for children, as helping them to prepare for adult challenges (Chua, 2011).

It can seem implausible to link discourse with emotional expression though historical and cultural differences regarding perceived emotions and well-being have varied in ways we associate with discourse. In Western countries, for example, upper-class children were expected to show their emotions in ways differing from children of lower economic classes (Gordon, 1989; Hochschild, 1979). Such upper-class children were separated from their parents, especially their mothers, and raised by nurses, private school masters, governesses, or nannies. Negative 'emotional' effects of this upper-class tradition, such as contemporary concerns about 'lack of attachment', or growing up 'depressed', 'frustrated', or 'angry', were not considered, even in relational terms.

It is only recently that relationships between children's 'well-being' and access to free and institutionalised education have been identified and accepted as a priority. The United Nations International Children's Emergency Fund (UNICEF, 2013) identifies five dimensions of children's well-being: material well-being, health and safety, education, behaviours and risks, and housing and environment. However, in the developing world, many children continue to go without education while UNICEF's other dimensions of children's well-being would seldom feature in public discourse.

The National Council for Curriculum and Assessment (NCCA, n.d.) proposes that well-being must focus on how children develop in two areas:

'psychological' and 'physical'. Accordingly, children's relationships and interactions with their families and communities are considered key areas contributing to their well-being. Warm and supportive relationships with adults contribute not only to well-being but to independence, creativity, exercising, good nutrition choices, spirituality, and hygiene (NCCA, n.d.). Other groups, such as The Children's Society (2014), identified six priorities for children's well-being: conditions for learning and developing, a positive self-regard and a respected self-identity, have enough of what matters (like food, education, health services), positive relationships with family and friends, a safe and suitable home environment and local area, and opportunities to take part in positive activities. Such priorities on these 'psychological' and 'physical' aspects have varied in literatures of children's well-being that we reviewed.

The changing concepts and discourses of childhood 'well-being' have had a profound influence on policy and practice in medical, psychological, or educational contexts, and within everyday life interactions occurring within families and among peers. It can be difficult to distinguish these varied notions of childhood well-being from the practices used to promote 'it', or from normative expectations that may have obscured differences between children and adults to begin with, as in the case of children working. Evolving and varied notions of 'childhood wellbeing' have been taken up with the assistance (or reinforcement) of truant officers, social workers, physicians, and other state officials. By the end of the Second World War, Dr Benjamin Spock (Spock & Needlman, 2012) became the voice of child-centred parenting practices in North America. Later, developmental psychologists advanced relational attachment as paramount to children's (and adults') emotional well-being (e.g. Bowlby, 1969). These evolving notions of children's well-being clearly reflect changing historical and cultural contexts, and the priorities adults have set within them. 'For their own good', a phrase once used to support the corporal punishment (and well-being) of children, is seldom publicly used these days.

Children's mental health

An increasingly nuanced discourse of children's 'mental health' (Arbuckle & Herrick, 2006) has been developing, accounting for a need to *prevent* childhood 'psychopathologies' (Harari, 2013; Hassim, & Wagner, 2013), 'mental disorders' (Frances, 2013), 'traumas' (Quosh & Gergen, 2008), as well as 'behavior disorderedness' (Graham, 2005, 2010). Like any discourse, this 'mental health' discourse enables and constrains different understandings and responses to children, extending to their 'needs', 'interventions', 'treatments', 'problems', 'disorders' across the helping and educating professions. Mental health discourse melds the contemporary knowledge and practices of psychology and psychiatry to promote the 'good life' (Ehrenwald, 1991; Rose, 1990); it does so in ways that recruit non-professionals seeking to apply these means to the 'good life' to their own lives, and to the lives of children.

While, broadly speaking, an adult 'mental health' literature developed over centuries, a children's 'mental health' literature was almost non-existent before the 18th century (Wolfe, 2014). This discourse arguably lives in the shadow of its counterpart: mental illness, a 20th-century discourse about children which we will elaborate later. The purportedly preventable kinds of concerns we now call 'mental illness' were then understood as outcomes of demonic possession, God's will, or biological imperfections (cf. Frances, 2013). Prior to the 20th century, society was unfamiliar with 'prevention' or 'educational' interventions that ensure children's 'mental health'. Thus, such practices as praying for children or exorcising their demons were among the ways adults tried to promote or sustain what we now refer to as children's 'mental health'. Without a concept of 'children's mental illness', many children with what we now term 'mental disorders' were kept at home to receive family care, when households permitted this practice (Foucault, 1976; Moran, 2001).

The pioneering insights and therapies of Freud and early behavioural psychologists ushered in an aspirational and prevention-oriented era we now associate with modern mental health. New individual, group, and family interventions, along with changing educational initiatives (Martin & McLellan, 2013), were developed to optimise the mental health of children. By the 1970s, a new mental health frontier opened up as newly diagnosed disorders, such as ADHD, were found to be responsive to pharmaceutical interventions (e.g. Conrad, 2007; Whitaker, 2010). A new kind of mental health logic accompanied these interventions: chemical *management* (as opposed to 'treatment') of disorders deemed incurable (Frances, 2013; Lakoff, 2008).

Children's 'mental health' as it is conceived today began with the mental hygiene movement in the early 20th century (Noll, 2011). This movement, applying research from developmental psychology, focused on the 'normal child's' (Crow & Crow, 1951; Truitt, 1926) expected (i.e. age-related) behaviours and responses. People adopting such a focus tended to pathologise children's behavioural problems, such as later delinquency, drug addiction, school performance and attendance concerns, and teen pregnancy. Parental care was targeted as the most relevant (and preventive) factor in children's 'mental health'. In the words of John Bowlby (1953), 'mother-love in infancy in childhood is as important for mental health as are vitamins and proteins for physical health' (p. 182).

By the 1970s, advocates for children's mental health engaged home and school cooperation in areas such as education, prevention, and remediation (Group for the Advancement of Psychiatry, 1972). School programmes, for example, were developed and delivered to enhance the social environments and experiences of students. These mental health programmes, through engagements in extracurricular activities, sought to improve the intrinsic motivation, self-esteem, and self-resilience of children who were away from their families and communities. Such programmes also trained children to deal with difficult situations, stressful transitions, by enhancing individual coping skills seen to increase strengths while reducing problems. Though the focus of such mental health initiatives was on individuals, the *social* contexts within schools and families were considered ideal spaces in which to intervene (Waller, 2013).

Recent efforts to optimise children's 'mental health' have expanded to address: risk and prevention concerns (Lawless, Coveney, & MacDougall, 2014), mental health awareness (Lamerichs, Koelen, & Molde, 2009; Wolfe, 2014), stigma (Corrigan & Watson, 2002; Harper, 2005), discrimination (Mossakowski, 2003; Stevens & Vollebergh, 2008; Tran, 2014), child neglect (Miranda, De la Osa, Granero, & Ezpeleta, 2013), or healthy identity (McAdams & McLean, 2013). Consensus on what children's mental health is has eluded scholars and practitioners though this has not stopped the recurring bandwagon approaches that continue to be found in school and public health campaigns (Stephan, Weist, Kataoka, Adelsheim, & Mills, 2007; Weist, Rubin, Moore, Adelsheim, & Wrobel, 2007). Even such purportedly field-unifying concepts as 'mental health literacy' lack consistent theoretical articulation (Francis, Pirkis, Dunt, Warwick Blood, & Davis, 2002). Instead, children's mental health discourse (and the practices to follow from it) seems to have developed in piecemeal fashion through media efforts and diverse programmatic interventions to target individuals and social practices within schools and families.

In children's daily social interactions, across family conversations, and activities on school playgrounds, it can be difficult to assess how children's mental health discourse has or has not been taken up. Are changes in health outcomes and practices (mental health's equivalent to brushing one's teeth) resulting from these varied efforts at understanding and enhancing children's mental health as we have been describing? Do families discuss and practice contemporary forms of mental health hygiene as identified within the evolving children's mental health discourse? Research answering such questions is equivocal (Browne, Gafni, Roberts, Byrne, & Mahumdar, 2004) reflecting the field's lack of conceptual and programmatic consensus, particularly when factoring in differential funding priorities across political jurisdictions. While focused on notions of healthy child development and its enhancement, mental health continues to be primarily understood as an absence of its counterpart: mental illness (Hutchby, 2007).

Children's mental illness

Children's mental illness discourse developed in tandem with psychiatry's focus on diagnostic classification and treatments related to diagnosed 'disorders'. It is a discourse of pathology, located purportedly 'in' the children. As the 'S' in the *DSM* title signifies, it can be based on statistically normative concepts of health and illness.

The medicalised construct of children's 'mental illness' has been contested by scholars and practitioners (Barry, 1998; Conrad, 2007; Foucault, 1976; Ineichen, 1979; Tausig, Michello, & Subedi, 1999). Family therapists, for example, have long challenged the notion that the behaviours associated with mental illness be assessed apart from the social and interactional contexts in which they find their meaning (e.g. Nichols & Schwartz, 2008; Strong, 1993). Such a focus on individuals' symptoms of mental illness can also obscure links between symptoms and the socially unjust circumstances in which they occur (Eriksen & Kress, 2005).

'Madness', as Foucault's (1976) historical review showed, eventually became an institutional issue. Specialised hospitals and asylums were built to house and treat 'mad' adults who had been suffering or behaving in ways contrary to accepted cultural 'norms' (Moran, Topp, & Andrews, 2007; Yanni, 2007), while children's mental health or illness received scant attention. Instead, children diagnosed with mental illness are increasingly 'treated' within their natural educational and home contexts. As suggested earlier, fuzzy boundaries between mental health and mental illness have been developing (Clarke, Mamo, Fosket, Fishman, & Shim, 2010; Lakoff, 2008) out of a growing view that all humans need pharmaceutical management to live 'healthy' lives. For example, people with diabetes are seldom considered ill these days. Extending such logic, diagnosing concerns about children can be seen as a step towards optimising their living 'healthy' lives through eventual pharmaceutical management (a 21st-century version of the hippie credo 'better living through chemistry').

Recent concerns have developed over the increase in diagnosed children as well as diagnosable disorders. The process leading to the publication of *DSM-5* (APA, 2013), for example, prompted Allen Frances (Chair of *DSM-IV* process, 2013) to protest *DSM-5*'s 'diagnostic inflation and exuberance' with an evocative plea for 'Saving normal'. His concern is with expanding and accelerating practices of *statistically* classifying mental illness according to normative expectations of healthy development (Burman, 1994; Rose, 1990). To that end, psychologists have developed many tests to assess children for normalcy or diagnosable disorders. New kinds of diagnostic reasoning are also developing that rely on responses to medication (if one responds expectedly, one has the disorder the medication was tailored for, Lakoff, 2008). This is perplexing when children, for example, are increasingly treated with anti-psychotic medication to address behaviours adults find emotionally disruptive (e.g. Rodday et al., 2014).

Children's 'mental illnesses' presumably require clinical 'treatments' (Barry, 1998; Kahan, 1971; Wilkinson, 2012) despite controversy over what can be seen as excessive medicalising of children's (or adult's views of children's) 'difficulties' as 'mental illnesses' (e.g. Conrad, 2007, 2010; Frances, 2013; Jahnukainen, 2010; Smith, 2010). Concerns over diagnosing such difficulties solely in biological terms prompted Mills (2014) to provocatively describe some children as having 'psychotropic childhoods'. In another aspect of addressing mental illness, the conversational work involved in making such diagnoses has interested discourse analysts, for whom such diagnostic conversations are linguistically constitutive of children's identities (Avdi, 2005; Hutchby, 2007). Being diagnosed as mentally ill or having a mental disorder can have profound positive or negative consequences within families and schools enabling diagnosis-related responses to the child, and particular kinds of self-understandings (Hutchby, 2007).

How one develops a childhood 'mental illness' is also still a matter of debate. The mother-blaming that occurred in the heyday of attachment-focused research and practice has rescinded somewhat (Caplan, 2013). Today many potentially influential or causal factors are considered relevant to developments associated with childhood mental illness: heredity, environment, personality, stress, family functional interactions, communication patters, ethnicity, community roles, cultural influences, gender differences, and financial circumstances (Barry, 1998; Gaw, 1993). Barry (1998), for example, emphasised how each culture defines local standards for acceptable and unacceptable behaviour. For Harré (1986), this line of thought has micro-interactional implications within families and cultures, in what he termed 'emotionologies' – ways that people perform emotions in contextually and culturally accepted ways. Returning to Barry, 'A major human task is learning to control emotions' (p. 143) in such accepted ways. 'Unacceptable' behaviours and emotional expressions that become possibly diagnosed as mental illness have concerned critics for decades (e.g. Szasz, 1970). Critics such as Alverson and colleagues (Alverson et al., 2007) propose instead that mental illness be understood in context-sensitive and context-informed ways since the associated discourses of mental illnesses differ across cultural contexts. Efforts to standardise a global discourse of mental illness have been problematic for people of non-Euro-American cultures (cf. Watters, 2010).

The identity-implicating features of children's mental illness discourse have alternatively been seen as sources of help and concern. Garro and Mattingly (2000) regard healing narratives of human suffering as a means to construct and reconstruct identity in the face of cultural constructions of mental illness. An example of evolving children mental illness' discourse is the case of Asperger's syndrome. First classified in 1992 in the International Classification of Diseases (ICD-10) and in 1994 by *DSM-IV*, a new Asperger's Disorder was included. Distinctions between *syndrome* and *disorder* seemed unimportant at the time, though adults began self-diagnosing as 'aspies' (organising themselves in such groups as *Aspies for freedom, American Asperger's United, Dude, I'm an Aspie)* and some parents understood their children's behaviour as part of 'Asperger' diagnostic discourse through culture (e.g. in literature Mark Haddon, 2003, with *The curious incident of the dog in the night-time*, or Margaret Atwood, 2003, with *Oryx and Crake)*, awareness movements (e.g. UN World Autism Awareness Day or Autism Society, 2014) and internet resources (e.g. *My Asperger's Child*, Hutten, 2014). Social media groups for parents with an Asperger's child developed on Facebook 'clubs' offering a venue to discuss different facets of children's 'aspie' life like *Parents of Children with Asperger's syndrome Club*, 2014, or *Parents of Children with Asperger's Support Club*, 2014.

Concurrently, funding for Asperger's-related research projects grew enabling academics and new journals to publish on this topic. And, at about the same time, drugs for treating children diagnosed with Asperger's syndrome became more available, such as *Risperidone* (an atypical antipsychotic drug used mostly for schizophrenia) and *Lisdexamfetamine* (prescribed for ADHD to control hyperactivity). To the surprise (and disappointment) of many, Asperger's Disorder was reclassified in *DSM-5* within a broader Autism Spectrum Disorder. Though still classified within ICD-10 but not in *DSM-5*, the Autism Research Institute (2014) continues to explore the implications of this change. What this means for children previously diagnosed with Asperger syndrome has sparked considerable discussion (e.g. Annear, 2013).

Mental illness discourse, the kind articulated and classified in *DSM-5* is relatively recent (*DSM-I* was published in 1952) and was originally developed as a language for researchers to communicate with each other. Long ago, however, this discourse became a language of professional mental health practice as well as mental health administration, and it has been increasingly taken up as cultural means of self-identification (Illouz, 2008). A public seeking expert knowledge to make sense of human concerns has found readily accessible self-diagnosis tools online, and plenty of expert testimony and advice as to what to do about these self-diagnosed 'disorders' (cf. Grohol, n.d.).

Final thoughts

Following the proliferation of children's mental illness discourse have been corresponding shifts in mental health, school, and parenting discourse. Where mental health efforts focused on prevention education, moral correction, and 'hygiene', diagnosable children are increasingly chemically managed (Frances, 2013; Whitaker, 2010). Arguably, more children are being medicated because adults' diminished tolerance for concerns about children (e.g. misbehaviour) now require lower and more medically diagnosable symptom thresholds (Conrad, 2007; Rapley et al., 2011). Unclear at this point is how this approach to addressing formerly non-medicalised concerns about children is being taken up within families – apart from pill-taking rituals. How are the social practices of families (cf. Dreier, 2008) affected once a combination of children's mental illness discourse and medication become interwoven into everyday family interactions?

Parents' concerns have been central to the three discourses (well-being, mental health, and mental illness) we have been reviewing. Children might be recruited into understandings, roles, and activities associated with each discourse but typically adults, usually parents, have initiated these differing discourses for optimising their children's well-being, sustaining their children's mental health, and treating their children's diagnosed mental illnesses. Children nowadays live with their parents longer, spend more time in school, and within most Euro-American families there are fewer of them than was the case decades ago. Mental illness is increasingly being inferred from children's behaviours and emotions in ways that parents and doctors would have deemed normal not so long ago (Frances, 2013). Following the alarming analyses of Eva Illouz (2008), we wonder if understanding and responding to children through mental health and mental illness discourses is on the rise – along with a greater vigilance for potential mental illness symptoms and treatments – and what such a focus may bring to families and childhoods.

Clinical and educational relevance summary

Any discourse enables some ways of understanding and responding to aspirations or challenges while constraining other understandings and responses (Martin & Sugarman, 1999). Evolving discourses of children's well-being, mental health, and mental illness influence personal, interpersonal, institutional, and even national efforts to address what is seen as best for children. Discursive capture, an often unrecognised over-commitment to a single constraining discourse, is not always obvious. It often becomes evident when an either/or-ness (a correct/incorrect binary) develops over how one is to move forward despite lack of success in doing so, particularly if this is to involve others (Lyotard, 1988). In our view, such discursive capture begs reflection, resourcefulness, and flexibility to overcome conflict or stuckness in inclusive ways.

Discourse analysts and other child-focused researchers have important roles to play in exposing such forms of discursive capture, and in highlighting where attempts to move forward institutionally or societally are not occurring in inclusive ways. Differences over how to understand and respond to children's well-being, mental health, and mental illness have yet to be overcome with political or scientific consensus. Thus exposing where such differences are unresolved, while bringing new understandings and ways of communicating to political and scientific dialogues has been, and remains, a key to moving forward. At a more immediate interpersonal level, recent 'discursive

Table 6.1 Clinical practice highlights

1. Consider distinctions you make between discourses of well-being, mental health, and mental illness – given how these discourses may influence your interactions with children and others.
2. Consider how you and the clients, students, parents, and other professionals with whom you interact may be constrained or 'captured' by discourses of mental health and illness, such that other ways of understanding and acting on children's concerns might be obscured.
3. Where there are situations of discursive capture or conflict, see questions as ways to invite reflective distance on the language and discourses currently being used. Help others see that they can become used by language as opposed to discerning users of language and discourse.
4. Mental health and especially mental illness discourses can obscure a focus on what kids do well. During any discussion of mental health or illness, concerns invite consideration of the child's resources and capacities in ways that acknowledge concerns while not losing sight of strengths.
5. How are the policies of your profession or site of practice shaped by discourses of children's well-being, mental health, and mental illness? How are the wishes of children, parents and others included in these policies, and what is enabled and constrained in terms of your own practice by these policies? Where you see people excluded or your own practices constrained, how do you invite distant reflection and generative conversations on the discourses animating these policies?

therapies' (Lock & Strong, 2012; examples: narrative, solution-focused and collaborative therapies) have been developing that help clients and professionals be discursively reflective, flexible, and resourceful in addressing constrained or captured circumstances. These therapies engage participants in rhetorically and reflectively distancing themselves from constrained or captured understandings and ways of communicating, while inviting new and preferred responses for getting beyond such forms of constraint and capture. The aim of these therapies is to help people distance themselves from constraining discourses to find new ways of moving forward – in this case, to address children's well-being, mental health or mental illness. For a simple summary of the implications for practice, see Table 6.1.

Summary

Our aim throughout this chapter has been to provide some historical context for understanding historical shifts in the ways adults' concerns about children's well-being, mental health, and mental illness have evolved to present-day understandings. Our point throughout has been to highlight how such discourse developments have played roles in changing understandings of and responses to children. Quite literally, new kinds of children and families (schools as well) emerge out of such discourse developments. In educational and therapeutic contexts, other changes co-evolve with these developments as new intervention programmes and therapeutic approaches follow. As Philip Cushman's (1995) historical research suggests, societies seem to be in constant cultural dialogue with the experts of the day over how to understand and respond to personal and cultural concerns. We hope we have helped to historically illustrate how concerns and hopes for children have been differently understood and responded to in ways still relevant today.

Acknowledgements

The authors would like to acknowledge the Social Sciences and Humanities Research Council and the Werklund School of Education for financially supporting this research.

References

Alverson, H. S., Robert, D. E., Carpenter-Song, E. A., Chu, E., Ritsema, M., & Smith, B. (2007). Ethnocultural variations in mental health discourses: Some implications for building therapeutic alliances. *Psychiatric Services, 58*(12), 1541–1546.

American Psychiatric Association (2013). *Diagnostic and statistical manual of mental disorders* (5th edition). Washington, DC: American Psychiatric Association.

Annear, K. (2013). *The Disorder Formerly Known as Asperger's*. Retrieved from http://www.abc.net.au/rampup/articles/2013/05/24/3766915.html.

Arbuckle, M., & Herrick, C. (2006). *Child and adolescent mental health: Interdisciplinary systems of care*. Mississauga, ON: Jones and Bartlett Publishers.

Ariés, P. (1962). *Centuries of childhood: A social history of family life* (R. Baldrick, Trans.). New York: Viking.

Atwood, M. (2003). *Oryx and crake*. Toronto, ON: Random House.

Autism Research Institute (2014). *Updates to the APA in DSM-V – What Do the Changes Mean to Families Living With Autism?* (Paragraph 4). Retrieved February 16, 2014 from http://www.autism.com/index.php/news_dsmV.

Autism Society (September 112014). *Asperger's Syndrome*. Retrieved February 16, 2014 from http://www.autism-society.org/about-autism/aspergers-syndrome/.

Avdi, E. (2005). Negotiating a pathological identity in the clinical dialogue: Discourse analysis of a family therapy. *Psychology and psychotherapy: Theory, research and practice, 78*(4), 493–511.

Bagchi, S. S. (2010). *Child labor and the urban third world: Toward a new understanding of the problem*. Lanham, MD: University Press of America.

Barry, P. (1998). *Mental health and mental illness* (6th edition). Philadelphia, PA: Lippincott-Raven Publishers.

Bowlby, J. (1953). *Child care and the growth of love*. Harmondsworth: Pelican.

Bowlby, J. (1969). *Attachment*. New York: Books.

Browne, G., Gafni, A., Roberts, J., Byrne, C., & Majumdar, B. (2004). Effective/efficient mental health programs for school-age children: a synthesis of reviews. *Social Science & Medicine, 58*(7), 1367–1384.

Burman, E. (1994). *Deconstructing developmental psychology*. London, UK: Routledge.

Caplan, P. (2013). Don't blame mother: Then and now. In M. Hobbs & C. Rice (Eds.), *Gender and women's studies in Canada: Critical terrain* (pp. 99–106). Toronto, ON: Canadian Studies Press.

Choi, K. H., Reddy, L. F., & Spaulding, W. (2012). Child abuse rating system for archival information in severe mental illness. *Social Psychiatry and Psychiatric Epidemiology, 47*(8), 1271–1279.

Chua, A. (2011). *Battle hymn of the tiger mother*. New York: Penguin.

Clarke, A. A., Mamo, L., Fosket, J. R., Fishman, J. R., & Shim, J. K. (Eds.) (2010). *Biomedicalization: Technoscience, health, and illness in the US*. Durham, NC: Duke University Press.

Conrad, P. (2007). *The medicalization of society: On the transformation of human conditions into treatable disorders*. Baltimore, MD: The Johns Hopkins University Press.

Conrad, P. (2010). *Deviance and medicalization: From badness to sickness*. Philadelphia, PA: Temple University Press.

Corrigan, P. W., & Watson, A. C. (2002). The paradox of self-stigma and mental illness. *Clinical Psychology Science and Practice, 9*, 35–53.

Crow, L. D., & Crow, A. (1951). *Mental hygiene*. New York: McGraw-Hill.

Cushman, P. (1995). *Constructing the self, constructing America: A cultural history of psychotherapy*. Cambridge, MA: Perseus.

Danziger, K. (1997). *Naming the mind: How psychology found its language*. London, UK: Sage.

deVos, J. (2012). *Psychologisation in times of globalization*. London, UK: Routledge.

Dickens, C. (1838). *Oliver twist or parish boys' progress by Boz*. London, UK: Richard Bentley.

Dreier, O. (2008). *Psychotherapy in everyday life*. New York: Cambridge University Press.

Ehrenwald, J. (Ed.) (1991). *The history of psychotherapy* (2nd edition). London, UK: Jason Aronson.

Eriksen, K., & Kress, V. E. (2005). *Beyond the DSM story: Ethical quandaries, challenges, and best practices*. Thousand Oaks, CA: Sage.

Foucault, M. (1975). *The birth of the clinic: An archaeology of medical perception* (A. M. Sheridan Smith, Trans.). New York: Vintage.

Foucault, M. (1976). *Mental illness and psychology*. New York: Harper and Row.

Frances, A. (2013). *Saving normal: An insider's revolt against out-of-control psychiatric diagnosis, DSM-5, big pharma, and the medicalization of ordinary life*. New York: William Morrow.

Francis, C., Pirkis, J., Dunt, D., Warwick Blood, R., & Davis, C. (2002). Improving mental health literacy: A review of the literature. *Centre for Health Program Evaluation*. Retrieved September 20, 2014 from http://www.hirc.health.gov.au/internet/main/publishing.nsf/Content/4649FF5B003BDD5BCA257BF0001E034F/$File/literacy.pdf.

Freeman, J., Epston, D., & Lobovits, D. (1997). *Playful solutions to serious problems: Narrative therapy with children and their families*. New York: W. W. Norton.

Furedi, F. (2004). *Therapy culture: Cultivating vulnerability in an uncertain age*. New York, NY: Routledge.

Garro, L. C., & Mattingly, C. (2000). Narrative as construct and construction. In L. C. Garro & C. Martingly (Eds.), *Narrative and the cultural construction of illness and healing* (pp. 1–49). Berkeley, CA: The University of California Press.

Gaw, A. C. (Ed.) (1993). *Culture, ethnicity, and mental illness*. Washington, DC: American Psychiatric Press.

Gergen, K. J. (1982). *Toward transformation in social knowledge*. Thousand Oaks, CA: Sage.

Gershoff, E., & Larzelere, R. (June 26, 2002). *Is corporal punishment an effective means of discipline?* Retrieved March 18, 2014 from http://www.apa.org/news/press/releases/2002/06/spanking.aspx.

Gordon, S. L. (1989). The socialization of children's emotions: Emotional culture, competence, and exposure. In C. Saarni & P. L. Harris (Eds.), *Children's understanding of emotion* (pp. 319–349). Boston: Cambridge University Press.

Gould, M. (January 9, 2007). Sparing the rod caning was abolished 20 years ago. But do some teachers still wish it hadn't been? *The Guardian*. Retrieved February 16, 2014 from http://www.theguardian.com/education/2007/jan/09/schools.uk1.

Graham, L. J. (2005). *Discourse analysis and the critical use of Foucault*. Paper presented at Australian Association for Research in Education. Annual Conference, Sydney, November 27–December 1. Retrieved from http://eprints.qut.edu.au/2689/1/2689.pdf.

Graham, L. J. (2010). *(De)Constructing ADHD. Critical guidance for teachers and teacher educators*. New York: Peter Lang Publishing.

Grohol, J. (n.d.). *PsychCentral*. Online website retrieved from http://psychcentral.com/.

Group for the Advancement of Psychiatry (February 1972). *Crisis in child mental health: A critical assessment* (Vol. VIII, Report No. 82). New York: Group for the Advancement of Psychiatry.

Haddon, M. (2003). *The curious incident of the dog in the night-time*. London: Jonathan Cape.

Harari, E. (2013). Adolescent psychopathology and the death of meaning in psychiatry. *Australian and New Zealand Journal of Psychiatry, 47*(7), 605–608.

Harper, S. (2005). Media, madness and misrepresentation: Critical reflections on anti-stigma discourse. *European Journal of Communication, 20*(4), 460–483.

Harré, R. (Ed.) (1986). *The social construction of emotions*. Oxford, UK: Blackwell.

Hassim, J., & Wagner, C. (2013). Considering the cultural context in psychopathology formulations. *South African Journal of Psychiatry, 19*(1), 4. Retrieved February 16, 2014 from http://go.galegroup.com/ps/i.do?id=GALE%7CA325496183&v=2.1&u=ucalgary&it=r&p=HRCA&sw=w&asid=af3faa65963fe7f72ae3d4e17c8a585f.

Hochschild, A. R. (1979). Emotion work, feeling rules, and social structure. *American Journal of Sociology, 85*(3), 551–575.

Human Rights Watch (2010). 'Corporal punishment in schools and its effect on academic success': Joint HRW/ACLU Statement. Retrieved February 16, 2014 from http://www.hrw.org/news/2010/04/14/corporal-punishment-schools-and-its-effect-academic-success-joint-hrwaclu-statement.

Hutchby, I. (2007). *The discourse of child counselling*. Philadelphia, PA: John Benjamins Publishing Company.

Hutten, M. (September 11, 2014). *My aspergers child*. Retrieved from http://www.myaspergerschild.com/.

Illouz, E. (2008). *Saving the modern soul. Therapy, emotions, and the culture of self-help*. Berkeley, CA: University of California Press.

Ineichen, B. (1979). *Mental illness*. New York: Longman.

International Classification of Diseases (ICD-10) (1993). *World Health Organization: International Statistical Classification of Diseases and Related Health Problems* (10th revision). Geneva, Switzerland: WHO.

Jacobson, L. (2008). Children's lack of playtime seen as troubling health, school issue. *Education Week, 28*(14), 1.

Jahnukainen, M. (2010). Different children in different countries: ADHD in Canada and Finland. In L. J. Graham (Ed.), *(De)Constructing ADHD* (pp. 63–76). New York: Peter Lang Publishing.

Kahan, V. L. (1971). *Mental illness in childhood: A study of residential treatment*. London: Tavistock Publications.

Kirby, P. (2003). *Child labour in Britain, 1750–1870*. New York: Palgrave Macmillan.

Lakoff, A. (2008). *Pharmaceutical reasoning*. New York: Cambridge University Press.

Lamerichs, J. M., Koelen, M. A., & Molde, H. F. (2009). Turning adolescents into analysts of their own discourse. Raising reflexive awareness of everyday talk to develop peer-based health activities. *Qualitative Health Research, 19*(8), 1162–1175.

Lawless, L., Coveney, J., & MacDougall, C. (2014). Infant mental health promotion and the discourse of risk. *Sociology of Health & Illness, 36*(3), 416–431.

Lock, A., & Strong, T. (Eds.) (2012). *Discursive perspectives on therapeutic practice.* New York: Oxford University Press.

Lyotard, J. F. (1988). *The differend: Phrases in dispute* (G. Van Den Abbeele, Trans.). Minneapolis, MN: University of Minnesota Press.

Martin, J., & Sugarman, J. (1999). *The psychology of human possibility and constraint.* Albany, NY: SUNY Press.

Martin, J., & McLellan, A. M. (2013). *The education of selves: How psychology transformed students.* New York: Oxford.

McAdams, D. P., & McLean, K. C. (2013). Narrative identity. *Current Directions in Psychological Science, 22*, 233–238.

Mills, C. (2014). Psychotropic childhoods: Global mental health and pharmaceutical children. *Children and Society, 28*, 194–204.

Milteer, R. G., & Ginsburg, K. R. (2012). The importance of play in promoting healthy child development and maintaining strong parent-child bond: Focus on children in poverty. *Pediatrics, 129*(1), 204–213.

Miranda, J. K., De la Osa, N., Granero, R., & Ezpeleta, L. (2013). Maternal childhood abuse, intimate partner violence, and child psychopathology: The mediator role of mothers' mental health. *Violence Against Women, 19*(1), 50–68.

Moran, J. (2001). *Committed to the state asylum: Insanity, the asylum and society in nineteenth-century Ontario and Quebec.* Montreal, CA: McGill Queen's University Press.

Moran, J., Topp, L., & Andrews, J. (2007). *Madness, architecture and the built environment: Psychiatric a paces in historical context.* New York: Routledge.

Mossakowski, K. N. (2003). Copying with perceived discrimination: Does ethnic identity protect mental health? *Journal of Health and Social Behavior, 44*(3), 318–331.

Muzychka, M. (2007). *An environmental scan of mental health and mental illness in Atlantic Canada.* Ottawa, ON: Public Health Agency of Canada. Atlantic Regional Office.

Nichols, M. P., & Schwartz, R. C. (2008). *Family therapy: Concepts and methods* (8th edition). Boston: Pearson, Allyn & Bacon.

National Council for Curriculum and Assessment (NCCA) (n.d.). *The early childhood curriculum framework: Well-being.* Retrieved February 16, 2014 from http://www.ncca.biz/aistear/pdfs/PrinciplesThemes_ENG/WellBeing_ENG.pdf.

Noll, R. (2011). *American madness: The rise and fall of dementia praecox.* Cumberland, RI: Harvard University Press.

Parents of Children with Asperger's Support Club (September 11, 2014). Facebook page retrieved from http://www.facebook.com/ParentsOfChildrenWithAutismInTheGV?fref=ts.

Parents of Children with Asperger's Syndrome Club (September 11, 2014). Facebook page retrieved from http://www.facebook.com/pages/Parents-of-Children-With-Aspergers-Syndrome/119213814792472.

Prince, M. J. (2008). Claiming a disability benefit as contesting social citizenship. In P. Moss & K. Teghtsonnian (Eds.), *Contesting illness: Processes and practices* (pp. 28–46). Toronto, ON: University of Toronto Press.

Quosh, C., & Gergen, K. J. (2008). Constructing trauma and its treatment: Knowledge, power and resistance. In T. Sugiman, K. J. Gergen, W. Wagner, & Y. Yamada

(Eds.), *Meaning in action: Constructions, narratives and representations* (pp. 97–111). Japan: Springer.

Rapley, M., Moncrieff, J., & Dillon, J. (Eds.) (2011). *De-medicalizing misery: Psychiatry, psychology, and the human condition.* New York: Palgrave Macmillan.

Roberts, G., & Grimes, K. (2011). *Return on investment: Mental health promotion and mental illness prevention.* Ottawa, ON: Canadian Institute for Health Information.

Rodday, A. M., Parsons, S. K., Correll, C. U., Robb, A. S., Zima, B. T., Saunders, T. S., & Leslie, L. K. (2014). Child and adolescent psychiatrists' attitudes and practices prescribing second generation antipsychotics. *Journal of Child and Adolescent Psychopharmacology, 24*(2), 90–93.

Rogers, A., & Pilgrim, D. (2010). *A sociology of mental health and mental illness.* New York, NY: Open University Press.

Rose, N. (1990). *Governing the soul.* London, UK: Routledge.

Rose, N., & Abi-Rached, J. (2013). *Neuro: The new brain sciences and the management of the mind.* Rutgers, NJ: Princeton University Press.

Szasz, T. (1970). *The manufacture of madness: A comparative study of the inquisition and the mental health movement.* New York: Harper & Row.

Schmitz, C. L., Collardey, E. K., & Larson, D. (Eds.) (2004). *Child labor: A global view.* Kindle Version: Greenwood.

Shotter, J. (1975). *Images of man in psychological research.* London, UK: Methuen.

Smith, M. (2010). The uses and abuses of history of hyperactivity. In L. J. Graham (Ed.), *(De) Constructing ADHD* (pp. 63–76). New York: Peter Lang Publishing.

Society for Adolescent Medicine (2003). Corporal punishment in schools. *Journal of Adolescent Health, 32,* 385–393.

Spock, B., & Needlman, R. (2012). *Dr. Spock's baby and child care* (9th edition, 65th anniversary edition). New York: Gallery Press.

Spring, K. W., Sheridan, J., Kuoc, D., & Carnes, M. (2007). Long-term physical and mental health consequences of childhood physical abuse: Results from a large population-based sample of men and women. *Child Abuse & Neglect, 31*(5), 517–530.

Starker, S. (2002). *Oracle at the supermarket: The American pre-occupation with self-help books.* New Brunswick, NJ: Transaction.

Stephan, S. H., Weist, M., Kataoka, S., Adelsheim, S., & Mills, C. (2007). Transformation of children's mental health services: The role of school mental health. *Psychiatric Services, 58*(10), 1330–1338.

Stevens, G. W. J. M., & Vollebergh, W. A. M. (2008). Mental health in migrant children. *Journal of Child Psychology and Psychiatry, 49,* 276–294.

Strong, T. (1993). DSM-IV and describing problems in family therapy. *Family Process, 32,* 249–253.

Strong, T. (2011). Approaching problem gambling with a discursive sensibility. *Journal of Gambling Issues, 25,* 68–87.

Sugaya, L., Hasin, D. S., Olfson, M., Lin, K. H., Grant, B. F., & Blanco, C. (2012). Child physical abuse and adult mental health: A national study. *Journal of Traumatic Stress, 25*(4), 384–392.

Tausig, M., Michello, J., & Subedi, S. (1999). *A sociology of mental illness.* Upper Saddle River, NJ: Prentice-Hall.

The Children's Society (2014, January). *Promoting positive well-being for children. A report for decision-makers in parliament, central government and local areas.* Retrieved March 19, 2014 from http://www.childrenssociety.org.uk/sites/default/files/tcs/promoting_positive_well-being_for_children_final.pdf.

Thompson, R. (2012). Screwed up, but working on it: (Dis)ordering the self through e-stories. *Narrative Inquiry, 22*(1), 86–104.

Tran, A. G. (2014). Family contexts: Parental experiences of discrimination and child mental health. *American Journal of Community Psychology, 53*(1), 37–46.

Truitt, R. P. (1926). The role of the child guidance clinic in the mental hygiene movement. *The American Journal of Public Health, 16*(1), 22–24.

United Nations Children's Fund (UNICEF) (2013). *Child well-being in rich countries. A comparative overview*. Florence, Italy: Self-published. Retrieved February 16, from http://www.unicef-irc.org/publications/pdf/rc11_eng.pdf.

Waller, R. (2013). *Mental health promotion in schools: Foundations*. Sharjah, UAE: Bentham Science Publishers.

Watters, E. (2010). *Crazy like us: Globalizing the American psyche*. New York: Free Press.

Weist, M. D., Rubin, M., Moore, E., Adelsheim, S., & Wrobel, G. (2007). Mental health screening in schools. *Journal of School Health, 77*(2), 53–58.

Whitaker, R. (2010). *Anatomy of an epidemic: Magic bullets, psychiatric drugs, and the astonishing rise of mental illness in America*. New York: Broadway Paperbacks.

Wilkinson, A. P. (2012). Mental illness in children. *Pediatrics for Parents, 28*(1/2), 12–14.

Wolfe, D. A. (2014). *Spotlight on children's mental health*. Retrieved from http://www.rbc.com/community-sustainability/_assets-custom/pdf/Spotlight-on-children's-mental-health.pdf.

Yanni, C. (2007). *Architecture of madness: Insane asylums in the United States*. Minneapolis, MN: University of Minnesota Press.

Recommended reading

- Conrad, P. (2007). *The medicalization of society: On the transformation of human conditions into treatable disorders*. Baltimore, MD: Johns Hopkins University Press.
- Frances, A. (2013). *Saving normal: An insider's revolt against out-of-control psychiatric diagnosis, DSM-5, big pharma, and the medicalization of ordinary life*. New York: HarperCollins Publishers.
- Hutchby, I. (2007). *The discourse of child counselling*. Philadelphia, PA: John Benjamins Publishing Company.

7
Child Mental Health: A Discourse Community

Stephen Gilson and Elizabeth DePoy

Introduction

Over the past three decades, discursive analytic epistemologies have become central tools within the interdisciplinary field of disability studies. Particularly within the United States and Western European academic discourses, theorising disability as linguistic artefact has been potent in wrestling atypical embodiment away from its medical deficit prison and repositioning it as a socially and culturally constructed phenomenon. While vigorous and compelling, we assert that delimiting communication to discourse, conversation, and humans interacting with humans is incomplete. Rather, we propose meaning making and its productions in cultural expectations of development, normalcy, health, and illness as complex interactive discourse communities in which language, object, materiality, and interpretation dance in synergy. Given the local nature of community and its intrinsic discourse, this work, while emergent from US and UK scholarship, can be extrapolated and applied to analysis of discourse communities across the globe.

Expanding on the classic work of Swales (1990), discourse communities are defined as collectives, in which truths, meanings, cohesion, and exclusions are created through diverse communication mechanisms:

> By continually reworking their artifacts, reconstructing their reality, discourse communities develop traditions, repetitive practices, and norms that become part of how that community defines itself [and is defined by others].
>
> (Krippendorf, 2011, p. 6)

Given the materiality of the body in action and under scrutiny and the relative significance of artefacts as privileged communicators over words, particularly in childhood, this synthetic theory holds great promise to advance a fuller and deeper analysis of child mental health and its alternatives, mental illness and disability. Viewed as a production of complex discourse communities, child mental health and responses to its absence are read (Candlin & Guins, 2009) as expressive and material construction and revision of legitimate childhood, healthy childhood, child mental disability, and response.

The chapter begins by elucidating a revised and expanded construct of discourse communities more in line with Krippendorf's approach (2011), its theoretical relevance to advanced capitalist economic contexts in which visual culture is omnipresent, and its constructive power. The mosaic of discourse community analysis, including human members, materiality, and the meaning made by their complex synergic communications, is then applied to a profound exploration of contemporary childhood and the construction of child mental health and disability. Particular attention is focused on the activated body as object, the objectified child (and parental) body under scrutiny, and the discursive construction and function of the developmental normative and non-normative. The epistemologies of nomothetic inquiry, clinical trials, and the visual of the bell curve which dominate post-Enlightenment contexts are discussed as convincing cultural narratives and imagery that serve to construct elegant and seductive notions of monistic truth (DePoy & Gilson, 2007), reifying the mentally 'disabled' child as an artefact for biopharmacological and professional revision. (Note that we use the term 'mental disability' as expansive, to include diverse diagnoses that are ascribed to children who think and behave in an atypical manner.) We conclude the chapter with the relevance of this analysis and the guidance it provides for professionals.

Discourse communities, history and contemporary expansion

Late 20th-century development and use of the term 'discourse community' is most frequently attributed to Swales (1990), who ascribed the essential characteristics of shared interests, goals, and genres to these groupings. In its original iteration, membership was crafted as stable, using a secret handshake, so to speak, shared through diverse communication genres, to circulate internal messages of content and meaning. Since its inception, the concept of discourse community has been indicted with significant limitations of essentialism and utopianism, to name just two, fraying its edges, yet providing opportunity for revision, updating, and application (Devitt, Bawarshi, & Reiff, 2003). Viewing classic discourse community theory through a post-postmodern facet provides a third major criticism: the absence of the material as communicator (Alaimo, 2010; Miller, 2010). While the foci on language and cultural construction have been crucial in challenging the hegemony of positivism, reintroduction of palpable matter disintegrates Cartesian dualism, adding 'stuff' and especially the corporeal body as discursive (Miller, 2010; Siebers, 2010).

We therefore propose texturising discourse communities to bring their productivity in line with 21st-century thinking and praxis. In particular, two interaction genres join linguistic transmission: contemporary material culture

Table 7.1 Organisational framework

Discourse community/ genre	Medicalisation industry	Theory and inquiry (normal development and aberrations)	The commodified family	Professionals (teachers, mental health providers)
The words Body as object Stuff Imagery				

and the visual. Both enter discourse communities reactivating organic and inorganic objects and imagery as agentic and communicative (Berger, 2009; DePoy & Gilson, 2014; Rose, 2012; Turkle, 2011). Table 7.1 depicts the organisational scaffold that frames the subsequent discussion.

The discourse communities

While diverse discourse communities across the globe are concerned with children, their health and its absence or diminishment, we focus the spotlight on the medicalisation industry in the US, Western Europe, and other advanced capitalist economies; theory and inquiry: the commodified family; and professionals. We recognise that each is internally diverse and not necessarily mutually exclusive. However, in the limitations of a single chapter to avoid falling prey to essentialism, only trends in genre and lexis are discussed within specific communities, whetting the appetite for further exploration (Gnaulati, 2013).

The medicalisation community

Brought into public recognition in the late 1970s (Conrad, 2007), the medicalisation discourse community that has taken hold in advanced capitalist systems engages the language, artefactual, and image genres of medicine, pseudo-medicine, and scientific inquiry to pathologise human experiences and conditions. Two terms bear further definition here: 'medicalisation' and 'pseudo-medicine'. We address scientific inquiry in its own discourse home.

Medicalisation

Medicalisation refers to translation of daily experience into pathology and then serving it up for treatment by the medical industrial complex (Conrad, 2007). Although the distinction between medical and medicalisation is not always clear, we suggest that intention is the factor which separates the two. The primary commitment of medical treatment is to heal, while the primary aims of medicalisation are profit and control (DePoy & Gilson, 2014).

Pseudomedicine

Pseudomedicine refers to treatments that claim to be working concepts of medicine that either have no objectively verifiable benefit or are incompatible with the current state of knowledge in the field of science-based medicine (Psiram, 2013, para. 1).

As we discuss below, while we see research and science as grand narratives supporting the medical industrial community and thus a discourse in its own right, psychiatric diagnosis is suspect as masquerader even within this discursive genre.

Some history

Not unexpectedly, medicalisation has 'histories' rather than a single history, given that scholars take divergent approaches to this well-travelled retrospective analysis. Perhaps, best known for indicting physicians and their intellectual partners as greedy and insatiable were writers such as Szasz, Laing, Goffman, Illich, and Foucault. As eloquently stated by Nye (2003):

> These writers accomplished that most beautiful of intellectual operations: a perfect inversion of the orthodox position. Criminals were victims of labeling, stereotyping, and racial and economic injustice; the mad were explorers of new psychological horizons of personal liberty; and the rebellion against sexual and gender norms was a form of self-emancipation. In this new scenario, the real villains were the doctors, psychiatrists, and behavioral scientists who had used their knowledge and authority to shore up the 'establishment' and to segregate and pathologize the recalcitrant. (p. 116)

Contemporary medicalisation has moved away from vilifying providers. We agree. For us, headless corporatisation and action designed for maximising profit are the primary movers of medicalisation as we expand below.

Looking even further back to the end of the Middle Ages and birth of Enlightenment thinking, as religious, moral, and sin-laden explanations were nudged aside by scientific causal schemes, biology took the discursive lead in Western thinking, both lexically and materially defining illness and remediating it, to the extent possible at that point in the development of medical knowledge and technology (DePoy & Gilson, 2011). This trend created fertile soil for seeding, growing, and harvesting the medicalisation industrial complex, its lexicon, objects, and its images.

While more extensive histories are beyond the scope of this discussion, we do want to attend to thematic fragments that were and continue to be instrumental in provoking the design of young humans into binary categories, such as mentally healthy–not healthy, normal–not normal, disabled–not-disabled, and thus in potentiating the entrance and growth of medicalisation discourse, its following, and its continued popularity among many audiences.

One critical fragment is the role of developmental theory and the imagery of the bell or normal curve in essentialising humans by age. We discuss this factor as a discourse community in its own right. A second thematic fragment that complements and is dependent upon the post-Enlightenment positivist discourse of the normal curve is the power of scientific narrative and image in establishing an accepted reality and thus in legitimating practices in concert with that ontology. Particularly relevant to mental health and mental disability discourses is the work of Schuster (2011), who activates the story of neurasthenia in early 20th-century history. In his studies linking industrialisation and modernity to human distress, Schuster details how happiness entered the criteria for necessary and sufficient conditions for mental health, despite its previous absence from human health and rights discourses. Thus, happiness became an essential attribute that humans were supposed to possess in order to be healthy, with its absence remanded to the pathological or disabled. Through apprehending knowledge and cultural power, the positivist watershed trend birthed neurasthenia, a vague diagnostic entity, as medical 'reality', ripe for assessment, quantification, and capitalist exploitation. Rather than looking to more complex consequences of modernity on human well-being, the neurasthenic mood state of an individual body became the object of diagnosis and revision, leaving contextual practices detrimental to human flourishing unchallenged and intact.

Despite the disappearance of neurasthenia as a formal diagnosis, Schuster (2011) analyses it as an important history and contemporary analogy within current and more expansive medicalisation texts. Analogies persist in the continued diagnosis of victims of child abuse and neglect, conditions which need to be addressed directly rather than diagnosing and treating the damages that result from the tolerance of such social offences.

A third instigator as well as consequence of medicalisation was the repositioning of social control and citizenship status, enacted historically through war and politics. More recently, scientific knowledge and skill power held by only an elite few, the professional discourse community, unseated battle as human control agent and thus quietly and unrecognised as subjugator took up residence in the social control palace (Conrad, 2007).

Back to the present

Although some discourse communities have splintered and attempted to exit systems of economically driven human assistance, advanced capitalism, as it makes its way across the globe (Salmon,1985), continues its ascendency as the prime shaper of our current mental health discourse (Dumit, 2012). In the United States, for example, despite the myth, argued by some individuals, that the Affordable Care Act (ACA) is socialised medicine, it is capitalism at its best. As Arel (2013) states, 'The ACA is capitalism, and it is going to make a lot of people very rich' (para. 7). Within the United States, the combination of the ACA and mental health parity creates a firm place for nurturance and growth of the mental health medicalisation industry.

Although there are many theoretical models and conceptual frameworks that characterise capitalism, the general use in this chapter refers to social structures in which production, consumption, and profit form its royal triumvirate (Howard & King, 2008). Powered by popular culture, mass media and communication, marketing and branding, and public schooling (Jameson, 1991), late capitalism creates opportunity for entities such as pharmaceutical companies, advertising and marketing, insurance, and other influential players in the medical industrial complex to construct postmodern abstract and/or material symbols of pathology or its potential risk and to further convince the public of the need for consumption of services, artefacts, and goods to maintain or regain the grand narrative of health as crafted by empirical discourse. Within this economic context, the current child mental health medicalisation industrial discourse community invokes numerous genres to assure its survival, contain its knowledge power, and realise profit.

First, consider the *Diagnostic and Statistical Manual-5* (*DSM-5*) as both artefact and lexicon. The *DSM-5* imbues legitimacy, branding the well-meaning and true-believing provider as expert by languishing on his/her bookshelf or coffee table. The object itself is an impressive and large volume housing the engorging lexicon of the industry (see *American Psychiatric Association*, 2013).

Within its interior, this narrative contains a guide for branding the objects and humans in action, through which the mental health enterprise profits. The lexicon identifies and situates human objects as sick or at risk, necessitating attention, products, and services that feed the medicalisation industrial complex discourse community.

Curiously, as exposed by Francis, rather than containing well-researched and evidence-supported knowledge, the *DSM-5* serves as a set of tools to initiate, logofy, and obtain payment for those who are inculcated as objects in this community. According to Francis, the *DSM-5* is simply an insider narrative genre, containing the branding lexicon with the failed goal 'to promote research that might replace our superficial descriptive method of diagnosis with one based on aetiological understanding… high ambition has been combined with an unfortunate penchant for secrecy, a poorly organised methodology' (Francis, 2009, p. 391).

It is not surprising that Davis (2013) refers to the *DSM-5* as a playbook, choreographing the thinking and action of the medicalisation discourse community membership and its fodder.

Creating further cognitive dissonance is the process for diagnosis of child mental illness. Diagnosis is made in essence on the basis of a negative biomarker finding.

> Diagnosis usually begins with a medical doctor who takes a lengthy history and examines the child to rule out physical reasons for the difficulties. Lab tests may also be done to test for side effects of medication, for allergies, or for other conditions that could produce symptoms. If no medical reason is found for the behavior, the doctor refers the child to a psychologist or psychiatrist who treats children and adolescents.
>
> (kidsmentalhealth.org, 2009)

In an effort to scientise diagnostic rhetoric beyond using it to fill a void left by non-pathological medical testing, Meehl (1999) developed taxometrics, 'a statistical procedure for determining whether relationships among observables reflect the existence of a latent taxon (type, species, category, disease entity)' (Meehl, 1999, p. 165).

This impressive linguistic invention known only by an elite few is so precise and specialised that Waller (2006) chose the Platonic saying, 'carving nature at its joints', as its apt descriptor. Yet, even the ascription of computational complexity to diagnosis has been insufficient to waylay well-deserved critiques that expose childhood mental pathology as mere conceptual consensus based on political and economic concerns of the medicalisation discourse community. Thus, there are differences in how to understand to children's mental health and mental illness, and these differences contribute to the political and scientific dialogue (see Strong & Sesma-Vazquez, Chapter 6, this volume).

It is therefore curious that in the absence of an organic basis, the brain has become the most popular artefact for surveillance and blame for many brands of mental illness (Legrenzi & Umilta, 2011). Consider the attention deficit hyperactivity disorder (ADHD) brand which is ascribed to children who exhibit 'developmentally inappropriate inattention, motor restlessness, and impulsivity' (Curatolo, D'Agati, & Moavero, 2010, p. 36). Despite the knowledge that the neurological causal basis for this behavioural style has not been substantiated, pharmacological intervention to mediate 'neurobiological liability' is the treatment of choice. To us the term 'neurobiological liability' sounds much like political spin or deceptive advertising.

Interiorising mental illness within the skull joins hands with *DSM-5* diagnoses in perpetuating the narrative of individual responsibility and locus for change. Beautiful and compelling imagery accompanies lexis to convince the public of the viability and reality of brain disorder as the basis for child mental illness and disability, priming the pump for brain interventions, specifically 'brainpharma' (see Figure 7.1).

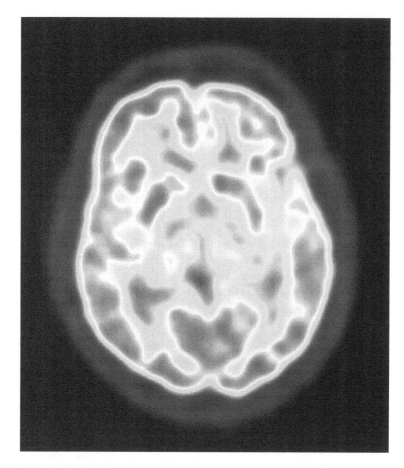

Figure 7.1   Brainpharma
Source: Jens Maus from http://commons.wikimedia.org/wiki/File:PET-image.jpg.

Dumit (2012) goes even deeper in his assertion that the pharmaceutical industry not only is an opportunistic drug dealer but also, more profoundly, acts as conniver using its power and wealth not simply to treat but to surreptitiously establish the parameters of health, its absence, and disability for the purpose of maximising profit. Two concepts are particularly nefarious in this agenda: risk and chronicity.

The concept of risk, Dumit (2012) reminds us, is an economic windfall, as drugs are sold to large populations who are not sick but are convinced through advertising, marketing, and well-intended provider guidance that they may become so without purchasing chemical intervention. Chronicity is often asserted, activating life-long medications, such as antipsychotics (Olfson et al., 2012).

Olanzapine (Zyprexa) and other antipsychotic medications are used 'off-label' for the treatment of aggression and other serious behavioral disturbances in children, including children with autism.

(National Institute of Mental Health, 2010, para. 4)

How is it possible that this discourse community has been so convincing in its medicalisation agenda? We step over the threshold into that supporting theory and inquiry discourse community in the next section.

Theory and inquiry, normal development and its aberrations

Developmental theory and its investigative practices provide a compelling discourse community serving the medicalisation industry. We begin with a brief discussion of theory and then move to its scrutiny.

Developmental theory

While diverse in their embodied domain of concern, even when posing as outside of this theoretical perimeter (e.g. neurological theories of human emotions, which still rely on the growth and development of the nervous system as the basis for emotive function), all developmental theories surveil and count the frequency of observations of the embodied object and its behaviour over chronological time and then translate the most frequent into the normative or 'what should be' at a particular age. What is gained by these theories is the power of prediction. However, on the proverbial flipside are the disadvantages of essentialism, homogenisation, and reliance on rules out of context.

As conceptual handmaiden to the medicalisation industry, this theoretical genre pronounces age-related normalcy, its opposite, the abnormal, and the explanatory schemes for revising the abnormal. Within the developmental theory discourse community, the desirable and healthy norms seek to create homogeneity by building a comparison matrix. Similarity to objects of one's age group and differences from those of other age groups are desirable in movement and growth on a hierarchy from diminutive to grand, from immature to mature, from non-skilled to skilled, and from internally disorganised and undisciplined to organised and self-controlled (Kail & Cavanaugh, 2013).

Classical human development discourse was initially concerned with describing how individuals and/or their parts unfold in an orderly and sequential fashion over the longitude of a life or part of it. These theories thus crafted life stages and distinguished their boundaries in the process. As they gained popularity, developmental theories have expanded their discursive girths, looking beyond the passage of the individual through phases of the lifespan to positioning themselves as prescriptive. As such, these theories articulate and visualise the normal embodied object in action, and its photographic negative, the abnormal disabled.

More recently, in the face of opposition to the normal/abnormal lexicon, the euphemisms 'typically and atypically developing' were coined particularly to sanitise references to children in special education. Alteration of the linguistic symbols, while a ubiquitous practice in the current politically correct climate, does little else. According to Younkins (2004), 'Political correctness is a...tool to limit thought in education, science, and culture' (para. 13). The requisite change in language, however, does unearth the devaluation of the abnormal objects renamed by the discourse community, while maintaining the developmental attitudinal and praxis status quo and serving to further institutionalise developmental theory as the basis for guiding judgement of and response to the embodied object.

In their eloquent criticism, Peters and Johansson (as cited in Olson, 2009) refer to the grammar of childhood, indicting developmental discourse as sparse and untextured. As such, these conceptualisations fail to historicise childhood, embedding it in a stifled and 'adultist' language of viewing the young as objects of 'not-yets' (Olson, 2009). Through the developmental lens, mentally disabled not-yets shift to delayed not-yets at best and 'nevers' at worst, as the atypically developing individual is surveilled and branded as unlikely to keep up or ever become equal in maturation with his/her age-group counterparts. Table 7.2 showcases just some of the popular culture lexicon of normal development and thus sets the stage for identifying delayed 'not yets' and 'nevers'. Lists such as the exhibit in Table 7.2 brand childhood normalcy and

Table 7.2 Developmental theory

Month	Description
Month 1	Makes eye contact Cries for help Responds to parents' smiles and voices
Month 2	Begins to develop a social smile Enjoys playing with other people and may cry when play stops Studies faces Prefers looking at people rather than objects Gurgles and coos in response to sounds around him/her First begins to express anger
Month 3	Starts a 'conversation' by smiling at you and gurgling to get your attention Smiles back when you smile at him, the big smile involving his whole body – hands opening wide, arms lifting up, legs moving Can imitate some movements and facial expressions (Finello, 2001)

Table 7.3   Words and text

Lexicon	Description
Developmental delay	Child development refers to the process in which children go through changes in skill development during predictable time periods called *developmental milestones*. Developmental delay occurs when children have not reached these milestones by the expected time period. For example, if the normal range for learning to walk is between 9 and 15 months and a 20-month-old child has still not begun walking, this would be considered a developmental delay (CASRC, 2011).
Underachiever	An underachiever is a person (as a student) who fails to achieve his or her potential or does not do as well as expected (Dictionary.com, 2013).
Autistic spectrum disorder (ASD)	ASD is a group of developmental disabilities that can cause significant social communication and behavioural challenges (CDC, 2014).

mental disability, scaring well-meaning parents and then advertising the road to redemption.

Now, consider the specific lexicon of child mental illness emerging from delayed acquisition or absence of the developmental markers showcased in Table 7.3.

Up until this point, we have discussed developmental theory and research as a lexicon for essentialising children. However, this discourse community does not discriminate according to age or familial role. Parents are equal objects of scrutiny and prescription (Tilsen, 2007). For example, Mowder (2005) proposed parent development theory (PDT) 'to assist professionals in organizing their thinking, practice, and research regarding parenting' (para. 1). This discursive framework suggested a model for universalising desired parent behaviour and its change over time, complete with a measurement artefact and acronym, the Parent Behavior Importance Questionnaire-Revised (PBIQ-R) for quantification and comparison (Mowder, Shamah, & Zeng, 2011).

Not only is parent development inscribed in the developmental lexicon, but the pressure for exhibiting the characteristics of an ideal parent is high. Look at Smith's statement appearing front and centre in her review of parenting research literature.

It could be said that the quality of parenting is the most important variable in a child's life.

(Smith, 2010, p. 690)

Gnaulati (2013), who criticises the over-diagnosis of childhood mental illness, still objectifies the normative child, albeit generous in the range of tolerated behaviours. However, an important part of his work references the pressure cooker in which many parents reside. The 'good parent plays multiple roles in the life of his/her child': event planner, sleep monitor, electronic media supervisor, and homework sergeant (p. 176).

It is therefore not surprising that for the parent who seeks to raise the good child, medical explanations for why a child as object is not equivalent to and following desirable developmental norms are seductive. Parents live with the National Institute of Mental Health (NIMH) pronouncement that 'once mental illness develops, it becomes a regular part of your child's behavior and more difficult to treat' (National Institutes of Mental Health, 2009, para. 2).

Medical texts can serve to exonerate parents who are charged with such offences and who measure themselves against the icon of the perfect parent and family presented in popular culture and promoted within the medicalised lexicon (Gnaulti, 2013). As stated by Solomon (2012), 'it is easier for parents to accept conditions arising from nature rather than nurture discourses' (p. 21), as they are not responsible for the abrogation. However, Solomon also reminds us that, 'nature v nurture is most often nature via nurture' (p. 21), acquitting well-meaning adults of charges of being bad parents when they fail to recognise themselves in their offspring.

In addition to the designed child of the 21st-century medicalisation discourse community in post–industrial societies, the parent, now named caregiver, is required to be the picture of mental health (Quinn, Briggs, Miller, & Orellana, 2014) and cannot falter too far from the constructed adult ideal, lest he or she becomes the family pariah.

According to Bringewatt (2013), children have been partially invited to evaluate the family function as well. Her retrospective study of people who had been given diagnoses of mental illness when they were children chronicled their acceptance of their conditions and their development as they came to embrace a medical story of their struggles.

Bringewatt provides the following interpretation:

> This study sheds light on the meaning-making processes of children, underscoring the power of discourse in shaping children's experiences and understandings, as well as the role children play in obtaining information and negotiating and interpreting these narratives. (p. 1224)

In her study, the computer makes an entrance as discursive object and expert, given its role in the information enterprise. Children could obtain their identities from the presentations of 'scientific' descriptions, many similar to the above-mentioned quote from NIMH. The family as commodity is thus completed, with any less than the theorised ideal unit becoming a potential profit centre for the medicalisation discourse community.

We are not discounting that children and parents have profound struggles that render family life difficult to intolerable. Nor are we suggesting that families do not need help. Rather, this discussion highlights the construction of the Westernised family ideal such that any aberration may be exploited for marketing.

Although perpetuating the medicalisation discourse community by reifying child mental illness, Solomon's (2012) recent tome provides some important insights to consider. In his quest to heal himself from an accepted diagnostic identity, and to engage in self-talk about his gendered identity, Solomon acknowledges the dissonance that occurs when children are dissimilar to parents in form and function.

The craft

Positivism is the epistemic craft used to verify and test the application of developmental theory to growing children and their family units. This research tradition, if enacted by carefully following its discursive methodological rules, is designed ostensibly to eliminate investigator bias and isolate factors that can then be theorised as causal of an observed object in action. As noted above, positivism and its related philosophical schools of thought posit the existence of a single reality apart from human thought that can be known objectively by enacting scientific method. The acceptance of logico-deductive discourse as the basis of scientific 'truth' apart from human preference is both simplistic and erroneous (Byers, 2011).

> It seems strange to call science a mythology since the story that science tells about itself is precisely that it, an activity pursued by human beings, is objective and empirical; that it concerns itself with the facts and nothing but the facts.... And yet science is a human activity. This is an obvious statement but it bears repeating since part of the mythology of science is precisely that it is independent of human beings; independent of mind and intelligence.... How do human beings create a system of thought that produces results that are independent of human thought?
>
> (Byers, 2011, ix–x)

Within this investigative discourse tradition, the conceptual and visual genres of the bell curve lie at its heart, forming the core object for positivism and thus monism, and the knowledge that is iced on top of this succulent thinking. Look at the elegance and symmetry of the image presented in Figure 7.2.

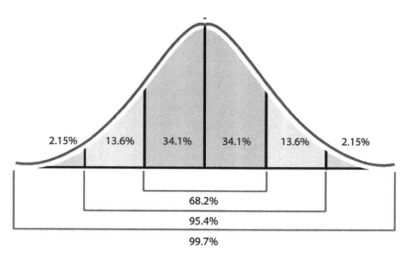

Figure 7.2   The normal curve

Historicising the bell curve provides longitudinal illumination of how it came to be iconic and take on truth proportions within the developmental theory and investigative discourse community. While Quetelet brought the bell-shaped curve to explanations of social phenomena, de Moivre actually is credited with its 'discovery' in the 18th century as an artefact of probability (Bellhouse, 2011). In his own words, de Moivre first elevated the bell curve into the realm of divine design.

> And thus in all cases it will be found, that although chance produces irregularities, still the Odds will be infinitely great, that in process of Time, those irregularities will bear no proportion to the recurrency of that Order which naturally results from Original Design.
>   (de Moivre in Gaither & Cavazos-Gaithe, 2012, p. 364)

Building on de Moivre's work, Gauss and LaPlace advanced the math of the normal distribution (Ernst, 2006), elucidating the typical symmetrical curvilinear shape that emerges from measures of natural occurrences (Goertzel & Fashing, 1981). Classical and contemporary scholars examining the application of mathematical theory to human surveillance suggested that Quetelet, in his effort to impose order on social and human experience, failed to distinguish equations of chance from a more concrete existence of an empirical world. Moreover, reflecting Aristotelian thinking, Quetelet assumed that God aimed for a single central point in creating humans without ever hitting the bull's eye. Conceptually, calculating the mean score of human attributes would therefore reveal the ideal human, with the extreme scores forming the numeric objects that depict the most distal limits of acceptance.

In his most recent work, Orrell (2012) suggested that the story told by the bell curve is one of deceptive beauty. Accordingly, the search for imagery and object depicting symmetrical order as an aesthetic seduces the investigative community into simplifying and limiting their understandings of the universe that they seek to detail.

While allowing for prediction, albeit only within an acceptable degree of probability defined through the lens of the investigative discourse community member, the numeric objects that feed the normal curve affirm a single 'reality' and thus concentrate their power. Words such as randomisation, manipulation, and control form a lexicon of elite literature which along with the bell curve and numeric objects seeks to verify group characteristics and distinguish them from non-group features, as in the case of normalcy in child mental development and ideal parenting.

Locating 'scientific inquiry' at the apex of epistemic and ontologic discourse nourishes the medicalisation industry despite the inability of its favoured child, logical positivism, to support its own claim of being the methodological owner of monistic truth (Keupink & Shieh, 2010).

Consider just a few basic words and their definitions to illustrate the recondite lexicon of the developmental theory and investigative discourse community.

Variable

Variable is a factor that differs among and between groups of people. Within the child development discourse, variables may include characteristics such as age, sex, and IQ or measurements such as activity level, academic outcomes, and standardised test scores. There can also be *treatment* or condition variables, for example, in a study of interventions for childhood schizophrenia, the severity of the disorder, and outcome variables.

Validity

Validity is the degree to which a result (of a measurement or study) is likely to be accurate, answer the research question, and reduce *bias* (systematic errors). Validity has several other meanings, usually accompanied by a qualifying word or phrase; for example, in the context of measurement, expressions such as 'construct validity', 'content validity', and 'criterion validity' are used. External validity positions a study as relevant to a broader group than that on which the study was done. Internal validity depicts rigour or sloppiness.

Normal distribution

We have already discussed this term as the visual of the normal or bell-shaped curve. The distribution is the set of scores or observations on which the imagery is based and which tells a story about a group. Related lexical terms which have their corresponding imagery are *mean*, standard deviation, stand error of the mean, and so forth. Now, look at just some of the symbols frequently found within the inquiry discourse community.

$\alpha = $ alpha
$\beta = $ beta
$\mu = $ population mean
$\rho = $ rho
$\sigma = $ sigma
$\chi^2 = $ chi-squared
$F = $ statistic for ration of variances

These images hold internal meaning to the community but doubtfully to those beyond its perimeter. But who is a member?

Are they the elite? The professional discourse community

This diverse and broad discourse community consists of those who are in service to children and their families, their educators, and their organisers. Included in this society are health and human service professionals (e.g. medicine, allied health, social work, and psychology), teachers in K-12, faculty who educate providers, and the professional organisations in which the lexicon is shared for a price.

In much of the literature, the professionals who deliver services are often cast as the villains of the medicalisation industry. After all, aren't these elites in the power seat, privy not only to the informed consumption of developmental and medicalised knowledge, its imagery, and its lexicon but more importantly to its generation as well?

While the classical medicalisation literature promoted this view, we beg to differ, suggesting that people enter the developmental theory and investigative discourse community with only the best of intentions, to help those who have fallen outside of or never made it into the precious space of the norm. Yet, the newbie as well as the seasoned professional or academic is compromised by many factors. In this section, we encounter two: academic capitalism (Munch, 2014) and professional organisations.

Academic capitalism is defined as the adoption of market strategies on the part of institutions of higher education. In so doing, colleges and universities have shifted their primary focus from the production and transmission of knowledge for the purpose of advancing civilisation to the commodification of knowledge for fiscal survival and growth within an advanced capitalist global context. We do not have the space to discuss academic capitalism in depth in a short chapter. However, this trend bears noting as a critical influence affecting knowledge discourse communities as they move into commercial discourse communities. In joining the medicalisation industry, higher education markets and advertises its wares.

> When students choose colleges, institutions advertise education as a service and a lifestyle... Colleges and Universities also regard students as negotiable to be traded to corporations as external resources... When students graduate they are presented... as products/outputs.
>
> (Slaughter & Rhoades, 2004, p. 1)

As faculty members move to become intellectual entrepreneurs, they are affected by and responsive to the marketplace and socialise students to follow suit (Park, 2011). Given the primacy of fiscal solvency and expansion, it is not surprising that faculty and future professionals who they teach are reactive to predefined market needs without the time or the socialisation to become sceptical. Once graduated, professionals are compelled to pay regular fees to, purchase objects from, and register for yearly conferences hosted by professional organisational discourse communities. Faced with competition and the market economy, the helping missions of the past are eclipsed by advertising and marketing survival tactics of the present (DePoy & Gilson, 2014).

Consider autism as exemplar. Mallett and Runswick-Cole (2012) have already exposed the body building effort that both constructed and muscled autism as a mental disability heavyweight with a full palette of colours, textures, and forms. It is therefore not surprising that organisational opportunism quickly appeared converting millions of followers as they are urged to reach into their pockets to grease institutional coffers. Consider the Autism Society. We are not singling out autism or the Autism Society nor are we indicting their handlers so to speak as malicious. Rather, we call on them as illustrative of a trend that we suggest needs to be exposed for analysis and revision.

More than a million individuals in our country (more than 50,000 in Ohio) and their families are affected by autism... the fastest growing serious developmental disability in our country. The Autism Society of Ohio is here to help them on the state and local levels by providing the following:

- Accurate information on autism
- Support services and programmes for individuals and families
- Educational resources to guide families and professionals
- Autism awareness to bring tolerance and understanding
- Advocacy to speak for those who cannot

Nowhere in its discourse are credit card and banking profits mentioned. Yet, like so many others, this organisation is immersed in market techniques such as cause branding, as reflected below:

> The Autism Society and UMB Card Partner have teamed up to launch the Autism affinity Visa® Platinum Rewards credit card. The Autism Society receives $50 when your Visa Platinum Rewards Card is activated and used as well as a small percentage of your purchases as a donation from UMB Bank! Our new Visa program provides a convenient way for you to make everyday purchases while contributing to a great cause.
>
> (The Autism Society of Ohio, 2012)

Clinical relevance summary

In this chapter, we have applied contemporary discourse community theory to the analysis of child mental health and the construction of child mental illness and disability with a particular focus on post-industrial contexts. Several discourse communities functioned to exhibit their lexical and material genres as means for crafting the commodified child and family and revising those who are not recognised as fitting within these landscapes. By theorising disability as artefact of language and positioning it as socially and culturally constructed has been helpful, but it is incomplete. Rather, it is important to consider health and illness as interactive discourse communities within which language, materiality, object, and interpretation are integrated. For a simple summary of the implications for practice, see Table 7.4.

Summary and conclusions

Our final words speak to the professional discourse community not as clinically prescriptive but rather as a call to healthy scepticism necessary to precede

Table 7.4   Clinical practice highlights

---

1. There are now important discourses of disability that recognise its socially and culturally constructed nature, but it is time to go further and think about the synergy between language, object, materiality, and interpretation. Practitioners could be reflective about this integration.
2. It is important to recognise the local nature of community, and although the emergent discourse work came from the United States and Europe, it can be applied to discourse communities across the globe. Thus, practitioners working in other countries may benefit from considering these discourses in their work.
3. In contemporary times, maximising profit tends to be the primary motivator of medicalisation.

---

profound positive change. While post-modern 20th-century portraiture of the professions painted members as perpetuators of medicalisation, we do not agree. Rather, we view the professional discourse community as in line to become the committed redesigners (DePoy & Gilson, 2014) of themselves, youth and families, all who are negatively influenced by their own commodification. Recall that we perceive professionals as committed and well intended but subjected to market strategies which are often skulking under their intellectual radar. It is precisely professional commitment that privileges and obligates this discourse community to become intellectual and praxis sceptics. By sceptic, we mean the questioner, not the cynic. The cynic is simply the critical pessimist who defaults to intellectual ennui, whereas the questioner holds high standards for theoretical discourses, their derivation, their advancement, and their relevant and virtuous application. As such, the questioner seeks to learn for transformative purposes without which destructive discursive and materialist trends that brand youth as deviant delayed 'not-yets' or 'nevers', accuse their parents of immoral deviancy, and indict professionals as greedy, go untamed.

The professional discourse community holds the discursive tools and resources to engage thoughtfully with those who can benefit from interaction, while also taking the lead in critical analysis and positive social change to release children and their families from pejorative branding and captivity in lifelong devaluing marketplaces. Scepticism provokes rethinking and redesigning not only responses to the mentally disabled child but to re-conceptualising childhood and family itself.

References

Alaimo, S. (2010). *Bodily natures.* Bloomington, IN: Indiana University Press.

Arel, D. (2013). *The Affordable Care Act Is not Socialism.* Retrieved December 17, 2013 from Speakout: http://www.truth-out.org/speakout/item/20806-the-affordable-care-act-is-not-socialism.

American Psychiatric Association (2013). *Diagnostic and statistical manual of mental disorders (DSM-5)* (5th edition). Washington, DC: Author.

Bellhouse, D. (2011). *Abraham de Moivre: Setting the stage for classical probability and its applications.* Boca Raton, FL: CRC Press.

Berger, A. (2009). *What objects mean.* Walnut Creek, CA: Left Coast Publishers.

Bringewatt, E. (2013). Negotiating narratives surrounding children's mental health diagnoses: Children and their contribution to the discourse. *Children and Youth Services Review, 35*(8), 1219–1226.

Byers, W. (2011). *The blind spot.* Princeton, NJ: Princeton University Press.

Candlin, F., & Guins, R. (2009). *The object reader (In sight: Visual culture).* London & New York: Routledge.

CASRC (2011). *What Is Developmental Delay and What Services Are Available if I Think My Child Might Be Delayed?* Retrieved July 5, 2014 from How kids develop: http://www.howkidsdevelop.com/developDevDelay.html.

Centers for Disease Control (CDC) (2014). *Facts About Autism.* Retrieved March, 2014 from Autism Spectrum Disorders: http://www.cdc.gov/ncbddd/autism/facts.html.

Conrad, P. (2007). *The medicalization of society: On the transformation of human conditions into treatable disorders.* Baltimore, MD: Johns Hopkins University Press.

Curatolo, P., D'Agati, E., & Moavero, R. (2010). The neurobiological basis of ADHD. *Italian Journal of Pediatrics, 36,* 36–79.

Davis, L. (2013). *The end of normal.* Ann Arbor, MI: University of Michigan Press.

DePoy, E., & Gilson, S. (2007). The bell-shaped curve: Alive, well and living in diversity rhetoric. *The International Journal of Diversity, 7.* Retrieved March 14, 2014 from http://ijd.cgpublisher.com/product/pub.29/prod.514.

DePoy, E., & Gilson, S. F. (2011). Studying disability: Multiple theories and responses. Thousand Oaks, CA: Sage Publications.

DePoy, E., & Gilson, S. (2014). *Disability design and branding.* London, UK: Routledge.

Devitt, A. J., Bawarshi, A., & Reiff, M. J. (2003). Materiality and genre in the study of discourse communities. *College English, 65*(5), 541–558.

Dictionary.com (2013). *Underachiever.* Retrieved March 21, 2014 from http://dictionary.reference.com/browse/Underachiever?s=t.

Dumit, J. (2012). *Drugs for life.* Durham, NC: Duke University Press.

Ernst, W. (2006). *Histories of the normal and the abnormal: Social and cultural histories of norms and normativity.* Abingdon, Oxon, UK: Routledge.

Finello, C. (2001). *Month-by-Month Guide to Baby's Emotional Development.* Retrieved March 21, 2014 from Parents: http://www.parents.com/baby/development/behavioral/month-by-month-guide-to-babys-emotional-development/.

Francis, A. (2009). Whither DSM–V? *British Journal of Psychiatry, 195,* 391–392.

Gaither, C. C., & Cavazos-Gaithe, A. E. (2012). *Gaither's dictionary of scientific quotations* (2nd edition). New York: Springer.

Gnaulati, E. (2013). *Back to normal: Why ordinary childhood behavior is mistaken for ADHD, bipolar disorder, and autism spectrum disorder.* Boston: Beacon Press.

Goertzel, T., & Fashing, J. (1981). The myth of the normal curve: A theoretical critique and examination of its role in teaching and research. *Humanity and Society, 5,* 14–31.

Howard, M., & King, J. E. (2008). *The rise of neoliberalism in advanced capitalist economies: A materialist analysis.* New York: Palgrave MacMillan.

Jameson, F. (1991). *Postmodernism, or, the cultural logic of late capitalism.* Durham, NC: Dike University Press.

Kail, R., & Cavanaugh, J. C. (2013). *Human development: A lifespan view* (6th edition). Belmont, CA: Cengage.

Keupink, A., & Shieh, S. (2010). *The limits of logical empiricism: Selected papers of Arthur Pap.* Dordecht, Netherlands: Springer.

kidsmentalhealth.org (2009). *How Is Mental Illness in Children Diagnosed.* Retrieved March 21, 2014 from Kids Mental Health Information Portal: http://www.kidsmentalhealth.org/how-is-mental-illness-in-children-diagnosed/.

Krippendorf, K. (2011). *Discourse and the Materiality of Its Artifacts.* Retrieved June 19, 2014 from Scholarly Commons: http://repository.upenn.edu/asc papers/259.

Legrenzi, P., & Umilta, C. (2011). *Neuromania.* Oxford: Oxford University Press.

Mallett, R., & Runswick-Cole, K. (2012). Commodifying autism: The cultural contexts of 'Disability' in the academy. In D. Goodley, B. Hughes, & L. J. Davis (Eds.), *Disability and social theory* (pp. 33–51). Basingstoke: Palgrave MacMillan.

Meehl, P. E. (1999). Clarifications about taxometric method. *Applied and Preventive Psychology, 8,* 165–174.

Miller, D. (2010). *Stuff.* Malden, MA: Polity.

Mowder, B. (2005). Parent development theory: Understanding parents, parenting perceptions and parenting behaviors. *Journal of Early Childhood and Infant Psychology,*

1. Retrieved June 19, 2014 from http://www.questia.com/read/1G1-220766550/parent-development-theory-understanding-parents.

Mowder, B., Shamah, R., & Zeng, T. (2011). Current measures for assessing parenting of young children. *Journal of Child and Family Studies, 20*(3), 295–302.

Munch, R. (2014). *Academic capitalism: Universities in the global struggle for excellence.* New York: Routledge.

National Institute of Mental Health (2010). Medications for autism. *Psych Central.* Retrieved August 7, 2014 from http://psychcentral.com/lib/medications-for-autism/0005716.

National Institute of Mental Health (2009). *Treatment of Children with Mental Illness.* Retrieved June 19, 2014 from National Institute of Mental Health: http://www.nimh.nih.gov/health/publications/treatment-of-children-with-mental-illness-fact-sheet/index.shtml.

Nye, R. A. (2003). *The Evolution of the Concept of Medicalization in the Late Twentieth Century.* Retrieved April 14, 2014 from http://onlinelibrary.wiley.com.prxy4.ursus.maine.edu/doi/10.1002/jhbs.10108/pdf.

Olfson, M., Gerhard, T., Huang, C., Lieberm, J. A., Bobo, W. V., & Crystal, S. (2012). Comparative effectiveness of second-generation antipsychotic medications in early-onset Schizophrenia. *Schizophrenia Bulletin, 38*(4), 845–853.

Olson, M. (2009). Democratic citizenship – A conditioned apprenticeship: A call for destabilisation of democracy in education. *Journal of Social Science Education, 8*(4), 75–80.

Orrell, D. (2012). *Truth or beauty: Science and the quest for order.* New Haven, CT: Yale University Press.

Park, T. (2011). Academic capitalism and its impact on the American professoriate. *Journal of the Professoriate, 6*(1), 94–99.

Psiram (2013). *Pseudomedicine.* Retrieved August 12, 2014 from Psiram: http://www.psiram.com/en/index.php/Pseudomedicine.

Quinn, A., Briggs, H. E., Miller, K. M., & Orellana, E. R. (2014). Social and familial determinants of health: Mediating effects of caregiver mental and physical health on children's mental health. *Children and Youth Services Review, 36*, 163–169.

Rose, G. (2012). *Visual methodologies: An introduction to researching with visual materials.* Los Angeles, CA: Sage.

Salmon, W. (1985). Profit and health care: Trends in corporatization and proprietization. *International Journal of Health Services, 15*(3), 395–418.

Schuster, D. (2011). *Neurasthenic nation.* Piscataway, NJ: Rutgers University Press.

Siebers, T. (2010). *Disability aesthetics.* Ann Arbor, MI: University of Michigan Press.

Slaughter, S., & Rhoades, G. (2004). *Academic capitalism and the new economy: Markets, state, and higher education.* Baltimore, MD: Johns Hopkins University Press.

Smith, M. (2010). Good parenting: Making a difference. *Early Human Development, 86*(11), 689–693.

Solomon, A. (2012). *Far from the tree.* New York: Scribner.

Swales, J. (1990). *Genre analysis: English in academic and research settings.* Cambridge: Cambridge University Press.

The Autism Society of Ohio (2012). *The Autism Society of Ohio.* Retrieved May 19, 2014 from The Autism Society: http://www.autismohio.org/index.php/information-mainmenu-118/autism-society-of-america-news/4535-sign-up-for-an-autism-society-visa-platinum-rewards-card.

Tilsen, J. (2007). We don't need no education: Parents are doing it for themselves. *Journal of Progressive Human Services, 18*(1), 71–87.

Turkle, S. (2011). *Evocative objects*. Boston: MIT Press.

Waller, N. G. (2006). Carving nature at its joints: Paul Meehl's development of taxometrics. *Journal of Abnormal Psychology, 1*(15), 210–215.

Younkins, E. (2004). *Political Correctness Threatens Free Society*. Retrieved March 15, 2014 from Rebirth of Reason: http://rebirthofreason.com/Articles/Younkins/Political_Correctness_Threatens_Free_Society.shtml.

Recommended reading

- Jameson, F. (1991). *Postmodernism, or, the cultural logic of late capitalism*. Durham, NC: Dike University Press.
- Kirk, S. A., Gomory, T., & Cohen, D. (2013). *Mad science: Psychiatric coercion, diagnosis, and drugs*. New Brunswick, NJ: Transaction.
- Munch, R. (2014). *Academic capitalism: Universities in the global struggle for excellence*. New York: Routledge.
- Park, T. (2011). Academic capitalism and its impact on the American professoriate. *Journal of the Professoriate, 6*(1), 94–99.

8

The Social Construction of Attention Deficit Hyperactivity Disorder

Sami Timimi and Lewis Timimi

Introduction

This chapter examines the social construction of attention deficit hyperactivity disorder (ADHD) by looking at its history, cross-cultural aspects, and how it is described on the Internet. The diversity of findings contrasts to its narrow construction in standard medical and Western public discourses. The chapter begins with a brief overview of the current dominant construction of ADHD as found in mainstream medical/psychiatric literature. In order to explore how ADHD came to occupy a hegemonic position in the modern child and adolescent mental health lexicon, its developmental history is outlined and a number of key drivers such as the impact of the pharmaceutical industry, commodification, and changing child-rearing patterns and expectations are discussed. Next, the cross-cultural research literature is summarised. Here there is a contrast between attempts to identify commonalities, in order to 'squeeze' diverse cultural beliefs and practices around children and child development into simplistic Westernised categories like ADHD, and more anthropologically informed research that explores these diversities. We then review findings from an Internet-search-based research project. This research used a standard search engine (Google) to search the themes of 'What is ADHD', 'What causes ADHD', and 'ADHD treatment'. This is followed by a thematic content analysis of the most commonly cited web pages for each question. The study helps us gain a glimpse into the type of discourses that members of the UK public are likely to encounter when carrying out their own searches on the topic of ADHD. Finally, we reflect on some potential clinical implications of our findings.

Construction of ADHD in mainstream psychiatric discourse

The standard construction of ADHD in the psychiatric literature goes something like this: According to the fifth edition of *Diagnostic and Statistical Manual of Mental Disorders* (*DSM-5*) (American Psychiatric Association, 2013), ADHD is characterised by a pattern of behaviour, present in at least two settings (e.g. school and home) and that can result in performance issues in social, educational, or work settings (see Horton-Salway & Davies, Chapter 9, this volume). Symptoms are divided into two categories of 'inattention' and 'hyperactivity and impulsivity' that include behaviours like failure to pay close attention to details, difficulty organising tasks and activities, excessive talking, fidgeting, or an inability to remain seated in appropriate situations. Children must have at least six symptoms from either (or both) the inattention group of criteria or the hyperactivity and impulsivity criteria, while older adolescents and adults (above age 17 years) must present with five. The descriptions of symptoms at different ages help clinicians better identify 'typical' ADHD symptoms at each stage of patients' lives. In *DSM-5*, several of the individual's ADHD symptoms must be present prior to age 12 years, compared to 7 years as the age of onset in *DSM-IV* (American Psychiatric Association, 1994).

ADHD is thought to be a common psychiatric disorder that affects between 3% and 5% of children, mainly boys, and which many do not grow out of (i.e. for many it will be seen as a lifelong disorder). It is often claimed that ADHD has a primarily genetic basis resulting in a chemical imbalance in the brain that can be treated using stimulant medication. However, evidence in support of a biological basis for ADHD is lacking, but this hasn't prevented ADHD from becoming a popular diagnosis with many children, adolescents, and increasingly adults, being diagnosed with what many are told is a primarily biologically based disorder (Timimi & Leo, 2009).

Culture and social construction

Before discussing some of the social, cultural and political dynamics that have contributed to how ADHD has come to be understood and how practice in this area is now conceptualised, it is worth understanding a little bit about what is meant when referring to 'culture' and then to 'social construction' (see Rafalovich, Chapter 5, this volume).

One definition of culture that we find useful is that culture is, 'The peculiar and distinctive way of life of a group, the meanings, values and ideas embodied in institutions, in social relations, in systems of belief, in customs, in the uses of objects and material life... the 'maps of meaning' that make things intelligible to its members' (Clarke, Hall, Jefferson, & Roberts, 1975, p. 10). In other words 'culture' refers to a set of beliefs and a 'way of doing things' that a certain group of people have in common.

We can examine and talk about 'culture' at many different levels as we can draw an imaginary boundary around groups of people that have something in common in many different ways, from small groups right up to sets of nations and whole regions. So, for example, we can talk about the culture of a particular family, a particular profession, a particular city, a particular country, a particular religion, and a particular continent even. Each of these groupings will contain 'something' in common to mark them out from other similar groupings (like a different family, a different profession, a different religion and so on). This, of course, also means we have to be very careful when we talk about 'the culture' of a particular group, as each group is made up of individuals who are going to possess significant differences between them and will simultaneously belong to several different possible cultural groupings. (For example, if we were to take a city as our starting cultural grouping, people within this city will also belong to different families, professional groups, religions, speak different languages, and so on.) This means we must always be careful not to 'tar everyone with the same brush' – not to stereotype.

The concept of social construction has a close affiliation to that of culture. Social constructs are essentially the by-products of countless human thoughts and choices that coalesce into cultural ideas and traditions. A major focus of social constructionism is to uncover the ways in which individuals and groups participate in the construction of their perceived social reality. It involves looking at the ways social phenomena are created, institutionalised, and made known to and by people. The social construction of reality is an ongoing and dynamic process. This process also introduces the possibility of change as what is considered a 'truth' or a 'correct' or 'moral' way of being and behaving shifts from one generation to the next (Hacking, 1999).

So, in essence 'social construction' refers to the process by which different cultures develop their sets of beliefs and values that guide the way they subsequently behave in the world and in relation to each other. Another way of putting this is that groups of people who share a common cultural set of values and beliefs are said to 'socially construct' their understanding of the world around them.

To help explain this further, let us give you a simple and relevant example of social construction in relation to a child who is perceived to be displaying persistent disobedience. As a generalisation (remembering the warning about stereotyping) we can see that different cultures will interpret the reasons for such behaviour differently and so go about the task of trying to modify the behaviours differently. In the West today, it may lead to a concern that the child has a medical condition and so a visit to the doctor may be organised that may eventually result in a referral for an assessment for ADHD. Among some religious groups in other parts of the world, because they have a different set of beliefs about children, families, and how behaviour is understood and dealt with, a concern that a child might be possessed by some supernatural entity may emerge. This belief will lead to a different strategy, for example a traditional healer or religious leader may be consulted and certain rituals may then be prescribed. For yet other groups, such displays of defiance may be interpreted as an expected developmental stage and no particular action or intervention deemed necessary. Here we are not making any value judgement about which of the above approaches are 'better', just illustrating that if we 'socially construct' our world differently then how we understand a problem and what we do about it will also be different.

A brief history of ADHD

Like all psychiatric diagnoses (apart from the dementias), ADHD was constructed in the imaginations of those who developed the concept rather than as a result of any new scientific breakthroughs. During pre-1970s, reports linking ADHD behaviours to a medical cause were rare and sporadic and referred largely to children likely to be afflicted with organic conditions.

Overactivity, poor concentration, and impulsivity in children were first conceptualised as possible medical phenomena over a century ago when the British paediatrician Frederick Still described a group of children who showed what he considered to be a poor capacity for sustained attention, restlessness, and fidgetiness. He went on to argue that these children had 'abnormal defects of moral control' (Still, 1902), although he generally assumed this was caused by pre-existing diseases affecting the brain such as cerebral tumours, meningitis, epilepsy, head injury, typhoid fever, or impairment of the intellect. Although Still lists behaviours that could be considered more appropriate to the category of conduct disorder (such as cruelty, jealousy, lawlessness, dishonesty) than ADHD, ADHD enthusiasts often paint his 1902 article in the *Lancet* as an early example of the identification of a medical syndrome of it (Barkley, 2006).

The next important link in the developing ADHD narrative was the chance discovery at an institution for neurologically impaired children that stimulant treatment (with the stimulant Benzedrine) improved the behaviour, concentration, and school performance of a group of these children. The children often presented with restlessness, personality changes, and learning difficulties with many having previously suffered from encephalitis (Bradley, 1937). Neither Still nor Bradley's papers received much attention at the time of their publication.

In the post-Second World War years, psychiatry became involved in the treatment of many traumatised men, women, and families and so began to expand its range of interests. The mental life of children – a group that had not drawn much interest from the psychiatric profession until then – became the focus of greater curiosity and interest. A number of doctors began to speculate that children who presented as hyperactive might have organic lesions in the brain that was causing their hyperactivity. Based on the observation that children who experienced identifiable brain injuries (e.g. from encephalitis, birth trauma, and epilepsy) presented with hyperactivity, the concept of Minimal Brain Damage (MBD) was developed as a hypothesis to explain the occurrence of hyperactivity in the absence of overt evidence of brain injury (e.g. Denhoff, Laufer, & Solomons, 1957; Strauss & Lehtinen, 1947).

By the 1960s, however, the term MBD was being criticised and losing favour, as evidence for underlying organic lesions in children who displayed hyperactivity was not being found. In addition, higher rates of brain insults were found to be present across most psychiatric categories (rather than any specific one) and many who had evidence of brain damage did not show hyperactivity. Instead, the Oxford International Study Group of Child Neurology suggested that MBD be redefined as Minimal Brain Dysfunction (Bax & MacKeith, 1963). The concept of Minimal Brain Dysfunction was a wider one and referred to children with learning or behavioural problems ranging from mild to severe, and presenting with 'deviations' of perception, conceptualisation, language, memory, attention, impulse, or motor function. As it was becoming recognised in the medical literature that the presence of hyperactivity couldn't be conceptualised as evidence of some sort of brain damage, it began to be thought of as being part of a behavioural syndrome that could arise from organic pathology, but could also occur in its absence. As a result, a movement away from aetiologically based definitions towards behaviourally based ones occurred. Thus, in the mid-1960s, the North-American-based *Diagnostic Statistical Manual*'s second edition *DSM-II* coined the term 'hyperkinetic reaction of childhood', to replace the diagnosis of MBD (American Psychiatric Association, 1966).

At the same time, growing interest from psychologists meant that psychological mechanisms were being hypothesised as the mediator between potential causal influences and subsequent behavioural manifestations. Thus the role of attention came to the fore as a new theory proposed that deficits in sustained attention and impulse control were the drivers of hyperactivity (Douglas, 1972). Thus, when *DSM-II* was replaced in the early 1980s by the third edition (*DSM-III*, American Psychiatric Association, 1980), the disorder was now termed Attention Deficit Disorder (ADD) reflecting this change in emphasis. This could be diagnosed with or without hyperactivity and was defined using three dimensions (three separate lists of symptoms): one for attention deficits, one for impulsivity, and one for hyperactivity. ADD now sets the scene for a revolution to take place in Western child psychiatric practice, as the drug Ritalin joins the growing popularity of using psycho-pharmaceuticals to deal with life's mental challenges. The three-dimensional approach was abandoned in the late 1980s when *DSM-III* was revised (and became *DSM-III-R*, American Psychiatric Association, 1987), in favour of combining all the symptoms into one list (one dimension). The new term for the disorder was 'Attention Deficit Hyperactivity Disorder' (ADHD), with attention, hyperactivity, and impulsiveness now assumed to be part of one disorder with no distinctions, and thus ADHD was born.

When the fourth edition of *DSM* (*DSM-IV*, American Psychiatric Association, 1994) reconsidered the diagnosis, the criteria were again changed, this time in favour of a two-dimensional model with attention deficit being one sub-category and hyperactivity-impulsivity the other. With each revision, a larger cohort of children is found to be above the threshold for diagnosis. For example, changing from *DSM-III* to *DSM-III-R* more than doubled the number of children from the same population diagnosed with the disorder (Lindgren et al., 1994). Changing from *DSM-III-R* to *DSM-IV* increased the prevalence by a further two-thirds, with the criteria now having the potential to diagnose the vast majority of children with academic or behavioural problems in a school setting (Baumgaertel, Wolraich, & Dietrich, 1995).

The recently published *DSM-5* criteria (American Psychiatric Association, 2013) are likely to expand numbers even further particularly as some key criteria have been broadened to allow more adolescents and adults to qualify for a diagnosis (e.g. increasing the required age of onset for symptoms to age 12 or earlier, which has increased from age 7 or earlier in *DSM-IV*). The meteoric rise in numbers diagnosed with ADHD has meant that in the United States now nearly one in five school-age boys and about 11% of school-age children have received a diagnosis of ADHD (Schwarz & Cohen, 2013).

The principal recommendation for treatment has long been that of prescribing stimulant medication – namely, the use of the amphetamine class of drugs (such as Ritalin and Dexedrine). The idea that ADHD has a specific medical treatment has acted as a powerful stimulant (pun intended) towards the popularisation of the concept, particularly in those countries that are dominated by a neoliberal value system where the perceived availability of a specific treatment has enabled commodification and commercialisation of the diagnosis.

Conrad (1975) attributed the shift in how hyperactivity is perceived, from being a relatively esoteric and rarely used diagnostic category, to becoming a well-known clinical diagnosis by the mid-1970s in the United States, to the emergence of the 'pharmaceutical revolution' in mental healthcare. The new ideology that mental disorders are the result of chemical imbalances for which there are specific medications was beginning to take hold. The marketing of medications directly to physicians coincided with the US government becoming interested in the merits of using psychoactive medications to treat hyperactivity. Furthermore, the expanding use of psycho-pharmaceuticals for hyperactivity happened at a time when 'scientific approaches' to parenting became fashionable, reaching their zenith in the mid-20th century in the United States (Apple, 2006). The ascendance of scientific discourses on children and parenting displaced more naturalistic approaches to childrearing characteristic of earlier eras, pushing the task of child rearing further towards ownership by professional groups. As a result, parenting came to be viewed as a rationalised and professionalised undertaking.

Previously in the United States, children were viewed as largely sturdy and resilient. An example of this can be found in the Fischer's late 1950s/early 1960s New England town study (Fischer & Fischer, 1966). According to the study, families understood their children's problematic behaviours as 'stages' that most children could be expected to pass through. Viewing problems in this way meant that parents did not feel an obligation to seek professional help for them. In fact, to do so would run counter to another prevailing sensibility, namely, that parents ought not to force children in case they damage their 'potential'. Thus children's 'bad' behaviour was interpreted through a normative lens as expectable and temporary.

These changing dynamics – moving childhood behaviour problems away from the parental 'common sense' arena towards ownership by a professional class – together with the greater emphasis on using psycho-pharmaceuticals to control emotions and behaviours have contributed to diagnosis and prescription of medications for childhood mental disorders increasing steeply in the past few decades in most post-industrial countries (particularly in North America, Northern Europe, and Australasia). While this fact is not disputed, it is subject to vastly different interpretations by scholars and clinical professionals, depending on their divergent theoretical assumptions. Those who believe that 'scientific' progress is behind this rapid change in practice argue that disorders such as ADHD were simply 'under-recognised' in the past. According to this depoliticised perspective, there have always been children 'suffering' from such disorders, but only as a result of recent clinical and scientific advances we have discovered these to be symptoms of medical conditions that can cause abnormal development. Critics of the view that conceptual and practical changes surrounding childhood mental disorders are result of new scientific discoveries (and we would count ourselves among these critics) point to the paucity of evidence supporting this contention (for a summary of these critiques, see Timimi, 2002, 2005, 2008, 2009, 2013; Timimi & Leo, 2009).

By the beginning of this century the concept of ADHD was migrating beyond that of a childhood developmental disorder to a lifelong disorder as the idea of adult ADHD took off, starting inevitably in the pharmaceutical marketing global capital, the United States. The emphasis in adult ADHD moves from observations of external behaviour to perceived failings of internal regulation, highlighting problems with 'self-concept' and largely disregarding hyperactivity in its diagnostic framework (Moncrieff & Timimi, 2011). As Western culture has heightened a narcissistic focus on the inner life and leans towards measuring the sense of self-worth through competitive accomplishments (as compared with more communally orientated cultures, where self-worth tends to be more tied to family/group well-being), diagnoses framed in ways that blame the problem on biochemical dysfunction can appear to provide an avenue of relief from the struggles involved in keeping a positive sense of self in the absence of achievement or personal satisfaction.

The focus on 'inner life' (such as depression and anxiety) that dominates adult psychiatry (as opposed to a focus on behaviour that dominates child psychiatry) has led to women being the predominant 'customers' for community psychiatric services. Not surprisingly perhaps, the shift in emphasis in definitions of adult ADHD from behavioural to internalised definitions has accompanied an increase in the numbers of women being diagnosed with adult ADHD, and in many countries women outnumber men (Castle, Aubert, Verbrugge, Khalid, & Epstein, 2007), despite the fact that the diagnosis of childhood ADHD is strongly associated with being a boy (Timimi, 2005). Advocates have argued that ADHD is merely under-recognised in girls (McGee & Feehan, 1991; Staller & Faraone, 2006), but the targeting of women by promotional material suggests that adult ADHD may be the latest framework offered to women through which their distress and dissatisfaction can be exploited.

This solution, however, remains fragile as it also fixes the notion of a permanent and biological disability and thus that of 'damaged goods', creating a more long-term problem with the diagnosed person invited into a potentially lifelong struggle to control and prevent their 'ADHD' from ruining their life.

The pharmaceutical industry appears to have also been instrumental in the rise of adult ADHD, especially over the last decade. In 2004, for example, pharmaceutical marketing companies explicitly identified adult ADHD as an 'expanding and lucrative market' for stimulants and related drugs (Lead Discovery, 2004). Several companies have run direct to consumer advertising campaigns in the United States, which market the disorder by suggesting that common behaviours (such as forgetting car keys) may be symptoms (Food and Drug Administration, 2005). Company websites also contain screening questionnaires that encourage people to seek help if they think they have the diagnosis. Moreover, it has been revealed that some of the researchers who most vigorously promoted the concept of adult ADHD, and conducted many of the drug trials, failed to disclose millions of dollars of income they had received from pharmaceutical companies (Harris & Carey, 2008).

Cross-cultural perspectives

The drive to 'discover' biomedical templates within which to place various children's behaviours that are considered socially difficult has resulted in the exclusion of cultural meaning, the local significance of particular behaviours, and local beliefs and practices for dealing with such behaviours. This adoption of a 'universalising' approach to the concept of ADHD can be seen in most mainstream academic publications on the subject. An example of this exclusion of the cultural perspective can be found in the 'Global Consensus on ADHD/HKD' written by a group of 15 prominent researchers and clinicians (Global ADHD Working Party, 2005). It concludes that ADHD is a genetically determined neurodevelopmental disorder that is found in all societies, with core aetiological and pathophysiological features, which are amenable to a uniform treatment approach (mainly medication), which can then be applied across the world, regardless of local beliefs and practices.

Literature that takes account of local beliefs and practices shows a much more interesting and varied picture. When viewed through the prism of local cultural beliefs, variations in practice can be found even within and between countries that recognise and have widespread availability of services for diagnosing and treating ADHD.

For example, in a context where self-control in response to aggression or provocation is highly valued, a lack of self-control is more likely to be interpreted as a disorder. Similarly, in contexts where school success is highly valued, poor school performance is more likely to be interpreted as being the result of a disorder. Thus a significant difference is found in how ADHD is conceptualised in real-world practice between the United Kingdom and the United States. Both epidemiological (Hart & Benassaya, 2009) and child/family interview (Singh, 2011) studies have found considerable differences in the sort of problems likely to lead to a diagnosis and treatment of ADHD between the United Kingdom and United States. In the United Kingdom, diagnosis and treatment is more likely to be given to children (mainly boys) from lower social classes and is associated with higher levels of behavioural problems, whereas in the United States middle-class children (again, mainly boys) particularly in areas of lower academic achievement are more likely to be diagnosed and treated (Hart & Benassaya, 2009). Interviews with young people diagnosed with ADHD in the United Kingdom and United States have found that their beliefs about ADHD mirror the epidemiological findings, with UK children much more likely to view ADHD as a disorder of self-control causing them to lose their temper and get into fights, whereas US children are more likely to believe that ADHD is a disorder that causes them to fail at school (Singh, 2011).

Despite attempts at standardising criteria and assessment tools in cross-cultural ADHD studies, major and significant differences between raters from different countries are apparent (e.g. Mann et al., 1992). There are also significant differences between raters when they rate children from different ethnic minority backgrounds (Sonuga-Barke, Minocha, Taylor, & Sandberg, 1993). One replicated finding is an apparently high rate of hyperactivity in China and Hong Kong (e.g. Luk & Leung, 1989). In these studies, nearly three times as many Chinese as English children were rated as hyperactive. A more detailed assessment of these results suggested that most of the 'hyperactive' Chinese children would not have been rated as hyperactive by most English raters and were a good deal less hyperactive than English children rated as 'hyperactive' (Taylor, 1994). One suggestion for such a consistent disparity in hyperactivity ratings between Chinese and English raters and children is that it may be due to the great importance of school success in Chinese culture which leads to an intolerance of lesser degrees of disruptive behaviour (Taylor, 1994). Whatever the reason(s), it demonstrates that hyperactivity and disruptiveness are highly culturally constructed entities.

Thus, whatever part of conditions such as ADHD is biological, how we construct meaning out of this is a cultural process. Brewis and Schmidt (2003) studied a middle-class Mexican school with over 200 pupils. Using standard diagnostic criteria, they found that about 8% of the children could be diagnosed as having ADHD, yet there was only one child in that school who had been given an ADHD diagnosis. Through interviews with parents and teachers of these children, they discovered that these carers regarded ADHD-type behaviours as within the boundaries of behaviours viewed as normal and expected for these children at those ages.

Within any culture/society, ADHD shows different patterns of distribution that may be traced back to social and cultural dynamics. Thus, in the United Kingdom, the social distribution of ADHD diagnosis follows the contours of a class-based gradient. Children exhibiting the symptoms of any emotional or behavioural disorder, including those with symptoms of ADHD, are much more likely to be poor, to be raised by single and/or unemployed parents, to grow up in underprivileged neighbourhoods, and to be exposed to stressful life events in their early lives. The highest excess of all is where a parent is in trouble with the law – a court appearance by a parent raises the risk by almost 200%. This class gradient is not found in the United States with the most economically advantaged children in the United Kingdom being much more protected against an ADHD diagnosis than their American counterparts (Hart & Benassaya, 2009).

Race and ethnicity is another area where within-country differences in diagnostic patterns can be found. For example, Carpenter-Song (2008) in her ethnographic study of families with children diagnosed with ADHD in a metropolitan area of the Northeastern United States found considerable differences between the beliefs and practices of 'Euro-American' families and 'African-American' families. She found that while Euro-American families voiced biomedical explanations and preferred to use a clinical lexicon of 'disorders' and 'conditions', or specific diagnostic categories to describe their children's behaviour problems, African-American families resisted pathologising their children's experience, which was reflected in using a more diffuse vocabulary of 'issues', 'challenges', and 'difficulties' to describe problematic behaviours and feelings in their children. These African-American families displayed active questioning of, and scepticism towards, mental health interventions whether these were medication or psychotherapeutic.

Project overview: An Internet search on ADHD

One of the authors (LT) carried out an Internet search on the subject of ADHD. The purpose of the research was to enable a 'snapshot' of the sort of discourses on ADHD members of the UK public may encounter if they (as many do these days) seek out information on ADHD from the Internet.

The research took place between December 2013 and May 2014. Using Google as the search engine, three separate search terms were entered: 'what is ADHD', 'what causes ADHD', and 'ADHD treatment'. The top 20 results were analysed for each question, although sponsored links and separate results from the same website were ignored. The analysis involved looking at the section of each web page which best addressed the particular question entered. The narrative on the relevant web page for each question was then categorised and entered into a table. The number of times a particular category appeared across the 20 search results for each of the three questions was calculated for each question separately.

What is ADHD?

The Google search on 'What is ADHD?' suggests that, on the Internet at least, ADHD is seen as a common biological disorder of childhood, characterised by a particular group of behaviours (see Table 8.1). Eighty per cent of the websites (16 out of the 20 looked at) included at least one of these four most frequent categories in their definition (a genetic/biological/neurological disorder, a common condition, a childhood/developmental disorder, a group of behaviours/symptoms). An example of this definition of ADHD can be found on the website Medical News Today. Their first paragraph says the following:

> Health experts say that ADHD (attention deficit hyperactivity disorder) is the most common behavioral disorder that starts during childhood. However, it does not only affect children – people of all ages can suffer from ADHD. Psychiatrists say ADHD is a neurobehavioral developmental disorder.

Table 8.1   What is ADHD? Google search – top 20 sites

ADHD is	Number of websites out of 20 (%)
A genetic/biological/neurological disorder	7 (35)
A common condition	7 (35)
A childhood/developmental disorder	7 (35)
A group of behaviours/symptoms	7 (35)
A psychiatric or medical disorder	3 (15)
An illness	1 (5)
A chronic condition	1 (5)

What causes ADHD?

The Google search for the question 'what causes ADHD' was categorised in a different manner to try and best capture the nature of the discourses found on the top 20 websites. What became apparent through examining the websites is that possible causes could be portrayed in positive (as in the particular factor mentioned is described as causing ADHD) or negative terms (as in the particular factor mentioned is described as not being involved in causing ADHD). It was felt that a more complete portrayal of the information found on these sites needed to capture both positive and negative causal attributions. In order to capture this a weighted system for classifying each factor was developed (see Table 8.2). Thus for each category of proposed cause found on the website a weighted score was given with +2 being for a 'clear or strong link', +1 for a 'probable link', −1 for an 'improbable or unlikely link', and −2 for a 'definite no link'. Thus the higher the score, the more likely that the factor is being portrayed as causal across the websites, and the larger a negative score, the more likely the factor is being portrayed across the websites as not being causally involved.

The results of the search are shown in Table 8.3. As can be seen, the websites very strongly favour biological causes as the primary explanation for what causes ADHD. The seven categories with the highest scores are all biological causes, with genetic factors and brain functioning leading the way. Only five of the 29 categories involved potential psychosocial causes, with four out of these five having negative scores, indicating that psychosocial causes were often discounted as being involved in causing ADHD. The lowest scoring category (and therefore most often mentioned as not being involved in causing ADHD) was a lifestyle cause of 'sugar consumption' followed by 'parenting/upbringing' and the 'family/social environment'.

Websites, therefore typically portrayed ADHD as a disorder caused by biological factors; for example, psychmedcentral.com says:

ADHD has a strong genetic basis in the majority of cases.

Table 8.2   Weighted scoring system for 'what causes ADHD'

Weighted score	Criteria
2	Page indicates that there is a clear or strong link
1	Page indicates that there is a probable link
− 1	Page indicates that a link is improbable/unlikely
− 2	Page indicates that there is definitely no link

Table 8.3   What causes ADHD? Google search – top 20 sites

ADHD is caused by ...	Weighted score
Genetic factors	32
Brain/central nervous system functioning	19
Pre-natal injury or exposure to toxins	16
Brain injuries/disease	15
Biochemical imbalance	12
Exposure to lead	12
Birth complications (e.g. low birth weight, premature birth)	10
Environmental factors	4
Exposure to pesticides	2
Anaesthesia exposure	2
Emotional or physical abuse	2
Exposure to industrial chemicals	1
Exposure to second-hand smoke	1
Infections during early life and childhood	1
Executive function difficulty	1
Exposure to chemotherapy	1
Epilepsy	1
Being male	0
Date of birth	−1
Variations in temperament	−1
Dietary factors	−2
Food additives	−2
Hormones	−2
Problems with vestibular system	−2
Large birth size	−2
Too much TV	−10
Family/social environment (including school environment)	−11
Parenting/upbringing	−13
Sugar consumption	−16

Whereas, at the same time, the websites discount theories of causation that include lifestyle and psychosocial/environmental features, for example myadhd.com says:

> No studies support the idea that ADHD is the result of poor parenting or other family environment variables.

ADHD treatment

In response to the top 20 website search for 'ADHD treatment', the various treatments mentioned by the websites were collapsed into three categories which cover the main classes of treatment mentioned (see Table 8.4 for the top 20 Google sites). The three categories are

Table 8.4   ADHD treatment. Google search – top 20 sites

Treatment	Number of websites out of 20 (%)
Medication	20 (100)
Psychotherapeutic approaches	15 (75)
Lifestyle interventions	8 (40)

1. Medication: included specific mention of stimulants, non-stimulants, and anti-depressants.
2. Psychotherapeutic approaches: included specific mention of behaviour therapy, parent management training, teacher training, classroom strategies, social skills training, family therapy, cognitive behaviour therapy, counselling, art therapy, working memory training, neurofeedback, therapeutic recreational programmes, and support groups.
3. Lifestyle interventions: included specific mention of diet, dietary supplements, exercise, sleep, relaxation, and personal tutoring.

As can been seen, of the 20 websites identified, all mentioned medication as a treatment for ADHD. The primacy of biological treatments on the identified websites follows on from the preference for biological explanations found when searching the question 'what causes ADHD?' Furthermore, 16 (80%) of the 20 websites make no mention of long-term outcomes for treatment with medication. Of the four websites that do mention long-term outcomes, all four refer to the lack of evidence on the long-term effectiveness of using medication to treat ADHD; for example, the website Wikipedia mentions:

> [T]hey have not been found to improve school performance and data is lacking on long-term effectiveness and the severity of side effects...indeed, after 14 months [in reference to a well-known large study] the medication group lost its advantage to the long discontinued behavior modification group. By year eight socioeconomic status and family structure were the only predictive variables for ADHD treatment.

It seems that across the three Internet-based searches a biological model predominates. This illustrates how the biological emphasis and the assumed neurodevelopmental basis for ADHD has reached the public and become part of the everyday discourse. Given the widespread use of the Internet and its global reach, these discourses are affected by and, in turn, affect the 'common sense' public conceptualisation, not just in the developed North but increasingly around the world. The continued spread of the biomedical explanatory model and the promotion of the use of medication present all manner of known and unknown risks to children and families as well as cultural beliefs and practices connected with childhood more broadly.

Clinical relevance summary

So far we have examined how ADHD is being socially constructed as an already-known brain-based biological disorder for which there is a specific and effective medication-based medical treatment. We have also reviewed the historical and cross-cultural research literature, which helped us understand how we have arrived at the current social construction of ADHD, and how across the world the behaviours that ADHD refers to are understood (and, therefore, socially constructed) differently. So what are the potential clinical implications of taking this 'social construct' as opposed to biomedical perspective on ADHD?

First, understanding that the dominant biologically deterministic model of ADHD reflects a social construct rather than a scientific 'truth' allows the possibility of 'deconstructing' the biological model in order to allow space for alternative perspectives.

Deconstructing the medical, bio-deterministic model is a relatively straight-forward process. 'Deconstructed ADHD' can simply be seen as a diagnosis that describes behaviours, but cannot explain them. Clinically, you may need to explain a 'truism' and repeat it whenever it becomes necessary when the young person and/or family you are working with slide back into inferring that ADHD explains any behaviour. Explaining the truism goes something like this:

> I would just like to explain something to help you understand why psychiatric diagnoses like ADHD are very different to those that we find in the rest of medicine. As you know there is no particular blood test, brain scan or any other investigation that I can do that will allow me to understand whether there is anything going on in your/your child's body and brain that can help explain why you/they are sometimes finding it difficult to stay focussed/get into trouble. You see psychiatric diagnoses are basically shorthand descriptions of whatever the presenting behaviours are. Shorthand descriptions can be helpful as some behaviours often go together such as in your/your child's case where difficulty concentrating often goes together with being hyperactive. This shorthand can be useful to help with communicating and with defining problems for research. What psychiatric diagnoses don't provide is an explanation of why those behaviours are happening and what might be helpful in terms of treatment, as what might be helpful for one person with a particular diagnosis may not prove to be the case for the next person with the same diagnosis. In other words diagnosis in psychiatry helps us describe a problem but doesn't help us understand the problem or what treatment might help. That's a much more individual thing.

Most people seem to understand and accept this sort of explanation and the subsequent implication that a 'one-size-fits-all' approach to helping that person is not particularly helpful.

Second, understanding that we have no way of discovering for certain what has 'caused' ADHD behaviours in any individual allows you to engage with 'reasons' rather than 'causes' (Drury, 2014). The difference here is that 'causes' tend to be expert derived, distal, and are thought of as being universal. In relation to ADHD, for example, the dominant bio-deterministic position conceptualises ADHD as being 'caused' by faulty genes leading to faulty brains. However, we have little evidence to support such a model, and no biological evidence is used to confirm what this supposed fault is, in any particular individual, before they receive the diagnosis. 'Reasons', on the other hand, are more everyday, proximal, and locally developed. Thus, most people will have their own 'theories' about why they/their child is behaving in a certain way. Becoming more attuned to and actively engaging with the child and/or families, 'reasoning' allows more flexible narratives that are potentially less dis-empowering to emerge in response to the 'why' question (as in why is he/she behaving like this). Reasoning can focus on immediate experiences (he is being bullied) or historical ones (she has witnessed much violence). It can vary from biological (he seems to react to certain foods) to environmental (she feels embarrassed about not being able to read like other students, so plays up so that nobody can see this). People usually have ideas about what might be 'going on', that is, resulting in certain behaviours, and these can be explored after/before/at the same time as deconstructing the bio-deterministic model of ADHD.

Third, privileging 'reasons' over 'causes' allows access to more diverse, creative, co-constructed (involving input into the treatment plan from both service users and practitioners), and individualised therapeutic approaches rather than 'one-size-fits-all' plans to emerge. Thus, plans that connect with the child and/or family's reasoning can encompass approaches such as individual therapy (such as anxiety management), family therapy (such as working with relationship insecurities), Parent Management Treatments (such as helping with parenting skills), working with schools (such as getting a learning assessment), dietary (such as trying certain supplements), and brief solution focused (such as a focus on emphasising existing strengths and positives). This can also include other specific programmes that promote a child's well-being (see e.g. Bradley & Butler, this volume). Medication can be used within this framework, but with a preference for temporary and short-term use (under one year) and with an emphasis on the ownership of change remaining with what the young person/family have done (with medication acting as an 'enabler' helping people get in touch with already existing skills they had, which they had lost touch with). For a simple summary of the implications for practice, see Table 8.5.

Table 8.5 Clinical practice highlights

---

1. Deconstruct the biological model of ADHD in order to allow space for alternative perspectives.
2. Privilege service-user 'reasons' over expert-derived 'causes'.
3. Co-construct individualised therapeutic approaches rather than 'one-size-fits-all' ADHD treatment plans.

Summary

In this chapter, we use the framework of 'social construction' to understand and reflect on how ADHD has come to be understood. We described the historical development of the concept of ADHD and then summarised the cross-cultural research literature. We also presented findings from an Internet-search-based research project that used a standard search engine to search the themes of 'What is ADHD', 'What causes ADHD', and 'ADHD treatment'. ADHD is being conceptualised in the mainstream literature and has now reached the public arena as an already-known brain-based biological disorder for which there is a specific and effective medication-based medical treatment. However, the historical and cross-cultural literature paints a more nuanced picture of changing and diverse interpretations and conceptualisations of children's hyperactivity and inattention, with changes occurring vertically as we move through the history of Western society and horizontally as we move from culture to culture. No scientific breakthrough that elucidates the proposed biological basis of ADHD has accompanied any of these noted changes and variations. This absence of evidence lends support to the hypothesis that ADHD is best conceptualised as a 'social construct' rather than a bona fide biologically based medical disorder. Finally, we explored some potential implications for practitioners of viewing ADHD as a social construct.

References

American Psychiatric Association (1966). *Diagnostic and statistical manual of mental disorders (DSM-II)* (2nd edition). Washington, DC: APA.

American Psychiatric Association (1980). *Diagnostic and statistical manual of mental disorders (DSM-III)* (3rd edition). Washington, DC: APA.

American Psychiatric Association (1987). *Diagnostic and statistical manual of mental disorders (DSM-III-R)* (3rd edition revised). Washington, DC: APA.

American Psychiatric Association (1994). *Diagnostic and statistical manual of mental disorders (DSM-IV)* (4th edition). Washington, DC: APA.

American Psychiatric Association (2013). *Diagnostic and statistical manual of mental disorders (DSM-5)* (5th edition). Washington, DC: APA.

Apple, R. (2006). *Perfect motherhood: Science and childrearing in American.* New Brunswick, NJ: Rutgers University Press.

Barkley, R. A. (2006). The relevance of the Still lectures to attention-deficit/hyperactivity disorder: A commentary. *Journal of Attention Disorders, 10*, 137–140.

Baumgaertel, A., Wolraich, M. L., & Dietrich, M. (1995). Comparison of diagnostic criteria for attention deficit disorders in a German elementary school sample. *Journal of the American Academy for Child and Adolescent Psychiatry, 34*, 629–638.

Bax, M., & MacKeith, R. (1963). *Minimal cerebral dysfunction.* Little club clinics in developmental medicine. London: Heineman.

Bradley, C. (1937). The behaviour of children receiving Benzedrine. *American Journal of Psychiatry, 94*, 577–585.

Brewis, A., & Schmidt, K. (2003). Gender variation in the identification of Mexican children's psychiatric symptoms. *Medical Anthropology Quarterly, 17*, 376–393.

Carpenter-Song, E. (2008). Caught in the psychiatric net: Meanings and experiences of ADHD, pediatric bipolar disorder and mental health treatment among a diverse group of families in the United States. *Culture Medicine and Psychiatry, 33*, 61–85.

Castle, L., Aubert, R. E., Verbrugge, R. R., Khalid, M., & Epstein, R. S. (2007). Trends in medication treatment for ADHD. *Journal of Attention Disorders, 10*, 335–342.

Clarke, J., Hall, S., Jefferson, T., & Roberts, B. (1975). Subcultures, cultures and class. In S. Hall & T. Jefferson (Eds.), *Resistance through rituals: Youth subcultures in post-war Britain.* London: Hutchinson.

Conrad, P. (1975). The discovery of hyperkinesis: Notes on the medicalization of deviant behavior. *Social Problems, 23*, 12–21.

Denhoff, E., Laufer, M. W., & Solomons, G. (1957). Hyperkinetic impulse disorder in children's behavior problems. *Psychosomatic Medicine, 19*, 38–49.

Douglas, V. I. (1972). Stop, look and listen: The problem of sustained attention and impulse control in hyperactive and normal children. *Canadian Journal of Behavioral Science, 4*, 259–282.

Drury, N. (2014). Mental health is an abominable mess: Mind and nature is a necessary unity. *New Zealand Journal of Psychology, 43*, 5–17.

Fischer, J., & Fischer, A. (1966). *The New Englanders of Orchard Town.* New York: John Wiley and Sons.

Food and Drug Administration (2005). FDA warning letter RE: NDA 21–411 Strattera (atomoxetine HCl). Retrieved June 3, 2014 from http://pharmcast.com/WarningLetters/Yr2005/Jun2005/EliLilly0605.html.

Global ADHD Working Party (2005). Global consensus on ADHD/HKD. *European Journal of Child and Adolescent Psychiatry, 14*, 127–137.

Hacking, I. (1999). *The social construction of what?* Cambridge, MA: Harvard University Press.

Harris, G., & Carey, B. (2008). Researchers fail to reveal full drug pay. *New York Times,* June 8.

Hart, N., & Benassaya, L. (2009). Social deprivation or brain dysfunction: Data and the discourse of ADHD in Britain and North America. In S. Timimi & J. Leo (Eds.), *Rethinking ADHD: From brain to culture* (pp. 218–254). Basingstoke: Palgrave MacMillan.

Lead Discovery (2004). Adult ADHD: Therapeutic opportunities. *Lead Discovery* [Online]. Retrieved May 4, 2012 from https://www.leaddiscovery.co.uk/reports/813/Adult_ADHD_Therapeutic_Opportunities.

Lindgren, S., Wolraich, M., Stromquist, A., Davis, C., Milich, R., & Watson, D. (1994). *Reexamining attention deficit disorder.* Paper presented at the 8th Annual Meeting of the Society for Behavioural Paediatrics, Denver.

Luk, S. L., & Leung, P. W. (1989). Connors teachers rating scale: A validity study in Hong Kong. *Journal of Child Psychology and Psychiatry, 30*, 785–794.

Mann, E. M., Ikeda, Y., Mueller, C. W., Takahashi, A., Tao, K. T., Humris, E., Li, B. L., & Chin, D. (1992). Cross-cultural differences in rating hyperactive-disruptive behaviours in children. *American Journal of Psychiatry, 149*, 1539–1542.

McGee, R., & Feehan, M. (1991). Are girls with ADHD underrecognised? *Journal of Psychopathology and Behavioural Assessment, 13*, 187–198.

Moncrieff, J., & Timimi, S. (2011). Critical analysis of the concept of adult attention deficit hyperactivity disorder. *The Psychiatrist, 35*, 334–338.

Schwarz, A., & Cohen, S. (2013). ADHD seen in 11% of U.S. children as diagnoses rise. *New York Times*, March 31.

Singh, I. (2011). A disorder of anger and aggression: Children's perspectives on attention deficit/hyperactivity disorder in the UK. *Social Science and Medicine, 73*, 889–896.

Sonuga-Barke, E. J. S., Minocha, K., Taylor, E. A., & Sandberg, S. (1993). Inter-ethnic bias in teacher's ratings of childhood hyperactivity. *British Journal of Developmental Psychology, 11*, 187–200.

Staller, J., & Faraone, S. V. (2006). Attention-deficit hyperactivity disorder in girls: Epidemiology and management. *CNS Drugs, 20*, 107–123.

Still, G. F. (1902). Some abnormal psychiatric conditions in children. *Lancet, 1*, 1008–1012, 1077–1082, 1163–1168.

Strauss, A. A., & Lehtinen, L. E. (1947). *Psychopathology and education of the brain-injured child*. New York: Grune & Stratton.

Taylor, E. (1994). Syndromes of attention deficit and over-activity, In M. Rutter, E. Taylor, & L. Hersov (Eds.), *Child and adolescent psychiatry, modern approaches* (3rd edition) (pp. 285–307). Oxford: Blackwell Scientific Publications.

Timimi, S. (2002). *Pathological child psychiatry and the medicalization of childhood*. London: Routledge-Brunner.

Timimi, S. (2005). *Naughty boys: Anti-social behaviour, ADHD and the role of culture*. Basingstoke: Palgrave MacMillan.

Timimi, S. (2008). Child psychiatry and its relationship to the pharmaceutical industry: Theoretical and practical issues. *Advances in Psychiatric Treatment, 14*, 3–9.

Timimi, S. (2009). *A straight talking introduction to children's mental health problems*. Ross-on-Wye: PCCS Books.

Timimi, S. (2013). No more psychiatric labels: Campaign to abolish psychiatric diagnostic systems such as ICD and DSM (CAPSID). *Self and Society, 40*(4), 6–14.

Timimi, S., & Leo, J. (Eds.) (2009). *Rethinking ADHD: From brain to culture*. Basingstoke: Palgrave MacMillan.

Recommended reading

- Timimi, S. (2005). *Naughty boys: Anti-social behaviour, ADHD and the role of culture*. Basingstoke: Palgrave MacMillan.
- Timimi, S., & Leo, J. (Eds.) (2009). *Rethinking ADHD: From brain to culture*. Basingstoke: Palgrave MacMillan.
- Timimi, S. (2012). Children's mental health in the era of globalisation: Neo-liberalism, commodification, McDonaldisation, and the new challenges they pose. In V. Olisah (Ed.), *Essential notes in psychiatry*. Rijeka: InTech.

9

Moral Evaluations in Repertoires of ADHD

Mary Horton-Salway and Alison Davies

Introduction

Attention deficit hyperactivity disorder (ADHD) is an inclusive category defined by the fifth edition of *Diagnostic Statistical Manual of Mental Disorders* (*DSM-5*) as a mental health disorder affecting children and adults (American Psychiatric Association, 2013). There is a history of scepticism about ADHD, and as Singh (2008) suggested, the controversy focuses on validity of its existence, diagnosis, causes, and the ethics of medicating children. Variation in prevalence rates, both between countries and within countries, suggests that ADHD is as much a cultural construction as a medical one (Davies, 2014). Nonetheless, much of the research into ADHD has focused on cognitive, neurobiological, and genetic explanations (Cooper, 2008), and much less attention has been given to the cultural constructions that define ADHD or the experience of families who are constrained by them (Davies, 2014).

In this chapter, based on our research findings, we focus on how the category of ADHD is produced in discourse and associated with moral evaluations produced in media stories and drawn on by parents who have a child with a diagnosis of ADHD. We examine how the social identities of these children and their parents are embedded in different explanations for the causes and management of ADHD, and we identify the dilemmas arising when parents talk about their children. Taking a critical discursive psychology approach, we treat social identities as fluid subject positions, taken up or attributed to others in interactions and textual reports, as part of versioned accounts, descriptions, explanations, and arguments. We demonstrate how biological and psychosocial explanations are used in competing explanations (interpretative repertoires) to construct the category of ADHD; to attribute cause, blame, or accountability; and to construct normality and abnormality in children and position parents and children within non-coping and failing families. Our analysis of parents' accounts of their personal experience will show how they refute negative identities and how they draw on competing repertoires to manage dilemmatic issues arising from the ADHD debate.

Project overview

Alison Davies' (2014) study of parents' accounts of ADHD and Mary Horton-Salway's study of ADHD in the media apply a critical discursive psychology approach to identify a patterned relationship between competing repertoires of ADHD, subject positions, and moral evaluations that constitute the meaning of ADHD. Using 13 audio-recorded interviews and 2 focus groups, recruited through ADHD support groups, Davies analysed parents' accounts of children who have a diagnosis of ADHD, their parenting practices, and their experience of interacting with schools and medical practitioners. Based on a qualitative analysis of her data set, Davies applied a synthetic approach bringing together an analysis of both the situated discursive constructions of participants and the cultural resources they use to construct meaning.

Clinical and educational relevance

Competing interpretative repertoires of ADHD invoke parental accountability (or lack of it) for their children's ADHD, placing them at the centre of a highly moralised debate. Encounters with medical and educational practitioners are therefore likely to be morally risky for parents. The identity of the good or failing parent is at stake in discourse about ADHD such that health and education practitioners need to recognise what dilemmas parents face in talking about their child's behaviour and their parenting. Key to this, for the reflective practitioner, is an appreciation of how parents are positioned differently by biological and psychosocial explanations of ADHD, as well as the routine forms of gendering that currently underpin the discourse of ADHD.

Moral evaluations in repertoires of ADHD

As much as changing historic definitions have contributed to the existing vagueness around the category of ADHD, a pervasive concern with ADHD as a moral category has threaded itself throughout this discourse in changing formulations of the meaning of hyperactive and impulsive behaviour (see Rafalovich, Chapter 5, this volume). Moral evaluations continue to be woven throughout current literature relating to ADHD (Schubert, Hansen, Dyer, & Rapley, 2009; Rafalovich, 2008), and the category of ADHD remains controversial due to competing explanations for the causes of ADHD and the debate regarding the parenting and medication of children (see Timimi &

Timimi, Chapter 8, this volume). Contrary biological versus psychosocial explanations of ADHD are treated here as argumentative resources in the moral discourse of ADHD, and through a discussion of research on media representations and parents' accounts, we set out to examine what social and discursive actions are accomplished by their use. We examine how these competing interpretative repertoires are used to perform a range of discursive actions such as making truth claims, causal attributions, and constructing social identities. Interpretative repertoires (Edley, 2001; Gilbert & Mulkay, 1984; Potter & Wetherell, 1987) are 'the building blocks used for manufacturing versions of actions, self and social structures in talk...resources for making evaluations, constructing factual versions and performing particular actions' (Wetherell & Potter, 1992, p. 90). As Billig et al. (1988) argued, when we explain events or phenomena, explanations perform moral evaluations, although these are not fixed attitudes. When competing versions of the causes of ADHD are used in everyday talk and text, they are versatile, indexical and are used to construct different explanations for ADHD on different occasions. Within such explanations, different subject positions for both children and parents are attributed and taken up and work to support different versions of the meaning of ADHD.

For example, in a study of UK newspaper stories, Horton-Salway (2011) identified two different repertoires of ADHD typically used in public discourse. The biological repertoire draws on a biological deficit explanation for children's bad behaviour to construct the abnormal child: this is typically constructed as factual using the language of science (Gilbert & Mulkay, 1984). For example, Henderson and Hawkes (2004) wrote in *The Times*: 'Children with attention deficit hyperactivity disorder...are suffering from a medical condition linked to abnormal development of the brain'. Abnormality is also constructed in detailed accounts of the extreme behaviours of children with diagnosis of ADHD; for example, Palmer (2006, p. 35) wrote in *The Express* of parents who described their anxiety and stress because of their son's extreme naughtiness. His 'abnormality' was described as an example of 'an estimated 400,000 children in the United Kingdom with ADHD, thought to be caused by an imbalance of chemicals in parts of the brain that deal with attention, impulses and concentration'. These accounts typically depict extreme disruption in families, out-of-control children, and a justification for medication treatment of 'abnormal' children. The biological repertoire is used to construct both 'abnormal' children and their non-coping parents as victims. Typically, in such accounts 'the subject position of abnormal child...is concurrent with the construction of the need for medical intervention and drug treatment' (Horton-Salway, 2011, p. 539).

However, the psychosocial repertoire is most commonly used to explain ADHD in UK newspapers providing for an entirely different set of subject positions for children and their parents, linked to a media critique of medicalisation and a claim that environmental factors rather than biological ones underpin the psychological and social causes of children's poor behaviour (Horton-Salway, 2011). In the psychosocial repertoire, the subject positions of sick or 'abnormal' child are contested through a critique of inadequate parenting, schools lacking discipline and accounts of environmental pollutants. The psychosocial repertoire links to contrasts between the 'good old days' of sound discipline and wholesome lifestyles and the declining moral standards of contemporary society (Horton-Salway, 2011). A 'meta-narrative' of decline used in the media constructs moral panics about the state of society (Seale, 2003), and in stories of ADHD, parents are represented as both perpetrators and victims of societal decline with the behaviour of children as symptomatic of that decline. Children with a diagnosis of ADHD are positioned as innocent victims of over-prescription and the interests of pharmaceutical companies, while their parents are positioned as lacking time, skill or willingness to provide adequate parenting (Davies, 2007). In such accounts, the category of ADHD is constructed as a bandwagon, serving as an excuse for ordinary but extreme naughty behaviour to be interpreted as individual pathology (cf. Horton-Salway, 2007).

Although the biological and psychosocial repertoires are competing explanations for the nature and causes of ADHD, they were both used to construct a prescriptive need for interventions with families, so they each have a moral function in discourse. Rafolovich points out that both repertoires work as 'disciplinary mechanisms', are well designed to give moral imperatives, and prescribe advice or interventions of some kind (Rafalovich, 2001, p. 373). The two repertoires draw on similar examples of children's extreme naughty behaviours to validate opposing claims: the facts of their behaviours or 'symptoms' do not speak for themselves, but are versioned through narratives that polarise into moral arguments, making parents accountable to family interventions. Interpretative repertoires on ADHD are thus both prescriptive and regulatory and are fluid in that they can be used in different ways, and they are associated with a problem to be solved, a solution to be found, and the accountability of parents. Key to understanding this is an appreciation of how the psychosocial repertoire aligns ADHD with ineffective parenting and, consequently, places parents of children with ADHD at the centre of a highly moralised debate.

Parenting and ADHD

The moral positioning of parents, and particularly mothers, in relation to their child's ADHD has been well documented. Research based on mothers' accounts shows an orientation to themes of responsibility and self-sacrifice (Austin & Carpenter, 2008; Singh, 2004), mothers are judged in relation to their children's behaviour and repertoires of maternal blame, and the subject position of the 'blameworthy mother' are frequently made relevant by the mothers of children with ADHD (Davies, 2014). Bennett (2007) suggested that while the 'blameworthy' mother is a prevalent subject position in psychological and popular discourse, the medical explanation will always be a desirable way to avoid being positioned in this way (see also Malacrida, 2001, 2004; Singh, 2002, 2003, 2004). Mothers' take-up of the biological repertoire is an inevitable consequence. However, research shows that fathers are more reluctant to accept the medical model and are more likely to normalise their children's behaviour (Singh, 2003). The following analysis of Davies' (2014) interviews with parents, examines how parents draw on biological and psychosocial repertoires in their accounts of ADHD, arguing that each repertoire offers distinct moral interpretations of the relationship between ADHD and parenting. In managing competing explanations for the causes of ADHD, parents produce very distinct social identities for themselves and for their children, which provide for distinctly different forms of solution and ADHD management.

Parents work to represent themselves as good and morally adequate parents, subject positions that are typically but not always associated with the biological repertoire. The biological repertoire works to attribute cause to the biological origins of ADHD and Davies found that it is the dominant one in mothers' discourse, which we notice is the opposite of what Horton-Salway found in media accounts, where the dominant repertoire is psychosocial. We note however, that the fathers in Davies' study position *other* parents as the ones lacking good parenting skills. The observers of family life (such as the media) tend to attribute blame to family members (the inadequate parent and the naughty child) while the parents themselves tend to attribute cause outside themselves. Such attributions are treated here as discursive, rhetorical phenomena produced on and for an occasion, rather than a cognitive phenomenon where such arguments represent fixed internal attributional bias. A further complicating issue in parents' accounts was the distinctly gendered subject positions that were commonly taken up by the mothers and fathers in relation to their child's health and discipline (Davies, 2014; Horton-Salway, 2013). We show how these positions are deployed by parents to represent themselves as a competent and knowledgeable parenting team.

The biological repertoire

In Extract 1, we see how Paula uses the biological repertoire to formulate ADHD as a condition that is caused by brain malfunction.

Extract 1: The Brain Explanation (Paula)

It's a problem with the brain it's something to do with the uhm frontal cortex or something of the brain and the frontal lobes or something not working properly [...] and it's a neural developmental problem and and er medication helps put those missing bits back together so they can get all of the message instead of part of it (laughs).

Paula locates the source of ADHD in the brain and draws upon clinical vocabulary to describe the areas of the brain that malfunction. The pathology of her son's condition is constructed here using a three-part list 'it affects the brain and it's a neural developmental problem and and er medication helps put those missing bits back together'. This is a rhetorical device to construct ADHD as a scientifically robust biological category and establishes the use of medication as appropriate treatment. Such explanations are commonly used in conjunction with genetic accounts.

Extract 2: The Genetic Explanation (Paula)

Yes I do believe it's genetic uhm [...] it must be in my side of the family because my sister's got a boy who's got the same [...] but when I look back at my mum [...] what I know about Asperger's and Autism and stuff I'm 99% sure that she has Asperger's [...] yeah I I di-I identified-me when I think back uhm I do think there's definitely the Autism is in our family and I even think part the ADHD is there as well although when I was a child I never sat still although I read a lot I would be I'd have music on or I'd be fiddling or I'd be doing something.

Paula describes a dual diagnosis of autistic spectrum disorder (ASD) and ADHD that she traces back through the family as evidence for its genetic origin. Examples of three family members with 'atypical' behaviour, her sister, her mother, and herself, are scripted as a three part list of predicates of ADHD that work to build a genetic claim. Her own claim to category membership adds further substance to the family history: 'I never sat still [...] I'd have music on or I'd be fiddling'. Words such as 'would' and 'never' are used to script these predicates of ADHD as typical for her but as anomalous to the 'normative base' of childhood behaviours and comparable with those of her child.

Extracts 1 and 2 show how brain and genetic explanations are constructed along with abnormality. A biological understanding of ADHD functions to construct children's behaviour as pathological 'abnormality', rather than the result of poor parenting. It is unsurprising therefore, that parents do rhetorical work to establish pathological abnormality to avoid being treated as bad parents. Extract 3 is an example of this.

Extract 3: Scripting the 'abnormal' child (Donna)

I used to take him to all the little pre-schools [...] I was the only parent running about after him (laughs) [...] I did all the right things you know [...] we did tumble tots that was another thing and I thought that would be great-the worst day I've ever had I took him-cos they're meant to line up [...] and he's running about off there and he's pulling it down and he's kicking it and he's pushing in [...] eventually I did get him diagnosed uhm that was really hard I had the same problems with my doctor 'this is normal this is what children do' [...] well my <u>other</u> son never did. Listen my other son's 13 and he's not got any problems apart from teenage attitude.

Normality and abnormality are produced in relational contrast with one another (Potter,1996) and here the 'abnormality' of Donna's son's behaviour is produced in contrast to the normal requirement 'to line up' at pre-school. A recognisable 'ADHD script' is also used to describe the boy's behaviour as dispositional and typical by the use of 'used to' and 'he'd' suggesting that these are not isolated events (Edwards, 1994). Evidence of his hyperactivity is produced in a detailed account 'I used to come back like that (demonstrates flagging) he'd still be on the go'. This constructs Donna's version of events as a factual and accurate account (Parker & O'Reilly, 2012) and the contrast between herself and other mothers is made using extreme case formulations: 'the worst day ever'; 'I was the only parent running about'; and in contrast to other children, 'my <u>other</u> son never did' (Pomerantz, 1986).

Donna's 'other' son provides contrasting evidence of her ability to distinguish between normal and abnormal behaviour (Potter, 1996), providing an alternative construct of recognisable normal childhood and 'teenage attitude'. The ability to recognise differences between normality and abnormality is important since the issue of medication is embedded within a debate in which biological abnormalities call for medical treatment while normal but naughty children call for better parenting. Parents typically work to justify their decision to medicate using such categories.

Extract 4: The Dilemma of Medication (Rachel)

We didn't go for diagnosis because I didn't see the purpose of it because I wasn't going to medicate because we could manage his behaviour [...] but it was when school started saying 'I'm concerned about his education cos he's really bright and not performing' and then we did the diagnosis [...] so we thought there's nothing to lose and once we as I say we started [....] the difference it made it's just his report at the end 'positive attitude to learning he's contributing' it wasn't that there was no problems but the difference in him.

The management of stake, or interest, is also used by parents to justify their decisions to medicate or not (Edwards & Potter, 1993; Potter, 1996). Parents' truth claims could be undermined if they were thought to have a personal interest in getting a diagnosis and medication for their child. Some media accounts construe ADHD and medication as a bandwagon (Horton-Salway, 2011), and parents need to demonstrate that they had an initial reluctance to opt for this solution. Such is Rachel's display of initial resistance to medication in Extract 4, above: 'I didn't see the purpose of it because I wasn't going to medicate because we could manage his behaviour'. This shifts the blame for lack of behaviour management and defends against accusations of using medication to avoid the demands of parenting. Rachel's confession that her initial viewpoint was incorrect is paired with positioning the school as the initiator of a referral, 'school started saying "I'm concerned about his education "cos he's really bright and not performing"'. She contrasts the effects of her own 'mistaken' decision not to medicate against the positive effects of medication demonstrated by her son's subsequent performance at school 'the difference it made it's just his report at the end, "positive attitude to learning"'.

The effect of medication on school performance is often given as justification for accepting medication, albeit reluctantly. Accounts of unanticipated, positive effects of medication, work as stake inoculation in parents' accounts confirming ADHD as a biological fact that was neither sought nor mis-attributed to their child by themselves. This example of a 'truth will out' device (Gilbert & Mulkay, 1984) is used by parents to validate a truth claim by citing positive outcomes that both validate the decision to medicate and strengthen the legitimacy of Rachel's argument. The credibility of her account rests upon her ability to demonstrate initial scepticism and position the school as the agent of referral.

Using the psychosocial repertoire

The extracts above illustrate how parents draw on biological and medical explanations when accounting for their child's ADHD. Although the parents in Davies' study typically used the biological repertoire, they also acknowledge psychosocial explanations for ADHD as demonstrated in Extracts 5 and 6.

Extract 5: Undermining the sceptics (Gill)

Gill: It seemed to be everywhere and lots of the newspapers and TV programmes at the time [...] 'is ADHD a real condition? [...] 'is it just bad parenting?' and I know that debate was going on then and I didn't read too much about it because I didn't have kids I was 21 and I thought oh ok it sounds like a you know you read a few excerpts and I said yes it sounds like bad parenting to me and I did probably think uhm think like that and it wasn't until later on when I was reading more of the scientific research [...] so I looked into a- 'ok actually I'm going to change my opinion'.

In Extract 5, Gill outlines how ADHD was debated some 10 years ago as being either a 'real condition' or 'just bad parenting'. She provides an account of her former sceptical self, who was not pre-disposed to see ADHD as a medical cat-egory. This allows her to set up the psychosocial explanation for ADHD as a mistakenly held former viewpoint and describe her new viewpoint as a con-version based on experience. She undermines the knowledge of the sceptics by aligning herself with them and then suggesting that her youth and lack of expe-rience with children was to blame for her former misinformed position (Billig, 1996).

Her original lack of interest in ADHD as a medical category follows the discursive pattern 'at first I thought, but then I realised', a structure used to construct the current rationality and neutrality of the speaker (Wooffitt, 1992). Thus, Gill's common sense explanation of ADHD as caused by bad parenting is discredited by unusual, but empirical facts, which are now impossible to ignore. Gill's naïve ignorance of ADHD is demonstrably altered by the scien-tific evidence and this process is contrasted to the lack of scientific rigour of the popular media 'it seemed to be everywhere and lots of the newspapers and TV programmes at the time'. She refers to 'people' 'writing about' ADHD, thereby attributing to them a lower status than scientists and whose accounts are less credible than her own. Here, non-scientific and popular misunderstandings are attributed to others, while she claims for herself properly considered scientific knowledge and experiential understanding of ADHD. The extracts above have indicated that parents' descriptions of ADHD perform two discursive functions. First, parents work up descriptions of ADHD constructing ADHD as a biological reality. Second, parents undermine sceptics whose psychosocial explanations of ADHD invoke environmental, social, and parental responsibility.

Disciplining fathers, expert mothers

There is a danger of setting up a dialogic polarity between the two repertoires of ADHD, suggesting parents only draw on the biological repertoire to undermine the psychosocial repertoire. This ignores the indexical, contingent nature of discourse. For example, parents may also invest in a psychosocial repertoire to make the point that there are differences between families in their skills and efforts to manage their children's behaviour. A contrast between the categories of 'interested' and 'disinterested' parents was drawn by one of the fathers in Davies' study (2014):

> [S]ome parents might-might not be interested [...] I'm very interested in his-his I mean his upbringing and his uhm doing the best so I'm interested in ADHD from that point of view and ... I mostly get it right.
>
> (Mick)

Although parents might concede that psychosocial factors play a part in ADHD, they also work, like Mick, to position themselves as good parents rather than bad ones. While the psychosocial repertoire suggests that ADHD is caused by an absence of discipline in the home, many take up the subject position of the 'disciplining father' to establish the identity of competent father. Extract 6 is a co-constructed, account in which the traditional gendered subject position of the authoritative father is made relevant.

Extract 6: The disciplining father (Ingrid and John)

Ingrid: yes 'cos sometimes [.....] you'll say ' ah well I'll let him get away with it because he has ADHD' but then I think 'well actually no it's still not acceptable' but it's really difficult.

John: 'yeah I'll set fire to the building but I've got ADHD'

Ingrid: yeah I find it very hard to draw the line [....]

John: [...] I've got I've got very strict boundaries of behaviour and this that and the other and I'm being right or wrong I don't care if you've got ADHD ABC or 123 [....]

Ingrid: but he definitely pushes a lot more with me

John: oh he does with you yeah

Discipline is a key topic for parents talking about their children with ADHD and deciding what can be accounted for by ADHD, and what is due to ordinary naughtiness. Ingrid implies that each requires a different parenting response. However, John follows Ingrid's turn with an ironic 'yeah I'll set fire to the building but I've got ADHD'. Following Ingrid's admission that she is more lenient because the child has ADHD, John's irony works to construct her position as mistaken and allows him to distance himself, making it clear that he will not use a medical label as an excuse for his son's behaviour: ' I don't care if you've got ADHD ABC or 123'. Throughout this extract, John takes up the position of a traditional disciplining father with firm boundaries in contrast to Ingrid's softer position. Although they work together to establish John as the stricter parent, Ingrid's softness is mitigated by John's strictness, just as the 'less knowledgeable father' is mitigated by the 'expert mother' elsewhere (Extract 7 below). These contrasts work in tandem to construct a skilled and effective parenting team to offset any suggestion that their parenting might be lacking. In Mick's account below and in Gill's it is expertise in ADHD and dealing with school authorities that is constructed as the important parenting attribute.

Extract 7: Expert Mother, interested father (Gill and Mick)

Gill: I seem to be uhm a project manager and I think you find that a lot with the parents it's-and I felt 'do you know what he's my son so it's my job to liaise with school and do my own research and do-find out from psychiatrists and I'm the one who pulls the information together for him.

Mick: I'm very interested in his-his I mean his upbringing and his uhm doing the best so I'm interested in ADHD from that point of view but if it wasn't for Gill's interest in the field in general we-we wouldn't be where we are and I think Gill's been instrumental in-really in his upbringing and I try to emulate what she does you know.

Gill describes herself as the parent responsible for liaison with the school and her role is produced using the language of the workplace, 'project manager' and 'my job to liaise'. Mothers often draw upon technical language in their accounts of managing interactions with schools and medical experts. This constructs their expert parenting of special children, who require special parenting approaches fostered by specialist knowledge. Gill indicates a privileged status in relation to her son in relation to the school and establishes her entitlement to assume responsibility and be an expert about him. Similarly, Mick positions Gill as the knowledgeable parent in combination with himself as an 'interested' father who is supporting her position.

We have seen how constructing abnormality reifies a biological explanation and provides for good parent identities; how parents also draw on the psychosocial repertoire in accounts of coping and managing challenging behaviour in good parenting teams. However, describing their children as 'abnormal' invites medical or social intervention and we can see how parents orient to this possibility in constructing the normality of their children and family life.

Challenging children in normal families

When describing their child, parents work to manage a number of dilemmas that are associated with the label, ADHD. One of these is the risk carried by defining a child as 'abnormal' in order to make a case for ADHD. Supporting previous findings by Singh (2003), Extract 8 shows how fathers normalise their sons' behaviours by identifying with them. Extract 9 demonstrates how mothers work to construct the challenge of having a child with ADHD, using their own specialised abilities to parent such children and talk about the normality of doing so. Accounts such as these typically emphasise the virtues of children with special needs and special gifts.

Extract 8: The hyperactive father (Mick)

For myself uhm I'm when I'm working there can be a million and one things [...] you know lots and lots of things going on in my mind and I I had to train myself to concentrate on 'do this' or 'do that' [...] you know back to the main task and is it it's it's a what-do-you-call it? a discipline you know and I've had to learn it [...] I I sometimes work on four computers at once uhm and I always have done [....]. but I've always got to discipline my mind to focus on what's important you know whereas... without that discipline I'd be a scatterbrain.

Extract 8 shows Mick expressing concern about medicating his son and scripting his own 'hyperactive' behaviour as typical and routine, rather than dysfunctional. Mick describes how he has managed to control the negative effects of his hyperactive behaviours through self-discipline and made these work to his advantage. His difficulties, such as lack of concentration, lack of focus, and hyperactivity, suggest identification with his son's ADHD. However, he mentions them to warrant a knowledge entitlement and informed opinion based on his own personal experience of having ADHD-type behaviours; Mick is careful not to claim the label for himself, instead suggesting that without discipline he would be a bit of a 'scatterbrain'. Mick's entitlement to knowledge of ADHD allows him to take up an informed position on the issue of medication and normalise his son's behaviour (see Sacks, 1995). Fathers often position themselves in this way and frequently define hyperactivity as an attribute of normal masculine behaviour (Davies, 2014; Horton-Salway, 2013). Such constructions also support a genetic explanation for ADHD-type behaviours thus undermining the psychosocial explanation that ADHD is caused by deficit parenting and environment. By normalising the predicates of ADHD and constructing them as useful attributes, fathers also construct normality and positive trajectories for their child's education and career.

Identification with their sons' behaviours can be seen as fathers' attempts to resist their sons being positioned within a pathological narrative and to enrol them in the more common category of normal active boy. Resisting a tragic narrative associated with abnormality also features strongly in mothers' accounts as in Extract 9.

Extract 9: Emphasising the positive (Kim)

uhm (pause) I don't know uhm I sometimes feel that it.. on the whole I feel quite positively about it really because I think we can deal with it uhm so in that way I'm sort of glad he's born into our family cos we're equipped to deal with it so I'm pleased for him because it would be so much worse if he (laughs) was in a chaotic uhm.

In this extract, Kim talks positively about the 'fit' between her son and the family. This family 'fit' resonates with Broberg's (2011) findings in relation to parental reactions to intellectual disability, in particular, one which framed acceptance of their children's difference within a repertoire of belonging. This,

Broberg (2011) suggested, is a way of resisting the social and cultural expectation that families who have 'atypical' children will necessarily feel sad and bereaved. Such families are frequently positioned within a dominant 'tragedy' narrative. As we have seen, the biological repertoire makes available a deficit account of a child that risks being associated with a 'tragic' subject position for the child and for the family. Medical abnormality and sickness inevitably confer sympathy and moral scrutiny of the family, especially of the mother where children are concerned. In the face of so much moral evaluation, the biological repertoire is clearly an appealing and possibly defensive one for mothers to take up.

However, the biologically 'abnormal' child becomes the focus of institutional regulation, scrutiny, and intervention, and this too extends to the family. Positioning individuals within a 'tragic' narrative, potentially risks formulating them as lacking in 'ordinary' worth; they are related to differently, and held to account. Some parents respond by resisting the tragedy narrative, emphasising the normality of their child and family life, despite the 'difference' of their child. Broberg (2011) considered that through emphasising ordinariness, parents assume an active subject position, one which permits them the possibility of 'normality, involvement and mastery' in relation to their parental duty of care. However, Kim's account of acceptance appears troubled by her use of hedging devices and modifying phrases: 'I don't know uhm I sometimes feel that it...on the whole I feel quite positively about it really because I think we can deal with it uhm so in that way I'm sort of glad he's born into'. The hesitancy and vagueness of Kim's account may indicate some difficulty in taking up a position of acceptance in relation to the child's ADHD. Although there is some moral imperative to think positively with regard to health and illness (Radley & Billig, 1996; Wilkinson & Kitzinger, 2000), acceptance of it could equally be defined as defeatism or ignorance of the special needs of their child. Furthermore, the moral imperative to think 'positively' does not easily permit a 'good' mother to express ambivalence about her child when the culturally dominant message is that a good mother loves her child unconditionally. Kim's description of accepting her son is therefore not to be understood merely as a transparent representation of a fixed attitude or accurate description of a coping family. Rather, all of the parents' accounts can be heard as making sense of their lives and experiences in the everyday negotiation of their own and their family identities. The identities constructed by the parents in this study are not static, but fluid and indicative of the very complex ways in which they 'negotiate and understand their children' within the shifting demands of the local interaction (Goodley & Tregaskis, 2006). As we have seen, parents move between constructing abnormal and normal parent–child dynamics to formulate different child and parental identities, all of which, within different discursive contexts, make relevant the good parent subject position and the normality of abnormal situations.

Clinical and educational relevance summary

The analysis in this chapter has several implications for both clinical and educational practice. First is that the moral positioning of parents is concurrent with the polarisation of biological and psychosocial explanations such that the ADHD debate turns upon an axis of moral evaluations about parenting. Families can be positioned in negative ways in the discourse of ADHD. In using biological and psychosocial explanations, parents are faced with the dilemma of positioning either their child as 'abnormal' or themselves as failing parents. However, these sometimes competing or entrenched explanations of ADHD need not be treated as fixed attitudes or beliefs held by parents, but attempts to manage dilemmas that arise in accounting for their children and their parenting.

Second, parents and professionals are equally constrained by existing culturally available repertoires of ADHD and practices that embody gendered and stereotypical cultural ideas, such as the focus on 'naughty boys', 'blameworthy mothers', and the relative 'absence' of girls and fathers in research on ADHD. This was a notable feature in Davies' data set that has its counterpart in the under-diagnosis of girls relative to boys and the 'absence' of fathers in clinical and educational encounters. Such gendering is embedded in research and practice and deserves more attention in research, medical, and educational contexts. Greater focus on parenting teams, fathers' involvement in parenting, and the consequences of undiagnosed 'inattentive type' ADHD in girls, would avoid focusing on the 'blameworthy mother' and 'naughty boys'.

Third the issues arising from interview research could be further developed into the contexts of medical, social work, and educational encounters to explore more fully the situated processes of referral, case conference, and diagnosis. Studying the micro politics of parent–professional encounters could inform reflective practice and facilitate a better understanding of how mental health problems and failing families are constructed, defined, and resisted by both parents and practitioners. For a simple summary of the implications for practice, see Table 9.1.

Summary

The accounts of parents in Davies (2014) contribute to our understanding of the dilemmas raised for them by the ADHD debate. We argue that mental health categories such as ADHD do not refer to objective and morally neutral scientific truths, but these are constructed through cultural representations and discourse

Table 9.1 Clinical and educational practice highlights

1. Clearly, families may be positioned in negative ways through the discourse of ADHD, and the use of biological and/or psychosocial explanations poses dilemmas for parents in positioning their child as 'abnormal' or positioning themselves as failing parents.
2. The culturally available repertoires of ADHD constrain both parents and professionals.
3. A greater focus on parenting teams, father's involvement in parenting children with ADHD, including girls, may avoid focusing on the relationship between 'blameworthy mothers' and 'naughty boys'.
4. The issues identified in this research could be applied to a wider range of contexts including medical, social work, and educational settings to explore in more detail the situated processes of referral, case conference, and diagnosis.

and are associated with moral evaluations and subject positions for children and their parents.

We have argued that dominant media understandings of ADHD align the category with ineffective parenting (Horton-Salway, 2011, 2013), while the accounts of parents overwhelmingly construct a different account of coping, caring, cooperation and expertise (Davies, 2014). The two different repertoires, biological, and psychosocial, hold parents accountable to a highly moralised debate, which is all too evident in their own responses. It was found that the biological repertoire is used both by parents and in UK media to construct the subject position of the abnormal child and the justification for medication, while the psychosocial repertoire is used to challenge the medicalization of normal-but-naughty children, and criticise inadequate parenting and declining standards of discipline. Both repertoires are used variably by parents and to justify or dismiss a need for interventions in family life. While this need is frequently recognised by parents, mothers construct themselves as experts in collaboration with schools and medical practitioners, and fathers work to dismiss the implications of individual pathology or deficient parenting. Fathers identify with their sons and represent them as normal, lively children, positioning 'other' parents as deficient, in contrast to their own effective family parenting team.

References

American Psychiatric Association (2013). *Diagnostic and statistical manual of mental disorders (DSM-5)*. Washington, DC: American Psychiatric Association.

Austin, H., & Carpenter, L. (2008). Troubled, troublesome, troubling mothers: The dilemma of difference in women's personal motherhood narratives. *Narrative Inquiry, 18*(2), 378–392.

Bennett, J. (2007). (Dis)ordering motherhood: Mothering a child with attention deficit hyperactivity disorder. *Body Society, 13*(4), 97–110.

Billig, M. (1996). *Arguing and thinking: A rhetorical approach to social psychology.* Cambridge: Cambridge University Press.

Billig, M., Condor, S., Edwards, D., Gane, M., Middleton, D., & Radley, A. (1988). *Ideological dilemmas: A social psychology of everyday thinking.* London: Sage.

Broberg, M. (2011). Expectations of and reactions to disability and normality experienced by parents of children with intellectual disability in Sweden. *Child: Care, Health and Development, 37*(3), 410–417.

Cooper, P. (2008). Like alligators bobbing for poodles? A critical discussion of education, ADHD and the biopsychosocial perspective. *Journal of Philosophy of Education, 42*(3–4), 457–474.

Davies, B. (November 19, 2007). The scandal of kiddy coke. *Daily Mail,* p. 19.

Davies, A. (2014). *'It's a problem with the brain': A discursive analysis of parents' constructions of ADHD* (unpublished PhD thesis). Milton Keynes: The Open University.

Edley, N. (2001). Analysing masculinity: Interpretative repertoires, ideological dilemmas and subject positions. In M. Wetherell, S. Taylor, & S. J. Yates (Eds.), *Discourse as data: A guide for analysis* (pp. 189–228). London: Sage.

Edwards, D. (1994). Script formulations: An analysis of event descriptions in conversation. *Journal of Language and Social Psychology, 13*(3), 211–247.

Edwards, D., & Potter, J. (1993). Language and causation: A discursive action model of description and attribution. *Psychological Review, 100*(1), 23–41.

Gilbert, G. N., & Mulkay, M. (1984). *Opening Pandora's box: A sociological analysis of scientists' discourse.* Cambridge: Cambridge University Press.

Goodley, D., & Tregaskis, C. (2006). Storying disability and impairment: Retrospective accounts of disabled family life. *Qualitative Health Research, 16*(5), 630–646.

Henderson, M., & Hawkes, N. (September 9, 2004). Brain scans show it's not always easy to be good. *The Times* (London), Home News, p. 11.

Horton-Salway, M. (2007). The ME Bandwagon and other labels: Constructing the authentic case in talk about a controversial illness. *British Journal of Social Psychology, 46*(4), 895–914.

Horton-Salway, M. (2011). Repertoires of ADHD in UK newspaper media. *Health (London), 15*(5), 533–549.

Horton-Salway, M. (2013). Gendering attention deficit hyperactivity disorder: A discursive analysis of UK newspaper stories. *Journal of Health Psychology, 18*(8), 1085–1099.

Malacrida, C. (2001). Motherhood, resistance and attention deficit disorder: Strategies and limits. *Canadian Review of Sociology and Anthropology, 38*(2), 141–165.

Malacrida, C. (2004). Medicalisation, ambivalence and social control: Mothers' descriptions of educators and ADD/ADHD. *Health: An Interdisciplinary Journal for the Social Study of Health, Illness and Medicine, 8*(1), 61–79.

Palmer, J. (December 19, 2006 1st edition). The softly-softly season; your health – This time of year used to be a nightmare for the Thomson family, whose son Robin has disruptive behavioural problems. Here, his mother Ruth tells JILL PALMER how they have learned to manage their boy. *Express,* Features, p. 35.

Parker, N., & O'Reilly, M. (2012). 'Gossiping' as a social action in family therapy: The pseudo-absence and pseudo-presence of children. *Discourse Studies, 14*(4), 457–475.

Pomerantz, A. M. (1986). Extreme case formulations: A new way of legitimating claims. *Human Studies, 9*, 219–230.

Potter, J. (1996). *Representing reality: Discourse, rhetoric and social construction.* London: Sage.

Potter, J., & Wetherell, M. (1987). *Discourse and social psychology: Beyond attitudes and behaviour.* London: Sage.

Radley, A., & Billig, M. (1996). Accounts of health and illness: Dilemmas and representations. *Sociology of Health and Illness, 18*(2), 220–224.

Rafalovich, A. (2001). Disciplining domesticity: Framing the ADHD parent and child. *The Sociological Quarterly, 42*(3), 373–393.

Rafalovich, A. (2008). *Framing ADHD children: A critical examination of the history, discourse and everyday experience of attention deficit/hyperactivity disorder.* Plymouth: Lexington Books.

Sacks, H. (1995). *Lectures on conversation.* Oxford: Blackwell Publishers Ltd.

Schubert, S., Hansen, S., Dyer, K., & Rapley, M. (2009). 'ADHD patient' or 'illicit drug user'? Managing medico-moral membership categories in drug dependence services. *Discourse & Society, 20*(4), 499–516.

Seale, C. (2003). Health and media: An overview. *Sociology of Health & Illness, 25*(6), 513–531.

Singh, I. (2002). Bad boys, good mothers, and the miracle of Ritalin. *Science in Context, 15*(4), 577–603.

Singh, I. (2003). Boys will be boys: Fathers' perspectives on ADHD symptoms, diagnosis and drug treatment. *Harvard Review of Psychiatry, 11*(6), 308–316.

Singh, I. (2004). Doing their jobs: Mothering with Ritalin in a culture of mother-blame. *Social Science and Medicine, 59*(6), 1193–1205.

Singh, I. (2008). Beyond Polemics: Science and ethics of ADHD. *Nature Reviews Neuroscience, 9*, 957–964.

Wetherell, M., & Potter, J. (1992). *Mapping the language of racism.* Chichester, NY: Columbia University Press.

Wilkinson, S., & Kitzinger, C. (2000). Thinking differently about thinking positive: A discursive approach to cancer patients' talk. *Social Science and Medicine, 50*(6), 797–811.

Wooffitt, R. C. (1992). *Telling tales of the unexpected: The organization of factual discourse.* London: Harvester/Wheatsheaf.

Recommended reading

- Gray, C. A. (2008). *Lay and professional constructions of childhood ADHD (Attention Deficit Hyperactivity Disorder): A discourse analysis* (Unpublished PhD thesis). Edinburgh: Queen Margaret University.
- Horton-Salway, M. (2011). Repertoires of ADHD in UK newspaper media. *Health (London), 15*(5), 533–549.
- O'Reilly, M. (2008). 'I didn't violent punch him': Parental accounts of punishing children with mental health problems. *Journal of Family Therapy, 30*(3), 272–295.

10
Leaving Melancholia: Disruptive Mood Dysregulation Disorder

Valerie Harwood

Introduction

This chapter provides a theoretical examination of the constitution of contemporary discourses of depression in childhood and adolescence, focusing on a new depressive disorder described in the fifth edition of *Diagnostic and Statistical Manual of Mental Disorders (DSM-5)* (American Psychiatric Association, 2013). Termed 'disruptive mood dysregulation disorder' (DMDD), this is a depression filled with energy and agitation, a new mental disorder characterised by 'temper outbursts' and that can only be diagnosed between the ages of 6 and 18 years (with an age onset of under 10 years). As such, this is not only a new disorder for children and young people, it is also a depressive disorder that appears to depart from the most commonly understood character of depression: the lifelessness or low energy characterised by the figure of Melancholia. Following this diagnostic formula, depression may be read into the temper outbursts of children.

In this chapter, I consider this change in the contemporary moment by using two striking literary and artistic figures, Melancholia and Orestes, as a means to bring to the fore the variations in the discourses of melancholia and depression. Melancholia is arguably familiar to us with a characteristic immobile and downward-looking figure. Orestes, on the other hand, is a figure that at times might remind us of melancholia and, at other times, is startlingly energetic and agitated.

DSM-5 has been met with considerable debate in relation to the new disorders added and those that have been changed (or unchanged) and those now omitted from the manual (Gitlin & Miklowitz, 2014). Notably for children and young people, the former chapter 'Disorders Usually Diagnosed in Infancy, Childhood, or Adolescence' in *DSM-IV-TR* has been removed, and *DSM-5* now has a chapter on neurodevelopmental disorders (Halter, Rolin-Kenny, & Dzurec, 2013).

DSM-5 has also arguably instilled child and adolescent depression with extreme agitation. While it is the case that 'child and adolescent' versions of depression in the previous DSMs (e.g. *DSM-IV-TR*) made links between depression and the disruptive disorders (American Psychiatric Association, 2000), these links are comparatively 'tame' when compared to the DMDD. Widely circulated comments by former chair of the DSM-IV Task Force, Allen J. Frances openly criticised DMDD: 'I very much oppose the inclusion of this new "disorder" – fearing that DMDD would medicalise temper tantrums in children and run the risk of exacerbating the already shameful overuse of antipsychotics' (Frances, 2011). While there is a slowly growing critical literature about DMDD, most of this takes issue with this new disorder, but appears to ignore its placement in *DSM-5* in a chapter on depressive disorders.

Project overview: Disruptive mood dysregulation disorder

It is difficult to ignore the presence of the child-focused DMDD in *DSM-5*. Producing this new disorder, which was originally called 'temper dysregulation disorder with dysphoria' (Wakefield, 2013), has prompted considerable reaction. Especially, as Rao (2014) pointed out, there is 'limited empirical data available' (p. 12) about this new disorder, and as Wakefield (2013) stated, it is a 'largely untested diagnosis' (p. 150). DMDD is justified as a means to restrict the alarming rates of bipolar disorder diagnosis in children. These rates have been reported to have had a '40 fold' increase 'in the past decade' (Rao, 2014, p. 3). This is explicitly declared in the introduction to the chapter 'Depressive Disorders' in *DSM-5*:

> In order to address concerns about the potential for the overdiagnosis of and treatment for bipolar disorder in children, a new diagnosis, disruptive mood dysregulation disorder, referring to the presentation of children with persistent irritability and frequent episodes of extreme behavioral dyscontrol, is added to the depressive disorders for children up to 12 years of age. Its placement in this chapter reflects the finding that children with this symptom pattern typically develop unipolar depressive disorders or anxiety disorders, rather than bipolar disorders, as they mature into adolescence and adulthood.
>
> (American Psychiatric Association, 2013, p. 155)

Emphasis on the 'correction' of incorrect diagnoses of bipolar disorder is then reiterated in the DMDD criteria section (pp. 156–160) of the 'Depressive Disorders' chapter, which includes the note that 'disruptive mood dysregulation disorder was added to *DSM-5* to address the considerable concern about the appropriate classification and treatment of children who present with chronic, persistent irritability relative to children who present with classic (i.e. episodic) bipolar disorder' (American Psychiatric Association, 2013, p. 157).

As these statements make clear, not only does this new disorder address concerns with overdiagnosis of bipolar disorder, it anticipates a trajectory of future adult disorder, namely 'unipolar depressive disorders'. A key stepping stone to DMDD was the proposal of 'severe mood dysregulation' (SMD) disorder. SMD was proposed by Leibenluft et al. (Leibenluft, Charney, Towbin, Bhangoo, & Pine, 2003) 'as an alternative diagnosis [to bipolar disorder] for those with chronic irritability' (Pliszka, 2011, p. 8).

DSM-5 describes major depressive disorder as the *'classic condition* in this group of disorders' (American Psychiatric Association, 2013, p. 155, emphasis added). Other disorders listed in 'Depressive Disorders' of *DSM-5* are persistent depressive disorder; premenstrual dysphoric disorder; substance/medication-induced depressive disorder; depressive disorder due to another medical condition; other specified depressive disorder; and unspecified depressive disorder (American Psychiatric Association, 2013).

Referring to major depressive disorder as the *classic condition* attests this is the category with which the colloquial term 'depression' is commonly associated. In *DSM-5*, major depressive episode 'is characterized by discrete episodes of at least 2 week's duration (although most episodes last considerably longer) involving clear-cut changes in affect, cognition, and neurovegetative functions and inter-episode remissions' (American Psychiatric Association, 2013, p. 155).

'Depressive Disorders' stands as a newly separated (and distinct) chapter in *DSM-5*, with 'the former "Mood Disorders" chapter...now divided into two chapters, "Bipolar Disorders" and "Depressive Disorders"' (Wakefield, 2013, p. 141). In the previous edition, *DSM-IV-TR*, 'Depressive Disorders' were placed under the category 'Mood Disorders', a category that also included 'Bipolar Disorders' (American Psychiatric Association, 2000). Indeed, as Wakefield (2013) surmised:

> Depressive disorders experienced the most changes and the most controversy of any chapter... [this] include[s] elimination of the major depression bereavement exclusion, introduction of the new category of disruptive mood dysregulation disorder in children, introduction into the main listing of premenstrual dysphoric disorder, introduction for further study of the new category of persistent depressive disorder, and introduction of a new major depression specifier, 'with anxious distress'. (p. 148)

Placed in this category, DMDD sits with a well-known contemporary, major depressive disorder, a disorder said to widely affect the populations of many countries. According to the US National Institute of Mental Health, '11% of adolescents have depressive disorder by the age of 18' (National Institute of

Mental Health, 2008). DMDD thus joins a category that houses some of the most commonly diagnosed – and popularly known – mental disorders.

Meeting melancholia

For several centuries, the image of physical lack has pervaded as a signature of melancholy, and arguably, depression. This image is famously portrayed in Albrecht Dürer's engraving, *Melancholia I* (see Figure 10.1).

Figure 10.1   'Albrecht Dürer: *Melancholia I*' (1514) (43.106.1)
Source: In *Heilbrunn Timeline of Art History*. New York: The Metropolitan Museum of Art http://www.metmuseum.org/toah/works-of-art/43.106.1. (October 2006) [OASC].

In *Melancholia I*, the central *adult* figure sits limp and forlorn, lacking energy or motivation, and unable to move. This is not the figure of a child; it is an adult, one placing the representation of depression as that of a grown figure. In this woodcut engraving, Dürer famously depicted the tension between 'melancholy, creativity, knowledge' (Sullivan, 2008). Strewn aside, the tools and implements surrounding melancholia tell of the vanished adult creativity, and the star on the horizon is suggestive of the role of divine inspiration. Drawing on Hippocrate's humoral theory, Dürer's engraving portrays immobility; the figure's potency lost from within, (which explained melancholia in terms of 'an imbalance in the humours: the more severe the imbalance, the more severe the symptoms of melancholia' (Lawlor, 2012, p. 26)).

The image of immobility echoes across the interpretations of melancholy, and again, these are adult images of the melancholic. Analysing a period of melancholy's immobility, Foucault's *The History of Madness* contributes instructive observations on melancholy. Based on his researches into 17th-century medicine, Foucault announces that 'melancholy never attains frenzy; it is a madness always at the *limits of its own impotence*' (2006, p. 266, emphasis added).

The Anatomy of Melancholy, first published in 1621, describes melancholy as either 'disposition or habit' (p. 83). This famous book was written by Robert Burton, who by his own admission busied himself writing his book as a means to avoid melancholy. Disposition refers to a 'transitory melancholy which goes and comes upon every small occasion of sorrow, need, sickness, trouble, fear, grief, passion, or perturbation of the mind, any manner of care, discontent, or thought, which causeth anguish, dullness, heaviness and vexation of spirit' (Burton, 1621/2004, p. 218). It is marked by its opposition to specific emotions, including feelings such as 'pleasure, mirth, joy, delight' and can cause 'frowardness in us, or a dislike' (p. 219) (in the *Oxford Dictionary* 'frowardness' is stated as having origins in Old English, meaning 'leading away from' (Soanes & Stevenson, 2013)). To describe melancholy of habit, Burton said, 'we call him melancholy that is dull, sad, sour, lumpish, ill disposed, solitary, any way moved, or displeased' (Burton, 2004).

The Anatomy of Melancholy, encyclopaedic in its references across literature, mentions violence, but it is overwhelmingly violence directed upon the self (and again, when this occurs it is an adult violence). Instances of violence towards others are few; the picture of melancholy is one of impotence. It is certainly not a melancholy that could house the likes of DMDD.

The idea that melancholy cannot attain the vigour possible in other ailments is clearly demonstrated in Foucault's recount of the descriptions provided by the 17th-century anatomist and physician Thomas Willis (1672, 1683). Foucault describes Willis' account of melancholy, in which 'the spirits are carried away by an agitation, but a *weak* agitation that lacks power or violence a sort of impotent upset that follows neither a particular path nor the *aperta opercula* [open ways] but traverses the cerebral matter constantly creating new pores' (2006, p. 266). This description draws a picture of movement without direction but with a telos of dissipation. In this movement, 'the spirits do not wander far on the new paths they create, and their agitation dies down rapidly, as their strength is quickly spent and motion comes to a halt' (2006, p. 266). This 'melancholic experience' extends from the physiological to the soul, a view that prompts Jeremy Schmidt (2007) to conclude that for Willis, the mind and the body are both involved in the melancholic condition. The melancholy described by Willis is one of diminishing strength, reduction in agitation. It is, again, one of impotence and not a reference to disruption, agitation, or tantrums.

Surveying medical accounts of the 18th century from the work of English physician Robert James (1743/1745) and Paris physician Anne-Charles Lorry (1765), Foucault points out that while certain explanations vary and symptoms shift, there is a conceptual unity that writes the story of melancholy. What we have is an organizational apparatus that assembles symptoms, one that crafts explanations and faithfully portrays the idea of melancholy. The image of melancholy as impotent pervades, one of immobility, reduction, and loss of power.

For much of the 'clinical history' of medicine over the last several centuries, impotence has been readily discernible in the imagery of melancholia. Indeed, what perhaps earmarks depression as appearing as though it has a 'continuous history' with melancholia (Foucault, 1977) is the association with recurring depictions of impotency. Coined in the mid-19th century, and replacing melancholy, the term 'depression' came from usage that was 'popular in middle nineteenth century cardiovascular medicine to refer to a reduction in function' (Berrios, 1995, p. 386). Under this name, depression was characterised as 'reflected loss, inhibition, reduction, and decline' (Berrios, 1995, p. 386). In *The Dictionary of Philosophy and Psychology* (Baldwin, 1901), Joseph Jastrow defined depression as '[a] condition characterized by a sinking of spirits, lack of courage or initiative, and tendency to gloomy thoughts' (1901, p. 270). Here the word 'sinking' conjures the distinct image of deflation. The sense of impotence is brought to the fore when Jastrow distinguishes depression from dejection: 'depression refers more definitely to the lowered vitality of physical and mental life, dejection to the despondency of the mental mood' (Jastrow, 1901, p. 270).

Foucault made the observation on melancholy's impotence with reference to the 17th century. While it is not the case that melancholy became, as it were, what was defined as depression in the various DSMs (American Psychiatric Association, 1987, 1994, 2000, 2013), this characteristic of impotence is a point to labour upon. The kinship of melancholy and depression over the last several centuries might be more usefully portrayed as reflecting their similar reliance on the idea of impotence.

While there is contention regarding the proposition that melancholy is the historical antecedent of depression, there is good justification for considering the cultural understanding attributed to the emblematic features of the two; namely, the notion of impotence. However, it is important to note that to consider these concepts together is not to stake a claim of continuity between them. For instance, the suggestion of a relationship between the concepts is rigorously analysed by Radden (2003), who differentiated between melancholy and depression on the basis of descriptive versus causal accounts, concluding that they are distinct. This view explicitly questions the attribution of melancholia as an historical precursor to contemporary depression. Acknowledging the significance of this distinction, I suggest it can be argued that it is the emphasis on *impotence* that enables a relationship between the two to be perceived.

What then happens when the 2013 *DSM-5* category of Depressive Disorders includes DMDD, a disorder that includes diagnostic criteria such as that

> [s]evere recurrent temper outbursts manifested verbally (e.g. verbal rages) and/or behaviourally (e.g. physical aggression toward people or property) that are grossly out of proportion in intensity or duration to the situation or provocation.
>
> (American Psychiatric Association, 2013, p. 156)

Changes such as this stand opposed to an image of impotency that has arguably dominated our discourses of depression over several centuries. This suggests that what we may be experiencing (or possibly about to more frequently experience) is a different object of depression, one that could mean leaving Melancholia.

Leaving melancholia

The idea that depression is impotent and lacking vigour or energy is challenged by DMDD. Likewise, DMDD challenges our conception of distinctions between depression in childhood and adulthood, curiously bringing these together through the vehicle of temper outbursts. Certainly, we might be wise to consider the proposal that we are now 'leaving Melancholia' and, with it, the sense of overwhelming immobility. While temper outbursts might be considered emblematic of a certain kind of 'immobility' that frustrates adults, this is not the same kind of immobility depicted in *Melancholia I*.

Yet, although DMDD stands out among the Depressive Disorders, it is not the case that 'anger' has been wholly absent from contemporary conceptualisations of depression. For instance, the previous edition of the DSM, *DSM-IV-TR* (American Psychiatric Association, 2000), stated that for 'prepubertal children' 'Major Depressive Episodes occur more frequently in conjunction with other mental disorders (especially Disruptive Behavior Disorders, Attention-Deficit Disorders, and Anxiety Disorders) than in isolation' (American Psychiatric Association, 2000, p. 354). In adolescents, the association between depression and other disorders is expanded to include the group of disruptive behaviour disorders as well as 'Anxiety Disorders, Substance-Related Disorders, and Eating Disorders' (American Psychiatric Association, 2000, p. 354).

It is also relevant to note that in relation to adults, the description for major depressive episode has criteria for the specifier 'Psychotic Features' that includes either delusions or hallucinations (American Psychiatric Association, 2000), and these may point towards violence to others. The Specifier is differentiated into either 'Mood-Congruent Psychotic Features' or 'Mood-Incongruent Psychotic Features', with the latter defined as 'content [that] does not involve typical depressive themes of personal inadequacy, guilt disease, death, nihilism, or deserved punishment' but that does include 'persecutory delusions, thought insertion, thought broadcasting, and delusions of control' (American Psychiatric Association, 2000, p. 413). Inclusion of persecutory delusions does render the possibility that, within a diagnosis of major depressive episode, there is scope for potency. That said, this has been a less emphasised characteristic of depression.

A means by which to conceptualise these changes might be to think in terms of 'control' and 'lack of control' (Toohey, 2004). As Toohey (2004) argued with reference to changes to 'depression' and 'melancholia' between Greek and modern representations, 'We witness in this evolutionary shift a movement from activity to passivity, from body to mind (and interiority), from complicity to estrangement, from public to private, from mark to sign, and paradoxically and above all, from lack of control to control' (2004, p. 56). In this view, depression and melancholy, as we have recently known it (along with immobility), demanded a sense of 'control' (perhaps this is a cue for the issue of control and temper outbursts). In much earlier representations dating from 400 BCE, this sense of control is far from evident; what we see is a 'lack of control' that is caused by none other than *agitation.*

A depression of children and young people that speaks of DMDD might thus be better understood as one that eschews popularised images of Melancholia, and instead embraces lack of control, where one is caught in the maelstrom of agitation. A legendary example of this form of agitation is the figure of Orestes who, interestingly, has been interpreted to represent not only a 'melancholy' but also an 'energetic madness'. In the following section, I focus on an adult portrayal of melancholy and agitation in order to consider the varying ways melancholia and depression have been historically construed. As I will show, this enables me to examine how the politics of truth is very much implicated in the production of diagnostic criteria that now connect a child's temper outbursts to depressive disorders, the most diagnosed of mental disorders in the world (see Giles, Chapter 12, this volume).

Meeting Orestes

Orestes is claimed to be an archetypal figure of melancholy. Take, for instance, Lawlor's assertion in *From Melancholia to Prozac: A History of Depression*: 'It has been argued that depression has existed since classical times, and the character of Orestes, in Aeschylus' tragic trilogy, the *Oresteia*, is proof' (2012, pp. 24–25). Here we can consider depression and melancholia might be thought of as forms of 'madness'. By murdering his mother (matricide), Orestes plunges into a 'melancholic madness', a story famously told in *The Eumenides* by Aeschylus (458 BCE) and later in *Orestes* by Euripides (408 BCE). Also described as myth (Ingham, 2007), the story of Orestes has variations in this melancholic madness as well as in the politics of its discourse – variations that remind us of the changes to how an anger verging on madness can come to be portrayed as depression (or, in *DSM-5*, as DMDD).

The presence of depression in Orestes is considered to be portrayed in an Apulian vase of the 4th century BCE:

> Orestes, as depicted on a fourth-century BC red-figure Apulian vase (now in the Louvre), is undergoing a rite of religious purification in order to rid him of the murder of his mother, Clytaemnestra, who had been party to the murder of his father, Agamemnon. Oreste's depression is manifested in his posture, the downcast eyes and drooping body, drained of all energy.
>
> (Lawlor, 2012, pp. 24–25)

Images of the Apulian red-figure bell-krater can be accessed on the website for The Louvre, Paris (http://www.louvre.fr/en/oeuvre-notices/apulian-red-figure-bell-krater).

A rather different depiction of Orestes can be found in the Greek play *Orestes* by Euripides, a play in which we see frenzy and anger. Take, for example, the description by Electra, the sister of Orestes, who states, 'my poor Orestes fell sick of a cruel wasting disease; upon his couch he lies prostrated, and it is his mother's blood that goads him into frenzied fits' (Euripides, 408 BCE). This 'cruel wasting' changes abruptly:

> 'Tis now the sixth day since the body of his murdered mother was committed to the cleansing fire; since then no food has passed his lips, nor hath he washed his skin; but wrapped in his cloak he weeps in his lucid moments,

whenever the fever leaves him; other whiles he bounds headlong from his couch, as a colt when it is loosed from the yoke.

(Euripides, 408 BCE)

'Bounding from his couch', a movement so vigorous it is compared to 'a colt loosed from the yoke' is vastly different from the reclining figure depicted on the Apulian bell-krater. It is an image more reminiscent of the painting by Adolphe William Bouguereau, *Orestes Pursued by the Furies* (1862) (see Figure 10.2).

This 19th-century painting depicts an Orestes with energy, attempting to escape the merciless anger of the furies. Bourguereau, responding to critics of the painting, commented on energy, stating 'I soon found that the horrible, the frenzied, the heroic does not pay' (Harrison, 1991, p. 111). This depiction of Orestes shows frenzy and, in so doing, is an energetic depiction of Orestes' experience of melancholia.

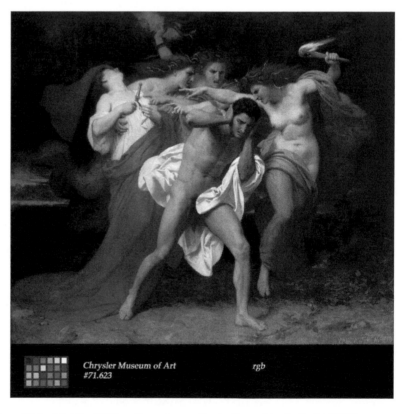

Figure 10.2 Adolphe William Bouguereau, *Orestes Pursued by the Furies* (1862)
Source: Chrysler Museum of Art, Norfolk, Virginia, USA.

The melancholic madness of Orestes

Understanding the figure of Orestes represented in the Apulian red-figure bell-krater demands knowledge of his story, a point underscored by Toohey (2004) in *Melancholy, Love and Time: Boundaries of the Self in Ancient Literature*. In his analysis of Orestes' 'madness' (which I draw on closely in the remainder of this chapter), Toohey (2004) maintains, 'If we did not know that it was Orestes and had not noticed that he had a sword in his hand, then we would say that the male seated in the center of the representation ... was bored' (p. 15).

That said, while this representation of Orestes could be assumed to be similar to that of Dürer's (1514) *Melancholia* (also appearing with lassitude), closer inspection shows a figure containing energy, which is, in Toohey's (2004) words, 'agitation':

> Orestes' face and much of his posture exhibit a patina of motor retardation. But there are clear signs of mental activity – of agitation. There is the sword in his right hand: that Orestes intends it for some form of violent use is apparent by the apprehensive index finger on his right hand. That the sword points in the general direction of the Furies suggests that it is intended for use against them, rather than as a symbol of his act of matricide (Podlecki, 1989; Shapiro, 1994; Sommerstein, 1989), as a symbol of suicidal thoughts ... or simply as a means for slitting the piglet's throat. (p. 17)

While this analysis appears to examine melancholia and depression as 'constructs', Toohey does not consider either as purely constructed. He views depression as 'a persistent cultural entity that not unexpectedly, certain eras find difficult to accommodate conceptually' (2004: p. 39). While Toohey's book is critiqued for its methodological and historical content (Whitmarsh, 2005), it does provide a useful analysis of the figure of Orestes and the varying representations of his 'melancholia', variations that strike a chord with DMDD and depressive disorders.

Greek interpretation of melancholia drew on 'the humoral theory of black bile, (μέλαωα χολή) from which the word melancholy proceeds. μέλαωα χολή or melaina chole was translated into Latin as atra bilis and into English as black bile' (Lawlor, 2012, p. 27). Thus, from a medical standpoint that used humoural theory, 'the individual in whom black bile predominates comes increasingly to be seen as "melancholic"' (Toohey, 2004, p. 28). Such is the medical means through which the madness of Orestes was understood.

Toohey (2004) offered an analysis that critiques the straightforward attribution of 'melancholia', noting the similar complexity of melancholia represented in the Apulian red-figure bell-krater by the Eumenides Painter of the 4th century BCE and that depicted by Aeschylus in *The Eumenides* (458 BCE). This melancholia, however, is differently portrayed in Euripides' *Orestes* (408 BCE). Recalling the excerpt from Euripides' *Orestes* cited above, it is clear that Orestes moves between 'waste' and energy; a vacillation that might prompt some (including Toohey, 2004) to retrospectively propose a diagnosis of bipolar disorder. This, however, is a contested notion, not only because 'retrospective diagnosis' of historical figures is problematic, but 'manic depression, as construed by post-nineteenth-century definitions, bears no relation to the classical forms of mania, in which mania (insanity and delirium) might emerge from melancholia if the melancholia became particularly severe' (Lawlor, 2012, pp. 26–27). Here, again, we see not only a stark difference of opinion, but also an indication of the complexity of representation of the melancholic.

Significantly though, melancholia was much more than 'wastage', with agitation playing a key part in this malaise. The point is that when we meet Orestes, we begin to see not only the presence of agitation (as opposed to Melancholia's impotence and lack), but how this agitation is variously taken up and portrayed. For example, referring to his comparison of Orestes in Euripides' play with that of the figure depicted on the Apulian red-figure bell-krater, we see connections with agitation, but startling differences between exteriority and interiority:

> What links both figures is that they are prey to an extreme agitation and an awful fear (even terror) that plays havoc in their lives. But the agitation and fear of Euripides' Orestes and his violent melancholia is driven from without – from the gods and the Furies. The fearful agitation of the Eumenides Painter's Orestes is driven from within – the Furies cannot have caused this, because they share the same facial expression.
>
> (Toohey, 2004, p. 23)

Euripedes' 408 BCE play did not engage with a focus on the affect of impotency. By contrast, this was represented in the Apulian red-figure bell-krater.

Differing representations of Orestes, such as the Apulian vase where an affect without energy is shown, are suggestive of the ways discourses of madness not only shift and change, but even more elementally, reveal how context impacts representation of experience. Certainly, Toohey's (2004) analysis (to which I have referred extensively) picked up differences between classical Greek medical and popular usages, arguing that while medical usage connects to a wasted, impotent figure reclining on the Apulian red-figure bell-krater, '[n]on medical usage... associates the term with violence and anger' (p. 27) and thus takes up the 'Euripidean (violent and angry) Orestes' (p. 27). For Toohey (2004) then, the different portrayals reveal much about how melancholy was engaged, with

'Euripides *Orestes* provides a sobering illustration of the inability or unwilling-ness of the literary tradition to represent melancholia in truly complex manner' (p. 20). Here we see quite simply the workings of discourse in the production of knowledge about melancholia.

Interestingly, following Toohey's (2004) line of argument, it becomes clear from his analysis that more attention was paid to exteriority than on interior-ity. Before proceeding to Toohey's conclusion, it is worth pausing to consider the 'neuro-interiority' that is clearly marked out in the *DSM-5*. For instance, the new manual is structured to follow a 'neurodevelopmental life span approach' (Halter et al., 2013, p. 33). At the same time, DMDD, formerly labelled 'temper tantrum dysregulation', has been criticised as 'a symptom of other disorders' (Welch, Klassen, Borisova, & Clothier, 2013, p. 168). As we see an emphasis on 'neuro-interiority', we can also see how internalising and externalising disor-ders, what might be considered to be separate can be brought together under the diagnosis of DMDD. For instance, DMDD is reported to be 'highly comorbid with internalizing (depression and anxiety) and externalizing disorders (atten-tion deficit hyperactivity disorder, oppositional defiant disorder, and conduct disorder)' (Axelson, 2013, p. 137)

Reflecting on his analysis, Toohey (2004) concluded that the differing repre-sentations of Orestes reveal difficulties with how melancholia was discursively engaged, arguing that while it was 'present' and 'periodically acknowledged, but its time had not come' (p. 42). As he outlined:

This simple, though astounding, fact has been little understood. *The discursive tradition, then, with which depressive melancholy had to contend was one taken up with the outer, the surface, the mark, and the body as it was perceived in society.* The passivity of depressive melancholia – for this period a mere epiphenomenon of mania – had little to offer such tradition.

(p. 42, emphasis added)

Following the argument set out by Toohey (2004), the story of Orestes presents a tantalising account of differences in the way melancholia is conceived, as well as the differences between popular and medicalised accounts. The for-mer account is closer to Euripedian 'anger' and 'colt free from the yoke', while the latter depicts the more benign, yet subtly agitated figure on the Apulian bell-krater. The two variations radically demonstrate different ways through which melancholia can be conceived, with both signalling agita-tion, albeit in strikingly different forms. In this sense, we could propose that the DMDD of *DSM-5* is an eerie 'resurfacing' of a very old interpretation that fills melancholia with agitation, and as such is not so strange after all. There is, however, another angle to consider: the question of the politics of truth.

Conclusion: The politics of truth

Foucault engaged with the story of Orestes in two significant places in his work: in *The History of Madness* (2006) and in his 1983 lectures 'Discourse and Truth: The Problematization of Parrhesia' (published as *Fearless Speech*; Foucault, 2001). In the former, Foucault drew on Racine's *Andromache* (1667) in his analysis of the 'literary experience of madness in the Renaissance' (p. 580), and for the latter, he worked with Euripides' *Orestes* (408 BCE) in his analysis of Greek parrhesia (Foucault, 2001). Both engagements with the story of Orestes offer a means to contemplate what we might call the politics of melancholy, a politics that has occurred with the creation and insertion of DMDD into a chapter 'Depressive Disorders' in *DSM-5*.

As Foucault explained in *Fearless Speech* (2001), 'Euripides' *Orestes* – a play written, or at least performed, in 408 BC, just a few years before Euripides' death, and at a moment of political crisis in Athens when there were numerous debates about the democratic regime' (Foucault, 2001, p. 57). The critical importance of this political moment, to the play (and for Foucault, to his interpretation of parrhesia) is evident in Foucault's extensive description of this political moment:

> And now we can see the precise historical and political context for this scene. The year of the play's production is 408 B.C., a time when the competition between Athens and Sparta in the Peloponnesian war was still very sharp. The two cities have been fighting now for twenty-three long years, with short intermittent periods of truce. Athens in 408 B.C., following several bitter and ruinous defeats in 413, had recovered some of its naval power. But on land the situation was not good, and Athens was vulnerable to Spartan invasion. Nonetheless, Sparta made several offers of Peace to Athens so that the issue of continuing the war or making peace was vehemently discussed. In Athens the democratic party was in favor of war for economic reasons which are quite clear; for the party was generally supported by merchants, shop-keepers, businessmen, and those who were interested in the imperialistic expansion of Athens. The conservative aristocratic party was in favor of peace since they gained their support from the landowners and others who wanted a peaceful co-existence with Sparta, as well as an Athenian constitution which was closer, in some respects, to the Spartan constitution. The leader of the democratic party was Cleophon – who was not native to Athens, but a foreigner who registered as a citizen.
>
> (Foucault, 2001, p. 70)

Orestes is also depicted in Racine's *Andromache* (1667), which Foucault discussed as an exemplar of the shift from a renaissance madness which might contain reason to one where madness is understood as unreason (Foucault, 2006):

> The gesture that banished madness into the dull, uniform world of exclusion is neither the sign of a pause in the evolution of medicine nor an indicator of a halt in the progress of humanitarian ideas. Its exact meaning comes from the simple fact that in the classical world madness was no longer the sign of another world, and became instead a paradoxical manifestation of non-being. In the final analysis, confinement was not overly concerned with suppressing madness or removing from the social order a figure which could not find its place there, and its essence could not really be described as the exorcism of any danger. It only manifested what madness is, in its essence: the unveiling of non-being. (p. 249)

Here Foucault's reference is to the classical world, roughly the 17th and 18th centuries and the period of the great confinement, where madness, as unreason, could be banished, or more exactly, excluded. Orestes, then, in Racine's *Andromache*, signals the fundamental shift in madness to non-being and, importantly for this analysis, the connections with politics.

As the above analysis reminds us, madness requires a politics and this very politics means that madness shifts and changes (see O'Dell & Brownlow, Chapter 16, this volume). A figure such as Orestes can be absolved by the casting vote of Athene (Aeschylus, 458 BCE), be cured by the divine Apollo (Euripides, 408 BCE), or can be expunged from society (Racine, 1667). In terms of the politics of our times, DMDD is, I venture to suggest, no different. This is a child mental disorder that has been created to meet the needs of a politics concerned with 'over diagnoses' and attendant issues of extremely high prescription rates. This is a disorder emerging from political concerns couched in politically aware ways, deploying terms such as 'false positives' (Leibenluft, 2011, cited in Gitlin and Miklowitz, 2014, p. 89). This can also be formulated in a manner evoking praise, such as 'the new diagnosis [DMDD] is viewed as an alternative to assigning a lifelong diagnosis of bipolar disorder, which often is accompanied by powerful drug treatment (Margulies, Weintraub, Basile, Grover, & Carlson, 2012)' (Halter et al., 2013, p. 34).

Reading Foucault's (2001) discussion, we can see how Euripides' *Orestes* picks up on the politics of the time: '[O]ne of the issues clearly present in Orestes' trial is the question that was then being debated by the democratic and conservative parties about whether Athens should continue the war with Sparta, or opt for peace' (Foucault, 2001, p. 71). The absolution of Orestes' murder of his mother connected with the political needs of the times. It did not, for instance, follow a course similar to that of Pierre Riviere's murder of his family, a case so closely analysed by Foucault (1978). Is it not the case that the very discourse that produces DMDD is likewise none other than a politics of our time? This is to say that the decision to diagnose this state of mental disorder onto a child with tantrums rests upon the concerns of a select few (arguably not even the diagnosticians, but rather the writers of diagnostic texts).

It is surprising how little the debate on DMDD has been concerned with the placement in the 'Depressive Disorders' chapter in *DSM-5*. It would seem that the debate has come to rest on the diagnosis of 'tantrums' as disorders, and in so doing it has overlooked the conceptual shifts that occur when depression becomes disruptive. Perhaps adherence to the notion that child DMDD flows into an adult unipolar depression absolves DMDD from, as it were, 'disrupting' our understanding of depressive illness. While this may be a convincing argument for some, it does present considerable problems for conceiving the experience of children as distinct from adults. Will it be the case that tantrums come to mean the harbouring of depression?

In his discussion of melancholy and the 'melancholic experience' in the *History of Madness*, Foucault (2006) emphasised:

> The key point is that this process did not go from observation to the construction of explanatory images, but that on the contrary images fulfilled the initial role of synthesis, and their organizing force made possible a structure of perception where symptoms could finally take on their significant value, and be organized into the visible presence of the truth.
>
> (Foucault, 2006, p. 277)

This emphasis on the 'structure of perception' supports a line of reasoning that takes as its object how depression (or melancholy) is perceived. Thus images of objects of impotence (or their tantrums) enable an 'organizing force' that, to paraphrase Foucault (2006), structures our perceptions and consequently gives value and weight to the symptoms that tell the truth of depression. So images of impotence or of tantrums all add weight to how we conceive of, as well as perceive, disorders such as depression. This may partly help to answer why having disruption and tantrums in a chapter on depression hasn't been the key issue of debate with DMDD. Quite simply, we may just be adjusting our view to take in Orestes as well as Melancholia.

The philosopher Ian Hacking (2002) pointed out how ontology and 'new names' are interrelated: 'With new names, new objects come into being. Not quickly. Only with usage, only with layer after layer of usage' (p. 8). In the case of depression, while it is not a matter of 'new names', we should not be fooled. Rather, we need to consider the effects of a discourse that shifts and changes with layers of usage such that a new object (albeit with the same name, depression) comes into being.

Table 10.1   Educational practice highlights

1.	DMDD is a problematic inclusion in the manual of psychiatric disorders.
2.	The inclusion of DMDD as a form of depression has not yet been substantively critiqued.
3.	The inclusion of 'tantrums' is contested.
4.	There is a risk of stigmatisation for children in the school environment.

Summary

This chapter outlines the importance of critically reflecting on the diagnostic criteria for DMDD now included in *DSM-5*. In so doing, it mounts the argument that DMDD is a new and problematic inclusion to the 'Depressive Disorders' in an extremely influential manual of psychiatric disorders. Significantly, the inclusion of this new 'disruptive' and 'energetic' disorder as a form of 'depression' has yet to meet with substantive critique. DMDD criteria include 'tantrums', a point that has been hotly debated. For instance, as Wakefield (2013) pointed out, 'Children tend to outgrow these temper tantrum problems, so treatment and stigma may be applied unnecessarily to large numbers of children' (2013, p. 150). It is unknown how this new child disorder will impact, positively or negatively or even if it will afford the clarity that it is hoped to deliver. As Gitlin and Miklowitz (2014) concluded, 'whether this new category will advance diagnostic clarity and/or more appropriate treatment is unknown' (2014, p. 89). The chapter demonstrates how historically informed analysis can be drawn upon to reflect on how interpretations and representations of melancholia and depression are very much connected to the political, the discursive, and, in the 21st century, to the authors of one manual of mental disorders. For a simple summary of the implications for practice, see Table 10.1.

References

Aeschylus (458 BCE). *The Eumenides*. Orange Street Press Classics.
American Psychiatric Association (1987). *Diagnostic and statistical manual of mental disorders (DSM-III-R)* (3rd revised edition). Washington, DC: APA.
American Psychiatric Association (1994). *Diagnostic and statistical manual of mental disorders (DSM-IV)* (4th edition). Washington, DC: APA.
American Psychiatric Association (2000). *Diagnostic and statistical manual of mental disorders, text revision (DSM-IV-TR)* (4th edition). Washington, DC: APA.
American Psychiatric Association (2013). *Diagnostic and statistical manual of mental disorders (DSM 5)*. Washington, DC: APA.
Axelson, D. (2013). Taking disruptive mood dysregulation disorder out for a test drive. *American Journal of Psychiatry, 170*(2), 136–139.

Baldwin, J. M. (1901) Dictionary of Philosophy and Psychology. New York: The Macmillan Company.

Berrios, G. E. (1995). Mood disorders. In G. E. Berrios & R. Porter (Eds.), *A history of clinical psychiatry: The origin and history of psychiatric disorders* (pp. 384–408). London: Athlone Press.

Burton, R. (2004). *The anatomy of melancholy* [1624]. Retrieved from http://www.gutenberg.org/files/10800/10800-h/ampart1.html.

Dürer, A. (1514). *Melancholia I.*

Euripides (408 BCE). *Orestes* (E. P. Coleridge, Trans.): Internet Classics Archives.

Foucault, M. (1977). Nietzsche, genealogy, history. In D. F. Bouchard (Ed.), *Language, counter-memory, Practice: Selected essays and interviews* (pp. 139–164). Ithaca, NY: Cornell University Press.

Foucault, M. (Ed.) (1978). *I, Pierre Riviere, having slaughtered my mother, my sister and my brother: A case of parricide in the 19th century.* London: Peregrine.

Foucault, M. (2001). *Michel Foucault: Fearless speech.* Los Angeles: Semiotext(e).

Foucault, M. (2006). *History of madness* (J. Murphy & J. Khalfa, Trans.). Abingdon, Oxon: Routledge.

Frances, A. J. (2011). Should temper tantrums be made into a DSM 5 diagnosis? *Psychology Today.* Retrieved from http://www.psychologytoday.com/blog/dsm5-in-distress/201110/should-temper-tantrums-be-made-dsm-5-diagnosis.

Gitlin, M. J., & Miklowitz, D. J. (2014). Psychiatric diagnosis in ICD-11: Lessons learned (or not) from the mood disorders section in DSM-5. *Australian and New Zealand Journal of Psychiatry, 48*(1), 89–90.

Hacking, I. (January 16, 2002). Inaugural lecture: Chair of philosophy and history of scientific concepts at the Collège de France, . *Economy and Society, 31*, 1–14.

Halter, M., Rolin-Kenny, D., & Dzurec, L. (2013). An overview of the DSM-5: Changes, controversy and implications for psychiatric nursing. *Journal of psychosocial nursing, 51*(4), 30–39.

Harrison, J. C. (1991). *The chrysler Museum handbook of the European and American collections: Selected paintings, sculpture and drawings.* Norfolk, VA: The Chrysler Museum.

Ingham, J. M. (2007). Matricidal madness in Foucault's anthropology: The Pierre Rivière seminar. *Ethos, 35*(2), 130–158.

James, R. (1745). *A medicinal dictionary, including physic, surgery, anatomy, chymistry and botany, and all their branches relative to medicine.* London: T. Osborne.

Jastrow, J. (1901). Depression. In Baldwin, J. M. (Ed.), *Dictionary of philosophy and psychology* (Volume 2). London: MacMillan.

Lawlor, C. (2012). *From Melancholia to Prozac: A history of depression.* Oxford: Oxford University Press.

Leibenluft, E. (2011). Severe mood dysregulation, irritability, and the diagnostic boundaries of bipolar disorder in youths. *American Journal of Psychiatry, 168*, 129–142.

Leibenluft, E., Charney, D. S., Towbin, K. E., Bhangoo, R. K., & Pine, D. S. (2003). Defining clinical phenotypes of juvenile mania. *American Journal of Psychiatry, 160*, 430–437.

Lorry, A. C. (1765). De melancholia et morbis melacholicis (2 volumes), Paris.

Margulies, D. M., Weintraub, S., Basile, J., Grover, P. J., & Carlson, G. A. (2012). Will disruptive mood dysregulation disorder reduce false diagnosis of bipolar disorder in children? *Bipolar Disorder, 14*, 488–496.

National Institute of Mental Health (2008). *The numbers count: Mental disorders in America.* Retrieved from http://www.nimh.nih.gov/health/publications/the-numbers-count-mental-disorders-in-america.shtml.

Pliszka, S. R. (2011). Disruptive mood dysregulation disorder: Clarity or confusion. *The ADHD Report, 19*(5), 7–11.

Podlecki, A. J. (Ed.) (1989). *Aeschylus: 'Eumenides'*. Warminster: Aris and Phillips.

Racine, J. B (1982 [1667]). *Andromarche: Tragedy in five acts* (R. Wilbur, Trans.). New York: Harcourt Brace Jovanovich.

Radden, J. (2003). Is this dame melancholy? Equating today's depression and past melancholia. *Philosophy, Psychiatry and Psychology, 10*(1), 37–52.

Rao, U. (2014). DSM-5 Disruptive mood dysregulation disorder. *Asian Journal of Psychiatry*, Vol.11, October 2014, pp.118–123.

Schmidt, J. (2007). *Melancholy and the care of the soul: Religion, moral philosophy and madness in early modern England*. Hampshire, England: Ashgate Publishing.

Shapiro, H. A. (1994). *Myth into art: Poet and painter in classical Greece*. London: Routledge.

Soanes, C., & Stevenson, A. (2013). Oxford University Press. Retrieved August 3, 2013 from http://oxforddictionaries.com/definition/english/froward.

Sommerstein, A. H. (Ed.) (1989). *Aeschylus: 'Eumenides'*. Cambridge: Cambridge University Press.

Sullivan, E. (2008). The art of medicine: Melancholy, medicine and the arts. *The Lancet, 372*(9642), 884–885.

Toohey, P. G. (2004). *Melancholy, love and time: Boundaries of the self in ancient literature*. Ann Arbor, MI: University of Michigan Press.

Wakefield, J. C. (2013). DSM-5: An overview of changes and controversies. *Clinical Social Work Journal, 41*, 139–154.

Welch, S., Klassen, C., Borisova, O., & Clothier, H. (2013). The *DSM-5* controversies: How should psychologists respond? *Canadian Psychology, 54*(3), 166–175.

Whitmarsh, T. (2005). Melancholy, love, and time: Boundaries of the self in ancient literature (review). *American Journal of Philology, 126*(2), 281–284.

Willis, T. (1672). *De anima brutorum*. London: Typist E. F. impensis Ric. Davis, Oxon.

Willis, T. (1683). *Two discourses concerning the souls of brutes* (De anima brutorum, English Trans). London: Thomas Dring.

Recommended reading

- Axelson, D. (2013). Taking disruptive mood dysregulation disorder out for a test drive. *American Journal of Psychiatry, 170*(2), 136–139.
- Frances, A. J. (2011). Should temper tantrums be made into a DSM 5 diagnosis? *Psychology Today*. Retrieved January 18, 2014 from http://www.psychologytoday.com/blog/dsm5-in-distress/201110/should-temper-tantrums-be-made-dsm-5-diagnosis.
- Gitlin, M. J., & Miklowitz, D. J. (2014). Psychiatric diagnosis in ICD-11: Lessons learned (or not) from the mood disorders section in DSM-5. *Australian and New Zealand Journal of Psychiatry, 48*(1), 89–90.
- Pliszka, S. R. (2011). Disruptive mood dysregulation disorder: Clarity or confusion. *The ADHD Report, 19*(5), 7–11.

11

Why Does a Systemic Psychotherapy 'Work'?

Sim Roy-Chowdhury

Introduction

Systemic family therapy is a key psychotherapy approach taken to work with children and their families within Child and Adolescent Mental Health (CAMHS) settings (see Kiyimba & O'Reilly, Chapter 30, this volume). Within publicly funded settings, psychotherapy has been increasingly scrutinised in the United Kingdom and elsewhere in order to determine whether support for this provision is warranted by the evidence. In so doing, the randomised controlled trial (RCT) has become positioned as the 'gold standard' for evidence of 'what works' in psychotherapy. Indeed, the question of what constitutes evidence in the psychotherapies has become highly politicised. In the United Kingdom, those therapies that have been shown in 'good quality' RCTs to be effective are selected by the watchdog National Institute for Health and Care Excellence (NICE) for public funding. Consequently, cognitive behavioural therapy (CBT) is the treatment of choice in the UK government's Improving Access to Psychological Therapies (IAPT) programme.

Where does this leave the evidence base for systemic psychotherapy? Carr (2014) found strong evidence from trials to support the effectiveness of family therapy and systemic interventions. However, according to Pilling (2009), 'more high-quality efficacy research to establish the evidence base for family/systemic interventions, particularly in adults, is needed' (p. 202). I would like to examine this premise through an exploration of the assumptions made about psychological therapies when conducting an RCT and the evidence used to support these assumptions. Put differently, one might ask the question, what is the evidence to support the use of RCTs as evidence for the effectiveness of psychotherapy? I will then consider the use of RCTs in constructing an adequate evidence base for systemic family therapy. I suggest that qualitative research methods, particularly discourse analysis (DA), offer researchers the tools to design studies that are more epistemologically congruent with systemic thinking in order to examine research questions and generate findings that illuminate aspects of psychotherapeutic practice. This allows us to make some cautious, provisional statements about what the actual practice of a systemic psychotherapy looks like and why it 'works'.

Assumptions made when conducting an RCT

Condition-based trials

Pilling (2009) informed us that NICE '...does not rely solely on...RCTs when coming to its conclusions' (p. 196). However, in assessing the efficacy of psychological therapies, outcome studies which conform to the conventions of an RCT are given greater credence by NICE than other studies. Pilling (2009) provided us with a helpful summary of those psychological therapies that have been shown to be effective for different 'conditions'. CBT is top of the class and 'works' with '...depression, childhood depression, OCD, PTSD, anxiety, antenatal and postnatal mental health' (p. 198). While family therapy has been shown to be effective for depression in children, and family interventions are recommended for schizophrenia, the overall verdict for studies demonstrating the effectiveness of systemic interventions is 'must try harder'.

The first assumption made by RCTs, then, is the reliance upon psychiatric diagnoses or 'conditions'. Although psychiatric diagnoses form the basis of the dominant discourse in the mental healthcare field, we should not skate over validity problems found by researchers. Bentall (2009) offered a good summary of these studies. Psychiatric diagnoses have poor reliability between psychiatrists, with one study finding that agreement was reached in only 50% of cases (McGorry et al., 1995). Factor analytic studies of symptom clusters do not support the validity of diagnostic categories (Liddle, 1987). Similarly, the predictive validity of diagnoses (prognosis and treatment effects based upon diagnosis) has been found to be poor (Tsuang, Woolson, & Fleming, 1979). Hence, from the perspective of empirical evidence, there are at the very least objections to incorporating psychiatric diagnoses within research designs: to base a study upon theoretical premises with a weak evidential base undermines the validity of the study.

For systemic therapists and researchers, there is of course the additional difficulty that systemic theory and practice is epistemologically antithetical to individual diagnosis. This is in terms of both its de-emphasising of individual psychopathology in favour of relational analyses and its theoretical affinity with relativist social constructionist discourses, providing at least one reason that there are comparatively few diagnosis-based systemic psychotherapy RCTs.

'Active ingredients' within an RCT

RCTs seek to control all aspects of the human interaction that we call psychotherapy, apart from the 'active ingredients' of the therapy itself. Thus, the aim is to strip out the relationship between provider and receiver of therapy and strip out the unique characteristics of interactants, in order to assess whether the therapy is effective. In so doing, RCTs employ a drug metaphor (Shapiro et al., 1994): A technician applies a dose of an active substance called, let us say, a systemic psychotherapy, to a passive recipient, which, rather in the manner of an antibacterial treatment for a bacterial infection effects a symptomatic improvement.

Careful readers may have noticed that these characteristics of an RCT differ somewhat to the ways in which we think about and do therapy. Indeed it might not be overstating the case to say that RCTs seek to remove from examination the elements of a systemic psychotherapy that make it a systemic psychotherapy. We might offer an account for our work that proposes that the therapist brings her own ways of seeing the world as do family members, who are a product of experiences, as well as dominant social and cultural discourses; that the therapist seeks to construct a helpful conversation that makes moment-by-moment adjustments in response to the specific and idiosyncratic requirements of participants. Hence, individuality, relationship, and context are not 'non-specific factors', which we must push to the margins in our examination of a (systemic) psychotherapy, but constitute the therapy itself. I have put the case elsewhere (Roy-Chowdhury, 2003, 2010), as have others (e.g. Kaye, 1995), that the version of psychotherapy proposed by RCTs bears no more than a passing similarity to our actual practice. Furthermore, Harper, Gannon, and Robinson (2013) provided a methodological critique of RCTs in examining the effectiveness of psychotherapies, not least because the essential double-blind element, where therapist and therapee are not aware of whether they are in a treatment or control group is seriously compromised in psychotherapy. It should come as no surprise then, to discover that meta-analyses of process–outcome research have found that only a relatively small amount of the outcome variance can be accounted for by specific therapist techniques, the change agents scrutinised within an RCT, and rather more by the quality of the therapeutic relationship, client, and therapist factors.

Why does a psychotherapy work?

The assumption contained within an RCT design, which is that the most important factor in creating successful change is the therapy model (and associated techniques), is founded on scant evidence. For example, Shapiro et al.'s (1994) meta-analysis found 2% of outcome variance to be accounted for by therapist techniques. Indeed, research has actually shown therapist effects to be much greater than treatment effects (Blatt, Sanislow, Zuroff, & Pilkonis, 1996; Crits-Cristolph et al., 1991; Wampold, 2001).

Blow, Sprenkle, and Davis (2007) found no evidence to suggest that fixed therapist traits had an impact upon outcome and only a relatively small effect size relating therapist experience to outcome. On the other hand several studies have found that the therapeutic relationship or alliance is a significant factor in successful therapy outcomes. Horvath and Symonds (1991) and Horvath, Del Re, Fluckiger, and Symonds (2011) found that ratings of the strength of the therapeutic alliance reliably predict outcome. Blow et al. (2007) concluded, '... it is in the therapeutic relationship that therapists make or break therapy' (p. 306). Asay and Lambert's (1999) review similarly asserts that '(t)he therapist-client relationship is critical' (p. 34), and that the assessment of the client of the strength of this relationship is more significant that any external ratings.

An alternative research paradigm

If quantitative methods, specifically the use of RCTs, have significant shortcomings for the meaningful examination of a systemic psychotherapy, what are the alternatives? Qualitative research methods are based upon epistemological assumptions closer to those that underpin a systemic psychotherapy and offer the means for the investigation of questions that would be meaningful and interesting from a systemic perspective. Sweeping statements are to be avoided; however, in general terms, it is possible to assert that qualitative methods, as for most versions of systemic psychotherapy, eschew positivism and naïve realism. Social constructionism has exerted a considerable influence over the field of systemic psychotherapy (see MacNamee & Gergen, 1992). In that sense, it may be asserted that systemic practice offers us a glimpse into the practice of social constructionist psychotherapy.

Discourse analysis is becoming increasingly popular as a research method within systemic literature (e.g. Kiyimba, Chapter 3, this volume; Kiyimba & O'Reilly, Chapter 30, this volume; Roy-Chowdhury, 2003, 2006; Stancombe & White, 1997). One reason for this increasing interest is that the researcher adopts an epistemological position that is in tune with the social constructionist turn in systemic therapy (for a description of DA, see the glossary, as well as Antaki, Billig, Edwards, & Potter, 2002; Potter, 1997). (For its application to family therapy, see O'Reilly, 2014; Roy-Chowdhury, 2003, 2006; Stancombe & White, 1997; Sutherland & Strong, 2011.)

How can DA be used to research systemic psychotherapy?

If the therapeutic relationship is key to effective psychotherapy, it follows that DA allows us to examine what stronger and weaker therapeutic relationships look like in the talk that takes place in therapy. For example, what are the conversational coordinates of a stronger therapeutic relationship? What can studying discourse tell us about the importance of relationships, the application of methods drawn from theory, and the interaction between relationship and technique? In seeking answers to these questions, what can we learn about the actual practice of a systemic psychotherapy, and indeed, what may be the potential pitfalls of a psychotherapy that eschews positivism and naïve realism? In order to consider these questions, we now turn to an analysis of family therapy session transcripts, drawing on an earlier paper examining the therapeutic relationship (Roy-Chowdhury, 2006) and a larger research project (Roy-Chowdhury, 2001).

Commonly used transcription notation was employed. This is summarised succinctly by Flick (1998). Additionally, double parentheses, (( )), were employed to indicate clarificatory information, for example, ((laughter)) and ((stands up)), as suggested by a number of authors (e.g. Potter, 1996; Sacks, Schegloff, & Jefferson, 1974). Each line was numbered for ease of reference. Billig's (1997) advice against making guesses of unclear passages was followed in order to avoid compromising accuracy.

Analysis of sessions

The first extract comes from a 'good' session, where there are good levels of participation and consensus concerning subject positions and appropriate conversational formats. There are few problems of speaking, hearing, understanding, or 'trouble sources' (Schegloff, 1992). Therapees bring 'troubles talk' to the interaction, without which psychotherapy could not be accomplished, and the therapist responds in a satisfactory way such that the belief that a psychotherapy is being enacted appears to be sustained on all sides.

Such a session is the third session between Liz, the therapist, and the family of which Paul and Anne are parents and Ian their son. Paul has been diagnosed as suffering from a bipolar disorder and has had a lengthy psychiatric 'career'. This is the first session, which Ian has attended with his mother only. In common with other sessions examined, the session follows an interview format for much of its duration with the exception of a message delivered at the end of the session following a break. However, there are some notable discursive features that seem to strengthen the engagement between participants and their collaboration in creating a psychotherapy. In the extract below, Ian has described at some length his father's behaviour when he is 'high'.

Extract 1

```
1  Anne: that's the thing that concerns me when he's like that he'd buy
2  the world (.)
3  Liz: Right
4  Anne: You know he'd go out and order (.) he'd go out... [
```

```
5   Ian: [And it's very difficult to stop him (.)
6   Liz: Right
7   Anne: No you couldn't stop him
8   Ian:=He's so adamant that he's fine, and everything's excellent
```

In the following turn, Liz asks a question regarding the responses of Paul's other son, Mike, to his father when Paul is behaving in this way. It is noticeable that Liz often offers only minimal responses over a number of turns by Ian and Anne. This response of 'right', in the extract above, fulfils her obligations as an interactant and acts both to acknowledge her understanding of the previous accounts given and a prompt for further elaboration. Stiles (1992) has found this to be a common response mode for the therapist within a client-centred therapy in which the principal therapeutic aim is to understand the client's account from her point of view. The effect of this speech act appears to be a legitimising of Liz's right to ask her questions when she does, not just due to her category entitlement as a therapist but also within the expectancies of troubles talk with an interested observer in non-psychotherapeutic settings. None of Liz's questions in the session are challenged or resisted, and any problems in the interaction are easily repaired.

Interspersed with questions and these minimal acknowledgements are reflections which are used by Liz to convey some attempt to understand and put into words the experiences of Ian and Anne. This is exemplified in the extract below. Ian and Anne have been talking of the effects upon them of Paul's acutely 'psychotic' periods from which he has eventually 'recovered', and that the present 'spell' has lasted for five years. Liz asks if they fear that this time he may not recover.

Extract 2

```
1   Ian: [=I think well you've said that to me before
2   Anne: I do, I think that way
3   Liz: Do you (.) do you say it or do you [just think it?
4   Ian:                                    [=Yes you've said it to me;
5   I don't think he's ever going to get better
6   Anne: [=Yes, I have said it yes, I have said it, I've said it to my
7   sister (.)and to my sister-in-law
8   Ian: I don't think anyone is actually expecting [
9   Anne:                                           [BECAUSE it, because
10  it seems to be the unknown, you know, I just I don't er (.)
11  Liz: That must be very upsetting for you.
12  Anne: Mmm
```

One or two aspects of this extract are worth remarking upon. There is the sense that both Ian and Anne are talking of matters of importance to them, demonstrated by high levels of participation, their use of volume, and tone of voice to add emphasis. Each is keen to add her or his own account as demonstrated by the number of occasions where one begins to talk before the other has fully completed her or his turn. The purpose of this overlapping talk is not to contradict the previous speaker but rather to present the therapist with a particular construction or to add a point of view.

Liz inserts a question in line 3, which noticeably mirrors the lexical choices of Anne and Ian in their previous turns, in enquiring whether or not Anne speaks of her thoughts. Ian and Anne both answer this question directly rather than pursuing their earlier responses to Liz's previous question. The extract concludes with a reflection by Liz, the purpose of which is to convey an understanding of Anne's emotional state as she talks of her husband's illness. This is a short turn which does not use the reflection as a vehicle for an interpretive comment, nor does it attempt a transformation of Anne's 'lifeworld' experience into a professionalised account. Indeed, through its lexical choice the intervention is located within the realm of everyday talk. Anne indicates agreement to Liz's comment.

In Extract 3, Liz employs an advisement format, where she will offer opinions and interpretations to Anne and Ian.

Extract 3

```
1  Liz: It is interesting isn't it because at some level you believe
2  he's got some control. I'm not saying whole control, I'm not saying
3  that he's (.) you know he's (.) you know he's putting it on or being
4  rude or anything like that but there's a couple of things that
5  you've said that you know (.) when he has to you've talked about
6  when he's been ashamed of his illness that (0.4) and your wedding
7  that when he has to he can make himself (.) and you know make
8  himself get a bit better, but at home he doesn't have to put the
9  effort quite so much.
```

Here, Liz is contradicting the earlier assertion made by both Anne and Ian that Paul had no control over his 'symptoms'. She initially introduces this contradiction with the claim that their previous descriptions indicate to her that Anne and Ian believe that Paul has 'some control'. The emphasis upon 'some' softens the challenge, as does the use of 'at some level'. The account of Paul's control, initially based upon Anne and Ian's belief, is skilfully elided into a claim made by Liz through her use of the first person from line 2 onwards. She further softens the challenge in lines 2 and 3 while at the same time pre-empting and rebutting counter-challenges to her own account. She bolsters the facticity of this claim through making use of Anne and Ian's accounts of times when he has appeared to be in control of what he says and does. Liz further modifies her turn in order to reduce the likelihood of resistance (see Heritage, 1997) by emphasising that Paul can exert this control only when he has to and even then can only 'make himself a bit better'.

Close attention to the lexical choice in this turn demonstrates a use of Anne and Ian's language to describe Paul, notably in the reference to his 'illness'. Liz once again chooses to couch her construction in the language of everyday talk rather than institutionalised rhetoric, for example, by her use of the phrase 'he's putting it on' in line 3. The effect of this claim is that Ian uses a subsequent lengthy turn to support Liz's construction by providing a description of times when Paul does indeed seem to be capable of controlling his illness. Anne does not comment directly upon whether or not she agrees with Liz and Ian; however, the overall effect of this fragment of the session is that Liz and Ian support each other's narrative concerning Paul, which goes unchallenged and yet appears to offer a quite different account of Paul's behaviour than that previously provided by the family. A successful 'restorying', if you will.

The attention to the use of the family's language that we noted in Extract 3 is also a common feature of the therapist's talk. This differs from the transformation of the everyday descriptions of patients into the decontextualised discourse of biomedicine that we have noted as a common feature of medical interviews and reduces the common psychotherapeutic practice of inducting the therapist into professionalised formulations (Hak & Boer, 1995). Where Liz is unsure of her lexical choice she will check its acceptability with Anne and Ian. For example, when referring to Paul's 'first breakdown', Liz asks, 'do you call it that because he was actually hospitalised that time?' This reflects the use by Anne and Ian of the word 'breakdown' with reference to other hospitalisations, but not specifically Paul's first period as an in-patient. Liz only proceeds when the use of this term is confirmed as being acceptable.

Finally, in looking to the discursive contours of a session where there are the features associated with a positive therapeutic relationship, one other aspect is discernable. On a number of occasions, Liz presents Anne and Paul with choices concerning the structure of the session. For example, she returns from the end of session break and asks the family whether they would like to hear the team's feedback or to arrange another appointment first. This is clearly a permissible question within the context of expectations established within the session as Ian promptly responds with the unembroidered request, 'Feedback first'. She concludes the session with a request for their views as to which members of the family might most usefully attend the next session.

We are now in a position to test out our inferences from this session against instances in family therapy sessions where trouble sources (see Glossary for a definition) appear in the talk, which for our purposes provide a discursive approximation to breaks in the therapeutic relationship. These may be defined as problems of speaking, hearing, or understanding (Schegloff,

1992). In identifying trouble sources, it is important to scrutinise the orientation of speakers themselves to these conversational moments that appear to be problematic. Typically, attempts will be made to correct a 'defective utterance' by the speaker (self-initiated repairs) or the listener (other-initiated repairs). Where the speaker does not accept the listener's invitation to correct the prior problematic utterance, the listener will typically undertake the repair herself in a subsequent turn (other corrections).

Jean is the therapist in this session. Vikram and Louisa are the parents of David, their adult son, who has been given a diagnosis of depression.

Extract 4

```
1    Vikram: Can you please tell me something? If he got himself in this
2    sort of present stage, can a person be changed, can a personality
3    (0.2) can he say, 'I'm not happy with myself, I don't want to, you
4    know, follow my dad's path', what can he decide on his own, what is
5    his personal life, that sort of thing, so can a person change his
6    personality at that stage, I mean [
7    Jean:                    [Well I suppose I'd like to hold
8    off from answering that question really, cause I suppose I'd like to
9    have an opportunity to hear some of my
10   (both Jean and Vikram talk at the same time)
11   team members' views as well [
12   Vikram:                     [Well, yeah, you are the professional
13   team, so say a personality(0.2) can a person break away from the
14   patterns of life which someone has decided to that duration (.) can
15   a person [
16   Jean:    [Before I take a break is there anything extra that you'd
17   like to add as well? ((to Louisa)).
```

Session extracts may be characterised by their tussles between therapist and family members over control of conversational formats, with family members seeking to manoeuvre the therapist into an advisement format, and meeting resistance from the therapist (Roy-Chowdhury, 2003). In conversational terms, this discursive shift is commonly signalled by a family member reversing the usual interview format and asking a question of the therapist. Studies have shown that the preferred response to such a request is to provide the information requested (Potter, 1996; Sacks & Schegloff, 1979). However, we see in Extract 4 that the therapist does not directly answer Vikram's question as to whether or not David can change his personality. In doing so, she gives a dispreferred response, which would commonly require a repair. The attempt by Jean to correct this trouble source is minimal, citing as she does her wish to hear the views of members of the therapy team and offering no explanation as to the reasons for this wish, conferring an inability to respond to Vikram's question. Her lexical choice repeated in lines 7 and 8 that she would '*like* to hold off answering that question' and would *like* to hear her colleagues' views imply not that she does not wish to answer the question but rather that she is unable to do so. The latter account, of a constraint to action, is commonly used to initiate a repair to a refused request (Potter, 1996). This might take the form of an account of the constraints upon Jean that makes her *unable* rather than *unwilling* to accede to the request for information. She fails to do so, and Extract 4 concludes with the recurrence of a trouble source that remains unrepaired as Jean leaves Vikram and Louisa's requests for advice unanswered and returns to her preferred conversational format.

The evidence from the talk within these therapy sessions is that the therapeutic relationship is under strain, and there are problems of speaking, hearing and understanding. Vikram's misunderstanding of Jean's turns on a number of occasions is symptomatic of these difficulties. Louisa and Vikram often do not locate themselves as troubled but rather separate David's problems from familial contingencies. In this way, they resist the suppositional basis for a family therapy. Their repeated attempts to elicit expert opinions from Jean are mirrored by their own marshalling of accounts for their son, which invoke pathology and personal deficiencies. The discrepancy between these accounts and those offered by Jean are rarely addressed and negotiated. The effects of differences in lexical choices between interactants is both symptomatic of and contributory to low levels of engagement and a therapeutic relationship that stumbles from one conversational crisis to another. Vikram and Louisa describe their son's 'deficiencies', his lack of self-confidence, his shyness; Jean carefully avoids the use of this language. She resolutely avoids labelling or blaming David and instead seeks to problematise his parents' actions in relation to him. In the absence of these conversational elements which serve to build a successful interaction, there is not a consensus achieved sufficient for participants to maintain the conditions for psychotherapy.

Given that Jean has encountered a family who are challenging her preferred conversational format and are seeking to position her as an expert, a legitimate question would be to ask: what can she do that might bolster the therapeutic relationship? Let us look at a further session extract, involving Liz, the therapist whose work we looked at in Extracts 1–3, working with another family for at least one possible answer to this question. From an analysis of Jean's work with Vikram and Louisa, together with other sessions, we have found that family therapists seek to talk into being a representation of themselves as, to put it within the terms of theoretical descriptions, 'non-expert'. Despite repeated appeals by therapees, they resist being positioned as expert advisors, and yet in doing so, they clearly demonstrate their professional control over permissible conversational formats. This then is a theoretically driven position taken by the therapist, which, when placed ahead of maintaining and strengthening the therapeutic relationship, appears to undermine the therapy. Let us consider a further example of this appeal for an expert opinion, but one that this time ends a little differently.

The following extract is taken from the end of the first session between the therapist, Liz; Tracey, a trainee family therapist; and the family comprising David, Julia, and their three children, John (the oldest), Peter, and Kathy (the youngest). David has hit Julia in the past. All three children have 'behavioural problems'.

Extract 5

```
1    Tracey: So, we're actually inviting you to come to further sessions
2    and we would like you to make a decision about whether that would be
3    helpful to you as a family=
4    David: I'd come if I knew the reason why I was coming and I don't
5    think it's fair to drag me and him ((indicating John)) up. This is
6    obviously another underlying problem I don't know about or I do know
7    about and none of the others do (.) I don't know.
8    ((Kathy opens and closes the room door)).
9    Tracey: I guess [
10   David:          [Unless I get an answer, I'm not coming back
11   Tracey: I guess this is what (.) there are no answers to that. This
12   (0.2) is where we try to move away from what is the reason, who is
13   to blame, etc., we are here to sort of (.) think about what is going
14   on.
15   David: When (.) when you are out to change something you gotta know
16   what you are trying to change. You don't go into Sainsbury's (.) [
17   Liz:                                                             [If
18   it meant that you were less angry, she was more happy and he was
19   working and the children stopped bedwetting would that be enough
20   reason? (0.3)
21   Kathy: Yeah.
22   David: Yeah.
23   Kathy: Yeah, yeah, yeah.
```

The interchange between David and Tracey up to the close of David's turn on line 15 has an air of familiarity about it. Tracey is putting to the family the offer of another appointment and is making an appeal to the family, which in this extract David has taken up. She uses her category entitlement as a psychotherapist to instruct them that they must take responsibility as a family for the 'decision' to take up the offer. David does not give the preferred response to an offer, that is, an acceptance (Sacks & Schegloff, 1979), but rather offers an acceptance that is conditional upon the provision of an expert opinion. He positions himself, alongside John, as being unfairly 'dragged' to the therapy unless he can be provided with a satisfactory explanation by the therapists of why they think it would be helpful to attend further sessions. He puts this bluntly in line 10. Tracey uses her subsequent turn to decline his demand for a professional opinion, as to do so would be to accept his positioning of her as 'the one who knows' and to undermine her earlier account of the decision-making power and responsibility lying with the family. David responds by supporting the reasonableness of his request through an appeal to common-sense idiomatic expressions (see Drew & Holt, 1989), including the invocation of the clarity and simplicity of a purchaser–provider interaction within 'Sainsbury's'.

At this point, we might have expected a further refusal to accept the positioning requested of them by the therapees, perhaps a move toward closure as we saw Jean make at the end of Extract 5. Instead, Liz intervenes (making use of her category entitlement as the more experienced family therapist) and accepts his request to provide an expert assessment of the ways in which, in her opinion, the family needs to change and hence the reasons for them to attend further sessions. The fact that she puts this opinion in the form of a question does not lessen its impact as it invites a response and does not disguise the advisement form of the turn that David has been seeking. The response is initially a rather stunned silence, which indicates that participants view Liz as acting outside the interpretive repertoire of the therapist that she and Tracey have established during the course of the session. Kathy and David then give the minimum monosyllabic response required by the question in order to effect a repair to the conversation, answering in the affirmative. This allows Liz and Tracey to move rapidly toward making the arrangements for the next session. Remarkably, given his previous problematising of future attendance, David says that he will come to the next session.

We can observe the difference between the similar struggle for control of the conversational format in Extract 5 and the therapist acceding to David's request for advice. In this extract, David repeatedly initiates openings for the therapists to repair their earlier refusal to accede to his request. In giving the preferred response in lines 16–18, Liz repairs the trouble source and rescues the therapeutic relationship from the risk of a breakdown. It seems that for the therapist to maintain flexibility regarding invitations to vary the conversational format and to provide an expert opinion is a stance that serves not only to enhance the therapeutic relationship but, in this instance, to save it.

In other words, it seems to be the case that the application of psychotherapy theory is unlikely to be effective without a strong therapeutic relationship. In those instances where the therapist maintains a theory-driven conversational format, for example a Milan-systemic style of questioning or a reluctance to be positioned as the expert, and places this before maintenance and repair of the relationship, the therapy can be seen to be under pressure. We saw in Extract 5 that Jean uses and seeks to maintain an interview format with David, Louisa, and Vikram. This format, where the therapist questions family members and provides minimal informative or reflective responses herself, is characteristic of a Milan-systemic model.

Let us look at another session where there are clearly difficulties in constructing a therapeutic relationship. This extract is taken from the first therapy session with Adam and Kate. John is the therapist.

Extract 6

```
1   Adam: Yes broadly speaking that's the situation I think.
2   John: Hm are there things that you've done or tried (.) to address
3   these issues as a couple
4   Adam: Are there things that we've? ((doesn't seem to have heard the
5   question))
```

Ostensibly this may appear to be a momentary lapse in hearing, but it is the first such occasion in the session, and we may ask, why here? Quite soon afterwards, John asks a similar question and this time Kate does not hear the question. A little later John again asks if they are doing anything that seems to make things better between them. This time Kate replies, 'I don't think that I have been doing anything different'. Taken together, it does appear that we are observing further problems in the therapeutic relationship which may reflect earlier unrepaired trouble sources but also an insufficiently negotiated move to transform Kate and Adam's experiences into an institutionalised frame of reference where solutions or (expressed within the theoretical language of solution-focused therapy) 'unique outcomes' are sought. There appear to have been too few occasions where the therapist acknowledges and demonstrates an understanding of the couple's experiences to allow the shift in the discursive register required in moving from experiences to questions eliciting 'solution-seeking behaviour'. Furthermore, the therapist's language, for example, of addressing issues, indicates professionalised rhetoric, rather than the use of everyday or idiomatic resources. Our findings demonstrating the conversational coordinates of a positive therapeutic relationship are again supported; without this engagement, difficulties appear in the use of theoretically informed conversational formats. There is an unwillingness of those designated as therapees to do therapy talk, and without this cooperation the therapist struggles to be a therapist. Following the second therapy session, Kate and Adam choose not to take up the offer of further appointments.

Discussion

In sessions where engagement is high, there is a demonstrable flexibility by the therapist as to the implicit or explicit demands of the family. This shows itself in a willingness to adopt a variety of conversational formats, including at times an advisement format. The impression is created in the therapists' talk that she is more interested in their experiences and their descriptions of these experiences than in transforming their accounts into a professional frame of reference. Her willingness to simply listen to the family while providing minimal responses in order to convey understanding and invite elaboration is one indication of this positioning as is her attention to the language used by the family and frequent recourse to everyday or idiomatic language. The therapist negotiates an alignment with the family by these and other means. She refers explicitly to asymmetries of knowledge and endeavours to clarify the extent of what she knows. She uses the privileges of her status as a psychotherapist to provide and bolster expert opinions and to entitle her to determine aspects of the session's structure. However, in providing opinions she takes care to do so in ways that acknowledge and 'fit' with prior constructions. At times, she presents the family with choices and opportunities to discuss session structure, which consolidates the impression that there are limits to her expertise.

The finding appears to be robust across therapy sessions, where, without good levels of engagement, difficulties appear in the use of theoretically informed conversational formats by the therapist. There is an unwillingness of those designated as therapees to do therapy talk, and without this cooperation, the therapist struggles to be a therapist.

Clinical relevance summary

In the chapter, there are several implications for practice. First, the analysis has pointed to a more dialectical process where unrepaired trouble sources between all participants emerge again and again and subvert the therapeutic aims of the conversation. The persistence of this finding even where the therapist is seeking to work within a model of creating behavioural change tempts one into generalising the importance of the therapeutic relationship irrespective of model.

Second, the detailed analysis of interactions that has been made here can serve as a useful method for examining the conversational coordinates of a successfully achieved therapeutic relationship. In the present analysis, these coordinates have been found to include an attention to lexical choices made by family members and an attention to turn design, a willingness to use everyday, non-institutionalised language, a flexibility in the use of a number of conversational formats including advisement in response to demands from the family, a willingness to negotiate session structure and content and to discursively manage asymmetries of knowledge and expertise. The therapist who will simply listen, acknowledge, and witness, if you will, the accounts offered by family members rather than be too quick to transform lifeworld descriptions into institutionalised rhetoric is more likely to generate an engagement necessary for the accomplishment of therapeutic goals.

Third, we see among the family therapists whose work we have encountered an ambivalence towards the power and presumptiousness accorded to them by virtue of their position in relation to those in therapy, a wish to deconstruct the dominant discourses of power, which is in tune with a postmodernist sensibility. Family therapists not only make use of category entitlements in positioning those in therapy as troubled and in need of help and in controlling the format and structure of sessions but also seek rhetorically to subvert their own power through evocations of doubt and uncertainty. The reasons for this are complex and, as we have seen, potentially confusing to family members. A social constructionist family therapy theory, with its emphasis upon the deliberately oxymoronic 'unknowing expert', undoubtedly contributes to this therapist positioning and also is a product of a deliberate attempt to induct the therapee into talking in a particular way, one that casts doubt upon familiar self and other descriptions and opens a space for alternative descriptions.

Fourth, given this complex and paradoxical positioning within discourses organising power relations, therapist persuasions are also put in a simultaneously knowing and unknowing format. That such persuasions, or, to put it another way, theory-driven techniques, are an intrinsic element of the therapy process is clear, and many of the transcript extracts include attempts at changing narratives, beliefs, behaviours, and so on. In Extract 3, for example, Liz, the therapist, seeks to persuade Anne and Ian that Paul has some control over his behaviour, which they had previously denied. In theoretical language, she is seeking to effect a reframe, in order to introduce new information, a new way of seeing the situation in the family. She does so skilfully making use of Anne and Ian's own representations of Paul's actions and constructing her turn carefully to soften the challenge and introduce ambiguity and uncertainty.

Fifth, the care taken in enacting such persuasions is in marked contrast to the less ambiguous evocation of expert status that is a feature of a similar process for medical practitioners and reflects the construction of the persuasion within the 'lifeworld' of the therapee rather than in the decontextualised domain of biomedicine (see Hak & Boer, 1995). The persuasion must make sense within the therapee's lifeworld, as it is within this domain of knowledge and experience that the change is being suggested, whereas in medicine the change is located outside this lifeworld in a parallel biological field of explanation. This explains also the therapists' repeated and effective use of idiomatic language in order to engage with the lived experiences of family members as they themselves would construct these experiences.

This engagement, which seeks transformations that are more subtle (the reasons for a doctor advising an overweight patient to exercise more are readily explained; the reasons for seeking to persuade family members that a husband and father has some control over his actions, less so) and yet profound, requires of the psychotherapist a capacity for interpreting from moment to moment the multiplicity of influences upon individual narratives provided by therapees. In the absence of such an analysis, persuasions are significantly less likely to be successful: therapees may actively discount such attempts at creating change or nod compliantly and fail to return to subsequent sessions.

The findings of this qualitative study support the findings of quantitative studies in that the successful use of theory-driven techniques have been found to be continuously undermined without a close attention to the need to build and maintain a strong therapeutic relationship. We have found that therapists accord to therapeutic encounters differing importance to relationship and technique. It is possible to tentatively assert that this is an important factor that distinguishes more from less successful therapy and that therapists who are able to manage this balance well and accord the therapeutic relationship the centrality warranted by the evidence are more likely to provide effective psychotherapy. This is one factor that is likely to account for therapist differences in outcomes as found in Crits-Cristolph et al.'s (1991) analysis. For a simple summary of the implications for practice, see Table 11.1.

Table 11.1 Clinical practice highlights

1. This study has found support for the finding that a strong therapeutic relationship is essential for a successful psychotherapy.
2. Close attention to the alliance in psychotherapy is warranted, and this should guide a collaborative approach to the use and timing of techniques and advice.
3. The self of the therapist and the power relations within the therapeutic encounter should be monitored by the therapist and made available for consideration within the therapy.
4. The taking up of a 'non-expert' position by the therapist requires careful handling, as this may, paradoxically, place the therapist in a more powerful position through control of the permissible conversational format.
5. Qualitative analyses of therapy sessions can provide valuable pointers towards successful practice at a detailed conversational level, and this chapter gives some indicators of the means by which the therapeutic relationship can be strengthened.

Summary

The evidence suggests that a successful psychotherapy resting upon a social constructionist assumptive base places importance upon conveying understanding and of locating the self of the therapist within the therapy, as proposed by client-centred therapists, psychoanalysts, and, more recently, systemic theorists. A discourse analyst's regard to inter-textual analysis and the subversion of a straightforward correspondence between talk and meaning and between signifier and signified appears to be warranted. This study has found support for the finding of quantitative studies that have assessed the relative importance of techniques drawn from psychotherapy theory as being less significant in creating therapeutic change than 'common factors' such as the strength of the alliance and collaboration between therapist and therapee. This applies as much to theory-driven attempts to control the conversational format by the therapist as to the use of specific techniques. Asymmetries of power and knowledge and the therapee's positioning as the unknowing expert, an aspect of the doing of a social constructionist psychotherapy, laden with ambiguities as it is, has been found to require expert handling and care in its application.

Using a qualitative research method, DA, research questions have been investigated that have more often been reserved for methods that quantify constructs and subject this quantification to a statistical analysis. Quantitative methods such as the RCT have a role in answering such questions as to whether or not a psychological therapy 'works' or is found to be helpful according to certain predetermined criteria. In this chapter, I have sought to answer the question of why a systemic psychotherapy is successful at the level of the interaction between participants within the therapeutic conversation, to understand which of the conversations are likely to be more helpful to those designated the client, service user or patient. A DA has been found to be a robust method for addressing these questions, which is epistemologically congruent with the therapy under scrutiny. One might make use of a similar qualitative methodology to examine where therapy works well or less well with specific populations, for example, cross-cultural therapy or therapy with children and young people.

As clinicians and researchers, it is important for us to seek, individually and collectively, to influence the political agenda that locates RCTs as the gold standard for psychotherapy research. RCTs themselves are ripe for reform to take better account of the evidence that a drug metaphor is of limited methodological utility. This might allow for a greater degree of epistemological congruence with systemic theory and practice. Without this, intrinsic aspects of systemic practice (and indeed other psychotherapy models) are occluded and hidden from the researcher's gaze. While we seek to increase the influence of a broader range of methodologies, we should also work towards building greater clinical complexity into the design of RCTs. This is, I believe, an urgent task. Allowing the present methodological hegemony to strengthen holds real risks for the continued availability of systemic and other psychotherapies within public healthcare systems such as CAMHS services.

References

Antaki, C., Billig, M., Edwards, D., & Potter, J. (2002). Discourse analysis means doing analysis: A critique of six analytic shortcomings. *Discourse Analysis Online, 1*(1), online.

Asay, T. P., & Lambert, M. J. (1999). The empirical case for common factors in therapy: Quantitative findings. In M. A. Hubble, B. C. Duncan & S. D. Miller (Eds.), *The heart and soul of change: What works in therapy* (pp. 33–56). Washington, DC: APA.

Bentall, R. (2009). *Doctoring the mind: Why psychiatric treatments fail.* London: Allen Lane.

Billig, M. (1997). Rhetorical and discursive analysis: How families talk about the Royal Family. In N. Hays (Ed.), *Doing qualitative analysis in psychology* (pp. 39–54). East Sussex: Psychology Press.

Blatt, S. J., Sanislow, C. A., Zuroff, D. C., & Pilkonis, P. A. (1996). Characteristics of effective therapists: Further analyses of data from the National Institute of Mental Health treatment of depression collaborative research program. *Journal of Consulting and Clinical Psychology, 64*, 1276–1284.

Blow, A. J., Sprenkle, D. H., & Davis, S. D. (2007). Is who delivers the treatment more important than the treatment itself? The role of the therapist in common factors. *Journal of Marital and Family Therapy, 33*(3), 298–317.

Carr, A. (2014). The evidence base for family therapy and systemic interventions for child-focused problems. *Journal of Family Therapy, 36*(2), 107–157.

Crits-Cristolph, P., Barancackie, K., Kurcias, J. S., Beck, A. T., Carroll, K., Luborsky, L., & Zitrin, C. (1991). Meta-analysis of therapist effects in psychotherapy outcome studies. *Psychotherapy Research, 1*, 81–91.

Drew, P., & Holt, E. (1989). Complainable matters: The use of idiomatic expressions in making complaints. *Social Problems, 35*, 398–417.

Flick, V. W. E. (1998). *Introduction to qualitative research.* London: Sage.

Hak, T., & Boer, F. (1995). Professional interpretation of patients' talk in the initial interview. In J. Siegfried (Ed.), *Therapeutic and everyday discourse as behaviour change.* New Jersey: Ablex.

Harper, D., Gannon, K. N., & Robinson, M. (2013). Beyond evidence-based practice. In R. Bayne & G. Jinks (Eds.), *Applied psychology: Research, training and practice* (pp. 32–46). London: Sage.

Heritage, J. (1997). Conversation analysis and institutional talk. In D. Silverman (Ed.), *Qualitative research: Theory, method and practice* (pp. 161–182). London: Sage.

Horvath, A. O., & Symonds, B. D. (1991). Relationship between working alliance and outcome in psychotherapy: A meta-analysis. *Journal of Counselling Psychology, 38*, 139–149.

Horvath, A. O., Del Re, A. C., Fluckiger, C., & Symonds, D. (2011). Alliance in individual psychotherapy. *Psychotherapy, 48*(1), 9–16.

Kaye, J. (1995). Postfoundationalism and the language of psychotherapy research. In J. Siegfried (Ed.), *Therapeutic and everyday discourse as behaviour change* (pp. 29–59). New Jersey: Ablex.

Liddle, P. F. (1987). The symptoms of chronic schizophrenia: A re-examination of the positive-negative dichotomy. *British Journal of Psychiatry, 151,* 145–151.

MacNamee, S., & Gergen, K. J. (1992). *Therapy as social construction.* London: Sage.

McGorry, P. D., Mihalopoulos, C., Henry, L., Dakis, J., Jackson, H. J., Flaum, M., & Karoly, R. (1995). Spurious precision: Procedural validity of diagnostic assessment in psychotic disorders. *American Journal of Psychiatry, 152,* 220–223.

O'Reilly, M. (2014). Blame and accountability in family therapy: Making sense of therapeutic spaces discursively. *Special issue: Qualitative Psychology, 1*(2), 163–177.

Pilling, S. (2009). Developing evidence-based guidance – implications for systemic interventions. *Journal of Family Therapy, 31,* 194–205.

Potter, J. (1996). Discourse analysis and constructionist approaches: Theoretical background. In J. E. Richardson (Ed.), *Handbook of qualitative research* (pp. 125–140). Leicester: BPS.

Potter, J. (1997). Discourse analysis as a way of analyzing naturally occurring talk. In D. Silverman (Ed.), *Qualitiative research: Theory, method and practice.* London: Sage.

Roy-Chowdhury, S. (2001). The language of family therapy: What we say we do and what we actually do in therapy. Unpublished PhD thesis, City University.

Roy-Chowdhury, S. (2003). Knowing the unknowable: What constitutes evidence in family therapy. *Journal of Family Therapy, 25*(1), 64–85.

Roy-Chowdhury, S. (2006). How is the therapeutic relationship talked into being? *Journal of Family Therapy, 28,* 153–174.

Roy-Chowdhury, S. (2010). IAPT and the death of idealism. *Clinical Psychology Forum, 208,* 25–29.

Sacks, H., & Schegloff, E. A. (1979). Two preferences in the organization of reference to persons in conversation and their interaction. In G. Psathas (Ed.), *Everyday language: Studies in ethnomethodology* (pp. 15–21). New York: Irvington.

Sacks, H., Schegloff, E. A., & Jefferson, G. (1974). A simple systematics for the organisation of turn-taking for conversation. *Language, 50,* 696–735.

Schegloff, E. A. (1992). Repair after next turn: The last structurally provided defence of intersubjectivity in conversation. *American Journal of Sociology, 97,* 1295–1345.

Shapiro, D. A., Startup, M., Bird, D., Harper, H., Reynolds, S., & Suokas, A. (1994). The high-water mark of the drug metaphor: A meta-analytic critique of process-outcome research. In R. L. Russell (Ed.), *Reassessing psychotherapy research* (pp. 1–35). New York: Guildford Press.

Stancombe, J., & White, S. (1997). Notes on the tenacity of therapeutic presuppositions in process research: Examining the artfulness of blamings in family therapy. *Journal of Family Therapy, 19*(1), 21–41.

Stiles, W. B. (1992). *Describing talk.* Thousand Oaks, CA: Sage.

Sutherland, O., & Strong, T. (2011). Therapeutic collaboration: A conversation analysis of constructionist therapy. *Journal of Family Therapy, 33*(3), 256–278.

Tsuang, M., Woolson, R. F., & Fleming, J. A. (1979). Long-term outcome of major psychoses: Schizophrenia and affective disorders compared with psychiatrically symptom-free surgical conditions. *Archives of General Psychiatry, 36,* 1295–1301.

Wampold, B. E. (2001). *The great psychotherapy debate: Models, methods and findings.* Manwah, NJ: Erlbaum.

Recommended reading

- Heritage, J. (1997). Conversation analysis and institutional talk. In D. Silverman (Ed.), *Qualitative research: Theory, method and practice* (pp. 161–182). London: Sage.
- MacNamee, S., & Gergen, K. J. (1992). *Therapy as social construction.* London: Sage.
- Wampold, B. E. (2001). *The great psychotherapy debate: Models, methods and findings.* Manwah, NJ: Erlbaum.

Part III

The Social Construction of Normal/Abnormal

12
Red Flags: The Social Construction of a Symptom

David C. Giles

Introduction

How does a 'behaviour' become a 'symptom'? A symptom is only meaningful in its context, as a kind of puzzle for an officially recognised expert to solve. While physical symptoms – rashes, sore throats, runny noses – are relatively easy for a medical practitioner to piece together to form a coherent, familiar picture, behavioural symptoms pose a much greater challenge. A swab cannot be taken that will confirm the doctor's diagnosis; as Szasz (1971) puts it, 'no-one knows for certain who is, or is not, mentally ill' (p. 34). The ambiguity of 'mental illness' is a fabulous breeding ground for symptoms, jostling for priority, yearning for interpretation. Compared to the neatly interlocking puzzle pieces available for constructing a common physical malady, the psychiatrist is confronted with a plethora of different shapes, most refusing to fit together, and probably deriving from several different boxes.

Now bring psychiatric diagnosis out into the open, plaster it all over the Internet, and see what happens: The symptoms (never mind the syndromes) grow and multiply; behavioural quirks and oddities are recognised by more and more people as symptoms of one or other condition. And when the people spotting those symptoms happen to be parents of the children demonstrating those behavioural quirks and oddities, and are quite reasonably seeking pieces to fill in the missing gaps in their own familial jigsaw puzzles, the context of interpretation becomes stretched ever wider. In the digital age, we are all analysts, performing our own interpretations, and we find symptoms everywhere.

Nowhere has this process become more elaborated than in the field of autism (see O'Reilly, Karim, & Lester, Chapter 14, this volume). In this chapter, I use the proliferation of autism symptoms (and their accompanying 'treatments') as a case study of how behaviours become socially constructed as pathological – as mental illness phenomena or, specifically, symptoms. I will do this by applying a blended analysis of discourse circulating through both the contemporary World Wide Web and through the vast academic literature on autism.

217

The proliferation of autism symptoms

It has been frequently noted that there is a contemporary 'scare' about autism that is aligned to increased diagnosis of the condition, as well as the emergence of a popular discourse around autism (Hacking, 2009). Autism is sometimes referred to as a modern 'epidemic' (Eyal, Hart, Onculer, Oren, & Rossi, 2010), and in the most recent edition of the American Psychiatric Association's *Diagnostic and Statistical Manual of Mental Disorders* (*DSM-5*, 2013), the scope of the condition has been stretched further by the absorption of related disorders into a broad 'autism spectrum' that encompasses conditions previously diagnosed as discrete disorders: Asperger's disorder, childhood disintegrative disorder, and the ragbag category pervasive developmental disorder – not otherwise specified (PDD-NOS). From 2013, individuals are able only to receive a formal psychiatric diagnosis of autism spectrum disorder (ASD).

At the same time, there is a healthy scepticism about the remarkable increase in autism diagnoses in recent years and the – often bizarre – psychiatric explanations offered to account for it.[1] Greater awareness and improved detection is the most common, although as Eyal et al. (2010) have shown, detection rates have more to do with diagnostic criteria than with 'awareness', and the salience of symptoms varies markedly across cultures. For example, Daley and Sigman (2002) found that Indian professionals were less likely to consider language delay as a central characteristic. Nadesan (2005) has suggested that autism be considered what Ian Hacking (1998) called a 'transient mental illness' – one that is intrinsically linked to a certain time and place and can just as easily vanish completely. Hysteria served as a useful construct for Freud and his contemporaries to explain mental distress in their middle-class female clients, but the diagnosis failed to make the second edition of the *DSM*. Shell shock ceased to be diagnosed when combat-related disturbance was found well away from the trenches of the First World War. Neither multiple personality disorder nor Asperger's disorder lasted longer than a single edition as *DSM* categories.

Along the same lines, Sami Timimi and colleagues (Timimi, Gardner, & McCabe, 2011; Timimi & Leo, 2009) have argued that autism and attention deficit hyperactivity disorder – and in particular the medicalisation of such conditions – are one of contemporary society's ways of producing and then controlling unruly boys (see Timimi & Timimi, Chapter 8, this volume). Indeed, there is plenty of material in the diagnostic tools used by clinicians that suggest that the main beneficiaries of autism diagnoses are not parents or children, but teachers and other educational practitioners, who are charged with supervising mealtimes or whose increasingly academic curricula require preschool children to gather and behave at regular intervals for practices like circle time. Among the 'target behaviours' listed for pre-schoolers receiving therapy for autism are 'sitting quietly', 'walking in line', 'waiting quietly', and 'participating in circle games' (Scott & Baldwin, 2005). In our ultra-competitive and materialistic world, diagnosing autism is a convenient means of producing obedient and compliant youngsters, who can attend educational institutions full-time while their parents contribute to the economic growth of the nation.

As the diagnostic criteria for 'autism spectrum disorder' become ever broader, the opportunity for additional signs and symptoms of the condition increases, and the diversity of research findings, theories, and explanations for different manifestations of autism (spectrum disorder) has led to an expanding set of associated behaviours that can, or may, be a potential source of concern for parents. Inevitably, the more symptoms that are displayed in the individual, the worse the prognosis that is given.

The ever-increasing plethora of symptoms is exacerbated by the extraordinary range of autism resources available on the World Wide Web. Most of these are aimed squarely at the parents of children with autism, or parents who are concerned, or even hopeful, that their children may have autism (I say 'hopeful' not in order to trivialise the parents' concern, but because there is a strong community of autism advocates who prefer to construct autism as a positive attribute). Some websites are run by the autism community itself (e.g. the hugely successful *Wrong Planet*, a US-based site with over 60,000 members and its own merchandise including T-shirts, bags, and hats) or by charitable bodies engaged in raising autism awareness; others have been set up by health professionals with the aim of educating the public or linked to more general health sites.

Project overview

In this chapter, I pull examples of online data fairly indiscriminately from the above sources. I am not claiming that these examples are typical, or representative, of any specific category of website, but simply present them as illustrative of the kind of discourse circulating online that parents and other interested parties may stumble across, or actively seek out, to inform their understanding of autism and its status as a medical disorder.

I refer to my analytic approach as 'blended' in the sense that it draws from the genealogical approach to discourse (identifying discourse as a cultural manifestation of underlying ideology, power structures, etc.), as well as the more systematic approach of discursive psychology – the microanalytic process of breaking down texts into analytic units, identifying recurrent patterns, with the emphasis more specifically on the way in which language is used as an interactional tool.

Kendall and Wickham (1999) outlined five steps for conducting Foucauldian discourse analysis, which I have broadly used in this chapter. First, a discourse is identified as 'a corpus of statements whose organisation is regular and systematic' (p. 42): Here, symptom talk (with specific reference to autism) would seem to fit the bill. The remaining four steps concern the identification of various sets of rules, governing: (1) the production of statements (here, when symptoms are added to existing lists of diagnostic indicators); (2) how these limit what can be said (here, about the behaviours that are said to constitute symptoms); (3) how they create spaces for new statements (by allowing further symptoms to be added to the list); and (4) practice as simultaneously material and discursive (the symptoms are subsequently used in order to label behaviour as indicative of autism, perhaps – at some later point – in formal diagnostic practices).

A key organising principle for the whole is Hacking's (1986, 1995) work on 'making up persons' and the 'looping effect of human kinds', which has influenced much of my previous research on the use of categories in online communication around mental health (Giles, 2006; Giles & Newbold, 2011, 2013).

How to identify autism symptoms

The term in use throughout the autism community for alerting parents to the presence of a possible autistic symptom is 'red flag'. A behaviour is said to be a red flag for autism if it functions as a potential sign or symptom of the condition. An example of this usage can be found on the website of *Autism Speaks* (*www.autismspeaks.org*), a high-profile US charity that describes itself as 'the world's leading science and advocacy organisation'. Their drop-down menu offers a page for 'symptoms' as well as 'signs', which restricts the former to broad diagnostic indicators such as 'social communication deficits' and the latter to the specific behaviours. On the 'signs' page, visitors are told: 'The following "red flags" may indicate your child is at risk for an autism spectrum disorder. If your child exhibits any of the following, please don't delay in asking your pediatrician or family doctor for an evaluation', followed by a list that includes 'No babbling by 12 months', 'No words by 16 months', and several rather more ambiguous behaviours.

The term 'red flag' in this context probably derives from a paper by Filipek et al. (1999) that features a list (Table III, on p. 452) entitled 'Parental Concerns that are RED FLAGS for Autism', including such things as 'Appears deaf at times', 'Gets things for himself', 'Tunes us out', 'Does things 'early', 'Lines things up', and 'Toe walks'. Although these descriptions are highly simplistic or colloquial, and at best entirely subjective, it must be noted that this paper was published in a highly respected, peer-reviewed scientific publication (*The Journal of Autism and Developmental Disorders*) and was authored by no fewer than 18 medical researchers affiliated to various American universities. Not surprisingly, many of the checklists of 'red flags for autism' found online are taken from this list in the many resources for parents. The list on the *Autism Speaks* website seems to be reproduced from one section of the Filipek et al. table, which was originally published under the heading '*Absolute indications* for immediate further evaluation' [*authors' italics*].

Elsewhere among autism-related online resources and in broader media coverage of autism, there are many more alarming manifestations of red flag discourse. A particularly unpleasant one can be found on the website of the American CBS news channel. Under the heading '*Autism: 9 Warning Signs Every Parent Should Know*', the visitor is presented with a clearly photo-shopped close-up of a sad-looking baby with deep blue marble-like eyes, followed by the text: 'Parents fear autism, and rightly so. The mysterious brain disorder devastates a child's ability to speak and interact with others...Is your child at risk for autism? Here are nine red flags to watch for', followed by the endorsement of a medical expert, and nine linked pages that stretch Filipek et al.'s list of behaviours, and other, newer additions to the standard list, into even more open-ended descriptions. For example, one page is headed 'Arm Flapping', with the text saying, 'Arm flapping is a well-documented red flag for autism. But other strange body movements and postures can also suggest trouble ahead. Does your child repeatedly stiffen his/her arms or legs? Keep twisting his/her wrists?'

Is 'arm flapping' really 'well-documented' as an autism red flag? The Filipek et al. (1999) list does include 'odd movement patterns', which is surely too ambiguous from which to infer 'arm flapping'. The *DSM-IV* criteria for 'Autism Disorder' include, as one of several possible symptoms, 'stereotyped and repetitive motor mannerisms (e.g. hand or finger flapping or twisting, or complex whole-body movements)', which seems to exclude 'arm flapping' as such, not to mention wrist twisting, or arm and leg stiffening. More ambiguously still, the *DSM-5* criteria for autism spectrum disorder simply offers 'stereotyped or repetitive motor movements', which is further qualified by the caveat that the list of diagnostic criteria is merely 'illustrative'. Therefore, stereotyped and repetitive motor movements – of any kind – are simply one instance of the plethora of possible 'repetitive patterns of behaviour' possibly indicative of autism.

Plumbing similar depths of ambiguity, another page is headed 'No name recognition', accompanied by a picture of a child covering his ears, which signals the following warning: 'By six months of age, a child should be quick to look up when someone calls his/her name. Does your child fail to look up when called – or look up only some of the time? That's a red flag'. Such a claim is not made even in the literature that advocates this test of autism detection. Nadig et al. (2007), the most frequently cited source, found that at six months only a 'nonsignificant trend' was observable between consistent own-name response and clinical outcome at 24 months; only from observations at 12 months could this pattern be supported by statistical analysis.[2]

How to build a symptom: Toe walking

In this section, I continue the analysis of symptom construction by exploring the discourse around a second behaviour: toe walking. Toe walking refers to the observed behaviour of a child moving around the environment with the feet bent at the toes (more commonly referred to as 'walking on tiptoes'). Toe walking is the kind of behavioural childhood quirk that one might initially expect to elicit delighted giggles from doting parents, evolving gradually into concern for the physical development of the feet. But the autism community paints a bleaker picture: toe walking is another red flag. It appears in Filipek et al.'s (1999) original list and has therefore been reproduced countless times on autism websites as something for parents to beware.

In the process of constructing an apparently quirky behaviour as a 'red flag' for autism behaviour becomes framed, not as a charming developmental eccentricity but as something almost akin to demonic possession. One *YouTube* video[3] shows an 18-month-old child exploring his parents' new apartment and apparently exhibiting certain 'red flags': The accompanying text reads, 'I didn't know it then, but looking back I can see the signs of autism beginning to emerge'. There are countless others which exhibit children spinning in circles, flapping hands, 'stimming' (unusual or repetitive behaviours that are believed to be stimulating), and generally failing or refusing to respond to their parents' demands. These often invite comments that amount to requests for expert advice. Following a similar *YouTube* video,[4] someone has posted[5]:

> I would love to send you some videos and see what you think [of] my 18 month old son Parker. I am seeking speech and OT but am worried he won't get the proper care or dx [*diagnosis*].

Childhood toe walking is raised as a concern in a number of online discussion threads on general parenting websites. In early 2012, a member of the large and influential UK parenting site *mumsnet.co.uk* posted the following query on the 'Children's Health' subforum[6]:

> My DS [*mumsnet acronym for 'darling son'*] has been walking on his toes for 2–3 months on and off. Although I noticed it, didn't really make a big deal about it, just thought it was a habit, and he would grow out of it. Anyway his teacher, has raised a concern, and has suggested we get him checked as he is also falling over quite a bit at school. Has anyone had any experience of this, or know anyone like this?

It does not take long for the word 'autism' to follow. The first respondent suggested dyspraxia; the second mentions that her daughter toe walked and had 'hypomobility and autism'. Three responses later, a parent recommends that 'it's worth getting checked out as it can be due to other underlying conditions such as autism'. Another says that her son has Asperger's and toe walks and that she calls it 'the aspie bounce'. The original poster replies that she doesn't think that her son has autism, but after a number of discussions about dance classes and physiotherapy a new discussant announces that 'toe walking can be a symptom of ASD'. Another replies that it is not necessarily a symptom, just 'something that a lot of children with ASD do, mainly due to sensory issues'. The previous poster responds by saying, 'toe walking is a red flag for autism'. There is some resistance to this interpretation, which is sensibly resolved by the original poster reassuring other members, 'It's very unlikely [my son] has autism, as he is showing no other signs'.

My aim in this analysis is not to claim that there are direct and measurable 'effects' of symptom talk of this kind, but to demonstrate how it permeates everyday interaction in a way that might raise unnecessary alarm about ordinary childhood behaviour: Fortunately, the original poster in this discussion had already ruled out (or had never reason to consider) autism as an explanation for her son's general behaviour; a more sensitive mother, or a mother of a child with delayed speech or poor social skills, who had suddenly started toe walking, could have reacted quite differently.

A more worrying thread appears on parenting website *thebump.com*[7], which opens similarly to the *mumsnet* one:

Okay, so daughter is 2.5 and has been toe walking for a while now. Dr is not concerned at all. I can't help worrying though that something isn't right. There has been a few instances where she will freak out about certain things. [*goes on to list panic reactions to various stimuli/situations*]

Although one or two responses suggest that it is a typical developmental phase and cite examples of people (including themselves) who toe walked without anything other than physical discomfort, a later member sounds a note of caution:

By 2.5 toe walking is a developmental concern – it could well be nothing, just a preference, but by that age it should be evaluated by a knowledgeable professional. I would get a second opinion.

This is followed by a post that ratchets up the discussion to the autism scare level, suggesting an explanatory link with (contested) theories about sensory processing deficits in autism and their associated 'treatments' (I explore the 'weighted vests' option in the next section of the chapter):

> This. Toe walking is a red flag for neurological differences. It might be nothing, especially since she's so young. But it's a red flag for ASD and sensory issues aside from the physical and gross motor complications it can cause. Sometimes people toe walk because it causes a significant joint compression in the large knuckles of the feet which can be calming to those with sensory issues. Sort of like weighted vests or Temple Grandon's squeeze machine.

There follows a succession of posts that reiterate the call for the original poster to seek additional professional advice, criticising her paediatrician for 'playing down' her concerns.

> I would not dismiss toe walking. It can be indicative of non neuro-typical development. If your pediatrician down plays your concerns I would look for a new pedi and schedule a evaluation through early intervention. You do not need the pediatricians permission to have your child evaluated for delays and it sounds like you have legitimate concerns.

Advice to seek professional advice is typical of discussion threads in online mental health communities (Giles & Newbold, 2011, 2013) and may act as a form of 'stake inoculation' (Potter, 1996) against making judgements that could turn out to be false, or merely contested. One later discussant even throws in a further disclaimer ('I always feel as though I am being an alarmist in these topics. I hate that. But honestly, I agree with [*the earlier post recommending a new paediatrician*]'. On a general parenting website like *the bump* it has the potential to alarm concerned members and undermine their confidence in their current practitioners (note that the posts are not applying critical judgement to medical discourse per se but to the judgement of a specific professional suspected of making a false-negative error).

The same 'stake inoculation' is practised by professionals themselves when addressing the toe walking issue online. On the US parenting site, *sixtysecondparent.com*, a paediatric therapist contributes a blog post entitled 'Toe-walking: what to do about children who consistently walk on tiptoes'. It starts by saying, 'children walk on their toes for various different reasons', and then the eye is drawn to a sentence in bold, two paragraphs below, which states, 'A very large number of children with autism spectrum disorder will persistently walk on their toes': the author goes on to say that she always carries out a general assessment in cases of toe walking that includes questions about language and social interaction.

It transpires that toe walking is not only seen as a 'red flag' for autism. On another US parenting site, *babycenter.com*, a member begins a thread by saying:

> My 2 year old has SPD [*sensory processing disorder*[8]] of the seeking variety. For the last month he has been walking on his toes. Two of our six specialists believe he is at risk for mild ASD. Does anybody know if this new toe-walking is related to ASD because it's a sensory-seeking behavior and kids with ASD often have sensory processing disorders as well?

The first response to this post begins: 'As far as I understand it, toe walking, if a physical cause has been excluded, is a red flag for SPD'. Subsequent responses offer further examples of toe walking children with a variety of psychiatrically recognised and unrecognised diagnoses. There is not room in this chapter to explore the various issues relating to the detection and diagnosis of SPD as such, but this example serves to illustrate how the 'red flag' discourse routinely applied to autism has begun to mutate and permeate talk around other conditions in their associated communities.

What scientific evidence is there for the status of toe walking as a symptom, or sign, of autism? As one might imagine, the literature on this topic is not vast, and the autism link seems to be taken for granted by most authors, without them necessarily being able to account for it in neurological terms, typically just reporting its association with 'language disorders'. What actually constitutes toe walking is hard to pin down because the vast majority of research relies on parents' anecdotal evidence that their child toe walked 'some of the time'. Despite referring to toe walking as 'an equinus gait' (Barrow, Jaworski, & Accardo, 2011, p. 619), these and other authors do not register abnormalities in their samples' walking or standing, apart from a minority exhibiting 'tight heel cords'. Overall, it seems that toe walking is a clinical matter largely because it has been observed by parents.

The prevalence of toe walking in autism is a source of considerable inconsistency: one older study estimated it at 62.9% (Accardo, Morrow, Heaney, Whitman, & Tomazic, 1982), while a more recent one puts it at 20.6% (Shattuck, 2006), the latter author explaining the decline as being masked by the widened diagnostic criteria for autism. Barrow et al. (2011) used the same argument to explain differing prevalence among subsamples of children with autism (20.1%) and Asperger syndrome (10%). They accounted for this discrepancy in terms of language disorder, Asperger syndrome being less likely to be characterised by speech delay in early childhood. The incidence in toe walking in their overall sample of children is recorded as 12.1%, though this figure is not broken down into ASD and non-ASD subsamples, and since about a third of the sample had an ASD diagnosis, it is hard to draw from the data a normative prevalence of toe walking. Nevertheless, one thing that can be confidently stated is that it is less prevalent in Asperger syndrome than in the general population.[9]

The relationship between language and toe walking is a strange one, not least because it takes us even further away from the sensory processing argument (not a compelling theory, but at least logically consistent). However researchers fail conspicuously to address the issue of why disorders in speaking or comprehension should manifest themselves in the behaviour of toe walking. Despite explaining their findings as a matter of language, Barrow et al. (2011) go on to suggest that it could result from a 'tonic labyrinthine reflex' arising from autistic infants' tendency to 'arch when cuddled' (p. 620), a reaction caused by their allegedly innate aversion to affection. This is wildly speculative (and many authors vehemently dispute the claim that children with autism shrink from affection), but at least has some logical consistency. The authors claim, however, that it would be extremely difficult to provide supporting evidence, since it would 'require identifying subtle neuromotor patterns consistent with this reflex pattern' (Barrow et al., 2011). Nevertheless, they conclude by recommending that all toe walking children be 'screened for language delay and autism spectrum disorders' (p. 621).

Proceeding from symptom to treatment: Sensory processing and blanket wrapping

Nothing powers a symptom like the first-hand account of a famous individual, and Temple Grandin is probably the most famous single person with autism (unless we count Einstein, whose reportedly late onset of speech led him to be adopted as an honorary 'aspie'). Grandin (1992) is the source of a popular therapy for autism, known as 'deep touch pressure'. This is rooted in the notorious experiments of Harlow and Zimmerman (1959), in which baby monkeys were presented with a cloth-covered 'surrogate' mother and spent more time clinging to it than to a wire surrogate with a feeding device attached. Harlow claimed this as evidence that infants of all types value touch, providing security and comfort over food.

One might expect a child displaying a 'tonic labyrinthine reflex' through aversion to cuddling to be the last to be seeking out the comforting touch of a parent, and yet Grandin (1992) claims that in autism, tactile stimulation is even more highly valued than in the general population, since the child often seeks out 'pressure' experiences, such as wrapping itself in blankets or bandages. This is, she argues, an attempt to cope with 'sensory processing deficits' resulting from neurological abnormalities in the cerebellum region of the brain. In the same paper, the author describes the development, and single-case trial with herself as subject, of a 'squeeze machine' – a padded device that allows the child to experience five minutes of 'deep touch pressure' that (based on anecdotal observation) results in 'a readily detectable calming effect'. She does qualify her findings by admitting that the squeeze machine may be of little use with some autistic children, since autism is by its nature a 'heterogeneous' disorder (somehow embracing aversion to cuddling at the same time as seeking out a hug).

Grandin's influence within the autism industry has inspired a host of other 'calming' instruments to be devised for use in occupational therapy in particular. A British company called weight-2-Go Blankets (*weight2goblankets.co.uk*) offers a variety of weighted blankets that function as bedtime quilts, as well as shoulder wraps and 'lap pads' that are claimed to reduce hyperactivity and keep children still (presumably at those important moments like circle time and lunch). The efficacy of their product range is outlined by reference to what they call 'sensory integration dysfunction', which incorporates both over- and under-sensitivity to sound, colour, touch, or movement, 'unusually high or low activity levels' and a host of other typical ASD symptoms and associated behaviours. It is hard to see how the weighted blankets, wraps, and pads could be helpful for a child with *over*sensitive reactions to touch, or unusually *low* activity levels, though, as Grandin (1992) pointed out, deep touch pressure is not a solution for *all* children on the spectrum.

It would be unfair to criticise a commercial website for overselling its product, but it is notable that, apart from a list of 'testimonials', the manufacturers are unable to provide any scientific literature that might constitute convincing evidence for the efficacy of their blankets (even in their 'FAQ' section, 'how do you know they work?' is not a question on their customers' lips). A US company, Mosaic Weighted Blankets (*mosaicweightedblankets.com*), with a similar product range, tries a little harder, by at least providing some neurochemical claims for their use with autism: The products stimulate the production of serotonin (that produces the calming effect) and melatonin (that helps children sleep, said to be a common problem in autism). The manufacturers' claims stretch far beyond autism and embrace a long list of psychiatric and sleep disorders; but, despite endorsements from occupational therapists, they still fail to cite any studies that support their effectiveness.

As Michael Billig (1999) argued, in the analysis of rhetoric one must pay as much attention to what is left unsaid, and one would imagine that businesses trumpeting the endorsements of practitioners and the activity of specific neurochemicals would be only too eager to present scientific citations to back up their claims. Alas, we are still awaiting the first conclusive controlled study that successfully demonstrates the efficacy of deep touch pressure therapy on any meaningful behaviour in autism. Most supportive studies are poorly designed (typically, failing to include a control group or condition), conflate statistics, or are confined to a single-case study (Stephenson & Carter, 2008).

When deep touch pressure studies are well designed, they are generally unable to find any significant decrease in the target behaviour that can be attributed to the pressure devices themselves, such as the recent study by Watkins and Sparling (2014) that compared behaviours while children were wearing 'snug vests' (at various levels of inflation) with those before and after the vests were attached. Even studies that have demonstrated some marginal benefits of weighted vests and similar devices are able to attribute those findings to more general aspects of therapy, such as increased attention and decreased demands (on the child) (e.g. Doughty & Doughty, 2008).

Can we conclude from the research that weighted vests, blankets, and other deep touch pressure appliances *do not work*? It all depends what constitutes effectiveness. Some of the authors of otherwise unsuccessful studies claim that the children began to ask to wear the vests and that they were able to act as 'reinforcers' (Doughty & Doughty, 2008). A discussant on the *autism.org.uk* forum says: '[my son's] got a blanket now and loves it... it doesnt always help but its another thing to help. anything that helps is a good thing'. But faith alone cannot guarantee effectiveness, and there have been at least two cases in non-clinical settings where children with autism have been suffocated when wrapped in blankets. One case concerned a Pentecostal 'storefront' church in a ritual intended to drive out the evil spirits responsible for the child's autism[10], and the other in a Canadian school specialising in learning disability where occupational therapists had demonstrated the technique previously.[11]

Clinical relevance summary

I hope it is quite clear from what I have said up to now that clinical practitioners should not deal with problematic behaviour by wrapping children in blankets. If clinical practice is intended to be evidence based, the evidence for sensory processing is poor: the evidence for considering 'toe walking' (however this is defined, and this is a problem in itself) as an indicator of autism is extremely limited and partial. It may be worth considering whether the checklist approach to autism – informed by such instruments as the Checklist for Autism in Toddlers (CHAT; Baron-Cohen et al., 2000) is appropriate now that the symptoms or signs of autism have proliferated to such an extent that two individuals with autism may not share any of the behaviours used to diagnose them.

Far better then, to focus on behaviour rather than *condition*, and so functional analysis would seem to be a starting point for any therapeutic intervention. This shifts the emphasis away from catch-all explanations that frame the individual child as a human kind (like 'autism') towards breaking their behaviour down into discrete elements: eye contact, which may or may not have anything to do with poor understanding of language, which may or may not be related to a tendency to toe walk, and so on. Stop worrying about what connects all these behaviours to each other; the answer may be 'nothing'. It also gives practitioners the opportunity to discriminate between different aspects of behaviour that might otherwise congregate under the autism banner. If a child is whacking other children in preschool, it is clearly a problem that requires attention, but lining up toy cars rather than racing them round the room only becomes problematic once the autism label is attached. Seeing autism as the conceptual thread linking the two behaviours together does not enrich our understanding of either one.

One way of looking at autism is as a 'black box': an input/output system whose workings are (perhaps deliberately) obscured in the process of using it to make a rhetorical point (Latour, 1987). This obscuration has allowed autism to contain, among a wealth of other inconsistencies, both under- and over-sensitivity to sensory stimuli, as well as claims that children with autism both avoid human contact and seek out tactile stimulation. However, even those researchers most committed to autism as a conceptual organising principle have been finally driven to admit that 'attempts at a single explanation for the symptoms of autism have failed' (Happé, Ronald, & Plomin, 2006, p. 1218), instead advocating breaking up the 'triad of impairments' of social deficits, communication deficits, and rigid or repetitive behaviours and abandoning the search for a 'gene for autism'. Has the lid finally come off the black box? (For a simple summary of the implications for practice, see Table 12.1.)

Summary: Watching the edifice being constructed

A final point I would like to make, from a methodological perspective, is that we cannot ignore the cultural and historical contexts in which discourse is produced and reproduced. As a microanalyst, I could have easily contented myself with exploring the conversation-like nuances of the thousands (millions?) of discussion threads on autism. They would have told us much about the way symptoms are constructed by parents of children with (or without) autism, but parents are only one of the groups with a stake in the autism industry. Alongside professionals from health, medicine, education, psychology, and the many therapies on offer, there is a flourishing autism culture that manifests itself in books (fiction and non-fiction), websites, and other resources and has

Table 12.1 Clinical practice highlights

1. Focus on behaviours as isolated behaviours, not as symptoms.
2. Only focus on behaviours that are intrinsically problematic.
3. If a child walks on her or his toes, chances are she/he isn't autistic.
4. Don't wrap children in blankets unless it's bedtime.

a profound influence on the way autism and related concepts are represented in contemporary society. Understanding the way this is constructed, brick by brick, requires a good deal of archaeology. We are fortunate to have the material culture so close to hand.

Notes

1. One of the stranger stories to have done the rounds concerns the claim that areas with clusters of employees in the IT industry, such as California's 'Silicon Valley', have higher levels of children diagnosed with autism (see http://news.bbc.co.uk/1/hi/health/2192611.stm). However, no research has thus far been published to substantiate the (anecdotal) claims, which are merely based on the belief that autism is genetically inherited (and the equally unsubstantiated claim that computer experts are 'more autistic' than everyone else!).

2. I am only citing this latter statistic in order to demonstrate the fallacy of CBS' claim. Further inspection of the Nadig et al. (2007) data reveals that only one child in the sample actually failed the 'own name' test at *both* 6- and 12-month intervals, suggesting that either this is a statistical artefact (i.e. not corrected for chance) or there is a developmental trajectory to own name recognition whereby children with autism progress typically until somewhere between 6 and 12 months and then, mysteriously, 'lose' this capacity. Like most autism symptoms, own name response is hopelessly under-theorised in the literature, therefore prone to all kinds of spurious interpretation.

3. http://www.youtube.com/watch?v=LWPEt6La8Gs

4. http://www.youtube.com/watch?v=Cq7Tq7qMWds

5. Note, all segments of data are copied exactly from their source and include all errors

6. Full thread can be found at http://www.mumsnet.com/Talk/childrens_health/a1383781-4yr-old-walking-on-his-toes.

7. *thebump.com*, a spin-off of wedding planning site *theknot.com*, is aimed at parents of babies and toddlers; the thread under discussion here can be viewed at http://forums.thebump.com/discussion/12328112/walking-on-tip-toes-at-2-5-years-old-and.

8. Sensory processing disorder (SPD) warrants a separate analysis of its own. In brief, it is a condition recognised largely within the field of occupational therapy that is said to arise from defective neurological 'integration' of sensory stimuli (typically, clothing fabric or food textures). SPD has failed as yet to acquire any psychiatric credibility and does not have an entry in either *DSM-5* or the World Health Organisation's Classification of Diseases (ICD-10). Although there is little scientific evidence in the literature for the use of sensory processing techniques in treating autism (Sniezyk & Zane, 2014), this has not prevented it from being a cornerstone of therapeutic practice in some fields: Indeed, one study found that 99% of US occupational therapists considered themselves to have a 'sensory-integration' orientation (Watling, Deitz, Kanny, & McLaughlin, 1999). Clearly, the SPD community is not going to give up without a fight. The website www.sensory-processing-disorder.com, in a revolutionary-style manifesto on its homepage, proclaims: '[SPD] is a REAL diagnosis and the time will come when we prove it scientifically to the medical profession'.

9. The authors refer to children with ASD and Asperger syndrome as discrete subsamples, although of course the latter is now included under ASD (no information is given about the diagnostic criteria applied in the particular clinic that these children visited).

10. http://news.bbc.co.uk/1/hi/world/americas/3179789.stm

11. http://www.canada.com/topics/news/national/story.html?id=410ebe27-db04-4ba9-99d9-39ec286f0b0f

References

Accardo, P., Morrow, J., Heaney, M. S., Whitman, B., & Tomazic, T. (1982). Toe walking and language development. *Clinical Pediatrics, 31*(3), 158–160.

American Psychiatric Association (2013). *Diagnostic and statistical manual of mental disorders* (5th edition). Arlington, VA: American Psychiatric Publishing.

Baron-Cohen, S., Wheelwright, S., Cox, A., Baird, G., Charman, T., Swettenham, J., ... & Doehring, P. (2000). Early identification of autism by the Checklist for Autism in Toddlers (CHAT). *Journal of the Royal Society of Medicine, 93*, 521–525.

Barrow, W. J., Jaworski, M., & Accardo, P. J. (2011). Persistent toe walking in autism. *Journal of Child Neurology, 26*(5), 619–621.

Billig, M. (1999). *Freudian repression: Conversation creating the unconscious.* Cambridge: Cambridge University Press.

Daley, T. C., & Sigman, M. D. (2002). Diagnostic conceptualisation of autism among Indian psychiatrists, psychologists, and paediatricians. *Journal of Autism and Developmental Disabilities, 32*, 13–23.

Doughty, S. S., & Doughty, A. H. (2008). Evaluation of body-pressure intervention for self injury in autism. *Behavioral Development Bulletin, 14*, 23–29.

Eyal, G., Hart, B., Onculer, E., Oren, N., & Rossi, N. (2010). *The autism matrix: The social origins of the autism epidemic.* Cambridge: Polity.

Filipek, P., Accardo, P., Baranek, G., Cook, E., Dawson, G., & Gordon, B., et al. (1999). Screening and diagnosis of autistic spectrum disorders. *Journal of Autism and Developmental Disorders, 29*(6), 439–484.

Giles, D. C. (2006). Constructing identities in cyberspace: The case of eating disorders. *British Journal of Social Psychology, 45*, 463–477.

Giles, D. C., & Newbold, J. (2011). Self- and other-diagnosis in user-led online mental health communities. *Qualitative Health Research, 21*(3), 419–428.

Giles, D. C. & Newbold, J. (2013). 'Is this normal?' The role of category predicates in constructing mental illness online. *Journal of Computer-Mediated Communication, 18*(4), 476–490.

Grandin, T. (1992). Calming effects of deep touch pressure in patients with autistic disorder, college students, and animals. *Journal of Child and Adolescent Psychopharmacology, 2*(1), 63–72.

Hacking, I. (1986). Making up people. In T. C. Heller (Ed.), *Reconstructing individualism: Autonomy, individuality and the self in western thought* (pp. 222–236). Stanford, CA: Stanford University Press.

Hacking, I. (1995). The looping effect of human kinds. In D. Sperber, D. Premack, & A. J. Premack (Eds.), *Causal cognition: A multi-disciplinary debate* (pp. 351–383). Oxford: Oxford University Press.

Hacking, I. (1998). *Mad travelers: Reflections on the reality of transient mental illnesses.* Charlottesville: University Press of Virginia.

Hacking, I. (2009). How we have been learning to talk about autism: A role for stories. *Metaphilosophy, 40*, 499–516.

Happé, F., Ronald, A., & Plomin, R. (2006). Time to give up on a single explanation for autism. *Nature Neuroscience, 9*, 1218–1220.

Harlow, H. F., & Zimmermann, R. R. (1959). Affectional responses in the infant monkey. *Science, 130*, 421–432.

Kendall, G., & Wickham, G. (1999). *Using Foucault's methods*. London: Sage.

Latour, B. (1987). *Science in action: How to follow engineers in society*. Milton Keynes: Open University Press.

Nadesan, M. H. (2005). *Constructing autism: Unravelling the 'truth' and understanding the social*. Abingdon: Routledge.

Nadig, A. S., Ozonoff, S., Young, G. S., Agata, R., Sigman, M., & Rogers, S. J. (2007). Prospective study of response to name in infants at risk for autism. *Archives of Pediatric Adolescent Medicine, 161*, 378–383.

Potter, J. (1996). *Representing reality: Discourse, rhetoric and social construction*. London: Sage.

Scott, K., & Baldwin, W. L. (2005). The challenge of early intensive intervention. In D. Zager (Ed.), *Autism spectrum disorders: Identification, education, and treatment* (3rd edition) (pp. 173–228). Mahwah, NJ: Lawrence Erlbaum Associates.

Shattuck, P. T. (2006). The contribution of diagnostic substitution to the growing administrative prevalence of autism in US special education. *Pediatrics, 117*, 1028–1037.

Sniezyk, C. J., & Zane, T. L. (2014). Investigating the effects of sensory integration therapy in decreasing stereotypy. *Focus on Autism and Other Developmental Disabilities*. Published online before print April 1, 2014, doi: 10.1177/1088357614525663.

Stephenson, J., & Carter, M. (2008). The use of weighted vests with children with autism spectrum disorders and other disabilities. *Journal of Autism and Developmental Disorders, 39*(1), 105–114.

Szasz, T. (1971). *The manufacture of madness: A comparative study of the inquisition and the mental health movement*. London: Routledge and Kegan Paul.

Timimi, S., & Leo, J. (2009). *Rethinking ADHD: From brain to culture*. Basingstoke: Palgrave Macmillan.

Timimi, S., Gardner, N., & McCabe, B. (2011). *The myth of autism*. Basingstoke: Palgrave Macmillan.

Watkins, N., & Sparling, E. (2014). The effectiveness of the Snug Vest on stereotypic behaviors in children diagnosed with an autism spectrum disorder. *Behavior Modification*. Published online before print April 28, 2014, doi: 10.1177/0145445514532128.

Watling, R., Deitz, J., Kanny, E. M., & McLaughlin, J. F. (1999). Current practice of occupational therapy for children with autism. *American Journal of Occupational Therapy, 53*, 498–505.

Recommended reading

- Accardo, P., Morrow, J., Heaney, M. S., Whitman, B., & Tomazic, T. (1982). Toe walking and language development. *Clinical Pediatrics, 31*(3), 158–160.
- Eyal, G., Hart, B., Onculer, E., Oren, N., & Rossi, N. (2010). *The autism matrix: The social origins of the autism epidemic*. Cambridge: Polity.
- Nadesan, M. H. (2005). *Constructing autism: Unravelling the 'truth' and understanding the social*. Abingdon: Routledge.
- Timimi, S., & Leo, J. (2009). *Rethinking ADHD: From brain to culture*. Basingstoke: Palgrave Macmillan.

13
The Production of the 'Normal Child': Exploring Co-constructions of Parents, Children, and Therapists

Charlotte Brownlow and Andrea Lamont-Mills

Introduction

The desire to 'be normal' is something that has been discussed in relation to a range of groups within society, including drug users (Nettleton, Neale, & Pickering, 2012), adults with mental health issues (Spondenkiewicz et al., 2013), and people with disabilities (Lee & Lin, 2013). The commonality shared by each of these is that individuals are being benchmarked against a shared societal understanding of what 'normality' looks like (see Lester & O'Reilly, Chapter 18, this volume). Nettleton et al. (2012) argued that such desires for normality are not necessarily a reflection of personal goals. Instead, they are contextualised and informed by a range of statistical and judgemental productions of normality, with the personal and the social being intertwined.

Understandings of 'normality' and 'normal' patterns of development for children is something that has received great interest in developmental psychology. Throughout the relatively short disciplinary history of developmental psychology, theorists have strived to provide theories of sequential development (e.g. Piaget, 1932), plot normative milestones of physical, social, and emotional development (e.g. Gesell, 1950), and provide measures through which professionals and parents alike can benchmark their own child's progress against the standardised 'norm'. Such norms have taken up a prioritised place within the discourse of developmental psychology, and both professionals and parents readily draw on such shared understandings of what the 'normal' child might look like. So powerful are these constructions that theorists, such as Rose (1989) and Burman (2008), have argued that childhood has become one of the most heavily governed periods of life development with practices closely observed by professionals, parents, and society.

Children who depart from this expected pattern of 'normal' development can be identified following comparisons, and psychologists have produced a suite of psychometric assessment tools based on assumptions of standardised abilities and behaviours that facilitate such identification. One of the most powerful tools available to and shared by a range of professionals is the *Diagnostic and Statistical Manual* (*DSM*) (American Psychiatric Association, 2013). The latest iteration of the manual has been recently released, and it is not without its critics. This is both in terms of pathologising what may have previously been considered differences or different expressions of particular concepts and the powerful constructive role that such a manual adopts for both individuals diagnosed and diagnosing professionals. Rose (2013) argued that the process of diagnosis has become a taken-for-granted feature of our society, where a patient expects to receive some sort of diagnosis following an appointment with a professional. Rose argued that such a diagnosis serves as an important framework for both the professional and individual to understand the condition and therefore make judgements about past experiences, and predictions about future ones.

Importantly, these frameworks rarely engage children as individuals. Niedel and McKee (2014), in their work with children with diabetes, argued that children need others to interpret their bodies and changes within these. We would argue that in the majority of cases, it is more powerful others who are the first to recognise and flag deviations in children's behaviour from the expected norm. Such 'others' can be parents, teachers, or other adults, all of who necessarily assume a more powerful position than the child in the production and performance of normative behaviours. Once identified as 'abnormal' or 'problematic', children then engage with professionals who seek to treat and potentially change behaviour in order to transform it to within more widely accepted normative parameters.

The (psychological) interventions to become more 'normal'

Given that the identification of problems in children is often done by (more powerful) others, the role that children play in the therapeutic relationship that seeks to change such behaviours is an interesting one. By default, children who attend therapy are there because they have been identified by others as being 'not normal' or 'having problems' and thus are in need of help through some sort of facilitated change process. Work by Parker and O'Reilly (2012) has focused on the social positionings of both parents and children within the therapeutic setting, and the roles that parents, children, and therapists play within such a setting. Parker and O'Reilly argued that in some instances parents seek to build alignments with professionals through techniques which serve to distance them from their child's behaviours – positioning problems as being within the child. This work highlights several examples of children's positions being constructed for them through the talk between parents and therapists. The location of the problem being placed within the individual child provides key examples of positionings *of* children within therapeutic contexts, rather that children taking up one of a number of available positions.

The complex role taken up by children in therapeutic interactions is not without tensions, and several researchers point to techniques used by children to both adopt and resist professional assertions (see Iversen, 2012). The particular challenge for children within therapy is what such resistances might look like, and how these may be interjected within the therapeutic discourse. One of the discursive devices available to perform such resistances to positionings is that of interrupting the therapist and/or parent. However, child interruptions within therapeutic settings are generally considered a negative event (O'Reilly, 2006). The unequal power relationships within such settings further compound the challenge of resistance for children, with not all members of the therapeutic relationship being afforded equal participatory rights in what is taken to be a shared discourse. Itakura (2001) has argued that such settings are very different from everyday exchanges, with participants in therapy both assigning and assuming particular roles. Everyday turn takings are therefore potentially problematic within a therapeutic setting as the therapist is typically the one afforded with the most power, followed by the parent, and lastly the child.

Project overview

Following on and inspired by some of the discussions that have come before, we seek to examine how such positionings of 'normal', 'problematic', and 'abnormal' are constructed, assigned, and adopted within a therapeutic exchange. In this chapter, we seek to provide illustrative examples of such concepts taken from real-life therapeutic exchanges. We examine the co-construction by children, parents, and therapists and also the subject positionings reflected in the adoption of such discourses. Our focus will be on two therapy sessions with a child who ends up being given an inclusive diagnosis of attention deficit hyperactivity disorder (ADHD).

In order to illustrate the production of normality and abnormality, we have situated our work within a framework of discursive psychology (DP) and critical discourse analysis (CDA). Given this, our methodological and analytic approach has been informed by the work of Potter (2012), Potter and Hepburn (2006), and Edley (2001). By adopting this approach, we are explicitly recognising the taken-for-granted understanding that psychological therapy is fundamentally a conversational activity through which therapeutic meanings, understandings, and actions are constructed and realised.

Data and analytical approach

The data that we have used come from Alexandra Street Press' *Counseling and Therapy in Video Volume III*. This video series is a training resource, and as such it includes recordings and verbatim transcriptions of real-life therapy sessions.

The general aim of the video series is to provide students and practitioners with real-life learning examples that capture the lived nuances and practices of therapeutic work. The decision to use this series as data reflects the second author's interest in exploring the nexuses between real-life therapeutic practices and psychological and counselling teaching and learning resources.

In order to identify data that illustrate the co-construction and negotiation of 'normality' and the problematisation of 'abnormality', we first searched the series for examples of ongoing therapy engagements between therapists, children, and parents. In this way, we wished to give a sense, albeit brief, of the longitudinal development of such discussions and positionings. The outcome of this search was the identification of a series of 10 counselling sessions with the one family that contained a combination of therapist–mother, therapist–child, therapist–mother–child, and therapist–therapist interactions. Therefore, we did not choose these sessions because we knew they contained such positionings and repertoires; rather, we chose these sessions and then examined these for potential positionings and repertoires.

Our analytic approach has been guided by the work of Larsson, Loewenthal, and Brooks (2012), where we first went through the transcripts in their entirety in order to get an overall sense of the role of the participants within the therapeutic session. After reading each verbatim session transcription, we selected two sessions as our main data source as these reflected discussions about the normative construction of the 'normal/abnormal child' rather than the 'normative construction of mothering'. The first is where Jonah[1] is not present and he is positioned by Mum and Therapist ('Counseling a Child with ADHD: Assessment and Interventions. A 10-part series. Intake Session', 2013), and the second is when Jonah is present and can participate in the positioning ('Counseling a Child with ADHD: Assessment and Interventions. A 10-part Series. Session 4', 2013).

The specific lens for this chapter is that of the production of normality within therapeutic practices, and therefore the approach to analysis was one that prioritised the repertoires that reflected this focus at the expense of others, which were evident in the exchanges. We therefore examined the transcripts for ideas that reflected the construction of normality and the ways in which these ideas shape dominant and shared interpretative repertoires concerning the production of normality. Within each session, we then selected sequences of interaction where the production of the normal, abnormal, and problematic child were constructed by drawing upon various subject positions and interpretative repertoires. These sections were then examined in a more microanalytical manner for their action orientated, situated, and constructed and constructive nature (Potter, 2012).

Each sequence was subjected to what we have called a 'light Jeffersonian' transcription', which entailed applying a reduced set of Jeffersonian (Jefferson,

1984a) transcription conventions to selected sequences. This decision reflects our interest in showing how in turn taking some voices and positions are privileged over others. This usage is also consistent with recent critical discursive work that has examined how the diagnosis of schizophrenia is constructed (see Larsson et al., 2012).

Analysis

Throughout the analysis we have prioritised two dominant interpretative repertoires that we argue reflect the production of the 'normal' child within these therapeutic exchanges. These are *Constructing the 'normal' child* and *The problematisation of behaviour*.

Constructing the 'normal' child

This first repertoire reflects the co-constructed and shared understandings of what a 'normal', and through contrast, an 'abnormal' child might be like. This draws heavily on behavioural and chronological age comparisons concerning the comparison of the child client, in this case Jonah, with taken-for-granted benchmarks of 'normality' and 'normal development'. Such developmental benchmarking is consistent with child and adolescent therapy practices where developmental history taking with reference to salient features is used by therapists to understand the presenting problem (Carr, 2006). This benchmarking was evident in several places within the sessions, and an example is provided below:

Extract 1: Intake Session Therapist 1 (T1), Mother (M)

```
T1:   .h an then h↑ow bout his um >developmental
      milestones< like walking talking did he have any
      delays as far as spee:ch (0.6) fine motor sk↑ills
      anyth↑ing
M:    he walk[ed late.
T1:         [°°abnormal°°
      (0.2)
T1:   °okay°
M:    but when he walked he just got up an walked
      [he didn't
T1:   [ok(h)ay
M:    toddle.
T1:   al(h)right
M:    um::
      (1.0)
M:    I think he's he started speaking (0.3) relatively
      early [in complete sentences
T1:         [°aha aha okay okay°
```

```
M:    you know he didn't
      (1.1)
M:    there wasn't a lot of babbling an' toddling =
T1:   =°okay°
M:    >it was almost like< he was just bo(h)[rn
T1:                                          [he just
      cl[imbed heh heh
M:      [gro(h)wn you know he left
T1:   heh heh [heh
M:            [he just CAME OUT and walked aw↑ay heh.
```

The therapist makes relevant the repertoire to be drawn upon by signalling that the upcoming talk will be concerned with '>developmental milestones<'. This request for developmental history is one where abnormal developmental information is given preference through 'delays as far as spee:ch (0.6) fine motor sk↑ills anyth↑ing' and the whispered '[°°abnormal°°'. The hearable emphasis on delays and speech and the elongation of speech signal for Mum that her response is to be a comparison of Jonah to the 'normal' child. The use of a three-part list (Jefferson, 1990), 'spee:ch (0.6) fine motor sk↑ills anyth↑ing', constructs the request as relating to Jonah's general development and its potential abnormality rather than being reflective of an individual or isolated event. Thus, the above prosodic and lexical features of the therapist's talk shape Mum's response to be one where shared understandings of the 'normal child' are made salient through comparison, and we see this in her response, 'he walk[ed late.' Exchanges such as this were evident in several places in the sessions.

What is interesting in such exchanges is that in addition to drawing on shared assumptions concerning the 'normal' child and what should happen in terms of standard development through benchmarking against typical development (see also O'Dell & Brownlow, Chapter 16, this volume), Mum also adopts an extreme position concerning Jonah. Here, Jonah's early behaviours are not constructed as being within the 'normal range', but rather as in advance of the normal expectations. Thus, Jonah is positioned not just as different, but in excess of normal expectations.

Mum does this by using a number of discursive resources. She first counters the possibility of Jonah's walking late as being perceived as deficit through the phrase 'but when he walked he just got up an walked'. The lexical choice of 'but' contrasts with the therapist's whispered overlap '[°°abnormal°°' and the walking late, hearably moving this behaviour to being advanced. Mum further constructs Jonah's walking as being in excess of normal expectations by contrasting what he did do 'just got up an walked' with what he did not do 'toddle' by using shared understanding of what 'normal' motor development looks like. The 'normal' child is further invoked by Mum when she states Jonah 'started speaking (0.3) relatively early' and that this was 'in [complete sentences', thereby contrasting Jonah's speech development, 'there wasn't a lot of babbling', to the taken-for-granted understanding that children babble before they talk.

What is interesting is at the end of this exchange Mum works to align the therapist with her positioning of Jonah as being advanced. Mum does this by drawing upon non-literal descriptions of Jonah as a baby where 'he was just bo(h)[rn gro(h)wn' and '[he just CAME OUT and walked aw↑ay heh'. This alignment is evidenced by Mum's production of within-speech laughter 'bo(h)[rn gro(h)wn', which acts as an invitation to laugh (Jefferson, 1979) and is taken up by the therapist as she responds in overlap by extending upon the first non-literal description and '[he just climbed heh heh' and then laughing. The therapist's response and continued laughter at her next turn at talk works to signal acceptance of this advanced positioning.

Such developmental comparisons continued throughout this session. On several occasions, the therapist asks Mum to consider Jonah's behaviour in relation to the behaviour of other comparable (normal) others.

Extract 2:

```
T1:   .h >and did you notice< as um far as his activity
      level or his ability to um (0.8) follow
      direct↑ions: (0.5) was that- (.) very different
      from the other kids:?
      (0.6)
T1:   °ah° similar? tell me bit about that say: (0.4)
      >an whatever< y-y- say when he was (.) five years
      old (.) kinda kindergarten ish (0.8) age. any
      diffe↑rence in um (1.3) >in ah< activity lev↑el
      or his ability to pay attent↑ion
      (1.1)
M:    he's always bin=
T1:   =Compared to the other kids.=
M:    =really really act↑ive
      (0.3)
T1:   °okay°
      (0.4)
M:    ah::=
T1:   =>so even more< so than the other kids his age at
      the time.
      (1.0)
T1:   >this of course w[ould five years old lot of
M:                    [I wouldn't say: y↑eah:
T1:   times they are running ar(h)ound<
M:    I don't know if he was more active
```

Comparisons with shared 'truths' highlights the assumption that childhood and child development can and should be considered in terms of developmental milestones, and the meeting of these, with a failure to do so being evident of development that is straying from the norm. While both professionals and lay individuals have a shared understanding concerning the importance of such milestones, professionals have an added layer of authority with respect to the outcomes of such benchmarking. This authority is displayed in intake sessions where developmental comparisons are used to inform diagnosis (Carr, 2006).

In this exchange, the therapist focuses on Jonah's levels of activity as a younger child. The therapist is exploring whether the reported levels of activity in the then five-year-old Jonah could be retrospectively considered abnormal. This retrospectivity is marked through the therapist's use of past tense (e.g. 'did you notice<', 'was that-', 'when he was') making clear that Mum is to reference back to Jonah's developmental history. What is different from the previous example is that here the therapist makes explicit the comparison to the 'normal' child or children and Jonah's potential difference from that norm. This is seen in 'was that- (.) very different from other kids:?' where the use of 'very', and its emphasis, works to make hearable the request for information about departures from normality. This utterance, as a yes-no interrogative, works to set up a difference comparison where acknowledgement of the difference from the norm is the preferred response. However, the preferred yes response does not emerge, despite there being opportunities where Mum could have joined the interaction (e.g. 'follow direct↑ions: (0.5)', 'from the other kids:? (0.6)', 'five years old (.)'), and provide a yes response. In fact, Mum does not provide any response.

This lack of response or confirmation of 'abnormality' is treated as problematic by the therapist as evidenced by her continuation with comparisons through the use of 'similar?', long pauses of 1.3 and 1.1 seconds, the um before the 1.3 second pause, her return to key diagnostic features of ADHD ('activity lev↑el or his ability to pay attent↑ion'), and her repeat of comparing Jonah to other five-year-olds who are expected to be active via 'running ar(h)ound<'. When Mum does respond, she acknowledges that Jonah was an active child, but rejects the therapist's positioning of Jonah's activity levels as being potentially 'abnormal' or 'even more< so than the other kids his age at the time'. Instead, she normalises Jonah's activity through normative judgements of his behaviour against other typical five-year-olds. While Jonah was 'really really act↑ive', Mum normalises this activity level through 'I wouldn't say: y↑eah:' and then 'I don't know if he was more active'. The use of I don't know should not be taken to infer an absence of knowledge here but rather as a means by which Mum can maintain her normalisation of Jonah despite the therapist's 'abnormal' positioning, thus protecting both her and the therapist's 'face-wants' (Tsui, 1991).

Such normalising comparisons are further invoked and supported through engaging in gendered positioning in the session:

Extract 3:

```
T1:   so even compared to his: classmates, he's
      >running around< and climbing on things more than
      them.
      (0.4)
M:    yeah:
T1:   even though they're all boys.
M:    ye[ah:
T:      [okay got it.
```

The comparison in the last example serves two important purposes. The first is invoking a comparison of Jonah and his behaviour against the benchmark of 'normal' 11-year-old boys. This is distinct from the previous developmental history benchmarking where it had been established that Jonah, as a five-year-old, was not necessarily more active than other five-year-olds. In this current developmental benchmarking, the membership category device of boys is invoked and, in doing so, shared understandings of gendered category-bound activities are introduced as the basis of comparison where Jonah is now being compared to the 'normal boy' child.

This comparison is hearable through the lexical choices of 'so even compared to his: classmates' with the hearable emphasis on 'even'. The benchmarking has also been unpacked where activities have now become running and climbing. The therapist's utterance is constructed in such a way that it sets up a preference for agreement with Jonah as now being more active than others his age. While Mum responds with 'yeah:', this comes after a pause at a turn relevant place (TRP). This is not understood by the therapist as conferring the position of 'abnormal boy' child as she herself introduces that membership category device. It is only at this point that Mum's second 'ye[ah:' is oriented to by the therapist as sufficient agreement with her positioning.

The problematisation of behaviour

The problematisation of behaviour interpretative repertoire is one that seeks not only to identify behaviour which strays from shared understandings of 'normal development' through comparisons to various agreed benchmarks but also to tag this difference as being problematic. While these comparisons may be based on the shared understandings between therapist and parent, they may also be evidenced in behaviour benchmarked against professionally constructed measures. Thus, unlike the previous positioning of Jonah as exceeding developmental milestones, the problematisation of behaviour seeks to categorise the behaviour, and by default the child, as being problematic and therefore requiring therapy. In these negotiations concerning the problematisation of behaviours, the site of contestation and rejection becomes Jonah and whether he displays this behaviour.

In the case of the exchange below, the therapist engages in an immediate problematising of behaviour, with clinical labels being suggested at the very start of the intake session.

Extract 4:

```
T1:  I'll be touching not just on attention problems
     and behaviour, .hh >but also< moo:d anxiety other
     things that (.) other difficulties kids sometimes
     have. (.) .hh >I really make sure< we kinda
     ↑cap↓ture °the big picture° so=

     ((lines omitted))

T1:  .hh your own so (.) .hh so >first of all< just
     tell me real briefly >you know we'll go in more
     det↑ails< but tell um me briefly (0.4) what are
     your ((inner)) primary concerns about Jo↑nah or
     what what kinda prompted you to contact the °the
     clinic°.
```

We see here how the therapist talks into being particular behaviours as being problematic. She does this by outlining what the upcoming discussion will focus on, 'attention problems and behaviours', as well as psychological states that are typically understood as being problematic (e.g. 'moo:d anxiety') and 'other difficulties kids sometimes have.'. Discussing and working with clients and their problems is part of the category-bound activities associated with being a psychologist. However, the point here is that by talking problems into being at the start of the session, problem discourse is immediately privileged and becomes the expected repertoire through which understandings will be co-constructed and negotiated.

What is interesting in the therapist's second utterance is the problematic behaviours and states that have been presented in the first part of the exchange are now expressed as concerns, only to be manifested as discursive 'problems' later in the session. While this may appear as a downgrade from problems, concerns is still hearable as meaning problems through the qualifier of 'primary' and the direct linkage of concerns to Jonah ('concerns about Jo↑nah'). This problematising continues with a 'what' interrogative question where Mum is explicitly requested to explain why she has contacted the clinic with the unsaid being why she is contacting for help.

Mum's initial response to the above questioning is to normalise and position Jonah's behaviour as one that is in excel of what would be considered typical. Thus, there is not an immediate uptake of the problematisation repertoire.

Extract 5:

```
M:   w↑ell hhh
     (1.2)
M:   Jonah is: (0.3) um (0.8) amazing.
     (0.3)
M:   he's ah (0.4) really (.) musical: an (.) and
     (0.4) artistic and he's (0.3) .hhh he's got this
     brilliant memory: and .hh um
```

To do this alternative positioning, Mum draws upon a number of extreme descriptors ('amazing', 'really (.) musical:', 'brilliant memory:'). These are presented in a three-part list to indicate the generality of Jonah being in excel of normal. However, this positioning does not come easily to Mum as her utterance contains a number of prosodic and lexical aspects of perturbed talk (Silverman, 2001), such as 'w↑ell', the numerous pauses in the utterance at non-turn construction units, the in- and out-breaths, and the ums. Mum's use of expressive caution is understandable, given that the delicate matter she is about to discuss is that of Jonah's problems. The therapist's utterance prior to this can be understood as a trouble-telling inquiry (Jefferson, 1984b) where this trouble telling is hearably difficult for Mum. This is unlike the first repertoire that we examined. That is, when referencing comparison with the 'normal' child, there is an absence of trouble-telling and perturbed talk features, thus the problematisation of the behaviour does not only talk into being the problem, it does so in a manner that reflects the troubled or delicate nature of expressing such a problem.

As with trouble-telling sequences, Mum acknowledges later on in the intake session that Jonah's behaviour could be considered problematic by some individuals, thus potentially invoking the problematisation repertoire. However, she counters this by continuing to normalise such behaviours through providing context in terms of their own family environment and her identity as mother.

Extract 6:

```
M:   he's (.) I-I stayed home with him for (.) a long
     time. >so he didn't go ta< (.) nursery school or
     kindergarten or any of that .hhh >an so it's
     funny how you< read these things that ask
     questions l↑ike (.) .hh you k↓now did the
     teachers complain: about him walking on the
     furni↑ture an .hh you know he's FIVE I'm like
     w[alk on the
T1:   [((°right°))
M:   furniture you know I bought furnit(h)ure you
```

```
        could walk o(h)n so heh .hh (.) >I did never
        thought of that ↑as:< being (0.6) overactive.
        [that's I just
T1:     [°°okay okay°° (.)
M:      that's (.) how kids are supposed to be. >y-you
        know I thought we did< (.)
```

The dilemma here is that what Mum goes on to describe is what could be interpreted by the therapist as problematic behaviour. Thus, in response to the therapist's problematisation of behaviour, Mum seeks to normalise Jonah's behaviour and place Jonah within the 'normal' child position through 'how kids are suppose to be'. She does this through her own comparison of Jonah to other children his age via her marked emphasis and loud production of his age 'FIVE'. Her use of 'you know I bought furnit(h)ure you could walk' and '>I did never thought of that ↑as:< being (0.6) overactive' rhetorically works to manage her accountability for not recognising Jonah's early behaviour as potentially problematic. In doing this, she draws upon her identity as mother that comes with it the shared understanding that if a mother knows their child's behaviour is problematic, they do something to change it, not facilitate it by buying furniture for a child to walk on.

What is interesting in these exchanges is the absence of Jonah in the session. Thus, all conversations and presentations of Jonah's behaviour in this intake session are co-constructed between Mum and therapist. In his absence, the therapist and Mum position Jonah in two quite different ways – ways which draw on extremes, which are at times incompatible and thus contestable. Several attempts are made by Mum to resist the therapist's positioning of Jonah as displaying problematic behaviour during the intake session, for example:

Extract 7:

```
T1:     now >did teacher ever complain that he< blurts
        out answers in the classroom without raising his
        hand.
        (1.2)
T1:     that he'll just shout out or do y↑ou does that
        happen with y↓ou: when you're in speaking
        relationship with him.
        (0.6)
T1:     that before you even finish the quest↑ion
        (0.7)
T1:     this is different than interrupting it's just
        like there's no:: (0.7) you know he's (0.8) >the
        brain is going before the< (.) .hh
        (1.1)
T1:     or the mouth is g[oing before the brain can
M:                       [no:::
        (.)
```

```
M:    I don't really (0.3) notice that as like a-a
      problem.
      (0.4)
M:    that's not something th[at I notice.
T1:                          [°okay all right.°
      (1.5)
T1:   °°okay .hh°° um is he if he's with a group of
      people, um if he's with a group of other kids .h
      >and there's a question that's asked< is he the
      one who has to be the first one to answer like
      he's the one that's gotta give y↑ou (0.3) the
      informat↓ion
      (0.7)
M:    yeah
T1:   and he'll just blurt it out?
      (0.6)
M:    yeah
      (0.5)
T1:   he is that kid
      (.)
M:    he likes to (.) yeah [he wants to
T1:                        [ok(h)ay
M:    be the one who knows.=
T1:   =all right .hh so but-but have teachers ever said
      though that that he that he, does he get in
      trouble for not raising his hand.
      (0.9)
T1:   that shout- he's shouting out in the classroom.
      (1.0)
T1:   >cause this is [one
M:                   [no
T1:   you would definitely hear about if it was
      happening i(h)n the class(h)room. [no
M:                                      [this this .hh
      this is not that's not something that she:
      she she could she's concerned about I mean she
      has told me a lot of concerns but I'm not getting
      that's a problem [I don't see
T1:                     [okay
M:    that as one problem.
```

From the above sequence, we can see how Mum appears to resist the positioning imposed by the therapist. This is done through her withholding of responses at TRPs where she could have but did not join the interaction (e.g. after 'raising his hand.', 'finish the ques↑tion', 'speaking relationship with him'). Such withholding of response has been considered passive resistance in other health interactions (Stivers, 2006). When passive resistance is encountered, health professionals typically continue with their turn at talk in order to elicit agreement with what is being proposed or asked. We see this here as the therapist continues her line of questioning until Mum engages in active resistance (Stivers, 2006). Here, Mum responds with a no and then declares 'I don't really (0.3) notice that as like a-a problem. (0.4) that's not something th[at I notice.' While Mum starts with the hedge of 'I don't really', she upgrades this assessment to 'not something', thus working rhetorically to strengthen her rejection of the therapist's positioning.

The therapist continues with a new line of questioning regarding Jonah's behaviour. Here, Mum acknowledges that Jonah does engage in the blurting-out behaviour being described, but she does so in a way that deflects and then de-problematises Jonah's behaviour. Here, she positions Jonah's actions as being a reflection of Jonah wanting to be seen as the child who knows the answers to the teacher's questions ('he likes to (.) yeah [he want to be the one who know.'). This then constructs Jonah as a child who is actually engaging in appropriate classroom behaviour, that of expressing and sharing knowledge when asked to do so.

While Mum has rejected the therapist's positioning of Jonah's behaviour as being problematic, we see the therapist continuing to draw upon the prob-lematisation repertoire. The therapist now elaborates on her questioning by using explicit examples of problematic behaviour ('not raising his hand.' and 'shouting out in classroom.') in order to elicit a response. This is again passively resisted by Mum with the therapist continuing until Mum actively resists the positioning by invoking the membership category device of teacher through her use of 'she' with stress and hearable emphasis ('that's not something that she:'). Invoking of this category allows for the unspoken, but shared, under-standings that are bound up with the category entitlement of teacher. That is, if the teacher has not mentioned this as problematic, then it is therefore under-stood and hearably accepted that Jonah is not engaging in this behaviour. Such exchanges in the therapy session indicate that not all of the contributors to the therapeutic exchange have equal status in terms of power, both in terms of the controlling of topics for discussion and positionings in terms of diagnostic labelling.

A recurring example of this is the power of the therapist to problematise behaviours, irrespective of the resistance from either Mum or later Jonah. The above exchange is an interesting demonstration of the power differ-entials between the therapist and Mum particularly in terms of topics and questioning. This is displayed in the question–answer sequences that perme-ate the intake session, where the therapist is the asker of the questions with Mum being sequentially shaped to be the response giver. There was only one instance in this intake session where Mum asks a question and that came at the end of the session. Thus, the person, who asks the questions, is able to sequentially shape and thus situate the interaction within preferred repertoires of understandings.

The above exchange is also an example of the continued problematisation questioning by the therapist and rejection of this by Mum. This contrasting positioning of Jonah's behaviour as being potentially problematic by therapist, and then typical by Mum captures the contested space in which repertoires are constituted. Despite the resistance by Mum, here and in other exchanges, the unfolding of the session indicates that she comes to share the construction by the therapist of Jonah's behaviour being different to what could be considered 'normal' and by default, problematic, and thus adopts the repertoire discourse that she has previously rejected.

There were also instances in the sessions where understandings of what constitutes a problematic behaviour was itself contested and negotiated. Thus, despite the introduction by the therapist in several places throughout the sessions of behaviour which could be considered 'problematic', the understanding of what constitutes such problematic behaviour is not always clear, and in several places we see both Mum, and Jonah in session 4, negotiating understandings of problematisation.

It is clear from the exchanges presented thus far that Jonah's problematic behaviours are being constituted between Mum and the therapist without reference to what Jonah himself may feel, think, or understand about his so called problems and in fact whether he even perceives that he has problems. We see in session 4 a similar pattern emerging where the second therapist asks Jonah what he would like from the sessions, not whether Jonah thinks he has a problem. Interestingly, in response to this Jonah states that he is not sure why he is attending the therapy session.

Extract 8: Second Therapist (T2), Jonah (J)

```
T2:   so. you know tell ↑me um:
      (0.5)
T2:   if I could be helpful to y↓ou: or your family
      what would you l↑ike out of this process how
      could ↑I (0.7) °°help with things°°
J:    um:: (.) I'm not really sure because (0.4) I'm
      not sure exactly: (0.5) why you're helping us
      with. >I'm not sure if you're helping< (0.9) me
      and my mom with disorganiza↑tion: (0.3) or if
      you're helping: me with schoo:l or:
```

We can see from the exchange above, and the examples presented thus far, that someone other than the child is always defining children's 'problems'. In the above example, Jonah states that he is not exactly sure why he is in the therapy session, yet at no point during this session does the therapist ask about this. This has important links to the relative power base adopted by therapists and others, with children typically being a marginalised force within such exchanges.

These power differentials are inherent in the identities of child and therapist. We see the therapist invoking his institutional identity in the first line of this example. He does this through 'tell ↑me', 'I could be helpful to y↓ou:', and 'how could ↑I (0.7) °°help'. These reference the category-bound activities of a psychologist where psychologists do askings and clients do tellings. The emphasis and sound stretching on you, the upward intonation on me, and the whispered help signal for the listener that this is a request for Jonah, as the child client, to inform the therapist what his therapeutic goals are. This again draws upon the problematisation of behaviour repertoire where a problem position is central and where help is offered to solve problems. This draws upon the understanding that service offers are not made if problems do not exist.

Jonah's response to this request is interesting in that it has aspects of hesitant talk within it. There is the starting 'um:: (.)' and the various pauses at non TRPs. Jonah's use of 'why you're helping us with.' may appear hearable as suggesting that Jonah is not clear why he and his mother, as indexed through the use of us, are in therapy. Jonah follows up with identification of two possible problems areas, 'disorganiza↑tion:' and 'schoo:l' that work up a display of a lack of understanding concerning whose problem behaviour will be addressed by the therapist, his or his mother's.

Clinical relevance summary

In this chapter, we have sought to examine the taken-for-granted assumptions surrounding what it means to be 'normal' and consider the positions taken up by and imposed on individuals through such discourse as it is enacted within a therapeutic setting. It is clear that a technique drawn upon by both therapist and Mum is the shared understandings of what a 'normal' 11-year-old boy would look like and how Jonah does or does not deviate from this benchmarked ideal. While all agents in the shared discussion have the opportunity to resist and disagree with various positionings, the therapists were consistently the contributors who demonstrated power in the exchanges in terms of both topic introduction and conclusions resulting from such discussions. Both ultimately culminated in the assignment of diagnostic labels for Jonah at the end of the intake session (data not shown). It is clear that of all the participants in this triadic clinical relationship, Jonah is the least powerful with him stating on more than one occasion that he was not sure why he was attending the therapy session. We therefore argue that the context of clinical consultations are an important place for the exploration of the negotiation

Table 13.1 Clinical practice highlights

---

1. The therapeutic setting is not a neutral environment and is one where conversational power sharing should be a key consideration.
2. The co-construction of 'problem' behaviour in children is something that needs to be shared with those positioned by it.
3. The category-bound activities of the therapist and the institutional context itself privilege some subject positions over others. This privileging is one that needs to be explicitly considered and debated in order to fully enact a child-client-centred approach to psychological therapy.

---

of positionings, not least given the ready framework provided by the DSM for abnormality/normality constructions, and the problematisation of abnormality that pervades most psychological encounters. For a simple summary of the implications for practice, see Table 13.1.

Summary

The data examined in this chapter suggest that the institutional context shapes the subject positions and interpretative repertoires that are drawn upon in clinical interactions. In this sense, we argue that there are institutional activities and membership categories that make some subject positions and interpretative repertoires more easily enacted. Following from this, there is the clear activity of identifying what the problem is, and we see this in the intake session. This occurs in a sequence, and within that sequence we argue that discourse is constructed in a way that makes the discursive identification of a problem the preferred response.

These data suggest that when powerful others constitute the problem for the child, it is difficult for the child and/or mother to reject such problem positionings. This is not to say that rejections are impossible as subject positions can and were contested by both Mum and Jonah, albeit in different ways. In this sense, we take the stance that positions are contestable and thus can be negotiated. Subject positions are constructed across a number of turns at talk. Thus, they are built up and contributed to by both interactants. This co-construction can occur in such a way that privileges certain positions over others. We found this was done through question construction whereby certain responses and information are preferred – problem responses and problem information.

We also argue that by working up a 'normal' child position across an interaction, extremes of behaviour are drawn upon. We see this in the intake session where Mum switches between Jonah as an amazing person to Jonah having behavioural difficulties. This allows Mum to display Jonah as having positive attributes but problematic behaviours. Finally, we argue that problems are not easy to give voice to. By this we mean that when Jonah's problems are spoken about by Mum, she does not talk about his problems in a way that is free flowing and untroubled. Instead, Jonah's problem is produced with delays, hesitations, and deflections to normal or exceptional activity, as if giving voice to the problem irrevocably makes it real.

Note

1. A pseudonym.

References

American Psychiatric Association (2013). *Diagnostic and statistical manual of mental disorders* (5th edition). Arlington, VA: American Psychiatric Publishing.

Burman, E. (2008). *Deconstructing developmental psychology* (2nd edition). London: Routledge.

Carr, A. (2006). *Handbook of child and adolescent clinical psychology: A contextual approach* (2nd edition). East Sussex, UK: Routledge.

Counseling a Child with ADHD: Assessment and Interventions. A 10-part series. Intake Session [Video] (2013). *Psychotherapy.net*. Retrieved May 13, 2014 from Counseling and Therapy in Video: Volume III.

Counseling a Child with ADHD: Assessment and interventions. A 10-part Series. Session 4 [Video] (2013). *Psychotherapy.net*. Retrieved May 12, 2014 from Counseling and Therapy in Video: Volume III.

Edley, N. (2001). Analysing masculinity: Interpretative repertoires, ideological dilemmas and subject positions. In M. Wetherell, S. Taylor, & S. Yates (Eds.), *Discourse as data: A guide for analysis* (pp. 189–228). London: Sage.

Gesell, A. (1950). *The first five years of life: A guide to the study of the pre-school child*. London: Methuen.

Itakura, H. (2001). Describing conversational dominance. *Journal of Pragmatics, 33*(12), 1859–1880.

Iversen, C. (2012). Recordability: Resistance and collusion in psychometric interviews with children. *Discourse Studies, 14*, 691–709.

Jefferson, G. (1979). A technique for inviting laughter and its subsequent acceptance/declination. In G. Psathas (Ed.), *Everyday language: Studies in ethnomethodology* (pp. 79–96). New York: Irvington Publishers.

Jefferson, G. (1984a). Notes on a systematic deployment of the acknowledgement tokens 'yeah' and 'mm hm'. *Papers in Linguistics, 17*, 197–216.

Jefferson, G. (1984b). On the organization of laughter in talk about troubles. In J. M. Atkinson & J. C. Heritage (Eds.), *Structures of social action: Studies in conversation analysis* (pp. 346–369). Cambridge: Cambridge University Press.

Jefferson, G. (1990). List construction as a task and resource. In G. Psathas (Ed.), *Interaction competence* (pp. 63–92). Washington, DC: University Press of America.

Larsson, P., Loewenthal, D., & Brooks, O. (2012). Counselling psychology and schizophrenia: A critical discursive account. *Counselling Psychology Quarterly, 25*, 31–47.

Lee, T.-Y., & Lin, F.-Y. (2013). Taiwanese parents' perceptions of their very low birth weight infant with developmental disabilities. *Journal of Perinatal & Neonatal Nursing, 27*, 354–352.

Nettleton, S., Neale, J., & Pickering, L. (2012). 'I just want to be normal': An analysis of discourses of normality among recovering heroin users. *Health, 17,* 174–190.

Niedel, S., & McKee, M. (2014). 'Is it normal?' A simple question that often lacks an easy answer. *The Royal Society of Medicine, 107*(2), 52–53.

O'Reilly, M. (2006). Should children be seen and not heard? An examination of how children's interruptions are treated in family therapy. *Discourse Studies, 8*(4), 549–566.

Parker, N., & O'Reilly, M. (2012). 'Gossiping' as a social action in family therapy: The pseudo-absence and pseudo-presence of children. *Discourse Studies, 14,* 457–475.

Piaget, J. (1932). *The moral judgement of the child.* London: Routledge & Kegan Paul.

Potter, J. (2012). Discourse analysis and discursive psychology. In Cooper, P. M. Camic, D. L. Long, A. T. Panter, D. Rindskopf, & K. Sher (Eds.), *APA handbook of research methods in psychology: Research designs: quantitative, qualitative, neuropsychological, and biological* (Vol. 2, pp. 119–138). Washington, DC: American Psychological Association.

Potter, J., & Hepburn, A. (2006). Discourse analytic practice. In C. Seale, G. Gobo, & J. F. Gubrium (Eds.), *Qualitative research practice* (pp. 168–184). London: Sage.

Rose, N. (1989). *Governing the soul: The shaping of the private self.* London: Routledge.

Rose, N. (June, 2013). *What is diagnosis for?* Paper presented at the Institute of Psychiatry Conference on DSM-5 and the Future of Diagnosis, London.

Silverman, D. (2001). The construction of delicate objects in counselling. In M. Wetherell, S. Taylor, & S. Yates (Eds.), *Discourse, theory and practice* (pp. 119–137). London: Sage.

Spondenkiewicz, M., Speranza, M., Taieb, O., Pham-Scottez, A., Corcos, M., & Revah-Levy, A. (2013). Living from day to day – Qualitative study on Borderline Personality Disorder in adolescence. *Journal of the Canadian Academy of Child and Adolescent Psychiatry, 22,* 282–289.

Stivers, T. (2006). Treatment decisions: Negotiations between doctors and patients in acute care encounters. In J. Heritage & D. Maynard (Eds.), *Communication in medical care: Interaction between primary care physicians and patients* (pp. 279–312). Cambridge: Cambridge University Press.

Tsui, A. B. M. (1991). The pragmatic functions of 'I don't know'. *Text, 11,* 607–622.

Recommended reading

- Goodley, D. (2014). *Dis/ability studies: Theorising disablism and ableism.* London: Routledge.
- Hepburn, A., & Wiggins, S. (2007). *Discursive research in practice: New approaches to psychology and interaction.* Cambridge: Cambridge University Press.

14
Should Autism Be Classified as a Mental Illness/Disability? Evidence from Empirical Work

Michelle O'Reilly, Khalid Karim, and Jessica Nina Lester

Introduction

Is autism a disability? Fundamentally, this question also raises additional questions, including: What is a disability? Who makes the definition valid? What function does such a label serve within society? The answers are perhaps dependent upon varied points of view and affiliations with particular theoretical frameworks. Indeed, there has been inconsistency regarding the terminology that has been utilised, with this terminology continuing to evolve. In the latest incarnation of autism in the *Diagnostic and Statistical Manual of Mental Disorders* (*DSM-5*), autism has been defined as Autism Spectrum Disorder (ASD) (American Psychiatric Association, 2013), while other terms are still used within the International Classification of Diseases (ICD). Controversially, the diagnostic label of 'Asperger's Syndrome' has been removed from *DSM-5*.

There are indeed some who would argue that autism in its broadest context has been misclassified and, in actuality, it is *not* in all its forms a disability or mental health condition. In fact, some individuals diagnosed with autism have contested this form of categorisation. For example, the term 'Aspies' is a self-referential term sometimes utilised by individuals who have been diagnosed with Asperger's syndrome who celebrate the individuality and strengths of the diagnosis in a positive way (see e.g. http://aspiesforfreedom.wordpress.com/about/). Accordingly, some people embrace autism as part of their identity (Baker, 2011) and actively oppose the search for a cure for a diagnosis that many presume to be a disabling impairment (Brownlow, 2010). However, there is also evidence from research that demonstrates that some individuals diagnosed with autism, and their family members, do find the condition disabling and the effects stressful (Huws & Jones, 2008), and therefore desire a cure and interventions (Bagatell, 2010). Additionally, there are concerns that the increasing diagnosis of autism is not so much related to 'new' scientific discoveries but to shifting cultural and social practices related to child development and what comes to be counted as 'abnormal development' (e.g. Brownlow & Lamont-Mills, Chapter 13, this volume; Timimi, Gardner, & McCabe, 2011).

Despite these tensions and seemingly diverse viewpoints, autism continues to be classified as a mental disorder by both the ICD and the *DSM*. Thus, at least in the clinical field, autism is very much viewed as a mental disorder. Yet, it is important to note that most of the diagnostic frameworks take up an objectivist, positivist, and realist framework, and position those with autism as disabled, mentally ill, pathological, and different. However, the tensions and arguments that exist cannot and should not be ignored, particularly in light of a whole range of alternative ideas that have been proposed over the decades, including the social model of disability and the political model of disability. These different models and perspectives have drawn upon a range of theoretical frameworks that have challenged thinking about mental disorders, with social constructionism in particular challenging the positivist positioning of autism.

As such, in this chapter, we draw upon a social constructionist framework and, more specifically, a discourse and conversation analysis approach to explore some of the arguments related to how autism is classified as a disability and how and why this might be challenged. Specifically, we draw upon three of our empirical papers (Karim, Cook, & O'Reilly, 2012; Lester, Karim, & O'Reilly (in press); Lester & Paulus, 2012) to develop an argument that seeks to illustrate that the disabling aspect of autism is in and of itself socially constructed. We begin the chapter by providing some context regarding the nature of autism and the different models of disability that have challenged biomedical discourses and practices associated with 'treating' autism. We then provide empirical works to illustrate some of the tenets of the arguments that challenge more mainstream thinking about this complex condition. Finally, throughout this chapter, we use the term 'autism', rather than 'autism spectrum disorders' or 'autistics'. In doing so, we acknowledge that autism is a social construct with multiple and shifting meanings, which has been *'developed and applied, not discovered'* (Biklen et al., 2005, p. 12, italics in original).

What is autism?

Autism has traditionally been understood in biomedical and psychiatric terms. The term 'autism' was first described by Leo Kanner in 1943 and was followed by a description of children with similar characteristics referred to as Asperger's syndrome by Hans Asperger in 1944. While the concept of autism continued to develop during the 20th century, there was a radical change in thinking following the work by Lorna Wing who coined the term 'Autistic Spectrum Disorder' (Wing, 1981). This recognised a broader range of impairments and has been classically defined as a 'triad of impairments', including:

1. Impairment in reciprocal social interaction;
2. Impairment in communication skills; and
3. Repetitive patterns of behaviour, and a rigid style of thinking.

(Wing, 1981, 1996)

Autism is often considered a complex neurodevelopmental condition, which presents with difficulties in daily functioning (Karim, Ali, & O'Reilly, 2014). Typically, autism is defined behaviourally, and in receipt of a diagnosis, professionals clinically judge the individual as demonstrating specific behavioural impairments across the triad (Muskett, Body, & Perkins, 2013).

While diagnostic criteria, such as ICD 10 and *DSM-5*, are essential in determining whether an individual has autism, these are open to interpretation, particularly as there is no definitive diagnostic test. This has led to concerns that the apparent 'epidemic' of autism has a sociological and cultural dimension rather than being a reflection of a true increase in prevalence (Eyal, Hart, Onculer, Oren, & Ross, 2010). Other arguments have been made that the increases in diagnosis are due to the developing diagnostic criteria, age of children, and an improvement in public awareness (Frith, 1989; Williams, Higgins, & Brayne, 2006). Consequently, this has had international ramifications, as the increase in prevalence has been seen in many nations (Bailey, 2008). Not surprisingly, therefore, autism is the most widely researched child mental health disorder (Wolff, 2004). However, even while diagnostic systems set the parameters of normality with a broadening of the diagnostic criteria, there has been little dialogue regarding the ways in which constructions of normality/abnormality in the context of autism are actualised (Lester & O'Reilly, in press). Indeed, some have suggested that with the ongoing shifts in diagnostic manuals, the 'pool of "normality"' is becoming 'a mere puddle' (Wykes & Callard, 2010, p. 302).

Debates and conceptualisations of autism are grounded in a rich and complex history of psychiatry, critical psychiatry, psychology, and alternative models of disability (Nadesan, 2005). Historically, autism has been conceptualised against the backdrop of medical discourses and practices that have constructed autism through a lens of disease and deficit, with metaphors of 'cure' commonly evoked within medical and popular media discussions of autism (Broderick & Ne'eman, 2008, p. 469). Autism, then, has most often been presented as an ahistorical, biological fact – one to be understood through positivist perspectives (Glynne-Owen, 2010), with the social and cultural processes inherent to naming and treating autism frequently ignored (Nadesan, 2005). Central to the dominating arguments that surround autism – in which it is conceptualised as a disabling medical condition – there has been an implicit reliance about cultural notions of 'normal' and 'abnormal' (Ashby, 2010), with autism positioned as an 'abnormal' identity when set against normative developmental trajectories. Yet, there are alternative ways to conceptualise autism and disability more generally, which we shall explore next.

Alternative ways of thinking about disability

Before we consider the different ways of conceptualising autism and 'autistic behaviour', we provide a (very) brief account of the broader disability literature into which individuals diagnosed with autism are often situated. Notably, disability theorists have not typically included psychiatric disability in their work. Yet, even still, these models and arguments apply well to mental health conditions, with many of the espoused ideas offering new perspectives and directions for those who experience mental distress (Mulvany, 2000).

In the field of disability studies, there have been several models of disability proposed, and, while they share some central characteristics, there are also important differences (see also Gilson & DePoy, Chapter 7, this volume). Many of these differences reflect the perception that the ideas are incompatible (Grue, 2011). Nonetheless, many scholars agree that a more critical approach is needed (Mulvany, 2000) and tend to suggest that:

- there is a necessary challenge to the conventional deficit view of disability (Barton, 1993);
- there should be a focus on the rights of those with disabilities, while there is a need for social change (Mulvany, 2000);
- disability is viewed as a disadvantage of activity caused by social organisation (Oliver, 1990), which is far different from impairment that refers to a bodily defect (Barnes, Mercer, & Shakespeare, 1999).

Perhaps most importantly, all of these critical approaches oppose the dominant biomedical discourse that is advocated by the medical model, and to a lesser extent the biopsychosocial model.

The medical model view is one that sees disability as a permanent biological impairment (Gilson & DePoy, 2000). Indeed, mental disorders have become increasingly viewed in genetic, biological, and scientific terms within contemporary psychiatry (LaFrance & McKenzie-Mohr, 2013). Thus, from a biomedical perspective, the 'problem' is positioned as an individual one, and one that may not be capable of being 'fixed' or cured by medical intervention; thus, many disability scholars have argued that such an orientation inappropriately positions the individual as deficient (Gilson & DePoy). It is argued by critics that this view condemns the disabled as second-class citizens (Mercer, 2002), positioning normalisation as the goal and denying the agencies of those identified as disabled (Eyal et al., 2010; Grue, 2011). For those with mental health conditions, this model has cast a judgement of laziness, weakness, or belligerence as explanations for why an individual is 'disordered' (LaFrance & McKenzie-Mohr, 2013). In slight contrast, the biopsychosocial model has tended to be employed as a way of attending to some of the criticisms of the medical model, recognising that psychological and social factors play a role in mental disorders (Santrock, 2007). Accordingly, this model is more closely related to care and considers the social aspect of the individual. However, the biopsychosocial model still orients to causation of illness as stemming from within the individual's body and is still underpinned by the position of positivism.

Critical perspectives on disability have therefore challenged these ideas with a range of different models being developed. It is beyond the scope of this chapter to provide details on all of the varied models and perspectives or to provide any of them in detail as this constitutes a substantial literature. For context, however, we briefly introduce three of the more common models: (1) the labelling model, (2) the political model, and (3) the social model.

The labelling model of disability

Labelling theory was developed mostly within the United States and stemmed predominantly from the writings of Lemert and the symbolic interactionist perspective (Petrunik, 1980). The focus for labelling theorists was on the social reaction of others to those who were labelled with a disabling condition as opposed to the perceptions of those who were labelled (Mulvany, 2000). In relation to mental health, the main argument of labelling theorists is that the role of those with conditions is consolidated by the social reactions of others (Goffman, 1968). They argue that society learns a stereotyped imagery of individuals with mental disorders and takes up negative language associated with characterising individuals diagnosed and labelled as 'abnormal' (Weinstein, 1983).

The political model of disability

The socio-political approach to disability argues that disability is the product of interactions between people and their environment (Hahn, 1985). The political model of disability moves disability into the domain of power and resources. Thus, from this perspective, disability is viewed as a condition that interferes with a person's ability to work, and by default, many people with disabilities are assumed to fail to make an economic contribution to society (Gilson & DePoy, 2000). From this perspective, disability is argued to be assessable by measures of visibility and labelling (Hahn, 1987). Notably, the political model of disability does have some connection to the social model of disability.

The social model of disability

The social model of disability orients to a Marxist sociology and was developed mostly within the United Kingdom (Grue, 2011). This model takes an alternative viewpoint to the medical model, arguing that the incapacity to function is positioned within a hostile environment (Gleeson, 1997) and thus disability is considered a form of economic and political oppression (Oliver, 1996). This model advocates that definitions of disability are based on non-disabled assumptions and therefore fail to reflect the personal realities of those who are classed as 'disabled' (Oliver, 1983). From this perspective, therefore, disability is seen as a diversity of the human condition rather than something that needs to be fixed, with disability positioned as socially constructed (Gilson & DePoy, 2000).

Social constructionism and autism

In this chapter, we take up a social constructionist position to notions of disability and autism specifically, which is inherent in the social model. We suggest that such a perspective is congruent with discourse and conversation analysis, as it assumes that the language defining disability is central to understanding the experiences and worldviews of people with autism. We thus offer a brief discussion of social constructionism, as this served to ground the argument we proffer.

Introducing social constructionism

Social constructionism is not a unified framework (Brown, 1995) and there are some differences in thinking between those who take up social constructionist positions. Nonetheless, social constructionism is an epistemological perspective that considers social and psychological phenomena as constituted through social and interpersonal processes (Georgaca, 2014).

Gergen (1985) described several important features of this perspective, including that:

- Social constructionists have instilled doubt in the taken-for-granted world.
- Social constructionists have suggested that knowledge is culturally, historically, and socially specific, as well as sustained by social process.
- Social constructionists have argued that explanations and descriptions are not neutral; rather, they constitute social actions that serve to sustain particular patterns and exclude others.

While some social constructionists have been centrally concerned with power structures, asymmetry, and the political aspects of disability, not all social constructionists take this approach (Burr, 2003). This different orientation reflects their particular ontological concerns, which are best illuminated by differences between micro- and macro-versions of social constructionism.

Macro-social constructionism has tended to be most concerned with the role that linguistic and social structures play in terms of shaping the social world (Gubrium & Holstein, 2008). This perspective is concerned with a focus on the constructed social forms and collective representations (Sudnow, 1965). Researchers who operate from a macro-perspective tend to focus on power relations and social positioning, with power and ideology being viewed as critically important (Gergen, 2009). Micro-social constructionism, however, tends to be more concerned with the micro-structures of language. Accordingly, research operating from this more micro-perspective tends to focus on talk, situated interaction, and local culture (Gubrium & Holstein, 2008). Micro-social constructionists, therefore, view reality as being constructed within daily discourses (Burr, 2003), place less emphasis on power, and privilege naturally occurring data for their research (Gergen, 2009).

Social constructionism and mental health

In mental health, social constructionist work began in the 1960s by examining psychiatric and community understandings of mental illness and by exploring the impact of labels (Mulvany, 2000). Its basic premise in this context is that professional practices of diagnosis and treatment are not based on objective scientific practices but rather are social constructions linked to the context by institutional, social, and practical considerations (Georgaca, 2014). The key message social constructionists convey in the field of health is that medical knowledge and practice is socially constructed and disease is an invention as opposed to a discovery (Bury, 1986).

In the context of mental health, then, social constructionist research has tended to focus on highlighting the socially produced character of categories of mental disorder and of the associated professional practices (Georgaca, 2014). It has attended to how biomedical categories in psychiatry have provided the foundation for defining pathological behaviour. In other words, the process of categorisation provided by psychiatry has given society the boundaries for normality and abnormality (Griffiths, 2001). Social constructionist work, therefore, aims to explore how these systems are accomplished in practice, as well as examining the consequences for mental health institutions and individuals experiencing mental distress (Georgaca, 2014).

Our position

In writing this chapter, we do not specifically subscribe to any particular model of disability. While we acknowledge that the social model of disability also adopts a social constructionist perspective, it does presuppose that power structures operate at different levels of society. From our position, we align more with a discursive framework (Edwards & Potter, 1992) and conversation analytic methodology (Hutchby & Wooffitt, 2008), which does not pre-suppose power or oppression and instead explores if and how these issues are made relevant through and in interaction. We therefore adopt a micro-social constructionist approach to understanding autism (see Gubrium & Holstein, 2008).

Our social constructionist argument therefore posits that, like other mental health disorders, autism is socially constructed. This is not an argument that denies the bodily realities of individuals with autism and their families, nor is it one that belittles the experiences, distress, and difficulties that some face when diagnosed with autism. Rather, this perspective simply provides an alternative view to the medicalised discourses that prevail in the rhetoric of 'autism as a disability'. Through this alternative view, we seek to give a mechanism for understanding how the prominent idea of autism as a disability is reified through medical discourses and open up the possibility to challenge this dominating storyline. From this perspective, autism is set against a construction of normality as defined by the diagnostic manuals and clinical parameters of the condition, rather than an ahistorical biological frame. We argue that the narrow interpretations offered through the discourses and practices that have medicalised autism and other mental health conditions may have a negative impact on those who are unable to conform to the prevailing standards of normality.

Indeed, autism has historically been positioned within discourses of deficit and 'fixing' the presumably broken, which has encouraged a general discourse and practice of treatment and cure (Broderick & Ne'eman, 2008). Many children with an autism diagnosis, as well as their family members, are expected to continuously negotiate what constitutes normal and abnormal behaviour, with parents and children often called upon to account for their presumably 'abnormal' behaviours (Lester & Paulus, 2012, 2014). Currently, society is organised around 'neurotypical' values and by contrast autism is considered a deficit (Brownlow, 2010); and typical forms of communication and behaviour are privileged in society (Lester, Chapter 24, this volume). Evidently the social construction of normality is only possible by comparison to something else – something which is presumably 'normal' (Lester & Paulus, 2012). Further, one's understanding of pathology or a pathological identity is based on a conceptualisation of the corresponding state of normality (Canguilhem, 1998). Thus, in this chapter, we argue that by utilising a social constructionist framework to the notions of normality and pathology, we are able to explore the ways in which the discourses that reify autism are legitimised by psychiatric rhetoric (Lester & O'Reilly, in press).

Project overview

The chapter draws from three papers and thus three separate projects in order to synthesise ideas related to arguments around the ways in which autism is constructed as a disability or not, and, more generally, the appropriateness and/or inappropriateness of autism being constructed as a disability.

Paper one (Karim et al., 2012): This social constructionist paper utilised a thematic analysis to explore the views and practices of a range of mental health professionals in diagnosing autism in children. The project utilised semi-structured interviews with practising psychiatrists, paediatricians, and educational psychologists who worked in the United Kingdom for the National Health Service (NHS).

Paper two (Lester & Paulus, 2012): This paper drew upon data from a larger ethnography of the everyday practices surrounding autism at a clinic in the United States. A discursive psychology approach was taken to study 14 interviews with parents of children with autism and eight interviews with physical, speech, and occupational therapists who work with children with autism. The paper focused on the ways in which autism, as a construct, was discursively performed and the boundaries of normality and abnormality were negotiated.

Paper three (Lester et al., in press): This social constructionist paper carried out a discourse analysis to explore the views and experiences of a range of stakeholders on varied issues affecting children with autism and their families. The project utilised focus groups with representatives from psychiatry, general practice, psychology, paediatrics, researchers, parents, and those with a diagnosis of autism from the United Kingdom.

Paper one: Arguments about diagnosis (*from Karim et al., 2012*)

One of the core problems plaguing the classification of autism is the limited number of standardised measures for diagnosis, as well the reliance upon the subjective judgements of practising professionals. The variability and discontent in the diagnostic process potentially means that the diagnosis is unstable, which has significant implications for labelling and the pathologising of the condition. Diagnosing and treating autism, indeed, is a challenge for services (Ridge & Guerin, 2011), and the complexity of the condition results in a range of presenting characteristics and behaviours. Typically, diagnosis of autism requires the input of a myriad of professionals and the use of a range of diagnostic schedules and tools; yet, despite this seemingly exhaustive approach, a diagnosis may remain unclear (Karim et al., 2012).

The variability in diagnosis is reflected in this paper (Karim et al., 2012) and thus illustrates that there are difficulties in the classification of children with autism. This noted variability suggests that the boundaries of 'normal' and presumably 'abnormal' autistic behaviours are not fixed or particularly clear, reflecting some difficulties in defining the condition. In this project, the problem of objectivity was one that was oriented to by the participants. The perceived idea that health conditions other than autism could be objectively measured was voiced often in relationship to the medical model. For example, one participant noted:

> *The diagnosis is not reliant on a blood test so you can't you know you can't use objective measures you have to objectify what are effectively subjective impressions.*
>
> (Participant 23 – Paediatrician; p. 4 of paper)

The orientation here was that medical tests are an objective way of determining the presence or absence of a condition. However, in the above quote, there was an acknowledgment that autism cannot be diagnosed according to such standardised measures and that there is a 'subjective' element to the diagnostic process. Of course, some tools for diagnosis are available, with diagnostic manuals be available. However, in the Karim et al. (2012) paper, it was noted that only 58% of the professionals included in the corpus utilised such a manual. For example, one participant stated:

> *Um, I don't actually use any standardised assessments, tools.*
>
> (Participant 9 – Psychiatrist; p. 4 of paper)

The above extract leads us to wonder whether this should be concerning. Should there be a greater drive towards more standardised measures of autism? Possibly no, considering the complexity of the condition and the range of presentation of symptoms that many children with autism display. Yet, for the families living the experience of having a child diagnosed with autism, the waiting time, the uncertainty, and the difficulties encountered during diagnosis are particularly concerning. Perhaps instead, there is a need to be more concerned with the loose boundaries that surround a condition that has great implications for the individual being constructed as disabled or mentally distressed. This reality is potentially worsened by the range of terminology that is associated with the condition, driven in part perhaps by the need for services and associated funding. For example, in the United States where funding is provided by private insurance companies, particular labels are essential to access such monies (Lester & O'Reilly, in press). Interestingly, although *DSM-5* has moved to a generic term of ASD, and removed other labels including Asperger's from *DSM-5*, society and indeed some professionals have not necessarily aligned with these changes. In this study (Karim et al., 2012), professionals were able to recognise the power of labels and discussed this in the interview. For example, two participants noted:

> *Some families feel that it is a very stigmatising term so we try to find a different terminology such as pervasive developmental disorder.*
>
> (Participant 6 – Psychiatrist; p. 6 of paper)

> *A lot of parents are keen on having a diagnosis of Asperger's I suppose because it has a connotation of their kid's bright and special kind of thing.*
>
> (Participant 7 – Psychiatrist; p. 6 of paper)

Language clearly matters and evidently has a central role in the way in which children diagnosed with autism are constructed not only by society but also by those professionals making the diagnosis. The language utilised has important implications for the disabling aspects that are frequently generated and sustained within everyday cultural and social practices.

Paper one: Clinical relevance

Diagnosis remains a central issue to the argument of whether autism is or is not a disability. The paper (Karim et al., 2012) we cited showed considerable variation in the diagnostic practices both within and between professional groups. Although there are guidelines and manuals, many of the participants noted that consistency was not yet being achieved. Terminology, that is the language used, remains a central source of confusion, with different terms being employed both professionally and in lay discourse. The perpetuation of different terms by professionals ostensibly to protect from stigma may actually be unhelpful.

Paper two: Arguments about labelling (*from Lester & Paulus, 2012*)

As parents and therapists were invited to offer accounts of the children with autism, it became clear that autism was not constructed as a static construct; rather, the participants oriented to it in shifting and even contradicting ways. The accounts around being named or identified as 'autistic' ranged from being linked to performances of 'abnormality' to being viewed as evidence of a 'gift' or unique ability.

Across the data, all of the therapists and parents oriented to autism as something that fundamentally required an explanation, particularly to those less familiarity with the varied performances of autism. For instance, Nicole, one of the participating parents shared:

```
Nicole:   When I when I when a stranger sees George I sometimes
          feel like I have to offer up if he isn't (.) uh if his
          behavior's a little bit different that he's autistic and
          sometimes I want
Jessica:  =hm=
Nicole:   =people um to know that um we had an incident yesterday
          at um at a business where he um had a meltdown in the
          bathroom because the bathroom door wasn't locking and it
          was very upsetting to him and
```

```
        then some employees came in and tried to get in and see
        if he was okay and it was you know you could hear him
        throughout the whole building so sometimes you know (.)
        I I want people to know=
Jessica: =mm hm=
Nicole:  that he has a a disability.
```
(p. 268 of paper)

Like many of the participant parents, Nicole oriented to her son as 'a little bit different', with this difference often demanding some type of explanation to outsiders. The use of the word 'sees' makes evident that there is something about her son that others orient to as being outside of normative expectations and therefore worthy of explaining. Descriptions of her son as 'disabled' were made, however, only in relationship to other's (mis)interpretations.

Many of the participating therapists also spoke at lengths about the 'look of autism', highlighting how this 'look' is often read by diagnosticians as evidence of 'abnormality' and therefore worthy of diagnosis. Megan, a speech therapist, stated:

```
Megan:   Chance that little boy his pediatrician diagnosed him
         (.1) and I'm sure to the pediatrician he looks very
         autistic 'cuz he has some of those sensory things going
         on and some of those behavior things and you know (.)
         language that hasn't developed (.) so I think (.1) even,
         I don't know, even psychologists sometimes they they will
         see a child for one evaluation visit and have to make a
         diagnosis based on that one visit (.) um (.) and I don't
         think they always know=
Jessica: =Mm hm=
Megan:   =Like if they look autistic and they meet the criteria
         (.) then they're gunna get a diagnosis basically.
```
(p. 264 from paper)

Garland Thomson (1997) highlighted that discourse functions to produce particular identities typically 'within a hierarchy of bodily traits that determines the distribution of privilege, status, and power' (p. 6). Thus, Megan's listing of 'sensory things', 'behavior things' and 'language that hasn't developed' aligned with the common ways in which autism is constructed, particularly within biomedical frameworks. Further, in the above excerpt, the child is positioned as looking 'autistic' or 'abnormal enough' to a diagnostician to be discovered as such. Hughes (1999) discussed how 'the nondisabled gaze is the product of a specific way of seeing which actually constructs the world it claims to have discovered' (p. 155). In this case, Megan illustrates how autism is something that 'must always be interpreted' (Biklen et al., 2005, p. 3).

While the participating parents spoke often of the ways in which they evoked a disability category in order to explain their child's behaviours to outsiders, they also went to great lengths to position their child as 'typical' and 'normal'. For example, Alisha, a participating parent, constructed her son as 'a very typical little boy'.

```
Alisha:    um (2) he is very affectionate he has feelings just like
           everyone else he has likes and dislikes just like any
           other child um (.) there's a lot of things about him that
           are very typical (.)
Jessica:   Mm hm=
Alisha:    =um (.) even how he expresses even though it sometimes
           doesn't always seem like it in many ways is a very
           typical little boy=
Jessica:   =Mm hm can you talk a little bit about that (.)
Alisha:    Um you know he enjoys he loves playing outside he loves
           the trampoline he'd probably live in it if we let him um
           he loves chocolate chip cookies and pancakes and he loves
           to watch cartoons he likes to he loves to draw (.) um
           and he you know he works really hard in school and he
           actually gets very good grades in school um he's an
           excellent speller he's a very good reader he reads above
           his grade level even though we can't have a conversation.
```

(p. 266 from paper)

Disability categories are frequently viewed as being synonymous with incompetence, which is particularly true for people who are nonverbal (Biklen et al., 2005), such as Alisha's son. Thus, it is striking that Alisha, like other parents, offered descriptions of her child that countered the dominant assumption that autism can be equated with some level of incompetence or significant abnormality. In fact, Alisha lists out activities that some might view as neurotypical (e.g. watching cartoons, drawing), as she navigates the fragile task of constructing her child as competent. Indeed, all aspects of what it means to be autistic were negotiated by the therapists and parents, as discourses and ideologies surrounding autism and disability were evoked and even resisted at times.

Paper two: Clinical relevance

Negotiating what it means to 'be autistic' is a fragile task that involves defining and redefining the bounds of normality and abnormality. Indeed, parents and therapists orient to the performances of autism as not being static, but fluid, layered, and even contradictory. It is therefore critical for clinicians to remain cognizant of the socially bound nature of autism, and the ways in which normative understandings of 'looking abnormal' or 'looking normal' play into the diagnostic process. Further, it is paramount that clinicians support families as they navigate the challenging task of explaining their child's behaviours and even reframing their child as a competent human being.

Paper three: Stakeholder views (*From Lester et al., in press*)

As noted, indeed the boundaries between normality and abnormality are not clearly evident, with medical sociology and disability studies in particular serving to question whether clear distinctions exist. Yet, beyond this literature base, stakeholders also frequently construct mental health constructs as fluid and shifting categories. Rocque (2010) argued that people diagnosed with autism are often positioned as 'deficient' most notably because they are 'assessed according to a model of embodiment that assumes one developmental trajectory' – one that is far 'too narrow to include them' (p. 12). As noted above (Lester & Paulus, 2012), the performances and behaviours of people diagnosed with autism are frequently deemed 'abnormal' and often expected to be accompanied with some type of account or explanation. In our research with stakeholders (Lester et al., in press), we found that the stakeholders went about negotiating what counted as autism in ways that made evident ideological dilemmas (Billig et al., 1988) inherent to naming this shifting construct. For example, one of the stakeholders, who also identified as being diagnosed with autism and a parent of two sons with autism, shared:

```
Pete:   I think th::e (0.4) there is (.) one key difficulties here
        (.) which is (.) that need t' be addressed and that's
        the generalisation (0.4) issue, the fact that (.) um (.)
        different individuals ↑a::re (.) affected t' different
        degrees in different realms different spheres at particular
        times (0.2) different times different >different different
        different< (0.4) an' that's (.) gonna be a key problem t'
        settin' up any (0.4) set (.) >sort of< (.) u::m material
        (.02) t' be accessed by peo↓ple.
```

(p. 7 of paper)

Like the other stakeholders in our study, Pete produced a dilemma around what constitutes the boundary between normal and abnormal, specifically in relationship to determining the materials or resources that might be needed by the person with autism. Distinguishing between 'normal' and 'abnormal' in the context of autism is incredibly complex, particularly in relation to diagnostic processes. In the excerpt above, Pete made evident this dilemma through his repeated emphasis on the way in which people with autism are 'different different different'. He positioned this difference as particularly relevant to 'settin' up any...sort of...material' or providing individuals with autism with support.

The stakeholders also made relevant a dilemma around whether autism should be regarded as a disability or ability, and therefore something that requires a cure, placing into question the very basis of how autism is frequently conceived (i.e. as a deficit to be cured).

```
Pete:   <But it also leads into what was ↑just ↑said> an' that's
        >↑you know< (0.4) ↑autism itself I'm a great believer that
        ↑autism itself actually isn't a disability in anyway at
        a::ll (.) in fact there are many areas where I would argue
        that my autism >is a s↑trength< (.) um
Joy:    >Its other people's ignorance< isn't it?
```

(pp. 17–18 of paper)

In the above extract, the dilemma of whether autism *is* in fact a disability was negotiated, with Pete orienting to autism as 'a strength'. Indeed, Pete has a stake and interest in this particular claim, as he positioned himself as autistic ('my autism') and must now manage this autistic identity against being positioned as 'abnormal'. More broadly, this notion that autism is *not* a disability has been contested (Nauert, 2011), and thus we were struck by Pete's personal belief claims of, 'I'm a great believer'. In interactions, people have the option of taking up positions offered by others or positioning their own identity (Davies & Harrè, 2001). Here, if Pete had taken up autism as a deficit or as something that needed to be cured, it would have positioned him as disabled. Joy, who identified as a parent of a child with autism, furthered this negotiation by casting the positioning of autism as a disability as being due to 'other people's ignorance'. As such, the disabling effects of autism were located squarely within society rather than within the individual diagnosed with autism.

Stakeholders also made evident the dilemma of who has the authority to speak about autism. Indeed, knowledge is linked to categories, with particular categories being oriented to as being entitled to know particular sorts of knowledge and therefore entitled to speak on a particular matter (Potter, 1996). Across the focus group data, stakeholders moved between different categories, working up their epistemic rights. In the extract below, for example, Joanne, a parent of a child with autism and ex-chair of her local branch of the National Autistic Society, took up two subject positions as she made evident her knowledge.

```
Joanne:   And if I can speak as a parent and someone who <provides
          support as well> erm (0.6) I've been listenin' t' this and
          I find it as well if I was a parent (0.2) who had just had
          a diagnosis erm (0.2) rather than have >an' I've heard
          these stories< (0.4) rather than have a pa::ediatrician or
          someone from CAMHS give me a bibliography on a bit of
          pap↑er (0.4) and say ↑there go and find these ↑books (0.4)
          if they said to me (0.6) this is a website...I think that
          would be absolutely ideal.
```

(p. 18 of paper)

Here, Joanne identified herself as both a 'parent' and 'someone who provides support as well'. She evoked her expertise on both a personal and professional level, with the legitimacy of her claims made on the basis of her positioning. This negotiation of possessing the authority or rights to speak about autism creates a dilemma around (1) who has the right to establish what constitutes autism, (2) who can claim to know what parents and others need to know, and (3) what counts as valid evidence about autism.

Paper three: Clinical relevance

This paper's findings offer important implications for professionals who work with families of children diagnosed with autism. First, they highlight the need to acknowledge that the fluidity around the diagnostic criteria of autism underscore the difficulties in establishing consensus around what 'truly counts' as autism. As such, it may be useful for professionals to acknowledge and be responsive to the full range of perspectives around the meaning(s) of autism, particularly as various stakeholders share varied and even conflicting views on what autism is and is not. Clinicians may be well served by extending their knowledge base to include the perspectives of those who may not typically be accounted for in professional literature around autism. Further, it may be useful to actively illustrate how autism can be understood as a social category, specifically as stakeholders go about the task of negotiating the fluid meaning(s).

Clinical relevance summary

It is critical for clinicians to remain aware of the subjective nature of the diagnostic process, as well as the ways in which diagnosis intersects with understandings around abnormality and normality. In addition, with stakeholders constructing autism in varied ways, it is paramount for clinicians to position their discussion about autism in context of these different perspectives. For a simple summary of the implications for practice, see Table 14.1.

Table 14.1 Clinical practice highlights

1. The diagnostic process is subjective and dependent upon the language used.
2. Diagnostic manuals have not necessarily resulted in the consistent diagnostic processes.
3. Stakeholders negotiate the meaning(s) and performances of autism and view autism as difficult to name as simply representative of 'abnormal' behaviour.
4. The bounds between abnormality and normality are fluid and shifting, frequently redefining what it means to be classified with autism.
5. Stakeholders may view autism as an 'ability', not simply a 'disability'.

Summary

In this chapter, we highlighted the ways in which stakeholders go about negotiating the meaning(s) of autism. This negotiation is pursued in relationship to normative discourses and practices, often resulting in stakeholders having to offer accounts that position autism as something other than a deficit.

References

American Psychiatric Association (2013). *Diagnostic and statistical manual of mental disorders* (5th edition). Washington, DC: American Psychiatric Association.

Ashby, C. E. (2010). The trouble with normal: The struggle for meaningful access for middle school students with developmental disability lab. *Disability & Society, 23*(3), 345–358.

Bailey, A. (2008). Autism as a global challenge. *Autism Research, 1*, 145–146.

Bagatell, N. (2010). From cure to community: Transforming notions of autism. *Ethos, 38*(1), 34–58.

Baker, D. (2011). *The politics of neurodiversity: Why public policy matters*. Boulder, CO: Lynne Rienner.

Barnes, C., Mercer, G., & Shakespeare, T. (1999). *Exploring disability: A sociological introduction*. Oxford: Polity press.

Barton, L. (1993). The struggle for citizenship: The case of disabled people. *Disability, Handicap and Society, 8*(3), 235–248.

Biklen, D., Attfield, R., Bissonnette, L., Blackman, L., Burke, J., et al. (2005). *Autism and the myth of the person alone*. New York: New York University Press.

Billig, M., Condor, S., Edwards, D., Gane, M., Middleton, D., & Radley, A. (1988). *Ideological dilemmas: A social psychology of everyday thinking*. London: Sage.

Broderick, A. A., & Ne'eman, A. (2008). Autism as metaphor: Narrative and counter narrative. *International Journal of Inclusive Education, 12*(5–6), 459–476.

Brown, P. (1995). Naming and framing: The social construction of diagnosis and illness. *Journal of Health and Social Behavior* (Extra issue), 34–52.

Brownlow, C. (2010). Presenting the self: Negotiating a label of autism. *Journal of Intellectual and Developmental Disability, 35*(1), 14–21.

Burr, V. (2003). *Social constructionism* (2nd edition). London: Routledge.

Bury, M. R. (1986). Social constructionism and the development of medical sociology. *Sociology of Health and Illness 8*, 137–169.

Canguilhem, G. (1989). *The Normal and the pathological*. Brooklyn, NY: Zone Books.

Davies, B., & Harre, R. (2001). Positioning: The discursive reproduction of selves. In M. Wetherell, S. Taylor, & S. J. Yates (Eds.), *Discourse theory and practice: A reader* (pp. 261–171). London: Sage.

Edwards, D., & Potter, J. (1992). *Discursive psychology*. London: Sage.

Eyal, G., Hart, B., Onculer, E., Oren, N., & Ross, N. (2010). *The autism matrix: The social origins of the autism epidemic*. Cambridge: Polity Press.

Frith, U. (1989). *Autism: Explaining the enigma*. Cambridge, MA: Blackwell Publishers.

Garland Thomson, R. (1997). *Extraordinary bodies: Figuring physical disability in American culture and literature.* New York: Columbia University Press.

Georgaca, E. (2014). Discourse analytic research on mental distress: A critical overview. *Journal of Mental Health, 23*(2), 55–61.

Gergen, K. (1985). The social constructionist movement in modern psychology. *American Psychologist, 40*(3), 266–275.

Gergen, K. (2009). *An invitation to social construction.* Los Angeles, CA: Sage.

Gilson, S., & DePoy, E. (2000). Multiculturalism and disability: A critical perspective. *Disability and Society, 15*(2), 207–218.

Gleeson, B. (1997). Disability studies: A historical materialist view. *Disability and Society, 12*(2), 179–202.

Glynne-Owen, R. (2010). Early intervention and autism: The impact of positivism and the call for change. *International Journal of Children's Rights, 18*, 405–416.

Goffman, E. (1968). *Stigma.* Hamondsworth: Penguin.

Grue, J. (2011). Discourse analysis and disability: Some topics and issues. *Discourse and Society, 22*(5), 532–546.

Grubrium, J., & Holstein, J. (2008). The constructionist mosaic. In J. Holstein & J Gubrium (Eds), *Handbook of constructionist research* (pp. 3–12). New York: Guildford.

Hahn, H. (1985). Towards a politics of disability: Definitions, disciplines and policies. *The Social Science Journal, 22*(4), 87–105.

Hahn, H. (1987). Civil rights for disabled Americans: The foundations of a political agenda. In A. Gartner & T. Joe (Eds.), *Images of disabilities/disabling images* (pp. 181–203). New York: Praeger.

Hughes, B. (1999). The constitution of impairment: Modernity and the aesthetic of oppression. *Disability & Society, 14*(2), 155–172.

Hutchby, I., & Woffitt, R. (2008). *Conversation analysis* (2nd edition). Cambridge: Polity.

Huws, C., & Jones, R. (2008). Diagnosis, disclosure, and having autism: An interpretative phenomenological analysis of the perceptions of young people with autism. *Journal of Intellectual and Developmental Disability, 33*(2), 99–107.

Karim, K., Ali, A., & O'Reilly, M. (2014). *A practical guide to mental health problems in children with autistic spectrum disorder: It's not just their autism!* London: Jessica Kingsley Publishers.

Karim, K., Cook, L., & O'Reilly, M. (2012). Diagnosing autistic spectrum disorder in the age of austerity. *Child: Care, Health and Development, 40*(1), 115–123.

Lafrance, M., & McKenzie-Mohr, S. (2013). The DSM and its lure of legitimacy. *Feminism and Psychology, 23*(1), 119–140.

Lester, J. N., Karim, K., & O'Reilly, M. (in press). 'Autism itself actually isn't a disability': The ideological dilemmas of negotiating a 'normal' versus 'abnormal' autistic identity.

Lester, J. N., & Paulus, T. M. (2012). Performative acts of autism. *Discourse and Society, 23*(3), 259–273.

Lester, J. N., & Paulus, T. M. (2014). 'That teacher takes everything badly': Discursively reframing non-normative behaviors in therapy sessions. *International Journal of Qualitative Studies in Education, 27*(5), 641–666.

Lester, J. N., & O'Reilly, M. (in press). Repositioning disability in the discourse of our times: A study of the everyday lives of children with autism. In G. Noblit & W. Pink (Eds.), *Education, equity, and economy.* Springer.

Mercer, G. (2002). Emancipatory disability research. In C. Barnes, M. Oliver, & L. Barton (Eds.), *Disability studies today* (pp. 228–249). Cambridge: Polity press.

Mulvany, J. (2000). Disability, impairment or illness? The relevance of the social model of disability to the study of mental disorder. *Sociology of Health and Illness, 22*(5), 582–601.

Muskett, T., Body, R., & Perkins, M. (2013). A discursive psychology critique of semantic verbal fluency assessment and its interpretation. *Theory and Psychology, 23*(2), 205–226.

Nadesan, M. H. (2005). *Constructing autism: Unraveling the 'truth' and understanding the social*. New York: Routledge.

Nauert, R. (2011). Viewing autism as difference, not just a disability. Retrieved from PsychCentral: http://psychcentral.com/news/2011/11/04/viewing-autism-as-difference-not-just-disability/31091.html.

Oliver, M. (1983). *Social work with disabled people*. London: MacMillan.

Oliver, M. (1990). *The politics of disablement*. Basingstoke: MacMillan.

Oliver, M. (1996). A sociology of disability or a disablist sociology? In L. Barton (Ed.), *Disability and society* (pp. 18–42). London: Longman.

Petrunik, M. (1980). The rise and fall of 'labelling theory': The construction and destruction of a sociological strawman. *The Canadian Journal of Sociology, 5*(3), 213–233.

Potter, J. (1996). *Representing reality: Discourse, rhetoric and social construction*. London: Sage.

Ridge, K., & Guerin, S. (2011). Irish clinicians' views of interventions for children with autistic spectrum disorders. *Autism, 15*(2), 239–252.

Rocque, B. (2010). Science fictions: Figuring autism as threat and mystery in medico-therapeutic literature. *Disability Studies Quarterly, 30*(1).

Santrock, J. W. (2007). *A topical approach to human life-span development* (3rd edition). St. Louis, MO: McGraw-Hill.

Sudnow, D. (1965). Normal crimes: Sociological features of the penal code in a public defender's office. *Social Problems, 12*, 255–276.

Timimi, S., Gardner, N., & McCabe, B. (2011). *The myth of autism*. Basingstoke: Palgrave-MacMillan.

Weinstein, R. (1983). Labeling theory and the attitudes of mental patients: A review. *Journal of Health and Social Behavior, 24*(1), 70–84.

Williams, J., Higgins, J., & Brayne, C. (2006). Systematic review of prevalence studies of autism spectrum disorders. *Archives Dis Child, 91*, 8–15.

Wing, L. (1981). Language, social and cognitive impairments in autism and severe mental retardation. *Journal of Autism and Developmental Disorders, 11*(1), 31–44.

Wing, L. (1996). *The autistic spectrum*. London: Constable and Company Ltd.

Wolff, S. (2004). The history of autism. *European Child and Adolescent Psychiatry, 13*(4), 201–208.

Wykes, T., & Callard, F. (2010). Diagnosis, diagnosis, diagnosis: Towards DSM-5. *Journal of Mental Health, 19*(4), 301–304.

Recommended reading

- Brownlow, C. (2010). Presenting the self: Negotiating a label of autism. *Journal of Intellectual and Developmental Disability, 35*(1), 14–21.

- Lester, J. N., & O'Reilly, M. (in press). Repositioning disability in the discourse of our times: A study of the everyday lives of children with autism. In G. Noblit & W. Pink (Eds.), *Education, equity, and economy*. Springer.
- Lester, J., & Paulus, T. (2012). Performative acts of autism. *Discourse and Society, 23*(3), 259–273.

15

Subjectivity in Autistic Language: Insights on Pronoun Atypicality from Three Case Studies

Laura Sterponi, Kenton de Kirby, and Jennifer Shankey

On est en présence d'une classe de mots, les «pronoms personnels», qui échappent au statut de tous les autres signes du langage. A quoi donc je se réfère-t-il? A quelque chose de très singulier, qui est exclusivement langagier : je se réfère à l'acte de discours individuel où il est prononcé, et il désigne le locuteur. [...] La réalité à laquelle il renvoie est la réalité du discours. C'est dans l'instance de discours où je désigne le locuteur que celui-ci s'énonce comme «sujet». Il est donc vrai à la lettre que le fondement de la subjectivité est dans l'exercice de la langue.

(Benveniste, *De la subjectivité dans le langage*, 1958)

We are in the presence of a class of words, the 'personal pronouns,' that escape the status of all the other signs of language. Then, what does *I* refer to? To something very peculiar which is exclusively linguistic: I refers to the act of individual discourse in which it is pronounced, and by this it designates the speaker. [...] The reality to which it refers is the reality of the discourse. It is in the instance of discourse in which I designates the speaker that the speaker proclaims himself as the 'subject.' And so it is literally true that the basis of subjectivity is in the exercise of language.

Introduction

In this chapter, we examine a clinical feature typically associated with the speech of children with autism: pronoun reversal and avoidance. Children

---

This chapter is dedicated to the memory of Jennifer Shankey, who prematurely passed away while we were completing the writing of the manuscript.

with autism are reported to use the second-person pronoun *you* or third-person pronoun *he/she* to refer to themselves, as well as to use the first-person pronoun *I* to refer to the person addressed. This behaviour is referred to as pronominal reversal. In addition, affected children make frequent use of proper names to refer to self or the addressee and sometimes deploy agentless passive constructions. These speech patterns are referred to as pronominal avoidance. These phenomena are located at the intersection of linguistic and social-relational processes, and as such they constitute a particularly interesting area of investigation. For the study of language acquisition generally, these phenomena reveal language's social underpinnings, as well as the relationship between language and the development of self. For autism research, atypical pronoun usage offers potential insights about core features of the condition.

We contend that a thorough understanding of the complex functioning of pronouns, and the mechanisms underlying pronoun use in normally developing children, can afford a more nuanced account of pronoun reversal and avoidance in childhood autism. Research on pronoun atypicality in children with autism is very often based on a partial view of how personal pronouns function. Specifically, it is grounded in what we refer to as the indexical-referential dimension of personal pronouns, articulated by linguist Benveniste (1971). Benveniste's analysis, well known among linguists, discerned the important distinctive status of person pronouns as linguistic signs, which 'do not refer to a concept or to an individual' (Benveniste, 1971, p. 226). The *I* and *you* in particular are referentially empty signs with respect to reality in that they designate something internal to discourse and inherently linguistic. In Benveniste's words: '*I* refers to the act of individual discourse in which it is pronounced, and by this it designates the speaker' (Benveniste, 1971). The first and second-person pronouns are deictic; that is, they enact their reference at the token level through actual contiguity between the utterance in which *I/you* occurs and the speaker/recipient of the utterance.

While personal pronouns cannot be properly understood without considering the indexical-referential function, scholarship in child language, conversation analysis, and linguistic anthropology has unveiled additional functional modes, not of secondary importance. We argue that more nuanced comprehension of the functioning of personal pronouns – one that appreciates these additional functions – has the potential to deepen our understanding of children with autism and their linguistic behaviour. In particular, the resulting insights provide support for a theory of autism centred around impairments in self–other relatedness, in identifying with the perspective of others (Hobson, 2010). At the same time, our analysis points to the inadequacy of considering atypical pronoun use as straightforwardly symptomatic of a condition residing within the affected child. For one, we point out that typically developing children display what has been referred to as pronoun reversal and avoidance in autism research, and we show that difficulties in self–other relations need not manifest in pronoun atypicality. Furthermore, we find attempts by the child with autism to overcome these very difficulties through the experiential affordances of language, as well as a subtle sensitivity to interpersonal positioning – both of which have been largely overlooked by mainstream autism research. In addition, we point to the critical role of adults' contribution to interaction in shaping the child's non-normative use of personal pronouns.

Pronoun atypicality in children with autism

Difficulties with the use of personal pronouns in children with autism are long recognised. Kanner documented occurrences of atypical personal references in his first description of infantile autism and considered the phenomenon as characteristic of the condition, along with echolalia (Kanner, 1943). In fact, the Austrian-American psychiatrist considered pronominal reversal as an epiphenomenal manifestation of echolalia (see also Bartak & Rutter, 1974), the tendency to repeat the utterances of oneself or others.

Further study in the following decades revealed the autistic child's difficulties with personal pronouns 'too complex and too deeply ingrained to be accounted for completely within the surface framework of "reversals" secondary to echolalia' (Fay, 1979, p. 248).

Fay contended that underlying personal pronoun errors are 'multiple developmental obstacles of social, cognitive and grammatical nature' (Fay, 1979, p. 247). Drawing from developmental research on self-differentiation, memory, and deixis acquisition in neurotypical and atypical populations, Fay identified three compounding sources of pronoun difficulties: (1) limited contact and involvement with others, which severely handicaps the child with autism in the development of a sense of self and the other; (2) restricted echoic memory,[1] which may limit access only to the most recent information within an interlocutor's utterance, leaving the segment with the person pronoun in subject position inaccessible; (3) a more general deictic impairment, which includes difficulties in the use of spatial and temporal deictic terms (e.g. here, there) as well as in pre-linguistic forms, such as pointing. Subsequently, other researchers emphasised and further delineated the psychosocial deficits that Fay had originally identified, by linking abnormalities in person pronouns to underdeveloped differentiation of self, atypical experience of self–other relations, and limitations in communicative engagement (e.g. Charney, 1980b; Oshima-Takane & Benaroya, 1989).

Noteworthy insight and nuance have been provided by Hobson and associates regarding the psychosocial factors underlying atypical pronoun use (Hobson, 1990; Hobson, Lee, & Hobson, 2010; Lee, Hobson, & Chiat, 1994). On the basis of a remarkably ample and varied set of experiments of perceptual viewpoint, shared agency, visuospatial role-taking, and photograph-naming,

Hobson's team has shown that, while children with autism can comprehend pronouns appropriately and only rarely produce pronoun reversals, they also manifest subtle but significant differences with respect to closely matched non-autistic subjects. These differences – such as a relative propensity to use the pronoun *I* rather than *me*, and the proper name rather than the first- or second-person pronouns – are interpreted as a reflection of the autistic child's difficulty in experiencing and understanding himself as self-in-relation-to-others, with a self of his own (Hobson, 2011).

Pronoun atypicality in typically developing children

Patterns of pronoun errors, however, have been documented also in typically developing children (see Bain, 1936 and Cooley, 1908, for early reports). Two different explanations have been put forward: (1) Pronoun atypicality results from an incorrect understanding of how personal pronouns function (i.e. treating them as if they have fixed referents, like names, rather than as deictic terms) – a problem of competence, so to speak (e.g. Charney, 1980a; Clark, 1978). This perspective anticipates a certain consistency in the children's errors such that, for example, *you* will systematically be used to refer to the child herself. (2) Instances of atypical pronoun use are performance errors, resulting from processing demands that exceed the child's cognitive and linguistic abilities (e.g. Dale & Crain-Thoreson, 1993). In this perspective, personal reference errors will be inconsistent (e.g. only sometimes would a child use *I/you* atypically) but they would correlate with factors such as the syntactic complexity of the utterances in which they are found.

A third explanation for pronoun errors in typical language development has been advanced by Chiat (1982) on the basis of a case study of a two-year-six-month-old boy. Chiat suggests that pronouns can be plurifunctional – that is, they might be used prototypically as speech-role-referring, but at the same time they might perform other non-adult functions. In other words, the child might use pronouns with more than one function, in ways that are both complex and predictable (albeit non-normative).

We find Chiat's perspective particularly insightful for three main reasons: First, it acknowledges that in discourse contexts personal pronouns are considerably more complex than the indexical-referential account would assume.[2] Second, consistent with this, it suggests that multiple mechanisms underlying pronoun use may coexist, particularly in early stages of language development. Third, at the same time, it argues that atypical uses of pronouns may be functional and orderly.

The semiotic functioning of pronouns in everyday speech

Insights from linguistic anthropology and conversation analysis corroborate and expand Chiat's analysis (albeit not engaging with the literature on the acquisition of personal pronouns directly or explicitly). Linguistic anthropologist Greg Urban (1989) has advanced a model of the semiotic functioning of person pronouns that expands the traditional indexical-referential analysis articulated by Benveniste (1971) by identifying additional functional modes of the *I/you* in discourse. For Urban (1989), the indexical-referential treatment of *I/you* is accurate but insufficient. He delineates two additional modes of functioning of *I/you* in speech production – the anaphoric and the de-quotative *I/you* – that are predicated on the idea that 'the I of discourse is not only an actual in-the-world subject, indexically referred to by means of the first person form' but 'can also be any being or entity, imaginary or not, capable of being reported as a speaker' (Urban, 1989, p. 29). The anaphoric *I/you* manifests in cases of reported speech, as in 'He said "I am going".' In such cases, the personal pronoun is not used indexically to refer to the utterer or addressee of the sentence. Rather, it functions as 'an anaphoric device, indicating the co-referential relationship between the subject of the two clauses' (Urban, 1989, p. 30). In other words, the first-person pronoun *I* of an embedded clause in direct quotation achieves reference, not indexically, but anaphorically – its meaning depends on another element in the discourse.

The other mode Urban delineates, the de-quotative mode, functions when the matrix clause of a quotation (e.g. 'he said', 'she said') is omitted. The de-quotative *I* does not point to the speaker of the utterance but to the person/character she is representing and animating. Thus, once again the ordinary indexical-referential functioning of *I/you* is suspended. In de-quotative speech, the speaker may deploy a range of indexical cues (notably pitch and voice quality alterations) to signal that she is animating someone else's utterances. The absence of the quoting frame however allows maximal projection of the speaker into another self (Urban, 1989).

Conversation analysis also offers a theoretical perspective that expands our understanding of pronoun use (see Fasulo, Chapter 1, this volume). Charles Goodwin (1981) has demonstrated that sentences emerge in conversation through the interaction between speaker and hearer. The interactional nature of turns in conversation cautions us from evaluating a child's pronoun use in isolation, and it compels us to think about child language not solely in terms of cognitive and linguistic development. A child's verbal contribution to a communicative exchange also relates to the specific contingencies of interaction in which she is engaged. By extracting utterances from the context of production and judging them as self-standing entities – that is, as individual outcomes that reflect an underlying linguistic capacity (or lack thereof) – we might attain inaccurate evaluations.

In order to gain additional analytic purchase for the examination of the inherent situatedness of talk-in-interaction, including the relationship between personal reference forms and their context of production, we draw on Wittgenstein's notion of 'language game' (Wittgenstein, 1953) The phrase refers, not to a game with words, but to the idea that language is inseparable from activity, and that utterances take on meanings within distinguishable courses of action. For our empirical analysis of children with autism in interaction, this notion offers a useful lens, alerting us to the way children's participation in these activities shapes their pronoun use.

Project overview

We now present empirical work with the intention to show that a more complete understanding of how personal pronouns function entails a more accurate picture of atypical pronoun use in children with autism and its relationship to autism more generally.

We structure our cases studies in order to evaluate the two categories of explanation for atypical pronoun use in each child's language. To evaluate the competence hypothesis, we start off with examining the range of linguistic resources each child mobilised to refer to self and others. This provides a general sense of the child's capacity to position himself as subject and in relation to others in conversation. We then investigate the overall frequency, distribution, and consistency of person reference atypicalities, in order to evaluate if the errors could be related to an erroneous semantic representation of pronouns. To evaluate the performance hypothesis, we next perform an in-depth qualitative analysis of the utterances presenting atypical pronoun use. In this analysis, we consider the linguistic makeup (i.e. the syntactic complexity) of utterances containing the atypicality in order to evaluate if cognitive/linguistic demand could account for the abnormal reference form. We then consider another form of performance explanation, according to which it is pragmatic complexity – rather than the syntactic complexity – that accounts for atypicality when it is produced. Accordingly, we characterise those utterances pragmatically in terms of their discourse function. Finally, we consider a third hypothesis, which is that the language games initiated by the child's interlocutors have a bearing on his personal reference and pronoun use. To evaluate this hypothesis, we draw on analytic tools from conversation analysis and philosophy of language to illuminate the sequential context in which the child's utterances are produced.

Data corpus

Our study consists of a comparative case analysis of three children with autism: Ivan, Benjamin, and Aaron (see Table 15.1). All three boys are similar in age, having just passed or approaching their sixth birthday. Their linguistic abilities, however, are different, ranging from an MLU (mean length of utterance) of 2.28

Table 15.1   MLU and percentage of pronoun atypicality of the three case studies

	Ivan	Aaron	Benjamin
Age	5;11	5;10	6;3
MLU	2.28	3.92	5.85
Pronoun atypicality (%)	35	28	1

for Ivan, to 3.92 for Aaron and 5.85 for Benjamin.[3] We purposefully selected for this comparative case study three autistic children with different verbal abilities in order to gain insight on the role of linguistic, psychosocial, and interactional factors in pronoun atypicality.

For each child, we have considered 90 minutes of video-recorded spontaneous verbal interaction in different contexts, meal, play with peer, reading/writing/drawing/music with parent/tutor, bath, and bedtime. The interactions were then fully transcribed according to the conventions of conversation analysis (Atkinson & Heritage, 1984).

Each person reference form was identified and coded as typical or atypical. In line with previous work on pronoun reversal in typically developing children (Dale & Crain-Thoreson, 1993; Evans & Demuth, 2012), frozen expression – such as *I love you, thank you, excuse me, let's, lemme* – were not coded. The coding was carried out by the second and third authors of this chapter. They first coded independently 10% of the data and reached agreement to nearly 100%. Then they proceeded with the coding of the remaining data.

The three case studies

We begin with a presentation of Ivan's case. Ivan has the lowest linguistic ability of the three children and the highest percentage of pronoun atypicality. This child's case demonstrates the potential limitations of both competence and performance explanations. It also points to the value of an interactional analysis – keyed to the contribution of the child's interlocutors – for shedding light on non-normative person reference forms. We next present Benjamin's case, whose linguistic ability offers a stark and useful contrast to Ivan's. We draw on Benjamin's data to show that difficulties in self–other relations need not manifest in pronoun atypicality. Finally, we present data from Aaron, who combines a high level of both linguistic ability and atypical pronoun usage. For this reason, Aaron's case is the richest and most interesting. We will argue that Aaron's use of de-quotative speech – entailing non-normative use of personal pronouns – subtly but clearly demonstrates social sensitivity and manifests attempts to overcome difficulties in identifying with others' perspectives.

Ivan

Firstborn of an upper-middle-class English-speaking family living in northern California, Ivan was five years and 11 months at the time of the video-recording. He had a sister who was three years old. He was in a regular kindergarten class 40% of the time and in special education classes the remaining 60%. Ivan was diagnosed with autism at age three. His language development was significantly delayed: at the time of video-recording his MLU was 2.28.

Ivan is an active participant in verbal interactions with family members and tutors. He often initiates conversational sequences, most typically basic adjacency pairs, and is fairly responsive to those launched by his interlocutors. However, the syntactic structure of Ivan's utterances is minimal and the range of verbal actions performed rather limited, consisting principally of requests, declaratives, protests, and responses to adult queries – most frequently yes/no interrogatives. Ivan's speech includes a noticeable component of repetitive phrases and frozen expressions. In his non-echolalic speech, grammatical errors are frequent, notably omission of copulae, verb inflections, and subjects (see Table 15.2).[4]

Table 15.2 shows that Ivan's person reference repertoire is rather limited and the rate of atypicality significant. Clearly, Ivan is capable of positioning

Table 15.2 Ivan's person reference repertoire

Person Reference	Typical	Atypical	Total
Proper/Role Name			
Subject	27	20	47
Object (direct and indirect)	–	1	1
Possessive	–	4	4
Pronoun			
First-person singular			
I	18	–	18
me	–	–	–
my	3	–	3
mine	–	–	–
Second-person singular			
you	2	2	4
you	–	–	–
your	–	–	–
yours	–	–	–
Third-person singular	–	–	–
First-person plural	–	–	–
Second-person plural	–	–	–
Third-person plural	–	–	–
Total	50	27	77

himself as subject and affirming relations of possession, as attested by the use of the first-person singular forms. His use of the second-person singular, on the other hand, is more vulnerable to mistakes. Third-person singular pronouns are absent from Ivan's repertoire, and so are the plural forms across persons and cases. This range of forms and the vulnerability to errors parallel the sequence of development of pronouns in the English-speaking neurotypical population: *my*, *mine*, and *I* are used consistently before *you*; *he/she* are normatively used last (Chiat, 1986). In addition, it is notable that Ivan's MLU is approaching that of normally developing children when they correctly use the first- and second-person pronouns, which is of 2.5. We'll return to this important point in the discussion.

The most significant area of atypicality in Ivan's person reference repertoire is the pronominal use of proper/role names – that is, the use of the proper name in self reference and proper or role names to refer to his interlocutor. As discussed earlier, these forms are referred to as 'pronoun avoidance'.

In order to shed light on these atypical forms, we examined the linguistic makeup of the utterances containing the pronominal use of proper/role names, as well as their pragmatic function. The linguistic structure of utterances with pronominal use of proper/role names could not be distinguished in syntactic organisation, and in semantic constituents and relations, from those without pronoun avoidance. At the pragmatic level, pronoun avoidance was found across speech actions performed by Ivan, except for protests, which overwhelmingly had the form of 'negation + complaint object' (for instance, 'no vacuum'). On this basis, we concluded that a performance explanation is inadequate in Ivan's case.

Given Ivan's significant delay in language development, it seems plausible that his tendency to 'avoid' personal pronouns is primarily a manifestation of his limited linguistic competence. We argue, however, that attention to the sequential context of atypical pronoun use, which includes a consideration of his interlocutors' contribution, adds important insights on why Ivan uses proper/role names *when he does*.

We noticed that the pronominal use of proper/role names frequently occurred in the context of baby talk. Parents and tutors often used the baby talk register when addressing Ivan. In addition to other forms of simplification, they used parental or proper names (e.g. Mommy, Daddy, Shelly) in self-reference when talking to the child.[5] They also sometimes addressed the child in the third person using the corresponding pronoun or his first name. In such occasions, we found that Ivan often adapted to the personal reference frame and syntactic construction of the interlocutor, which resulted in his 'avoidance' of personal pronouns.

Consider the following extract, in which the child is with his after-school tutor Shelly. Ivan and Shelly are cleaning the board to begin drawing shapes. Ivan takes the initiative and indicates that he wants to draw a heart (line 1). His formulation of intent is grammatical and contains the first-person singular pronoun *I*. We shall see that Shelly's subsequent clarification sequence introduces a shift in personal reference to which Ivan adapts in his response.

Extract 1 – Drawing shapes with tutor (Ivan, Tape#1)

```
1      IVAN     I want to make a hea:rt. ((sits in front of
                board with back to Shelly; holds a marker in
                hand))
2      SHELLY   okay. mmh. ((of effort: moving Ivan's chair
                closer to the board))
                you want to make a heart?
3      IVAN     make heart ((turns to look at Shelly))
4      SHELLY   who makes a heart. Ivan or Shelly?
5      IVAN     Ivan Shewy ((hands marker to Shelly))
6  →            Shewy make a heart.
```

In turn 2, Shelly utters an understanding check that is in keeping, syntactically and in personal reference format, with Ivan's opening utterance. Ivan's echoic response in line 3 seems to be only partially satisfying for Shelly (Ivan's response confirms the action being projected but not the agent). In line 4, Shelly formulates an open wh- interrogative ('who makes a heart') and then appends to it an alternative question ('Ivan or Shelly?'), which shifts personal references from pronouns to proper names (thereby instantiating a typical baby talk feature). The alternative question format projects a response that contains a partial repetition (i.e. one of the provided options) (Raymond, 2003). Ivan's reply in line 6, which if taken in isolation would be treated as an occurrence of pronoun avoidance, thus emerges as appropriate response to his tutor's simplified alternative question.

One could object that Ivan's response in line 6 indicates that his utterance in line 1 was actually an instance of pronoun reversal, referring to Shelly as *I*. We would argue, however, that if the analytic focus is not limited to talk but includes also the moment-by-moment use of non-verbal semiotic resources, such as gaze, embodied action, and object use, this interpretation can be easily refuted: When Ivan opens the sequence (line 1) by announcing his plan of action, which linguistically constructs him as subject/agent through the indexical-referential *I*, he faces the board and holds the marker in his hand. He continues to hold the marker and orient to the board during the first clarification sequence (lines 2 and 3). When the tutor poses the second clarification question, in line 4, Ivan turns to look at her, shifts his torso sideways to make room for Shelly to access the board and then hands the marker to her, just prior to uttering his response in line 6. So Ivan's bodily orientation and object use support an interpretation of the child's referential forms as appropriately used.

Benjamin

An only child of an upper-middle-class English-speaking family, living in an urban area in northern California, Benjamin was six years and three months old at the time of the video-recording. He was in a regular first-grade classroom. Benjamin's parents reported noticing something unusual in his development as early as age one, when he begun to show a precocious interest in literacy activities – first with spelling of words and composition of multi-digit numbers, then with books, maps, and different counting systems. At age three, Benjamin could not only read fluently and spell difficult words correctly, he also manifested a tireless preoccupation with them. His speech had been developing at a comparably exceptional rate, but his parents soon realised that Benjamin was not using language to communicate with them as much as to sustain his involvement with written texts. When Benjamin was diagnosed with autism at age three years and six months, he was also identified as hyperlexic. Since receiving the diagnosis, Benjamin's parents carefully crafted learning environments and intervention strategies that aimed to engage Benjamin's proclivity with textual practices in a way that could support his social and communicative development.

At the time of the data collection, Benjamin's MLU was 5.85, which exceeds the level of typically developing children. His turns often consisted of complex sentences, exploiting a wide range of subordinating conjunctions. Indeed, the pragmatic scope of Benjamin's speech was wide, including assertions, questions, requests, directives, and assessments. While his speech was always addressed to an interlocutor, minimally one that was spatially co-present but most frequently participating with him to the ongoing activity, Benjamin rarely pursued joint narratives or spontaneously inserted himself in the verbal interactions of his family members. The verbal interactions comprised by the study present Benjamin constantly, almost incessantly communicating with others through language, yet perceivably without a strong orientation towards his interlocutors as other selves – that is, towards the attitudes of the other as a central dimension to explore and to relate to his own (see Table 15.3).

As Table 15.3 indicates, Benjamin's person reference repertoire comprises all person forms, at least in the nominative case, always used in normative manner. Ego-centred forms (i.e. first-person pronouns across cases) largely predominate, representing 47% of the pronominal references produced by the child. It is also worth noticing, however, the use of the first-person plural, which is not

Table 15.3 Benjamin's person reference repertoire

Person Reference	Typical	Atypical	Total
Proper/Role Name			
Subject	7	–	7
Object (direct and indirect)	1	2	3
Possessive	1	–	1
Pronoun			
First-person singular			
I	110	–	110
me	4	–	4
my	13	–	13
mine	2	–	2
Second-person singular			
you	61	–	61
you	8	–	8
your	9	–	9
yours	3	–	3
Third-person singular			
he/she	12	–	12
him/her	3	–	3
his/her	3	–	3
his/hers	–	–	–
First-person plural			
we	12	–	12
us	6	–	6
our	1	–	1
ours	–	–	–
Second-person plural			
you	3	–	3
you	–	–	–
your	2	–	2
yours	–	–	–
Third-person plural			
they	9	–	9
them	2	–	2
their	–	–	–
theirs	–	–	–
Total	**272**	**2**	**274**

infrequent. We take this as an indication of the child's capacity to position himself as part of a multi-person unit.

The only few cases – two to be precise – of person reference atypicality in Benjamin are pronominal uses of the first name, in object position. They both occur within an exchange between the child and his mother, which we present here below. After some spelling games on an erasable board, Benjamin projects the shift to a next activity, the watching of some home movies:

Extract 2 – Playing with Mom (Benjamin, Tape#2)

```
 1   BEN          I want to- (0.4) and watch four t- (0.4) two t- (.)
                  three tapes.
 2   (3.0)
 3   MOM          on tv?
 4   BEN          three tapes.
 5   MOM          (we         ) to play and watch tv later=
 6   BEN          =three tapes I want to watch.
 7   MOM          which ones.
 8   (2.0)
 9   BEN          I meant four- I mean I mean one mo- one more- I mean
                  one more than three.
10   MOM          one more than three.
11   (2.0)
12   BEN          it's=
13   MOM          =four
14   BEN   →      yes. four tapes. and of the f- (.) of the four tapes I
                  wanna pick the tape that's about Benjamin.
15   (1.6)
16   BEN   →      it will be Benjamin sweep=sweep=sweep (.)
                  creep=creep=creep (.) meow=meow=meow and (.) and
                  hanoo=hanoo=hanoo. ((sic))
17   MOM          home movies he wants to watch ((to researcher behind
                  camera))
18   BEN          yes.
```

After obtaining his mother's assent to play home videos, Benjamin expresses his intention to watch a tape about himself (line 14). Both in that turn and in the following (line 16), which specifies what is featured on the tape, the child refers to himself with the proper name. The normative person reference in that context is the accusative form of the first-person pronoun *me*. The reflexive pronoun *myself* would also be acceptable. Although our data corpus for Benjamin does not comprise occurrences of reflexive pronouns – and in typical language development, these pronominal forms are the latest to be acquired – we have no evidence to infer that the child does not master this specific pronominal case: we did not encounter instances in which he used (atypically) non-reflexive forms in lieu of reflexive pronouns. In addition, it is evident that Benjamin has no problem with *me*. Thus, this instance cannot be accounted for by the competence hypothesis. Regarding a performance explanation, the syntactic structure and pragmatic scope of the two utterances containing the pronominal use of the proper name are neither infrequent nor challenging for Benjamin.

In light of these considerations, we would argue that the atypical use of the proper name in this sequence is to be related to the child's pronounced focus on texts: each home video tape has a label that includes the child name and a title. In producing the utterances in lines 14 and 16, Benjamin is *reading* from memory, so to speak, the label of the tape he wants to watch.

Aaron

Aaron was a five years and 10 months old boy at the time of the video-recording. He was living with his English-speaking parents in an affluent residential area in northern California, where he was also attending a fully inclusive kindergarten class. His parents reported that Aaron's language development was initially significantly delayed: He was only babbling until well beyond 18 months and his first utterances were highly repetitive and formulaic. Since receiving the autism diagnosis at age three, Aaron underwent a wide range of interventions, from physical and speech therapy to applied behavioural analysis (ABA) therapy, pivotal response training (PRT), and floortime. At the time of the video-recording, he was receiving one-on-one Floortime tutoring. He was also taking music lessons at home. The parents reported significant progress in language and social skills accomplished by their child over the course of the last 18 months. As a matter of fact, at the time of the data collection Aaron's MLU was 3.92, which is only slightly below that expected in typically developing children of his age. Aaron was an active participant to verbal interactions with his parents and familiar interlocutors. While his verbal interactional bids are not frequent, Aaron is responsive to his interlocutors' initiatives and is able to build on their turns to remain engaged and expand the conversational exchange.

Interestingly, while Aaron's MLU is higher than the level at which typically developing children exhibit full mastery of the pronominal system, his rate of atypicality in personal reference is not insignificant, amounting to 28% of the utterances comprising personal reference forms (see Table 15.4). This child's interface between linguistic abilities and personal reference repertoire is thus particularly interesting.

Table 15.4 shows that Aaron is able to deploy a range of person pronouns in subject position and correctly at least once. Not surprisingly, the first- and second-person singular are the most recurring pronominal forms; but we also notice in Aaron a frequent use of proper or role names to refer to himself or his interlocutors. While the first-person singular form is always used correctly, except once, for the second-person pronoun we find more atypical occurrences than typical. These are cases of pronoun reversals, in the traditional characterisation of the phenomenon. Another area in which atypical forms are frequent is that of proper and role names, in subject and object position. These are cases of pronoun avoidance.

Table 15.4 Aaron's person reference repertoire

Person Reference	Typical	Atypical	Total
Proper/Role Name			
Subject	27	15	42
Object (direct and indirect)	1	4	5
Possessive	9	–	9
Pronoun			
First-person singular			
I	39	1	40
me	–	–	–
my	1	–	1
mine	–	–	–
Second-person singular			
you	5	12	17
you	1	1	2
your	–	3	3
yours	–	–	–
Third-person singular			
he/she	3	–	3
him/her	–	–	–
his/her	–	–	–
his/hers	–	–	–
First-person plural			
we	7	–	7
us	–	–	–
our	–	–	–
ours	–	–	–
Second-person plural	–	–	–
Third-person plural	–	–	–
Total	93	36	129

In order to shed light on these two areas of person reference atypicality, we turned to a qualitative analysis of the utterances and sequential contexts in which the errors were lodged. The largest portion of pronoun atypicality in Aaron is found within utterances that are distinguishable in sequential, pragmatic, and acoustic terms as follows: (1) the utterance is addressed to the interlocutor and solicits a response from him/her; (2) it is produced with perceivable pitch alterations, notably a sing-song voice; and (3) it is recognisable as part of a familiar language game between the child and the interlocutor. In fact, in the prototypical format of the language game, the child's utterance *belongs to the adult* – it is proffered *by* the adult to invite a response *from* the child.

Thus, these utterances of the child containing pronoun atypicality – most frequently *I/you* reversals – can be characterised as ventriloquisations, which are utterances animating the interlocutor's voice. Put a slightly different way, borrowing from linguistic anthropologist Greg Urban, these utterances are de-quotative, they are quotations wherein the matrix clause has disappeared. Despite the absence of a quoting clause, there are indexical cues that frame the utterance as ventriloquisation, thus negating the (unmarked) indexical referential functioning of *I/you*. These cues are the pitch and voice quality alterations, and the synecdochic valence of the utterance itself, which by virtue of belonging to a familiar language game, with scripted turns and speech roles, points to the game as a whole.

We have ethnographic evidence for this interpretation. However, compelling evidence also comes from the interlocutor's uptake of these utterances: in most cases the interlocutor repeats the child's utterance with no pronoun adjustment, thus ratifying it as ventriloquisation and as animating his/her own voice.

As an illustration, we examine an episode from a dinner that Aaron and his mother are having together. Mom has been attempting to make small-talk with her son about the events of the day, but Aaron has offered no or minimal responses. After another of Mom's open-ended prompts, Aaron produces an utterance that, while not actually answering his mother's question, is responsive to its implicit aim of re-engaging the child in the verbal exchange:

Extract 3 – Dinner with Mom (Aaron, Tape#4)

```
1    (12.0)
2    AARON       mh-uh-uh ((looking away from mom))
3    MOM         what are you thinking about.
4    AARON       ((turns further away from mom looking behind him))
5    MOM         uh-oh.
6    AARON    →  you're looking at the barista ((singsong voice))
7    MOM         you're looking at the barista. we are not together.
8    AARON       ((turns immediately and rapidly))
9    MOM         uh ((of surprise)) now we are together.
10   AARON       ((turns away from mom again and then turns back))
11   MOM         now we are together.
12   AARON       ((turns away rapidly))
13   MOM         uh-oh.
14   AARON       ((laughs and turns back toward mom))
15   MOM         do you want to be together?
16   AARON       yes.
     [...]
17   AARON       ((turns away again and laughs))
18   MOM         uh-ho.
19   AARON       ((laughs and turns back))
20   MOM         ((laughs))
21   AARON    →  now we're together.
22   MOM         m:h. ((nods and looks at Aaron))
23   AARON       now we're together.
24   MOM         uh huh.
25   AARON       ((turns away and laughs))
26   MOM         ((laughs))
```

After a 12-second silence, in which Aaron progressively turns his torso and gaze away from his mother and then mumbles at low volume, Mom proffers a question (line 3, 'what are you thinking about') that she frequently uses to reorient her son's attention when he seems to be disengaged from their dyadic exchange. Aaron responds with a familiar phrase (line 6) that ventriloquises another of Mom's typical re-engagement devices, this one specifically related to a playful interactional format of rapid shifts between withdrawal and engagement. Typically, Aaron's interlocutor establishes the format by noting aloud that the child is not making eye-contact or facing him/her. The opening phrase usually deployed is: 'Uh-oh. you're/Aaron's looking at the ___', uttered with a very distinct prosody and voice quality (notably elongated vowels and sing-song voice) and frequently followed by another phrase, 'we are not together', with the same suprasegmental characteristics. The playful exchange then unfolds with Aaron turning to face his interlocutor (who signals his/her satisfaction with the expression 'now we are together') before abruptly turning away again, thereby triggering another round of the same exchange.

Aaron's de-quotative utterances in lines 6 and 21 are prosodically marked as re-animations of Mom's voice and familiar expressions. They are produced with a sing-song voice and a prosodic contour similar to Mom's characteristic re-engagement device. Interpreting Aaron's utterances as ventriloquisations recasts the pronoun form contained within the utterance as appropriate to the pragmatic scope of the turn, rather than reversed.

Indeed, in both utterances with atypical pronoun forms, the mother confirms that Aaron was right, the first time by repeating Aaron's de-quotative turn herself, with a sing-song voice and remarkably similar prosodic contour, and adding the next typical phrase ('we are not together', line 7), and the second time assenting, verbally and gesturally (line 22). Aaron's laughter and repeated initiations of new rounds of the game attest to his pleasure in ventriloquising his mother and having her confirm that his guess was correct.

Thus, what superficially could be labelled as pronoun reversal within a formulaic, echolalic utterance emerges here as complex layering of voices and processing of another's perspective. Ventriloquisation offers the child the possibility of making conjectures about the other and submitting them to that very other, for ratification or revision.

The second area of atypicality in Aaron is the pronominal use of the proper name to refer to himself and his tutors, and role names to refer to his parents when they are the addressees of his utterances. We have observed pronoun avoidance to occur prevalently in two contexts: the first is the same as that of pronoun reversals we just discussed; the second is a context that requires the child to shifting perspectives and relate to different speech roles.

In the first context, i.e. de-quotative speech, the utterances containing pronoun avoidance present the use of the child's proper name for self-reference and can be characterised pragmatically as ventriloquisations. The second context in which we observed the pronominal use of the proper/role name to occur is that of sequences where there is a high demand on the child in terms of shifting perspectives or weaving them together in his speech. These utterances are thus onerous not only in cognitive processing but also in terms of linguistic composition.

The following extract is illustrative of this context: In this sequence, we find Aaron with his music tutor Sarah. Sarah offers Aaron the chance of a break from music lesson and the child proposes a variation on the game we saw him play with his mom (in Extract 3). Aaron deploys quotative framing clauses and direct reported speech to instruct Sarah on what she should say as participant to the language game.

Extract 4 – Music lesson with Sarah (Aaron, Tape#4)

```
1    SARAH      do you need some silly time?
2    AARON      no. (0.2) yeah.
3    SARAH      yeah? some silly time? okay.
                ((lower volume)) >don't make a face or else I'm gonna
                tickle you.<
4               ((Aaron opens his eyes and mouth really wide))
5    SARAH      OHHHgghhhhh. ((tickling Aaron))
6    AARON      ((laughs)) if we're NOT toge- (.) if- I will say if
                we're not together. (1.0) if we're not together,
7    SARAH      then what.
8    AARON      I don' know:.
9    SARAH      how about, um:,(3.0) if we're not together, (1.0) then
                I'm gonna (.) tweak your nose.
10   AARON      ((gets up)) no::. ((laughing and walking away))
11   SARAH      OH:: NO:::. WE'RE NOT TOGETHER::::.((running after
                Aaron, out of frame))TWEAK.
12   AARON      if we're- (0.4) if we're by ourselves, (0.2) sca::ry.
13   SARAH      how about-
14   AARON  →   if we- if Sarah says- if we're by ourselves scary
                (Sarah says). how if Sarah says that. (0.4)
                Sarah will say that.
15   SARAH      tell me again.
16   AARON  →   if Sarah will say (.) if you: (1.2) if we (.) if
                you::, (0.2) how about Sarah says if ↑you go be by
                yourself Shelly's gonna get you. yeah?
17   SARAH      do you wanna play (.) with Shelly ((co-present
                Floortime specialist)) a little bit Aaron?
18              ((Aaron moves towards the door and enters the house))
```

While Aaron's music tutor Sarah has some acquaintance with the 'we are together, we are not together' language game, she does not frequently engage in it (in our data corpus this is the only occasion). Aaron seems sensitive to Sarah's relative unfamiliarity with the game and in launching it (in line 6) he self corrects to insert a framing clause – a pronoun and *verbum dicendi* ('I will say'), which pragmatically serve as quotation marks for what is said next.

In attempting to have Sarah produce certain game's moves in subsequent turns, Aaron repeatedly uses framing clauses (lines 14 and 16). In those clauses, however, the child avoids the use of the person pronoun and refers to his tutor in the third person using her name. By referring to his interlocutor as Sarah instead of *you*, a shift in reference of the pronoun *you* between the matrix clause and the quotation is avoided (in line 16). While the linguistic and cognitive demand of such sentences might result in the simplifying strategy of pronoun avoidance, the trouble in pronoun use might also relate to a difficulty in relating to the other as *you*. The game offers a frame for experiencing speech roles and perspectives. Yet, at the same time this scripted scaffold seems to remove the child from the more immediate *I–you* relationship so that the tutor becomes a sort of character on the game stage.

Discussion

In this study, we examined the person reference forms used by three children with autism, who, while being similar in chronological age, exhibited different language abilities. A comparison of the range of person reference forms deployed by the children and the patterns of atypicality in pronoun use revealed differences, both quantitative and qualitative. These differences suggest that multiple interrelated factors might underlie the difficulties in person reference forms in children with autism, belying a single explanation in terms of either competence or performance. Two explanations in particular illuminate the data from our comparative case study. The first is the idea that central to autism is a difficulty in self–other relations, in identifying with the perspective of others (Hobson, 2011). Since language is inherently interpersonal, we maintain, such difficulties would be expected to manifest in talk, very likely in the use of personal pronouns specifically. While our work confirms such difficulties, it also points to ways that children with autism may deploy language-based resources for processing others' perspectives, an insight unavailable from a deficit-oriented approach. The second explanation cautions us from over interpreting the psychological significance of atypical pronoun use. Our data show that atypical personal reference by children with autism may often be chiefly an interactional outcome – the result of specific language games or practices in which children participate with their adult interlocutors.

Ivan, whose language ability is the lowest of the three children we studied, displayed the most limited range of personal reference forms and the highest rate of pronoun errors (35%). Typically developing children of Ivan's language level are close to full mastery of the first- and second-person pronouns. The absence of third-person pronouns and plural forms in Ivan's repertoire might thus be assimilated to the normative progression in acquisition of pronominal forms in combination with his language delay. However, the anomalies in first- and second-person references require another explanation: Ivan frequently refers to himself and his interlocutor in the third person and reverses half of his second-person singular pronouns. We conjecture, on the basis of this finding, that in children with autism the acquisition of the pronominal system might be to some degree dissociable from the general process of language development. Put a slightly different way, the mastery of person reference form is not only a linguistic accomplishment. We thus consider these findings as indicative of the child's difficulty in representing himself in relation to others. The frequent occurrence of pronominal use of proper and role names might be indicative of the child's proclivity to see himself and his interlocutors from distance, a propensity related to a difficulty in engaging in the *I/you* of the conversational interchanges. At the same time, our qualitative analysis has revealed that Ivan's pronominal use of proper and role names was partly an interactional outcome. It sometime resulted from his interlocutors' use of baby talk, which included 'pronoun avoidance,' to which Ivan adapted (see Extract 1). As such, the child's pronoun non-use was contextually sensitive. In simplifying speech addressed to the child with autism, the interlocutor can constrain him to use simplified forms himself. Thus, a clinical feature such as pronoun atypicality appears not solely as instantiating an underlying syndrome, but as a response to contextual conditions.

Benjamin, whose linguistic skills exceeded those of typically developing children of his age (in terms of length and complexity of utterance), displayed a wide range of person reference forms, deployed normatively except in one sequence (Extract 2). Yet, Benjamin's speech remains markedly atypical in its weak orientation toward others' attitudes and the co-construction of shared stances. This finding suggests that pronoun atypicality is neither pathognomonic of autism nor the single linguistic manifestation of the children's difficulty in self–other relatedness. Moreover, this finding indicates that language skill can afford the child mastery of the pronoun system despite persistent difficulties in self–other relatedness.

Aaron's linguistic abilities and rate of errors in pronoun use (28%) were between those of Ivan and Benjamin. Compared to typically developing children, who at Aaron's MLU level exhibit full mastery of the pronominal system, Aaron's rate of atypicality in personal reference was still considerable. Aaron's resorting to proper names and third-person constructions in the articulation of shifts in perspective and speech roles further corroborates an interpretation of the child's person reference atypicality that connects it to difficulties in sense of self and others in relation to one another. Such interpretation adds insight to Hobson and associates' claim that 'even when autistic individuals have achieved the potential for adequate speech-role-referring pronoun use, they might be subject to lapses in the propensity to identify with others in role-appropriate ways... and they might be prone to experience themselves in a relatively "uncommitted manner" ' (Lee et al., 1994, p. 174). Our analysis suggests that those 'lapses' overwhelmingly occur when the child is attempting to weave together multiple perspectives – that is, when he is addressing his own weaknesses.

We demonstrated that the child's non-normative use of pronouns and proper/role names was also systematically related to discernible language games. In fact, we determined that many of Aaron's *I/you* reversals and pronominal uses of his proper name were de-quotative pronouns, within utterances that animated the voice of the interlocutor. We suggest that such a propensity to reproduce the words of another is both evidence of the child's difficulties in identification with others, and, at the same time, an indication of the child's efforts to overcome them. Because of the intimate connection between language and experience (Ochs, 2012), the talk of Aaron's interlocutors affords a point of entry into their experience, perspective, and orientation. (For more on this, see Sterponi, de Kirby, & Shankey, 2014.)

Clinical relevance summary

As a final note, we would like to suggest that our study has important implications for intervention (see Table 15.5). Indeed, the perspective that we as researchers bring to understanding autistic language has direct implications for how we are likely to conceptualise the process and goal of clinical intervention. If we regard stereotypical features of autistic language as manifesting deficits alone, we are likely to support effort to encourage the child to suppress or replace them. By contrast, from an appreciation that these linguistic features often represent efforts to marshal the affordances of language to overcome difficulties, we may conceive of interventions that could support these efforts. Furthermore, if we regard talk-in-interaction as joint accomplishment of participants in the verbal exchange, we are moved to consider the quality of the child–clinician interaction and the clinician's role in shaping the child's talk. For a simple summary of the implications for practice, see Table 15.5.

Summary

As we have argued in this chapter, a perspective on language that foregrounds its interactional and experiential dimensions has the potential to reframe our vision of pronominal reversal and avoidance in the communication of children with autism. On the one hand, our work confirms that pronoun atypicality

Table 15.5 Clinical practice highlights

1. In order to promote exposure to and opportunity for usage of pronominal contrast, it may prove helpful to involve the child in triadic interactions (for instance, the child with two clinicians, or with one clinician and a family member). Such interactional context can bring the child to attend to the talk not addressed to him and to track pronominal functioning therein; for instance, the use of *you* to refer to a person other than himself.
2. We would encourage activities that mobilise the articulation of different voices, through a range of semiotic means including de-quotation. Within these activities, the clinician could scaffold the use of person pronoun while encouraging the child to engage in stance- and perspective-taking, that is in identifying with himself and the others as selves. Play with puppets or dolls can provide engaging opportunity for voicing and perspective-taking moves.
3. Across activities, the specific characteristics of the clinician turns may go a long way in offering the child opportunities to exercise and improve his communicative potential. While baby talk's person reference forms may be justified in the earliest intervention stages, as an attempt to introduce self/other designation without the complications of indexical reference, once the child demonstrates to be able to differentiate self from others, even at a basic level, it is opportune to begin introducing language forms that make salient the inherent relational aspect of communication and aim towards experiencing perspective-taking.

is salient in autism and related to difficulties in self–other relations. On the other hand, a single, deficit-oriented interpretation of the phenomenon does not find support in our analyses – for two primary reasons. First, the heterogeneous manifestation of person reference atypicality in our three case studies calls for a model that includes multiple explanatory components, which reflects the inherently multifaceted nature of personal pronouns use in general. The interaction between different explanatory factors has yet to be investigated and represents a potentially fruitful line of research. Second, we have suggested that the child's difficulties are often manifested most precisely in the child's efforts to overcome them. Indeed, children with autism can go a long way towards mastering the pronominal system, engaging language to explore and express the subjectivity of self and the other. In this sense, Benveniste's emphasis on the exercise of language as basis of subjectivity (1971) still represents a valuable insight for autism research and intervention.

Notes

1. Echoic memory, also called *auditory store*, retains the raw auditory stimulus for 2–3 seconds (Neisser, 1967).
2. Chiat points out that in adult language we can observe the impersonal use of the second-person pronoun and additionally hypothetical and perspective-shifting functions.

3. The MLU for typically developing children of six years of age is 4.5 (Rice et al., 2010). Typically, developing children come to master the pronoun system, in comprehension and production, when their MLU is 3 (Brown, 1973).
4. Ivan's omission of subjects might be related to the phenomenon of pronoun avoidance. However, we did not code them as pronoun avoidance as we found it too difficult to systematically judge whether a given case of subject omission was an instance of grammatical error or genuine pronoun avoidance.
5. On the basis of a study of five parent–child dyads, Willis (1977) has articulated a systematic account of pronoun in child directed speech. She has observed that 'most BT pronouns are conventional pronouns used grammatically but deviantly in regard to participant role, number, or gender' (Willis, 1977, p. 273). Willis has also provided a classification of the baby talk deviations in pronominal use. The most frequent deviation in her data was reference to the speaker by means of a third-person form such as the role name, in place of the first-person singular pronoun.

References

Atkinson, M. J., & Heritage, J. (Eds.) (1984). *Structures of social action: Studies in conversation analysis*. Cambridge: Cambridge University Press.

Bain, R. (1936). The self- and other-words of a child. *American Journal of Sociology, 41*, 767–775.

Bartak, L., & Rutter, M. (1974). The use of personal pronouns by autistic children. *Journal of Autism and Childhood Schizophrenia, 4*, 217–222.

Benveniste, E. (1971). Subjectivity in language (Mary Elizabeth Meek, Trans.). *Problems in general linguistics* (pp. 223–230). Miami, FL: University of Miami Press. (original – 1958 – Journal de Psychologie, 55 – July–September).

Brown, R. (1973). *A first language: The early stages*. Cambridge, MA: Harvard University Press.

Charney, R. (1980a). Pronoun errors in autistic children: Support for a social explanation. *British Journal of Disorders of Communication, 15*(1), 39–43.

Charney, R. (1980b). Speech roles and the development of personal pronouns. *Journal of Child Language, 7*, 509–528.

Chiat, S. (1982). If I were you and you were me: The analysis of pronouns in a pronoun-reversing child. *Journal of Child Language, 9*, 359–379.

Chiat, S. (1986). Personal pronouns. In P. Fletcher & M. Garman (Eds.), *Language acquisition: Studies in first language development* (pp. 339–355). Cambridge: Cambridge University Press.

Clark, E. V. (1978). From gesture to word: On the natural history of deixis in language acquisition. In J. Bruner & M. Garman (Eds.), *Human growth and development: Wolfson College lectures 1976* (pp. 85–120). Oxford: Oxford University Press.

Cooley, C. H. (1908). A study of the early use of self-words by a child. *Psychological Review, 15*, 339–357.

Dale, P. S., & Crain-Thoreson, C. (1993). Pronoun reversals: Who, when, and why? *Journal of Child Language, 20*, 573–589.

Evans, K. E., & Demuth, K. (2012). Individual differences in pronoun reversal: Evidence from two longitudinal case studies. *Journal of Child Language, 39*, 162–191.

Fay, W. H. (1979). Personal pronouns and the autistic child. *Journal of Autism and Developmental Disorders, 9*(3), 247–260.

Goodwin, C. (1981). *Conversational organization: Interaction between speakers and hearers*. New York: Academic Press.

Hobson, P. R. (1990). On the origins of self and the case of autism. *Development and Psychology, 2*, 163–181.

Hobson, P. R. (2010). Explaining autism. *Autism, 14*(5), 391–407.

Hobson, P. R. (2011). Autism and the self. In S. Gallagher (Ed.), *The Oxford handbook of self* (pp. 1–15). Oxford: Oxford University Press.

Hobson, P. R., Lee, A., & Hobson, J. A. (2010). Personal pronouns and communicative engagement in autism. *Journal of Autism and Developmental Disorders, 40*, 653–664.

Kanner, L. (1943). Autistic disturbances of affective contact. *Nervous Child, 2*, 217–250.

Lee, A., Hobson, P. R., & Chiat, S. (1994). I, you, me, and autism: An experimental study. *Journal of Autism and Developmental Disorders, 24*, 155–176.

Neisser, U. (1967). *Cognitive psychology*. New York: Appleton-Century-Croft.

Ochs, E. (2012). Experiencing language. *Anthropological Theory, 12*(2), 142–160.

Oshima-Takane, Y., & Benaroya, S. (1989). An alternative view of pronominal errors in autistic children. *Journal of Autism and Developmental Disorders, 19*, 73–85.

Raymond, G. (2003). Grammar and social organization: Yes/no type interrogatives and the structure of responding. *American Sociological Review, 68*, 939–967.

Rice, M., Smolik, F., Perpich, D., Thompson, T., Rytting, N., & Blossom, M. (2010). Mean length of utterance levels in 6-month intervals for children 3 to 9 years with and without language impairments. *Journal of Speech and Language Research, 53*(2), 333–349.

Sterponi, L., de Kirby, K., & Shankey, J. (2014). Rethinking language in autism. *Autism*, first published online as, doi: 10.1177/1362361314537125.

Urban, G. (1989). The 'I' of discourse. In B. Lee & G. Urban (Eds.), *Semiotics, self and society* (pp. 27–51). Berlin: Mouton the Gruyter.

Wills, D. D. (1977). Participant deixis in English and baby talk. In C. E. Snow & C. A. Ferguson (Eds.), *Talking to children: Language input and acquisition* (pp. 271–298). Cambridge: Cambridge University Press.

Wittgenstein, L. (1953). *Philosophical investigations*. Oxford: Blackwell.

Recommended reading

- Chiat, S. (1986). Personal pronouns. In P. Fletcher & M. Garman (Eds.), *Language acquisition: Studies in first language development* (pp. 339–355). Cambridge: Cambridge University Press.
- Goodwin, C. (1981). *Conversational organization: Interaction between speakers and hearers*. New York: Academic Press.
- Sterponi, L., de Kirby, K., & Shankey, J. (2015). Rethinking language in autism. *Autism*, 19(5), 517–526.

16
Normative Development and the Autistic Child

Lindsay O'Dell and Charlotte Brownlow

Introduction

As critical scholars, we are interested in the social construction of 'normal' development in children and how this operates discursively through the construction of 'other' childhoods that are seen to be different from the norm. Ideas about 'normal' development are naturalised in and through psychological descriptions of children's behaviours, particularly in the appeal to childhood as biologically constituted, and produced through an evolutionary process (Brownlow & Lamont-Mills, Chapter 13, this volume; Burman, 2008; Morss, 1990; Rose, 1989a).

In this chapter, we explore how ideas of 'normal' child development work to shape our understandings of children with Autism Spectrum Disorder (ASD), abbreviated to 'autism' for simplicity throughout the chapter. We aim to interrogate the discursive practices that construct 'normal development' against which children with a diagnosis of autism are seen to be different, and therefore developing inappropriately. The inclusion of autism in the *Diagnostic and Statistical Manual of Mental Disorders* (*DSM*) (which draws largely on medical discourse) positions autism as a mental health 'disorder'. However, this is a much-debated issue among people with autism, parents, autistic activists, and some professionals (see O'Reilly, Karim, & Lester, Chapter 14, this volume). As critical developmental theorists, we consider an alternative understanding of autism, which takes as a starting point valuing people with autism and autistic traits. This, in the United Kingdom, has largely been taken up through a discourse of neurodiversity, which critiques a biomedical approach to understanding autism as a pathology or a mental health 'disorder'. Our position is that the concept of 'normality' and its measurement is framed largely within a 'mental health' frame through psychiatric diagnostic criteria and, as such, is a powerful mechanism through which these debates are framed and people's lives experienced. Thus, we feel that it is important for clinicians and practitioners to engage with a more positive and enabling view of autism, hence choosing to discuss these issues in this book.

Measuring/understanding normal development

Understanding 'normal' child development as a (bio)logical progression in which development occurs through a series of stages is a powerful discourse in contemporary research and public imagination; it is also embedded in a history of psychology where 'psy' professionals (Rose, 1989b) have worked to produce a body of knowledge in which the standardisation of developmental abilities are inscribed through psychometrics (Rose, 1989a). By employing the concept of the normal distribution, human variability is constructed in a particular way, enabling 'appropriate' action to be taken by expert psychologists for any individuals falling outside of the 'normal range'. The focus by some psychologists, psychiatrists, and other healthcare professionals is therefore on the identification and treatment of individuals whose functioning is seen to be out of the 'normal' range.

Burman (2008) and Rose (1989b) have argued that childhood is the most intensively governed area of personhood, with agents such as health practitioners and psychologists monitoring to ensure 'normal' development and actively promoting certain attributes such as intelligence, education, and emotional stability. Critical scholars such as Nik Rose and Erica Burman have observed that the concept of normality can be used to simultaneously provide a means for identifying abnormality and construct an image of the 'normal' child and family. The proposition of developmental stages has been implicit in healthcare advice, everyday talk, and clinical practice for many years. Critical developmental psychology has offered a critique of the assumptions inherent in developmental goals, stages, and milestones. Burman (2008) argued that developmental 'norms' serve to regulate parents, particularly mothers, invoking ideals of (particular kinds of) mothering and the appropriate practices required to ensure optimal child development. For example, in charting the experiences of a group of new mothers, Urwin (1985) discussed the role of service providers such as health visitors in normalising developmental goals for new mothers. Recent work by critical disability scholars has also begun to examine the influence of developmentalism and ideas about 'normal' childhood on the experience of parents and their disabled babies (Goodley & Runswick-Cole, 2012). We share the view of critical disability scholars that 'impairment' is produced discursively rather than a 'hard biological fact' (Goodley & Runswick-Cole, 2012, p. 54), and we seek to draw on critical approaches in our understandings of autism in this chapter.

Defining autism as 'abnormal' development

The mainstream understanding of autism is shaped through diagnostic tools such as *DSM*, currently in its fifth revision. The new *DSM-5* has encountered much criticism in its introduction of a whole new set of 'disorders', which, it is argued, serves to pathologise everyday aspects of childhood such as temper tantrums which is categorised in *DSM-5* as Disruptive Mood Dysregulation Disorder, producing an increase in the number of children falling outside of the norm (Frances, 2012; see also Harwood, Chapter 10, this volume for an interesting critical discussion).

The positioning of autism within *DSM-5* is, however, different in that commentators have pointed out that while other disorders may rise in their diagnostic prevalence, the changes in the definition of autism within *DSM-5* may actually lower the rates of diagnosis (Frances, 2012; Jabr, 2013). Frances (2012) estimated that the rates of diagnosis of autism may fall between 10% and 50%, raising concerns among advocates about service provision and support.

The powerful way that *DSM* enables a construction of autism is reflected in the creation and maintenance of distinct disorder groups. In the revised *DSM-5*, autistic disorder, Asperger's disorder and pervasive developmental disorder not otherwise specified (PDD-NOS), which have previously all been distinct disorders, are now subsumed into one category called autism spectrum disorder (ASD) (Jabr, 2013). Jabr (2013) warns that the debates surrounding the number of characteristics needed as evidence of particular disorders and the arrangement of disorders into groupings are not just about semantics. Jabr points to the very real effects that these categories can have in terms of people with autism accessing various medical, social, and educational services. Matson, Hattier, and Williams (2012) echo this point observing that the long-standing triad of impairments evidenced as characteristic of autism has now become a dyad, with social communication and social interaction being collapsed into a single category. Matson et al. observe that under the changes to *DSM-5*, individuals who would have previously been diagnosed with Asperger's syndrome or PDD-NOS would now receive a diagnosis of autism spectrum disorder. However, they note that the bar for receiving a diagnosis has now been raised, with more severe and numerous symptoms needing to be demonstrated in order to receive a diagnosis (Matson et al., 2012). This changing of both disorder groupings, and severity of symptoms threshold, has significant consequences for both the construction of autism, and also the experiences of individuals receiving such a diagnosis. However, what is not under discussion is the inclusion of autism as a series of behaviours that require a classification as a 'mental health issue'. That is of course not to suggest that others, including some professionals, would contest such an assertion. Critiques of the new DSM have arisen from many practitioners who work with children and adults with autism. There is a growing move, in the United Kingdom and other contexts, to consider autism as an identity or as a difference, rather than as a 'mental health' issue.

A counter discourse to a medicalised construction of autism as abnormal development is offered by advocates and adults with autism. Issues regarding what are considered 'normal' and 'abnormal' are key concerns within autism (advocate) literature, and terminology has developed that reflects these concerns. 'NT' or 'Neurologically Typical' is a term that has been crafted by people with autism as an alternative to the term 'normal' to describe people without autism. The term has been taken up by the wider autism community and some professionals and academics working within the field of autism. This chapter will use the term 'NT' to refer to those individuals who have not been assigned or do not identify with a label within autism spectrum disorders – herein referred to as 'autism'. While we recognise the many debates about language with respect to describing both NT and ASD, we are using people first language in this chapter, that is, 'child with autism' as a way of recognising that the child is more than their diagnosis.

Project overview

This chapter seeks to interrogate the discursive production of autism, the impacts of this on professionals and families, and how this sets up assumptions concerning 'autistic' and 'normal' childhoods. The data contributing to this chapter was part of a larger-scale research project examining the construction of autism in online discussion lists (Brownlow, 2007). The decision to focus on online discussion lists reflects a body of literature that has demonstrated the positive engagement with internet technologies by people with autism, and the possibilities to develop more empowering identities online (see e.g. Brownlow, Bertilsdotter Rosqvist, & O'Dell, 2013; Davidson & Orsini, 2013).

Following approval from a university research ethics committee, online discussion lists were joined and contributions to these over a three-month period were collected. The discussion lists comprised of four groups: two for people with autism, one primarily for parents of people with autism, and one primarily for professionals working with people with autism. The participants in the project were all members of asynchronous online discussion groups and ranged from those with an 'official' diagnosis to those self-diagnosed and non-autistic members of the lists. All quotes that appear in this chapter are reported verbatim, and all names of contributors are pseudonyms.

Approach to data analysis

The data was analysed using critical discourse analysis informed by Edley (2001) and Fairclough (1992, 2001). Like other forms of discourse analysis, critical discourse analysis goes beyond perceiving language as just a tool for communication, instead viewing language as actively constructing what is understood (Scior, 2003).

The focus of the analysis reported in this chapter is on the role of developmental assumptions inherent in the diagnosis and classifying of 'normal' and 'different' childhoods. Power is an important concept in critical discourse analysis and is a key theme running through our analysis, with a focus being placed on the power to define, diagnose, and treat/modify autism.

Observations from data analysis

From our reading of the data produced in the online discussion forums in relation to practices around child development, two key themes were prioritised for analysis. These were issues surrounding the definitions of abnormality and the goal of achieving 'normalisation' through therapeutic interventions.

Defining (ab)normality

A normative understanding of development is evident in many of the discussions by parents contributing to the online discussion lists in comparisons of their child with autism to 'normal' children. Parents often report suspecting their child may have autism long before an 'official'/formal assessment has taken place. This assertion is commonly made by the parents comparing the behaviour of their child to what they expect 'normal' children to be doing at particular developmental stages. In doing so, an age-graded discourse is invoked in which numerical age is given a marker of developmental maturity and assumptions about proficiency. For example:

> My daughter […], is now 11 years old, and diagnosed with Asperger's syndrome.
>
> Since […] is an only child, we really didn't notice some of her behavior differences until preschool. She was not a baby who liked to be hugged. She preferred to be carried around and to look at things. She would quickly lose interest in her bottle or food and was always underweight. She spoke at 9 months, didn't walk until 15 months and wasn't potty trained until almost 4! But she knew all the letters in the alphabet before she was two!
>
> (Rosie)

The mother in this excerpt above illustrates the norms of development, which she uses to compare her child against a 'normal' developmental pathway. She clearly cites developmental milestones, such as walking and potty training, as being delayed in her child, leading her to recognise her child as 'different' to the norm. The comparison made by parents of their child, to what is portrayed as 'normal' behaviour, was a common theme in the postings. The distinction between normal and abnormal behaviour is evident to parents over time and becomes part of the history of the child as told to professionals and others. It is also evident in the discussion forum, for example, Jennie (a mother of a child with a diagnosis of autism) discusses her son's behaviour:

> my eldest son is 3 & 1/2 yrs & has just begun assessment for asd. he had prob-lems at birth but seemed 2 develope normally til about 2 yrs. obsessive rituals began, speechprobs, violence etc. he had mmr vaccine b4 this btw . . . he went downhill a while, then slowly improved & in last month speech has bcome almost normal & only aggression when annoyed & hyperactivity remain! surely thats not autism!?
>
> (Jennie)

In the first part of the posting, Jennie reports initially being confident that her son was developing 'normally' until around his second year, at which point she identifies a series of 'abnormal' behaviours that have developed. This is a common history of a child given by parents when describing the development of their child with autism. It is often linked to developmental theorising which cites two years as the age at which children who are developing 'normally' show a significant increase in social behaviours.

Discussions surrounding normality are also evident in parental reactions to a diagnosis of autism for their children. Common parental responses have been described as an initial grief reaction, sometimes characterised as mourning for the 'loss' of their 'normal' child (see e.g. Case, 2000; Konrad, 2005). Once a child is seen to be developing in a way that is out of the 'normal' range, it is common that parents look for a cause for the identified 'deviant' (i.e. autistic) behaviour, which is often constructed as not a natural or inevitable aspect of some children's development. Thus particular forms of development are nat-uralised and others, such as autistic behaviour and traits, are constructed as a deviation from the norm. For example, a mother of a child with autism reflects on her experience:

> No one around me knew what AS was and brushed it off as if I was talking about the common cold. I felt lost, useless and drained. I kept saying to myself what did I do that made my son this way? Carefully going over my pregnancy and thereafter. Did I not pay enough attention to him? Maybe I should have held him more or nursed him longer.
>
> (Liz)

The role of mothers in the causation of autism has been a dominant discourse since discussions of the 'refrigerator mother' (Bettelheim, 1959). While this understanding has been critiqued and seen to lack foundation, there are new ways in which mothers are seen to be implicated in their child's diagnosis of autism. For example, in debates about the safety of MMR vaccination, the link between autism and MMR was powerfully (and incorrectly, in our view) made (O'Dell & Brownlow, 2005). However, MMR is still drawn on by parents in the discussion (as in Jennie's except above) to make sense of the development of autism in their child. The power of the discourse was in part because of the need for an explanation or a causal agent for development that does not conform to the (assumed to be) normal trajectory. We have argued elsewhere (O'Dell & Brownlow, 2005) that the MMR debate, as it took hold in the British media, particularly focused on the role of mothers (rather than 'parents') and their ability to understand the scientific advice and be responsible for vaccinating their child, was a commonly articulated public discourse.

The pervasiveness of the discourse of normative child development is evident in the mobilisation of concepts such as developmental 'milestones' and age-specified expectations by parents in their descriptions and comparisons of their autistic children. The construction of a 'normal' child and its associated behaviours serves to render deviant the child with autism, positioned in deficit terms rather than viewing autism as a similar, but different, developmental trajectory. Furthermore the deficit discourse problematises parents, particularly mothers, of children with autism.

'Normalisation' through therapy

Psychometrics and standardised tests, through which behaviours and abilities are understood and articulated, draw on the principles of normal distribution. Assessment process are constructed which are assumed to be able to help professionals identify child who are not fitting within the 'normal' range of development. The use of psychometrics and a discourse of a normal distribution of abilities were drawn on in the discussion forums to frame and provide a language for understanding the difficulties children were manifesting. For example, the excerpt below is from a professional who reflects on a particular individual's centile placing on the WISC III intelligence test:

> some of his responses on the wisc III did indicate difficulty understanding appropriate behavior in social situations...and serious weakness 16% tile inferential thinking and social cognition may have difficulty comprehending concepts and determining correct behavior in social situations.

> (Charlene)

The tools used by professionals to make a diagnosis of autism are frequently reflected on in threads of the discussion lists by professionals as in the quote above and are also used by parents and people with autism themselves. One common thread surrounds challenging the use of tests which are based on the ability to communicate effectively. For example, the following extracts are a discussion on the forum between two professionals working within the field of autism.

> The school wants to evaluate him using the Stanford Binet IV. I do not know much about this instrument. Is anyone familiar with it? Have any of you had this test done on your child? Is it appropriate for an autistic child??...What I'm looking for are parents who's children have been administered this test. Is it appropriate for a child with Fragile X/Autism? A child that does not communicate very well? A child that doesn't write?
>
> (Sandra)

> I agree with you [Sandra], The problem with these tests is that they are based on the ability to communicate, which is the impairment in autism, which means that you are measuring someone with the one factor that you already know is the problem.... thus they are not valid.
>
> (Gemma)

In the above examples, the validity of the tests is questioned due to their reliance on communicative abilities – an area highlighted by posters as being impaired in children with autism. However, despite questioning the use of existing tests with children with autism due, for example, to their heavy reliance on communication skills, there is still an implicit belief in the standardised tests as a meaningful and effective way to assess children and to differentiate between 'normal' development and autistic traits.

In addition to professionals questioning the appropriateness of some tests for children with autism, there is also a thread of discussions that focus on parental experiences of the tools of diagnosis. Through these postings, parents present a challenge to professionals by questioning their tests and resources. For example:

> The school is a joke on the assessments. They don't work with him enough to get to know him, before they do the test. They take him to a strange place (to him) to do the test and is always uncooperative. I took him one place on my own to get him tested, and that was a joke. She didn't get to know him or let him get used to the new place either. Done all the test in 3 one hour visits. She told me that I need to take some parenting classes and learn how to raise my child. She also told me there was something wrong with his brain in the same breath. I got a good mad laugh out of her. But anyway the school used her testing to write one years IEP.
>
> (Natalie)

Here the main focus of questioning is on the procedure with which the assessment takes place, not necessarily the actual tests used. The validity of the test results is therefore questioned due to their inappropriate administration; the person administering the tests does not know her child. The parent is not questioning the constructs underlying the tests. The expert role of professionals in the identification and management of (ab)normal behaviours, through the employment of standardised tests developed within the professional scientific community, therefore remains unquestioned. The results of such tests are significant for parents of children with autism; they are translated into the support offered to the family, including the child's individual education plan (which, in the UK, forms the basis of the individual's educational support).

Echoed again in discussions of therapeutic intervention is a separation of the concepts of normality and abnormality, with clear expectations of behaviours associated with 'abnormal' development. Once a child has been identified as not developing 'normally', therapy and other interventions can be utilised to attempt to 'normalise' their behaviour. Autistic advocates argue that the goal of therapy is therefore often to change a child's behaviour by making them resemble the more socially acceptable behaviours of NT children. For example, in the excerpt below, Tegal, a contributor to the ASD discussion forum, writes:

> I will not discredit any therapy because while it may not be good for one child, it may be doing a wizz-bang job for another, giving that child a chance at a maybe normal life. Isn't that the whole reason why people became professionals in the first place? To help people with Autism and their families?
>
> (Tegal)

Therapeutic intervention is clearly positioned as a means of making an individual more 'normal' and, therefore, implicitly 'less autistic'. It positions 'normal life' as a non-autistic life and hence intervention is not concerned with developing a positive identity and position within society for children (and adults) with autism.

The importance, and potential achievement, of the goal of making an individual's behaviour more 'normal' is highlighted in discussions of people with autism who are seen to have 'recovered', as for example in the reference to the famous individual below:

> Raun K. Kaufman is a 26 year old diagnosed as severely autistic when he was 2. He now *bears no traces of the affliction* thanks to the Son-Rise Program.
>
> (Kim, emphasis added)

The use of terminology, an 'affliction', highlights the construction of autism as a negative trait, and consequently something that has to be changed in some way. The 'success' of the process of change for the particular individual cited in the posting is clearly attributed to the type of intervention undertaken rather than the individual's ability to manage and negotiate difference. The dominant focus in such a discourse is on making the behaviours of a child with autism more closely reflect or align to non-autistic/NT traits. The negative perception of autistic traits reflects the dominant construction within professional discourse of autism as a deficit or deviance rather than a difference. Clinical discourse and the tools of assessment reflect this position, being concerned with identifying children who deviate from the prescribed norm and intervene where it is deemed necessary and appropriate. The voice of individuals with autism has until very recently largely remained silent in such debates.

Discussion

The construction of 'abnormal' behaviour and assumptions about 'normal' child development are very much in evidence in the analysis of online discussion groups. We argue that this is a reflection of the power of such discourse in public imagination and clinical debates and is based on the dominant NT worldview. Within this discourse, autism is not constructed in an equal position to NT, and therefore is positioned as a behavioural pattern that deviates from the taken for granted 'natural' course of normal child development.

An alternative discourse of development, marginal but evident in the discussion postings, serves to question the taken for granted 'natural' course of child development and the place that autism occupies within this construction. If we take NT to equate to typical development, then this has been 'normalised' by professional discourse. This is evident through practices such as child developmental milestones, where 'normal' development has been mapped out partly in order to identify the 'abnormal' behaviours and enable the regulation of development (see e.g. Burman, 2008; Rose, 1989a). Drawing on critical developmental psychology, and the newly emerging critical disability studies, we argue that neurotypicality is seen as an implicit goal in developmental psychology, maintained by close checks made by regulating bodies such as health visitors and teachers. These bodies can highlight any differences and, by definition, problems early in order to increase the chances of the maintaining normal/NT behaviour. This has important implications for the development of a positive identity for people with autism. Children with autism and their families are in a marginal position (and often feel powerless) with regard to speaking about therapeutic interventions and the goals of such interventions. The dominance of professionals in therapeutic interactions has been reflected upon in academic literature (see e.g. Todd, 2006). Drawing on Morgan (2000), Todd argues that an alternative role is possible for professionals to adopt, which engages in practices that assist people to develop their own preferred identities viewing autism as a difference rather than a deviation from a supposed norm.

The autistic community, particularly online, are inverting this assumption by 'othering' neurotypicality, and thus, as a consequence of their new definition, rendering NT worthy of study. Psychologists and psychiatrists have the power to define autism as other. However, the autistic community are politicising the label and inverting the othering to give a syndrome of NT, constructed in a similar manner to *DSM* descriptions and suggested therapeutic interventions for autism (see e.g. Brownlow & O'Dell, 2006). This remains a marginalised voice due to the power of the medical model of disability, particularly with specific reference to autism, however the neurodiversity discourse is becoming a more visible and powerful explanation for autism in particular cultural contexts such as in the United Kingdom.

Clinical relevance summary

The findings presented above have several important points for consideration in clinical practice. First, the findings highlight the power of normalising discourse in the lives of children and adults with autism. We argue that there is potential for clinicians to work within a framework of neurodiversity rather than a binary of normal/abnormal development, a nuanced approach that would suit the sensitive and complex practice of many clinicians. This shifts the focus of the 'pathology' away from the individual and enables families and clinical practice to consider the construction of impairment in a wider society.

The powerful way that discursive lenses shape the production of understandings of the child is considered by Goodley and Runswick-Cole (2012). In 'Reading Rosie: the postmodern disabled child', Goodley and Runswick-Cole interpreted Rosie, a child who has a diagnosis of autism, through a number of different discourses. They note different readings allow different interpretations of Rosie's life to be storied, including viewing her through the 'autism cannon'. Goodley and Runswick-Cole argue that; 'In this reading, there is no space for pleasure or joy, only a reading that confirms that Rosie has a "life-long developmental disorder"' (p. 62) through which all her interests and activities are seen and become problematic. However, viewed from other positions, Rosie's life and experiences can be read in different and more enabling ways; for example a Nordic relational account highlights Rosie as part of her family; as Goodley and Runswick-Cole argue, 'At home Rosie is, variously, "disabled" and "non-disabled", depending on the extent to which Rosie's "autism" and the environment match one another. At home Rosie is non-disabled as she is supported by her parents. Disability is normalised by the family' (op cit., p. 62).

Goodley and Runswick-Cole (2012, p. 63) argue that critical theorists need to resist the 'totalising autistic narrative' in order to explore the child's life in the whole rather than as made up entirely of their 'disability' and impairment. For clinicians, the significance is that it remains key to explore the resources of the family and the society/community in which families live. We share Goodley and Runswick-Cole's concern that 'the quirky quality, creative and personhoods of disabled children are merely understood as signs, symptoms and signifiers of pathology' (op cit., p. 64).

The discussions in this chapter also highlight issues of power within the therapeutic relationship and the importance of professionals engaging with this power imbalance. In discussion lists, parents and individuals with autism were drawing on the specialist language of clinical discourse in their exchanges with professionals, however many parents find the specialist language of diagnosis and classification alien and exclusionary. Engagement of people with autism and their parents would also serve to engage professionals in an abilities discourse which works with people's capabilities rather than their pathology and limitations.

The relatively new movement within autism advocacy drawing on a discourse of neurodiversity may contribute to shaping alternative understandings of autism which prioritise difference and diversity rather than deficit and impairment. The concept of neurological difference has been a long-standing issue for autism advocates and the neurodiversity movement has become a significant critical voice in the past 5 to 10 years of disability studies and disability advocacy (see e.g. Brownlow & O'Dell, 2013). Such a discourse must compete however with the powerful discourse of autism as a medicalised 'disorder'. For a simple summary of the implications for practice, see Table 16.1.

Summary

The analysis presented in this chapter has demonstrated how constructions of 'normal' development operate in discussions about autism held in online discussion sites for people with autism, parents of people with autism, and

Table 16.1 Clinical practice highlights

1. Clinical practice should work with abilities and strengths of a child with autism rather than assume a binary of 'normal' development/autistic development as two distinct developmental trajectories.
2. An awareness of the wider social context in which children and adults with autism live is essential.
3. A consideration of issues of power within the therapeutic relationship is crucial, and the importance of professionals engaging with this power imbalance so as to enable alternative understandings and interpretations of autism to become more prevalent within professional practice.
4. It is important to engage in a dialogue that acknowledges the role of particular discourses in shaping understandings and positionings of people with autism. It is also important to consider what possibilities are available within representations of autism for the crafting of positive autistic identities.

professionals working with people with autism. We have cited examples of parents, drawing on dominant constructions of child development, who compare their children against the idealised prescribed normal benchmark.

From our discussions, we have presented two dominant discourses surrounding the engagement of people with autism and their parents with professions: one drawing on the definition of (ab)normality, and the other drawing on the goal of normalisation through therapy. We argue that such dominant discourses limit opportunities for individuals to fashion positive self-identities, and through drawing on alternative discourses of neurodiversity, people with autism can be positioned as 'different' rather than 'deficient'.

References

Bettelheim, B. (1959). Feral children and autistic children, *American Journal of Sociology, 64*(5), 455–467.

Brownlow, C. (2007). *The construction of the autistic individual: Investigations in online discussion groups.* Unpublished PhD thesis. University of Brighton, UK.

Brownlow, C., Bertilsdotter Rosqvist, H., & O'Dell, L. (2013). Exploring the potential for social networking among people with autism: Challenging dominant ideas of 'friendship'. *Scandinavian Journal of Disability Research,* doi: 10.1080/15017419.2013. 859174.

Brownlow, C., & O'Dell, L. (2006). Constructing an autistic identity: AS voices online, *Mental Retardation, 44*(5), 315–321.

Brownlow, C., & O'Dell, L. (2013). 'Hard-wired from the factory'? Autism as a form of biological citizenship. In J. Davidson & M. Orsini (Eds.), *Worlds of autism: Across the spectrum of neurological difference* (pp. 97–114). Minneapolis, MN: University of Minnesota Press.

Burman, E. (2008). *Deconstructing developmental psychology* (2nd edition). London: Routledge.

Case, S. (2000). Refocusing on the parent: What are the social issues of concern for parents of disabled children? *Disability and Society, 15*(2), 271–292.

Davidson, J., & Orsini, M. (2013). The shifting horizons of autism online. In J. Davidson & M. Orsini (Eds.), *Worlds of autism: Across the spectrum of neurological difference* (pp. 285–304). Minneapolis, MN: University of Minnesota Press.

Edley, N. (2001). Analysing masculinity: Interpretative repertoires, ideological dilemmas and subject positions. In M. Wetherell, S. Taylor, & S. J. Yates (Eds.), *Discourse as data: A guide for analysis* (pp. 189–228). London: Sage.

Fairclough, N. (1992). *Discourse and social change.* Cambridge: Polity Press.

Fairclough, N. (2001). The discourse of new labour: Critical discourse analysis. In M. Wetherell, S. Taylor, & S. J. Yates (Eds.), *Discourse as data: A guide for analysis* (pp. 229–266). London: Sage.

Frances (2012). DSM 5 is a guide not Bible – Ignore its ten worst changes [Blog message]. Retrieved from Psychology Today: http://www.psychologytoday.com/blog/dsm5-in-distress/201212/dsm-5-is-guide-not-bible-ignore-its-ten-worst-changes.

Goodley, D., & Runswick-Cole, K. (2012). Reading Rosie: The postmodern disabled child. *Educational and Child Psychology, 29*(2), 53–66.

Jabr, F. (May 7, 2013). No one is abandoning the *DSM*, but it is almost time to transform it [Blog message]. Retrieved from scientific American: http://blogs.scientificamerican.com/brainwaves/2013/05/07/no-one-is-rejecting-the-dsm-but-it-is-almost-time-to-transform-it/.

Konrad, S. C. (2005). Mothers of children with acquired disabilities: Using the subjective voice to inform parent/professional partnership, *Journal of Death and Dying, 51*(1), 17–31.

Matson, J. L., Hattier, M. A. & Williams, L. W. (2012). How does relaxing the algorithm for autism affect DSM-V prevalence rates? *Journal of Autism and Developmental Disorders, 42*(8), 1549–1556.

Morgan, A. (2000). *What is narrative therapy? An easy-to-read introduction.* Adelaide, South Australia: Dulwich Centre Publications.

Morss, J. R. (1990). *The biologizing of childhood: Developmental psychology and the Darwinian myth.* Hove, Sussex, UK: Laurence Erlbaum Press.

O'Dell, L., & Brownlow, C. (2005). Media reports of links between MMR and autism: A discourse analysis. *British Journal of Learning Disabilities, 33*, 194–199.

Rose, N. (1989a). Individualizing psychology. In J. Shotter & K. J. Gergen (Eds.), *Texts of identity* (pp. 119–132). London, Newbury Park and New Delhi: Sage.

Rose, N. (1989b). *Governing the soul: The shaping of the private self.* London: Routledge.

Scior, K. (2003). Using discourse analysis to study the experiences of women with learning disabilities, *Disability and Society, 18*(6), 779–795.

Todd, L. (2006). Enabling practice for professionals: The need for practical post-structuralist theory. In D. Goodley & R. Lawthom (Eds.), *Disability and psychology* (pp. 141–154). Basingstoke: Palgrave MacMillan.

Urwin, C. (1985). Constructing motherhood: The persuasion of normal development. In C. Steedman, C. Unwin, & V. Walkerdine (Eds.), *Language, gender and childhood* (pp. 164–202). London, Boston and Henley: Routledge and Kegan Paul.

Recommended reading

- Davidson, J., & Orsini, M. (2013). *Worlds of autism: Across the spectrum of neurological difference.* Minneapolis, MN: University of Minnesota Press.
- Lawson, W. (2008). *Concepts of normality. The autistic and typical spectrum.* London: Jessica Kingsley.
- Armstrong, T. (2010). *Neurodiversity: Discovering the extraordinary gifts of autism, ADHD, dyslexia and other brain differences.* London: Da Capo.

Part IV

Situating and Exploring Child Mental Health Difficulties

17

A Conversation Analysis of the Problem Presentation Phase of Initial Assessment Appointments in a Child and Adolescent Mental Health Service

Victoria Stafford and Khalid Karim

Introduction

In this chapter, we focus on the ways in which children present their problems at a Child and Adolescent Mental Health Service (CAMHS) in the United Kingdom (UK). A conversation analytic approach is used to explore the interaction between the clinicians and children/young people during the problem presentation phase of their initial assessment appointments. This phase of the appointment is one of the only opportunities a patient has to vocalise their interpretation and understanding of the concerns which have resulted in their attendance (Heritage & Robinson, 2006). In a child-centred service, such as CAMHS, it is important to develop an understanding of the variety of ways children perceive and present the problems affecting them, particularly as attendance tends to be driven by the parents. In this chapter, we explore the ways that children describe the difficulties they are experiencing, when asked by clinicians why they think they are at the appointment.

CAMHS

Mental health problems in children are addressed by a wide range of professionals from those in primary care to more specialist secondary care. If these problems become enduring and are significantly affecting the child or family, they can be referred to a specialist CAMHS. In the United Kingdom, CAMHS are specialist services provided by the National Health Service (NHS) for children and young people aged approximately 3–18 years of age. This service sees children with a broad range of difficulties including emotional, behavioural, and neurodevelopmental disorders. Children and young people are referred to CAMHS by their General Practitioner (GP), paediatricians, and other specialist

313

agencies such as social services. Assessment and intervention are provided by a multi-disciplinary team including child psychiatrists, psychologists, and psychiatric nurses. The initial assessment appointment at CAMHS is of particular importance. Children are seen by clinicians who take a family and medical history, establish the child's and parents' understanding of the referral, and make a decision as to whether CAMHS is the right service to address the child's needs. Depending on the outcome of this assessment, children are further assessed by the service, treatment is initiated, or they are discharged from the service, and so this appointment is an important episode for both the families and clinicians.

Child-centred practice

The exploration of the interactions within a CAMHS setting needs to be appreciated within the contemporary framework of child-centred care. These principles have become increasingly important when considering interactions with children, and while this encompasses an enormous range of situations, it has considerable implications for services that work with children such as CAMHS.

Although there is no universally agreed definition of child-centred care, within health settings, child-centred practice is considered to be the right of the child to receive care that is adapted to their needs, to be treated with respect, and to be an active participant in decision making surrounding their care (Department of Health, 2003). Consequently child-centred practice has been increasingly recognised as synonymous with good patient care and best practice. Underpinning this position is the children's rights movement and the United Nations Convention on the Rights of the Child (UNICEF, 1989). The translation of these principles and the subsequent national and international policies remain challenging, but there has been an increased focus on making services more child-centred, with the interests of the child considered to be the focus of any therapeutic work. This emphasis is also reinforced within the United Kingdom with legislation such as the Children Act (1989).

The lack of any clear definition on child-centred care is further complicated by the complex dynamics which exist around any child. When accessing services, parents and family members have to play an active role in the attendance and engagement of the child throughout any therapeutic process. Practically this makes having a truly child-centred service more challenging, as the adults involved tend to be more vocal than the children. Children are often attributed what can be described as 'half-membership status' in many health settings (Hutchby & O'Reilly, 2010), with the contributions of parents and other adults being sought more readily and given greater credence. This can be for many reasons, as it may be due to perceived differences in the child's competence in relation to the adults (Lobatto, 2002), or the assumption that adults give more reliable accounts than children (Day, Carey, & Surgenor, 2006). Considering the importance of child-centred care, there has been surprisingly little research in this area and there remains no clear methodology on assessing whether it is truly occurring in practice. The available evidence demonstrates that children have very little input into their healthcare conversations (Stivers, 2002), which may not be advantageous for longer-term outcomes. It does become difficult to identify the ideals of child-centred care within a particular discourse and there is limited evidence available at this time. Indirectly, studies from fields such as family therapy have shown how children are managed in a complex therapeutic environment and have to adopt particular strategies to get involved (Hutchby & O'Reilly, 2010; O'Reilly, 2006; O'Reilly & Parker, 2013).

Additionally, children are likely to have a range of attitudes towards their appointments and can describe feeling peripheral to the process, although this may also depend on factors such as their age (Ross & Egan, 2004). What is particularly striking about Ross and Egan's findings is that it was younger children who felt more peripheral. This could indicate that clinicians need to do more to engage younger children and that there is perhaps a reliance on these children's parents and carers for information and experiences. The contrast in opinions about care across ages highlights the complexity of working with children and young people, and in turn the difficulty in creating services that are conducive to all ages. In order for services to become more child-centred, clinicians must recognise that children's views, opinions, and ideas are as valid as those of an adult (Dogra, 2005).

First assessments

In any specialist medical setting, the first appointment is often the most crucial part of the therapeutic process and has a number of functions including building a rapport, understanding the reason for attendance, and establishing expectations. When considering therapeutic interventions with children, it is important to build rapport and engage the child in the first appointment as they tend not to have been the initiators in seeking treatment or the main determinants of attendance at the appointments (Wolpert & Fredman, 1994). Evidently, therefore, they may already feel alienated from the therapeutic process. In some cases, the appointment has been actively resisted by the child. There are also instances where parents or carers arrange appointments on their child's behalf for concerns that the child may be unaware of. In circumstances such as this, the first appointment also serves the purpose of explaining to the child why they are there and what to expect.

In this data set, the initial assessments recorded were for non-urgent referrals and were attended by the child and their parents or caregivers. There were four possible outcomes of the appointment:

1. The child is assessed and CAMHS intervention is not deemed necessary.
2. The child requires CAMHS intervention and is placed on a waiting list for treatment.

3. The child requires CAMHS intervention and is seen immediately.
4. The assessment cannot be completed and an additional appointment is given.

Problem presentation

In any consultation there are different phases to the encounter which serve as a function in the negotiation between a patient and clinician. The problem presentation phase is one of the most important stages in a consultation and is the area of interest in this chapter. This phase of the appointment is also known as the 'complaint' phase (ten Have, 2001) and is the point at which patients describe their concerns. Its importance is highlighted by Heritage and Robinson (2006) as being the only occasion the patient is actively given the opportunity to describe their reason for attendance and offer a candidate understanding. However, in some instances, children actively seek to avoid participation in certain parts of the appointment, with the problem presentation phase being one of these (O'Reilly & Parker, 2013).

The presentation of the problem by participants is an important aspect of the patient-centred approach (Vanderford, Jenks, & Sharf, 1997). In order for the patient to do this successfully, the interaction must be facilitated by the clinician and cooperated with by the patient (Pomerantz, 2002). Patient-centred care is reciprocal as the patient has to respond appropriately for the interaction to be successful, as well as the clinician creating an interactional environment in which they are able to do this comfortably.

When patients attend medical appointments, they face the dilemma of presenting their reason for attendance as being appropriate for medical attention (Heritage & Robinson, 2006), while maintaining the position of the patient as a lay person and the doctor as the expert. When patients are offered a chance to explain the reason for their visit, they have choices to make about how they are going do that (Pomerantz, 2002), including explaining their symptoms, describing their experience of the problem, or making a diagnostic claim. This observation is as relevant when a child is the focus of the appointment as when an adult is the focus, as we will discover over the course of this chapter.

Project overview

In this chapter, we explore how children respond during the problem presentation phase of initial assessment appointments at CAMHS. We investigate how this is achieved through the interaction between the child and clinicians. This chapter is based on an empirical paper on a similar topic (Stafford, Hutchby, Karim, & O'Reilly, 2014), drawing on some of the data from that paper.

The setting

The data extracts discussed in this chapter are taken from a wider study comprising video-recordings of naturally occurring routine initial assessment appointments at a CAMHS in the United Kingdom, which aims to shed light on communication between families and clinicians in this setting, and the decision-making processes that occur. Children who present to CAMHS through the initial assessment process recorded are those deemed to be non-urgent cases. These cases cover a range of mental health concerns from behavioural and emotional, to neurodevelopmental. Twenty-eight families consented to having their appointments recorded, resulting in approximately 37 and a half hours of video footage. National Research Ethics Service (NRES) approval was sought and granted for the project.

Families and clinicians

The appointments were attended by the child and family members, which included one or both parents, grandparents, members of the extended family, and other professionals such as a family support worker. The service this data was taken from is open to children from birth to 18 years of age. The range of children captured in our sample is representative of the population attending CAMH services, with ages ranging from 6 to 17 (Mean $= 11.21$, $SD = 3.10$), of whom 18 were male and 10 were female (64% and 36%, respectively).

The initial assessment appointments recorded were conducted by a minimum of two clinicians from a multidisciplinary team, with one exception where only one clinician was present. A total of 29 clinicians were captured across the 28 appointments recorded, with the number of times being recorded ranging from 1 to 5 (Mean $= 2.17$, $SD = 1.37$). Where more than one clinician attended the appointment, their combinations varied across the range of appointments they were recorded in; that is, they did not have static pairings that they worked in. Whether they led the appointment or took notes also varied across the appointments. The team consisted of consultant, staff-grade, and trainee child and adolescent psychiatrists (4, 2, 4, respectively), clinical psychologists (3), assistant psychologists (2), community psychiatric nurses (CPNs) (5), a learning disability nurse (1), occupational therapists (4), and psychotherapists (2). Medical and nursing students (1, 1) were also present in two of the appointments recorded.

Data analysis

Collecting video-recorded data allows for the examination of a detailed record of the interactions collected. Conversation analysis (CA) was the approach utilised to explore the interactions in detail. CA is the study of talk in interaction (Schegloff, 1987) and is primarily interested in how talk is organised and how social actions are carried out through talk (Antaki, 2011; Mazeland, 2006).

CA has been applied successfully to a variety of institutional talk (Antaki, 2011) and is particularly applicable with this data where a variety of epistemics are at play as it gives equal importance to all turns by all participants in the interaction (Pilnick, Hindmarsh, & Gill, 2010). Of significant value to this research is the potential for CA to provide both practical and theoretical contributions to the area being studied (Barnes, 2005; Fasulo, Chapter 1, this volume). The video-recorded data was transcribed using the Jefferson notation system (Jefferson, 2004), in accordance with the analytic method employed (see the Preface for table of symbols).

'Why Are You Here?'

Establishing the reason for attendance at CAMHS during initial assessment appointments happens in four different ways within the sample of 28 appointments:

1. The child is asked by the clinician why they are there, and gives a reason (16 instances).
2. Either the child or parent is asked why they are there, and the parent gives a reason (6 instances).
3. The clinician tells the family why they are there (2 instances).
4. The information is mutually exchanged during the course of the appointment (4 instances).

The following analysis pays particular attention to the reasons given by the children as to why they are presenting at CAMHS from the 16 cases available.

The clinicians in 16 of the 28 appointments recorded asked the child directly why they thought they were attending an appointment at CAMHS. This was generally addressed with variations of the question, 'why are you here?' The children who responded did so with one of three types of response:

1. They gave a 'medicalised' response: they gave a technical mental health reason for their attendance (4 instances);
2. They gave a 'lay' response: they gave a general explanation of the difficulties they were experiencing (4 instances);
3. They gave a 'don't know' response: they responded with variations of 'I don't know' indicating they had an insufficient understanding to give an explanation (8 instances).

We explore how the children and clinicians managed each of these types of responses. Issues of clinical relevance will be discussed at the end of each section as there are particular areas of consideration specific to each type of response given by the child. General clinical considerations are discussed at the end of the chapter.

'Medicalised' responses

In a number of cases when children were asked why they need an appointment at CAMHS, they responded by giving a medical reason for their attendance. These technical mental health terms ranged from the very specific, such as OCD and Tourette's Syndrome, to terms that while considered to be medical mental health terminology are also general conversational terms used to describe a behaviour, such as self-harm. When probed further about their understanding of the terms they used, two of the three children described behaviours appropriate to the label they gave them. In the third case, the child stated that she could not remember what her teacher told her, suggesting that the use of the mental health term is the result of its being used by other adults.

Extract 1: Family 1 (Prac = clinical psychologist) (from Stafford et al., 2014)

```
1    Prac:    ↑ Do you kno:w (0.31) why you've c↑ome here toda↓y?
2    Child:   erm because (0.39) I- keep (0.94) doin my- (0.41) I
3             thi↑nk it's ↓O- C- D-
4    Prac:    Ri:ght (0.78) ↑ok:(0.92) um (0.52) °that is a (.)
5             important word you use° (.) m↑eaning when you say O-
6             C- D-
7    Child:   °pard[on-]°
8    Prac:         [ah ] wh- when you say o- c- d- what does it
9             me:an?
10   Mum:     whad'ya think it me:ans when you say o- c- d-
11   Child:   um-(1.10) Ah: can't remember what the teacher to↑ld
12            me-
```

The clinician in Extract 1 began by asking the child to tell him why she thinks she was at the appointment. This was then followed by the child giving a technical reason for attendance '↓O- C- D-' (line 3), and the clinician asking for an explanation of her understanding of the term (lines 4–9). The clinician directly questioned the child's understanding of the term: 'when you say o-c- d- what does it me:an? (lines 9–10). The child however was unable to offer an explanation, and also attributed the term to a third party 'can't remember what the teacher to↑ld me-' (lines 11–12). The clinician emphasised the importance and relevance of using a technical term in his exclamation '°that is a (.) important word you use°' (lines 4–5). Although the term 'OCD is a technical mental health term, it is one that is being increasingly used in general conversation. Tourette's however still maintains a status of specialised medical terminology.

Extract 2: Family 20 (Prac = Clinical psychologist) (from Stafford et al., 2014)

```
1    Prac:     Uh::m (2.0) well without kind of putting the
2              spotlight on you Daisy uh-d'you know why you're here
3              today (1.8) ↑O↓kay (0.4) uhm if you just want to
4              whisper it out.(0.8) Uh:.
5    Mum:      >D'you wanna< say why you're here?
6              (5.0)
7    Mum:      °You can say°
8              (1.9)
9    Nan:      °Speak up.° (0.6)
10   Mum:      °'Cause of ma,°
11   Child:    °Tourette's.°
12   Mum:      'Cause of her Tourette's. (0.7)
13   Prac:     Oka:y. (0.8) Uh:::m did somebody- (0.9) did somebody
14             tell you that (1.4) Did they (0.6) did anybody
15             explain to you what that is.(0.9) (We'll ask the
16             doctor to explain to you) Think just for our
17             understanding.= What does that me:an =What happens
18             when somebody has Tourette's.
19             (3.3)
20   Mum:      What do you do.
21             (2.2)
22   Child:    °Move my head.°
23   (Lines omitted)
24   Mum:      What else do you do?
25             (1.7)
26   Child:    °Make noi:ses°
```

Despite the child in this extract being younger than the child in Extract 1, she used the medical terminology more successfully in describing her difficulties. Again, once she had provided the reason for attendance as her Tourette's, the clinician probes her knowledge of the term 'What does that me:an =What happens when somebody has Tourette's' (lines 17–18). When prompted by her mum, the child began to describe her symptoms (line 22). Initially, however, the clinician verbally doubted the child's own knowledge of the condition by asking if somebody else had told her that (lines 13–15). It becomes evident later in the session that the term 'Tourette's' had been used in previous medical encounters at which the child had been present, and that her use of the term stemmed from that interaction.

It is not always as easy to identify when a child is using a technical mental health term or lay terminology. Due to an increase in the use of mental health terminology in general parlance, there are terms that are both lay and technical, for example 'self-harm' and 'phobia'. In the following excerpt, the clinician is faced with this issue.

Extract 3: Family 2 (Prac = Psychotherapist) (from Stafford et al., 2014)

```
1    Prac:    o↑k (1.25) y- you were ref↑erred by your (0.33) G↓P
2             (0.36) um ↑do you ↑know (0.88) why you're here
3             ↓tod↑ay (0.83) can you tell me a bit ab↓out that
4    Child:   (er) it's ab↓out self-↓harming
5    Prac:    ab↑out self-↓harm (0.63) ok↓ay (1.77) i- and what do
6             you mean by ↓that Call↑um °in what ↑way°
7    Child:   what(0.42)em:(0.38)it's (mainly) ↓I self-harm
8    Prac:    you self-↓harm (1.03)°ok° (.) c- can you say
9             s↑omething about that ↓is it i- ↓do you cut yourself
10            ↓or hurt yourself in a ↑different way
11   Child:   cut
```

In this extract, the child used a technical term that overlaps in to everyday language to describe the reason for attendance: 'self-harming' (line 4). The term 'self-harming' can be used to describe a behaviour and as a diagnostic term. It was marked as a diagnostic term by the clinician in her response by asking both what the child *means* by self-harm and *how* he self-harms (lines 5–6).

Following the child giving his reason for attendance, the clinician asked him for an explanation of the term used: 'can you say s↑omething about that' (lines 8–10). In using a technical term the child was assuming a level of knowledge about the symptoms he was experiencing; that he has enough knowledge of both the symptoms and the term used to enable him to link the two and suggest that it is what he was experiencing. The explanation the child gave in answer to the clinician's probing potentially enables the clinician to establish the actual level of the child's knowledge, as well as to align their vocabulary and expectations for the rest of the appointment. The alignment of knowledge is an important part of the problem presentation stage as the remainder of the appointment is built upon the clinician trying to understand the child's difficulties.

Clinical relevance

Within our data, the use of a candidate diagnosis was sometimes the result of contact with third parties such as teachers or other health professionals. While children may use a term syntactically correctly, it does not mean that their symptoms are congruent with the diagnosis offered. The clinicians in the extracts presented all clarified the child's understanding of the terms they used, which gives them the opportunity to find out what the child's problem behaviours actually are, rather than relying on the candidate diagnosis. Additionally, the concerns patients presented during the appointment may have been discussed and thought through with family and friends before the appointment occurred, enabling the child to situate their concern within the context of their own and other people's prior knowledge and experience (Pomerantz, 2002; ten Have, 2001). This is an important exercise for patients, and valuable for doctors to bear in mind, as the discussion of concerns with others will inevitably shape the account patients give doctors.

It could be argued that the use of more technical terms such as 'Tourette's' and 'OCD' are an indication of an increased ease of use of mental health and illness vocabulary in lay conversation. The ease of use of mental health discourse in conversation may also bring with it increased expectations of a child's condition and the treatment they will receive from CAMHS on the part of both the child and families. This requires careful management by the clinician not only to be able to assess the child from a neutral perspective but also to manage the expectations of the child and family, especially in cases where they may be incorrect.

'Lay' responses

An alternative way children responded to being asked why they were attending the appointment was with a lay reason. The children who responded in this way gave predominantly descriptive accounts of why they were there. The extracts below demonstrate the descriptive nature of these accounts, in contrast to the more specific terminology used by the children in Extracts 1–3.

Extract 4: Family 17 (Prac = Consultant Child Psychiatrist) (from Stafford et al., 2014)

```
1    Prac:     °↓alright° (0.74) ok do YOU know why you're here
2              to↓day
3    Child:    urm: (0.21) no not really (0.41) something 'bout
4              difficulties or ↓some'ing like that
5    Prac:     about your difficulties
6    Child:    ye:ah (.) some'ing like th[at]
7    Prac:                                [alr]ight
```

This extract follows a similar sequence to those above, and began with the clinician asking the child why they thought they were there, eliciting a child response (lines 1–4). The child began their answer by stating that they do not really know the reason. However, following a pause, continued with a possible explanation (lines 3–4). The initial 'no not really' (line 3) suggests limited knowledge of the reason for attendance. This is then corrected with the child's explanation of 'something 'bout <u>diffi</u>culties or ↓some'ing like that' (lines 3–4), acknowledging the presence of a problem. In contrast to this, some children responded straight away with an explanation.

Extract 5: Family 27 (Prac = CPN) (from Stafford et al., 2014)

```
1   Prac:    Okay.(0.7) pt Ri:ght so .hhh shall we sorta sta:rt
2            with hhh ↑why (0.3) you think y'he:re. Why d'you
3            think y'he:re [°Nicholas?°]
4   Child:                [.hhh      ] Because m'mum and dad
5            are tryna to sort out fings for me.
6   Prac:    Okha::y. What sorta things are they trynta sort ou::t
7            d'you think?
8            (0.9)
9   Child:   Li::ke to help me at school
```

In this extract, the child responded immediately with a reason for being in attendance, which was attributed to his parents needing to 'sort <u>out</u> fings' for him (line 5). From this response it is not clear what the 'fings' that need to be sorted out are. The clinician pursued this further and the child's following explanation of some of the problems he was experiencing suggests that he does recognise that a problem exists: 'Li::ke to help me at school' (line 9).

Extract 6: Family 21 (Prac = Community Psychiatric Nurse)

```
1    Prac:    a:nd I s'pose the obvious ↓thing would be to ask you
2             (0.62) why you've come back (1.07) why you've been
3             referred back to us
4             (3.93)
5    Child:   er
6             (4.67)
7    (Lines omitted)
8    Child:   I don know it's just getting bad
9    Prac:    okay so it's b[een getting]g wo:rse
10   Child:                 [yeah]
11            yeah
12   Prac:    what's been getting wo:rse
13   Child:   the touchin' (an) everyfink
```

The child in Extract 6 was older than most in the sample at 17 years. It was also not his first appointment at CAMHS, having had a referral for related concerns previously, which resulted in a diagnosis. He was vague about his reason for referral when questioned (lines 1–8), despite his age and an assumed knowledge of the both the problems he was experiencing and the technical terminology for it. He attributed a certain level of knowledge about his problems to the clinician with his explanation of 'it's just getting bad' (line 8), and only expanded on this when she explicitly asked 'wha<u>t's</u> been getting wo:rse' (line 12). Despite having a diagnosis of OCD, he still only responded with a lay explanation of 'the touchin' (an) everyfink' (line 13).

Clinical relevance

Lay conceptualised responses tend to be a general account for attendance, rather than a specific description of the symptoms. It is just as important for clinicians to probe the child's responses as it is when they offer technical terminology, only looking for elaboration rather than a level of understanding. While technical terminology tends to originate from a third party, the lay descriptions offered by the children in our sample all seem to originate from within themselves, which may explain their general nature and hesitant responses.

The child in Extract 7 had already received a diagnosis of OCD prior to this appointment, but interestingly he responded with a general description rather than saying he was there because of the OCD he suffered with. It is worth clinicians bearing in mind that just because a child answers with a lay conceptualised reason and not a technical term, it does not mean they do not have an understanding of the cumulative effect of the difficulties they are experiencing, or a possible diagnosis. This is potentially something that is worth pursuing in situ.

'Don't Know' responses

The third response children gave to the question of 'why are you here' was to give an answer indicating limited understanding, for example 'I don't know'. Claiming limited understanding was the most common way for children to respond in this sample. They displayed this lack of knowledge in two ways when asked whether they knew why they were there:

1. By responding with variations of 'I don't know'.
2. By responding with 'no' or a shake of the head.

Extract 7: Family 6 (Prac = Consultant Child Psychiatrist) (from Stafford et al., 2014)

```
1    Prac:    wha̲t did you th↓ink you'd co̲me here f↓or w- what was
2             yo̲ur under↓standing of why you'd ↓co̲me here to̲day
3    (Lines omitted)
```

```
4    Prac:     whAt ↓do you think ↓you're- what do you think you've
5              come for ↓here tod↓ay w- what do you ↓think (.) why
6              do you ↓think your: mum and ↓nana bought you here
7              to↓day
8    Child:    ↓don't know
9    Prac:     don't ↓know (0.40) do you ↑think it's to ↓do with
10             your behavio↓ur
11   Child:    don't ↓know
12   Prac:     don't ↓know ok do you think it's to ↑do with your
13             ↑feelings
14   Child:    ↓don't know
15   Prac:     o↓k (1.06) that makes me ↓think you do kind of ↓know
16             but not saying (0.35) ok (0.92) ↓so did you did your
17             ↓mum discuss with ↓you (0.23) that you were ↓coming
18             here today
```

Here the clinician used several different lines of questioning to elicit a response from the child, but each time the child responded with 'don't know'. The clinician continued to pursue the line of questioning rephrasing it slightly each time, but is continually met with 'don't know'. After several attempts, the clinician treated the child's answer as resistance rather than as not knowing (Hutchby, 2002), which is reflected in the clinicians statement of 'that makes me ↓think you do kind of ↓know but not saying' (line 15–16). It appears to be rare that clinicians pursue answers when 'don't know' responses are given with as much rigour as in this extract.

Extract 8: Family 3 (Prac = Trainee Child Psychiatrist) (from Stafford et al., 2014)

```
1    Prac:     o↓k alright I will ↑ask ↓Mum and Dad ↓a little bit
2              ↓as well do you know>↓why you are< why you- why you
3              are ↓here
4    Child:    ((shakes head))°no°
5    Prac:     oh (0.59) ↓w:ell the the were they were a bit ↓of
6              concern ab↓o:ut you that you are a little ↓bit
7              anxio↓us (0.80) so we will ask ↓Mum and Dad ab↓out
8              tha:t and then ↓we will get ba I if you want to tell
9              me any↓thing in in in in between (.) ↓just let me
10             kn↓ow ↑alr↑ight
```

In this extract, the child's claim to a lack of understanding came in the form of a quietly spoken 'no' (line 4). The difference in response may be a reflection of the difference in questioning style, from 'why are you here' in previous extracts to the more closed 'do you know why you are here' (lines 2–3). The response is treated with surprise from the clinician, using the change-of-state token 'oh' (line 5) (Heritage, 1984). Whereas the clinician in Extract 7 pursued the child's answer, this clinician responded by informing the child of why they are there (lines 5–7). It is interesting to note that the child is then side-lined from the conversation and only afforded half-membership (Hutchby and O'Reilly, 2010; Shakespeare, 1998), as the clinician went on to refocus the conversation on the child's parents, 'so we will ask ↓Mum and Dad ab↓out tha:t' (lines 7–8). The clinician has attributed knowledge of the child's difficulties to the parents instead of to the child, and in doing so has closed down communication with the child. Although invited to contribute to the following conversation 'if you want to tell me any↓thing in in in in between (.) ↓just let me kn↓ow ↑alr↑ight' (lines 8–10), it is difficult for children to interject in conversation between adults, especially in the context of therapy (Parker and O'Reilly, 2012).

On occasion the child may use an alternative physical or vocal representation of 'no' to answer the clinician's question, such as a shake of the head.

Extract 9: Family 8 (Prac = Trainee Child Psychiatrist) (from Stafford et al., 2014)

```
1    Prac:    o↑kay 1.32) erm: (1.74) do ↑you (0.37) d- do you
2             ↑know why ↑you're ↓here by the way
3             ((child shakes head))
4    (Lines omitted)
5    Prac:    °n↑o° (1.47) shall ↓we ask your ↑mum (.) wh↓at she
6             thinks (0.90) why- ↓why you are ↑here
```

The clinician in Extract 8 uses a closed questioning style to elicit a response from the child (lines 1–2) and is met with a non-verbal response of a shake of the head (line 3). The clinician treats the child's response as dispreferred by pursuing an answer, but each time is met with the same response. The clinician then continues by redirecting the question to the child's mum, 'shall ↓we ask your ↑mum (.) wh↓at she thinks (0.90) why- ↓why you are ↑here' (lines 5–6).

Extract 10: Family 16 (Prac = consultant child psychiatrist) (from Stafford et al., 2014)

```
1    Prac:    Kolomban do you know why (0.39) you've come (0.75)to[↓day]
2    Child:                                                        [no]
3    Prac:    you ↓don't (.) okay who have you bought with you today?
4    Later in the session
5    Mum:     Kolomban does know why he's ↓come today
6    Prac:    Right
7    Mum:     he does know what the ↓process i[s all abo]ut but=
8    Prac:                                     [yeah]
9    Mum:     =we're a little bit ↓silly this morning
```

In the first part of this extract, the clinician followed what appears to be a standard line of questioning in asking the child if he knows why he has come to the appointment (line 1). On receipt of a 'no' response (line 2), the clinician did not pursue a further answer but rather sought to keep the child engaged by moving the questioning on to who else was attending the appointment with him (line 3), all the while keeping the child as the focus of the interaction and maintaining his position as a knowledgeable participant. Slightly later in the appointment, the child's mother refuted the child's lack of understanding regarding attendance and claimed that he was aware, 'Kolomban <u>does</u> know why he's ↓come today' (line 5), and accounted for the claim of a lack of knowledge to him being 'a <u>little</u> bit ↓silly this morning' (line 9). In making this claim about the child's behaviour and its implications on his participation in the interaction, the child's mother was redefining his epistemic position in this phase of the appointment.

Clinical relevance

Claiming limited understanding was the most common way for children to respond to being asked why they were attending CAMHS. Not only was this demonstrated in a number of ways by the children in the sample, but the response was also treated in a number of ways by the clinicians: as representing a lack of knowledge, as resistance to the question asked, or as insufficient as a response. This indicates that it is difficult to interpret what the child means by 'don't know'.

From the perspective of child-centred practice, it is important to treat resistance to answering the questions or a claim to a lack of knowledge as a valid contribution to the interaction and to keep interacting with the child. This is in contrast to Extracts 7, 9, and 10, where the clinician deferred to the child's parents knowledge. In Extract 8, the clinician ended the line of questioning with the child following the 'don't know' response and continued the assessment with the parents. From examining the complete vide-file, this has implications for the level of the child's engagement in the remainder of the assessment. There are cases where the child responds with 'don't know' but the clinician kept engaging with the child and later in the appointment the child provided a reason for their attendance where they had not done so before, such as the family in Extract 7.

Clinicians sought the children's understanding through different types of questions and this may reflect in the form of the response by the child. The composition of the question varied and there seems to be a tendency to respond with a 'don't know' when the question from the clinician was phrased as a 'wh' question, for example 'why are you here today', 'what has brought you here today'. When phrased as 'do you know...', the question tended to elicit either a closed 'no' response or a shake of the head. Although 'no' and 'don't know' may be equivalent responses when taken in the context of the above questions, the response itself may have an implication for whether the clinician goes on to engage the child for the rest of the appointment.

Lack of engagement with the child

It is interesting to note that despite CAMHS advocating being a child-centred service and the child being present at the initial assessment appointments, the child was not always asked for their understanding of why they were there or what their concerns were. In six appointments the clinician directed the 'why are you here' question to the parents instead of the child, in two appointments the clinician described to the family why they were there instead of asking either the child or additional adults what their concerns were, and in four cases there was no overt inquiry as to why they were there. There does not seem to be an age-related reason for the clinician not seeking the children's accounts in these situations, or competency issues relating to the children's problems, and there is little consistency across clinicians to suggest it is the consequence of a particular clinician's therapeutic style.

Clinical relevance summary

Children enter a CAMH service with varying degrees of knowledge of the concerns that have triggered an assessment. During initial appointments it is important for clinicians to establish what the child's understanding is in order to carry out a comprehensive and effective assessment.

Types of response

When asked to explain why they have an appointment at CAMHS, children all seemed to respond in one of the three ways discussed above. Each of these responses provides the clinician with an insight into the child's understanding of the problems they may or may not be experiencing, whether the child responds with a candidate diagnosis, or a 'don't know'. How the appointment then evolves, whether with the child being the focus, or the parents, seems to be a direct consequence of the clinicians reaction to the child's response, rather than of how the child responds in the first instance.

A 'child-centred' approach to assessments

Within the 16 appointments where the child was asked for their understanding of why they were at CAMHS, eight children responded with an indication of a lack of knowledge, with some of these revealing later in the session that they did know why they were there. The other eight children gave a comprehensive reason for them being there, whether in the form of mental health

Table 17.1 Clinical practice highlights

1.	It is important for clinicians to establish the origins and understanding of medical terms when used by a child.
2.	The interpretation of 'don't know' responses takes skill and experience from the clinician.
3.	Maintain child engagement throughout the appointment, even if they don't know why they are there.

language or a lay conceptualised description of their symptoms. With a variety of ages included in both groups, it is evident that children are capable of making a significant contribution to their assessments and that their insights provide valuable information for clinicians about the difficulties as experienced by them, and their understanding. For a simple summary of the implications for practice, see Table 17.1.

Summary

We have outlined three different strategies children use to describe their reasons for attendance at CAMHS and looked at how these are initiated, presented, and responded to by the children and clinicians in the appointments. It is important to note that while all of the data presented in this chapter includes the child, there are many occasions evidenced in the wider data set in which the child is not included in the problem presentation sequence and is never asked what their understanding is of why they have been referred to CAMHS. While it is important for clinicians to ask parents for their understanding of their child's problems, in a child-centred service there should always be time dedicated in appointments to finding out what the child's understanding of their situation is and to probing their responses for their true understanding. As we have discovered here, children often have valuable insights into the problems they are experiencing, when given the opportunity to proffer them.

References

Antaki, C. (2011). Six kinds of applied conversation analysis. In C. Antaki (Ed.), *Applied conversation analysis: Intervention and change in institutional talk* (pp. 1–14). Hampshire: Palgrave MacMillan.

Barnes, R. (2005). Conversation analysis: A practical resource in the health care setting. *Medical Education, 39*, 113–115.

Children Act (1989). Retrieved May 17, 2014 from http:// www.legislation.gov.uk/ukpga/ 1989/41/contents.

Day, C., Carey, M., & Surgenor, T. (2006). Children's key concerns: Piloting a qualitative approach to understanding their experience of mental health care. *Clinical Child Psychology and Psychiatry, 11*(1), 139–155.

Department of Health (2003). *Children's National Service Framework*. London: DH.

Dogra, N. (2005). What do children and young people want from mental health services? *Current Opinion in Psychiatry, 18*(4), 370–373.

Heritage, J. (1984). A change-of-state token and aspects of its sequential placement. In J. M. Atkinson & J. Heritage (Eds.), *Structures of social action: studies in conversation analysis* (pp. 299–345). Cambridge: Cambridge University Press.

Heritage, J., & Robinson, J. D. (2006). Accounting for the visit: Giving reasons for seeking medical care. In J. Heritage & D. W. Maynard (Eds.), *Communication in medical care: Interaction between primary care physicians and patients* (pp. 48–85). Cambridge: Cambridge University Press.

Hutchby, I. (2002). Resisting the incitement to talk in child counselling: Aspects of the utterance 'I don't know'. *Discourse Studies, 4*(2), 147–168.

Hutchby, I., & O'Reilly, M. (2010). Children's participation and the familial moral order in family therapy. *Discourse Studies, 12*(1), 49–64.

Jefferson, G. (2004). Glossary of transcript symbols with an introduction. In G. H. Lerner (Ed.), *Conversation analysis: Studies from the first generation* (pp. 13–31). Amsterdam: John Benjamins.

Lobatto, W. (2002). Talking to children about family therapy: A qualitative research study. *Journal of Family Therapy, 24*, 330–343.

Mazeland, H. (2006). Conversation analysis. In *Encyclopaedia of language and linguistics* (2nd edition, Vol. 3, pp. 153–162). Oxford: Elsevier Science.

O'Reilly, M. (2006). Should children be seen and not heard? An examination of how children's interruptions are treated in family therapy. *Discourse Studies, 8*(4), 549–566.

O'Reilly, M., & Parker, N. (2013). 'You can take a horse to water but you can't make it drink': Exploring children's engagement and resistance in family therapy. *Contemporary Family Therapy, 35*(3), 491–507.

Parker, N., & O'Reilly, M. (2012). 'Gossiping' as a social action in family therapy: The pseudo-absence and pseudo-presence of children. *Discourse Studies, 14*(4), 457–475.

Pilnick, A., Hindmarsh, J., & Gill, V. T. (2010). Beyond 'doctor and patient': Developments in the study of health care interactions. In A. Pilnick, J. Hindmarsh, & V. T. Gill (Eds.), *Communication in healthcare settings: Policy, participation and new technologies* (pp. 1–16). West Sussex: John Wiley and Sons.

Pomerantz, A. (2002). How patients handle lay diagnosis during medical consultations. *Texas Linguistic Forum, 45*, 127–138.

Ross, N., & Egan, B. (2004). 'What do I have to come here for, I'm not mad?' Children's perceptions of a child guidance clinic. *Clinical Child Psychology and Psychiatry, 9*(1), 107–115.

Schegloff, E. (1987). Analysing single episodes of interaction: An exercise in conversation analysis. *Social Psychology Quarterly, 50*(2), 101–114.

Shakespeare, P. (1998). *Aspects of confused speech: A study of verbal interaction between confused and normal speakers*. Mahwah, NJ: Lawrence Earlbaum Associates.

Stafford, V., Hutchby, I., Karim, K., & O'Reilly, M. (2014). 'Why are you here?' Seeking children's accounts of their presentation to child and adolescent mental health service (CAMHS). *Clinical Child Psychology and Psychiatry*. Advance online publication, doi:10.1177/1359104514543957.

Stivers, T. (2002). Presenting the problem in pediatric encounters: 'symptoms only' versus 'candidate diagnosis' presentations. *Health Communication, 14*(3), 299–338.

ten Have, P. (2001). Lay diagnosis in interaction. *Text, 21*(1–2), 251–260.

UNICEF (1989). *United Nations Convention on the Rights of the Child.* London: UNICEF.

Vanderford, M. L., Jenks, E. B., & Sharf, B. F. (1997). Exploring patients' experiences as a primary source of meaning. *Health Communication, 9*(1), 13–26.

Wolpert, M., & Fredman, G. (1994). Modelling the referral pathway to mental health services for children. *Association of Child Psychology and Psychiatry: Newsletter, 16,* 283–288.

Recommended reading

- Dogra, N. (2005). What do children and young people want from mental health services? *Current Opinion in Psychiatry, 18*, 370–373.
- Parker, N., & O'Reilly, M. (2012). 'Gossiping' as a social action in family therapy: The pseudo-absence and pseudo- presence of children. *Discourse Studies, 14*(4), 457–475.
- Pilnick, A., Hindmarsh, J., & Gill, V. T. (2010). Beyond 'doctor and patient': Developments in the study of health care interactions. In A. Pilnick, J. Hindmarsh, & V. T. Gill (Eds.), *Communication in healthcare settings: Policy, participation and new technologies* (pp. 1–16). West Sussex: John Wiley and Sons.
- Stafford, V., Hutchby, I., Karim, K., & O'Reilly, M. (2014). 'Why are you here?' Seeking children's accounts of their presentation to child and adolescent mental health service (CAMHS). *Clinical Child Psychology and Psychiatry.* Advance online publication, doi: 10.1177/1359104514543957.

18

The Discursive Construction of Problem Behaviours of Children with Autism in Therapy

Jessica Nina Lester and Michelle O'Reilly

Introduction

The focus of this chapter is on the social construction of 'problematic' and 'acceptable' behaviours within interactions between therapists and children diagnosed with autistic spectrum disorder (ASD). Drawing from a larger corpus of data, we explore how the children and their therapists go about negotiating and renegotiating problem behaviours. More specifically, in the data drawn upon, therapists used a social skills curriculum, 'Superflex...A Superhero Social Thinking Curriculum' (Madrigal & Winner, 2008), which included a shared vocabulary for talking about problematic behaviours. Drawing upon discursive psychology (DP) (Edwards & Potter, 1987) and conversation analysis (CA) (Sacks, 1992), we give particular attention to the lexical items related to the *Superflex* curriculum, as the character and his associated behaviours were often used to position particular behaviours as acceptable or unacceptable, within or outside of a child's identity, and capable of being altered. With a broad focus on the ways in which 'appropriate' social behaviours were negotiated and constructed within this setting, we attend to how the lexical items (i.e. *Superflex* and the *unthinkables*) functioned to *define* and *mark* what counted as an appropriate behaviour, who was accountable for changing the 'inappropriate' behaviour, and what or who was responsible for causing inappropriate behaviour.

We begin this chapter by discussing ASD more generally and then move to explore literature related to social skills interventions and the concepts related to the normality/abnormality binary. We then draw together the analytical messages, which illustrate how the participating children often exhibited behavioural challenges that were named socially 'inappropriate' and frequently identified as needing to be 'remedied' or 'fixed'. In the examples included here, we highlight how the therapists went about negotiating the naming in a delicate, face-saving way.

Defining autistic spectrum disorder

To contextualise our analysis, we begin by providing an overview of the disorder under focus. In doing so, we recognise that diagnosis is a complex process with different professionals having different ways of working (Karim, Cook, & O'Reilly, 2012) and that families often remain in need of accessible information about this condition (O'Reilly, Karim, & Lester, in press). We also recognise that there are a range of descriptive terms for the condition, including autism, autism/autistic spectrum disorder, pervasive developmental disorder, Asperger's syndrome, and high-functioning autism (Karim, Ali, & O'Reilly, 2014). While each of these labels is loaded with category-rich inferences and has different degrees of preference by different groups, the generally accepted term, as advocated by DSM-5, is that of 'autism/autistic spectrum disorder' (American Psychiatric Association, 2014). Thus, from this point forward, we use the term ASD, the abbreviated version of the accepted terminology.

The concept of ASD is a relatively contemporary one, stemming from the work of Paul Bleuler in his attempts to describe Schizophrenia (Karim et al., 2014). The term 'autism' was initially used by Leo Kanner (1943) to describe children he characterised as displaying extreme aloneness, preservation of sameness, and a range of affective, cognitive and behavioural symptoms. During this same time period, Hans Asperger (1944) also described a similar syndrome that included symptoms such as a lack of empathy, clumsy movements, and a limited ability to form friendships. Because of particular historical events, such as the Second World War, this research was not broadly disseminated (Karim et al., 2014). It was widely believed that these symptoms were due to the children having problems with their mothers, that is the 'refrigerator mother' (Bettelheim, 1967), which fitted with the broader 'mother-blaming' discourse prevalent within the field of mental health at that time. While this mother-blaming ideology has been challenged by scientific rhetoric (Blum, 2007; Phelan, 2005), the discursive situating of blame has not fully disappeared.

Nonetheless, scientific advances and new research have promoted a better understanding of ASD, influencing the terminology and classification of the disorder. The term 'ASD' was first introduced in 1981 by Lorna Wing, which was designed to encapsulate the range of different presentations. In contemporary medicine, ASD is described as a neuro-developmental condition, which causes difficulties in various areas of daily functioning and is characterised by a 'triad of impairments', including qualitative impairments in social interaction, qualitative impairments in communication, and restrictive and repetitive patterns of behaviour. Currently, ASD is classified as a mental health disorder (although we recognise that this classification has received some challenges). ASD is typically defined behaviourally; thus, for a child to be diagnosed with the ASD, s/he must be clinically assessed to demonstrate specific impairments across the triad of impairments (Muskett, Body, & Perkins, 2013). With two of the core characteristics of ASD being social difficulties and communication difficulties, it is perhaps unsurprising that there has been a recent focus on developing interventions that specifically target modification and/or correction of social and communication problems.

Social skills interventions and behavioural therapy

While there is a broad range of medical, psychosocial, and therapeutic interventions used to 'treat' ASD, in this chapter, it is not our intention to explicate the detail of these. Rather, we focus specifically on those interventions with a social skills and behavioural training element. First, we do this as two elements of the 'triad of impairments' focus specifically on social skills and communication. Second, the data drawn upon later in this chapter is taken from a therapeutic setting in which the interactions were broadly classified as social skills training. Finally, this area of intervention is fairly contemporary and the evidence-base for it is slowly beginning to emerge, creating some tension and contention around what such interventions should entail and their intended outcomes; thus, we suggest that discourse techniques can offer an alternative way of exploring this particular therapeutic process.

In modern life, social skills encompass a range of verbal and non-verbal behaviours that are required for interpersonal communication (Rao, Biedel, & Murray, 2008). These behaviours include such acts as smiling, making eye contact, responding to questions, and giving/acknowledging compliments (Beidel, Turner, & Morris, 2000). Notably, children with ASD experience difficulties in initiating and responding to verbal cues, interpreting non-verbal cues, initiating and maintaining eye contact, exhibiting appropriate emotional reactions, and using non-verbal behaviours as maintenance techniques (Attwood, 2000). Thus, social impairments among children diagnosed with ASD may include interpersonal interaction, speech and linguistic conventions, impairments in social pragmatics, poor speech prosody, and problems in understanding non-literal language (Williams-White, Keonig, & Scahill, 2007). These impairments can be problematic for children diagnosed with ASD, as research indicates that social skills are linked positively to developmental outcomes, including academic achievement, positive mental health, and peer acceptance (Hartup, 1989). Such impairments highlight, therefore, that interventions specific to social skills and communication are essential in addressing social impairments (Matson, Matson, & Rivet, 2007).

Due to this need, there have been an increasing number of interventions developed that focus specifically on social skills. These social skill interventions broadly fall into two categories: (1) those delivered within a traditional behavioural model in a clinic or classroom and (2) those that include a generalisation component, such as community practice work (Rao et al., 2008). These interventions can be further delineated into three types: (1) adult mediated, whereby the interventions are mediated by clinicians' and/or teachers' instructions, (2) peer mediated, whereby the interventions are mediated by peers, and (3) combination approaches, whereby the social skills group has peers and an adult(s) present (Paul, 2003).

Over time, social skills interventions have steadily increased, with considerable progress being made in terms of technique identification and manual development (Williams-White et al., 2007). For example, some reviews of literature noted that common techniques included modelling and reinforcement, peer-mediated intervention, reinforcement schedules, scripts and social stories (Matson et al., 2007), perspective taking, and play-based conversation skills training (Scattone, 2007). While research in this area continues, progress in determining positive outcomes for children diagnosed with ASD has been hindered by several methodological and conceptual weaknesses. For example, many articles report small or poorly characterised samples (Williams-White et al., 2007) and a lack of a common definition of social skills, with some appearing to be universally framed and others more idiosyncratically developed (Rao et al., 2008). Further, literature suggests that social skills impairments likely increase over time, which means that when a child reaches adolescence, and their social milieu becomes more complex, the issue becomes more complicated (Tantam, 2003).

Regardless of these limitations and the mixed views within the evidence-base, it is widely agreed upon that children with ASD may benefit from some type of social skills intervention. This, therefore, positions the impairment as being located within the child, with the child defined as behaving 'inappropriately' and 'abnormally' in comparison to their typically developing peers. Behavioural-based social skills therapy is thus one of the main ways by which children diagnosed with ASD are taught to orient to particular behaviours as 'problematic', while often being offered alternative patterns of behaviour to consider taking up. The interactions between therapists and children with ASD in these therapy sessions are one of the primary sites within which specific behaviours are named 'problematic' and potentially reframed by both the child and therapist (Lester & Paulus, 2014). Further, such therapy sessions are often sites in which the boundaries of 'normality' and 'abnormality' are (re)negotiated (Lester, 2014). This widespread assumption has constructed a dichotomy of normal and abnormal behaviours, attracting the attention of social constructionists and critical disabilities studies scholars who consider the limitations of such individualistic perspectives (see Brownlow & Lamont-Mills, this volume; O'Dell & Brownlow, this volume).

Normality versus abnormality

The normality/abnormality binary has been steeped within medical discourses and ways of thinking. The medical presentation of normality suggests that those who deviate from predefined notions of 'normal' have aspirations to return to this state, with the elimination of conditions focused on symptom reduction (Baker, 2011). From this viewpoint, parents of children diagnosed with ASD have been encouraged to seek treatments that promote the potential for a cure or more 'normal' behaving on the part of their child (Chamak, 2008). At least in part, this emphasis on cure and 'normal' behaving is due to the ways in which society is organised around neurotypical values, with ASD being constructed as a deficit (Brownlow, 2010). Importantly, though, the social construct of 'normal' is only feasible in comparison to something else (Lester & Paulus, 2012), with some versions of 'normal' being privileged over others (Brownlow & Lamont-Mills, this volume). In this way, normality and abnormality are mutually constituted entities, as normality is not possible without the construct of pathology or abnormality being invoked as a comparative frame (Canguilhem, 1989). An understanding of the normality/abnormality dichotomy has important implications for professional interventions, which typically focus on 'correcting' autistic behaviours and changing the social interactions and communication skills that people with ASD display.

More specifically, like many mental health disorders, ASD has been categorised as a disorder which is considered to be best understood by professionals. The majority of research focused on ASD and its associated therapeutic interventions has assumed that researchers and trained professionals know exactly what it means to behave 'appropriately' or 'inappropriately' (Nadesan, 2005). This presumption has often functioned to entitle the 'experts' to teach the person with ASD how to behave 'normally' or more similar to society (Biklen et al., 2005). Further, a common feature of many interventions has been the focus on changing people with ASD to fit the presumed norm (Brownlow, 2010). Donnellan, Hill, and Leary (2010) suggested that within the professionalisation of the 'interactions with people with autism, we have trained professionals, parents and others to interpret what happens in terms of simple, binary views of behaviour (i.e. good/bad or positive/negative)' (p. 2). As such, the practices and discourses of particular professions have served to define what constitutes normal and abnormal behaviour and to some extent this normality and abnormality are tied to definitions of appropriateness and deviance (Foucault, 1965).

Related to this, Becker (1963) noted that 'social groups create deviance by making the rules whose infraction constitutes deviance, and by applying those rules to particular people and labeling them as outsiders' (p. 9). Discourses of 'good' or 'bad' behaviour are particularly pertinent to the field of autism, as there are clear arguments from critical disabilities advocates that it is not the child who is disabled, but the society around him/her that is disabling (Oliver, 1996); that is, disability is positioned as socially constructed (Gilson & DePoy, 2000). Furthermore, particular groups of individuals diagnosed with ASD, such as Aspies, strongly advocate for the recognition of the positive aspects of the condition (Bagatell, 2007). For example, in Rosqvist's (2012) study of an educational setting designed for adults diagnosed with Asperger's, she noted that the participants expressed ambivalence towards the purposes of 'being trained', with many describing training or intervention as meaningless if it required 'that they should be someone other than who they are' (p. 5). In contrast, the participants described training focused on practical skills (e.g. learning how to park in a garage) as meaningful and functional. Rosqvist suggested that interventions should be developed in relationship to the neurodiversity movement, which has reconceptualised ASD as being more than a source of problematic symptomology.

To date, however, little research has focused on how interventions go about marking particular behaviours as socially appropriate or as unacceptable. This has been particularly true in relation to therapeutic interactions, as much of the research centred on behavioural therapy specific for children with ASD has focused on the effectiveness (or not) of a particular approach (e.g. Kenworthy, Anthony, Naiman, Cannon, Wills, Luong-Tran et al., 2014). As such, in this chapter, we consider how the social actions of children with ASD are constructed, marked, and negotiated as 'good' or 'bad' in the context of behavioural therapy. We highlight how attending to such interactions at a micro-level offers important insights to clinicians, particularly as they navigate the fragile task of celebrating the unique characteristics of children diagnosed with ASD, while also teaching them ways to more successfully interact with others (Lester, 2014).

Project overview

The data drawn upon was part of a larger ethnography focused on the everyday practices of children diagnosed with autism and their parents and therapists. Methodologically and theoretically, this discourse analysis was informed by discursive psychology (Edwards & Potter, 1992), and, to some extent, conversation analysis (Sacks, 1992). Further, we oriented to disability and ASD specifically, from a critical perspective, viewing it as always already located at the intersection of culture and biology and bound within discursive practices.

Table 18.1   Participating therapists' demographic information

Pseudonym	Professional Title	Total Years at the Site
Bria	Occupational Therapist	4
Drew	Speech Pathologist/Clinical Director	4
Jennifer	Speech Pathologist	2
Megan	Speech Pathologist/Clinical Director	4
Michelle	Teacher/Autism Specialist	4
Patricia	Physical Therapist	1
Samantha	Medical Secretary	½
Seth	Occupational Therapist	½

The setting and participants

Data was collected at The Green Room (self-selected pseudonym) – a paediatric clinic located in the Midwest region of the United States. The clinic was designed to support children diagnosed with developmental disabilities and their families. Specifically, at the time of this study, more than 80 families in a bi-state area sought support at the clinic.

The Green Room employed a total of eight therapists, all of whom participated in the study. In total, there were three speech and language pathologists, two occupational therapists, one physical therapist, one teacher/autism specialist, and one medical secretary/social skill support (see Table 18.1).

Families at The Green Room with children with ASD were also invited to participate in the study. A total of 12 families agreed to participate, resulting in the participation of 12 children with autism labels (see Table 18.2), aged 3 to 11 years, 6 fathers, and 11 mothers.

Data Sources

In the larger ethnography, data included interview data with therapists, parents, and a state advocate, relevant documents, and observational data. For the purposes of this chapter, we focused our analysis on the 175 hours of video- and audio-recordings of the therapy sessions and waiting rooms conversations. More specifically, we focused on those therapy sessions in which the *Superflex* curriculum was made relevant.

Data Analysis

Informed by discursive psychology, we completed a four-step, data analysis process. First, we engaged in intensive listening of the 175 hours of naturally occurring data (Wood & Kroger, 2000), noting specifically those segments of the talk that focused on delineating between 'good' and 'bad' behaviours in relation to *Superflex* and the *unthinkables*. Second, Lester transcribed the segments of the

Table 18.2 Relevant information for participating children

Child's Pseudonym	Age	Gender	Race	Diagnostic Label
Billy	6	Male	Caucasian	Asperger's
Chance	3	Male	Caucasian	Autism
George	4	Male	Caucasian	PDD-nos[a]
Noodle/Nancy	7	Female	Caucasian	Chromosomal Deletion/Autism
Picasso	9	Male	Caucasian	Autism
Saturn	7	Male	Native American	Autism
The Emperor/TE	7	Male	Caucasian	Autism
Thomas	5	Male	Caucasian	Autism
Tommy	7	Male	Caucasian	PDD-nos/ADHD[b]
T-Rex/TR	7	Male	Caucasian	Autism/ADHD
Diesel Weasel/DW	11	Male	Caucasian	PDD-nos/Bipolar/ OCD[c]/ADHD/ Tourette's
Will	8	Male	Caucasian	Mental Retardation/Autism[d]

[a] Pervasive developmental disorder, not otherwise specified.
[b] Attention deficit/hyperactivity disorder.
[c] Obsessive compulsive disorder.
[d] The state in which this study occurred does not yet apply the term 'Intellectually Disabled' in lieu of Mental Retardation. Will's mother indicated that his school diagnosed him as 'Mentally Retarded' and a doctor told her 'I think he has autism.'

talk (1) that focused on *Superflex* and/or the *unthinkables* and (2) that occurred within the context of a therapy session. A modified Jeffersonian transcript was produced for all segments of talk included for further analysis (Jefferson, 2004). Third, we read the synchronised transcripts and noted key, discursive patterns across the data. Fourth, we engaged in a 'more intensive' study of the patterns and began to develop an overarching understanding. Throughout, ATLAS.ti 7, a computer-assisted qualitative data analysis software package, was used, specifically the coding and memoing features.

Ethics

Prior to beginning this study, ethical approval from the institutional ethics board was acquired. Pseudonyms were used throughout to protect the anonymity of the participants.

Findings: discursively negotiating 'good' and 'bad' behaviours

Within the therapy sessions, both the therapists and children often spoke about the superhero character, *Superflex*, as capable of defeating a group of *unthinkables* – which were those characters representing social and behavioural problems. For example, the therapists often referred to 'Glassman', who represented a common unthinkable, when talking about making a small problem into a big problem that resulted in a meltdown or an 'earth shattering reaction'. Superflex, in contrast, was described as being capable of defeating Glassman and preventing him from making small problems into big problems. Glassman, then, functioned to mark the meltdown as inappropriate, while Superflex marked the appropriate or desired behaviour or reaction, providing a contrast between two dichotomous positions. Thus, with a broad focus on how behaviours were constructed as appropriate or inappropriate within this setting, the lexical items (i.e. Superflex and the unthinkables) functioned to mark what was considered an appropriate versus an inappropriate behaviour, while also positioning the child as being capable of dealing with 'inappropriate' behaviours. We shall consider these points further in the next two sections, focusing on how normality was negotiated and blame for 'bad' behaviour was achieved.

Marking 'good' and 'bad' behaviours

Across the data, we found the use of lexical items, such as Superflex or one of the unthinkables, to be a means by which the therapists and children delicately negotiated the fluid bounds between normal and abnormal behaving. Extract 1 illustrates how the therapists' employed lexical items associated with the Superflex curriculum, intertwining these lexical choices within the very doing of therapy. The extract begins just after Billy, one of the participating children, had 'calmed down' after having yelled that he wanted chewing gum and did not want to do any of the other suggested tasks. Together, he and Bria, his therapist, created a list of tasks that needed to be completed before the end of their session, with chewing gum being included on the list, as well as swinging, buttoning a shirt, tying shoes, and handwriting. Chewing gum was listed as a task that would be completed near the end of the session. This extract begins just after this reality was established.

Extract 1

```
1  Bria:   First you pick and then I get to pick (2) kay=
2  Billy:  (Climbs on to the swing) (3)
3  Bria:   (Pushes swing) thank you for being Superflex and using a
4          nice voice (1) for a second there I thought Glassman [was
5          gunna come out
6  Billy:                      Whoa] (.)
7  Bria:   You were almost being kind of rude but [then↑
8  Billy:                                            ah] (.)
9  Bria:   out comes Superflex↑
```

Bria began by inviting the child to 'pick' the next activity to be completed (line 1). This initial move was a common pattern across the data, as the therapists invited the children to 'pick' an activity first and then they would pick the next activity ('then I get to pick'). This particular pattern highlighted the institutionality of the talk, as one of The Green Room's goals (or institutional goals) was to assist children in turn-taking practices. After being invited to select the next activity, Billy responded non-verbally by climbing onto the swing (line 2). Billy, like many of the participating children, often used his body, rather than spoken words, to respond to the therapist's questions, prompts, or requests. While the non-speaking or non-verbal body has historically been associated with non-thinking or limited cognition (Borthwick & Crossley, 1999), in this case, Bria oriented to Billy's physical movement as meaningful and connected to 'Superflex' (line 3). Bria marked his movements to the swing and 'nice voice' as being associated with Superflex (line 3), a figure that they were presumably striving to embody. She then moved to offer a contrast between Superflex and one of the unthinkables – 'Glassman' (lines 3–4). This contrast set up a distinction between appropriate behaving and the behaving associated with one of the unthinkables, in this case 'Glassman'. Generally, abnormality is constructed in direct contrast to that which is defined as normal (Canguilhem, 1989). Here, Bria positioned particular behaviours with normal or preferred behaving ('nice voice') and abnormal behaving ('being kind of rude'). Further, both Glassman and Superflex were positioned as being within the child's power to control. The child, then, was positioned as having the ability and agency to shape how they behaved, with 'Superflex' marking the preferred way of being in the world.

Extract 2 provides an additional example in which Drew, a participating therapist, attempted to transition a child from the waiting room to a therapy room in the upstairs level of The Green Room.

Extract 2

```
1  Tommy:  I wanna go (.) I wanna [go↑
2  Drew:   Yeah let's] go↑ let's go=
3  Tommy:  =Can I go up TOO↑ I won't be Mean J[ean
4  Drew:                               Okay] if we have
5          any Mean Jean moments (.) you'll have to come back
6          downstairs let's go
```

Extract 2 began with the child, Tommy, making relevant his desire to go to the therapy room (line 1). The repeated request ('I wanna go') added emphasis to his request, making evident the immediacy of his request. The therapist responded by overlapping her speech with his second request (line 2), acknowledging the request. The child moved immediately to mark his (potential) behaviour with a reference to one of the unthinkables, 'Mean Jean' (line 3). Mean Jean was described as an unthinkable that was one of the most difficult nemeses of Superflex, as she often said mean things to others or took toys from friends. Interestingly, in this particular extract, it is the child who marked his behaviour and even his identity ('I won't be Mean Jean') in relationship to the unthinkable. Drew responded by indicating that 'Mean Jean moments' would result in returning to the waiting room. While the child positioned 'Mean Jean' as something he could (or could not) be, the therapist referred to Mean Jean as 'moments', implying a momentary and passing element to inappropriate behaving. Generally, across the data, the therapists positioned the presence of the unthinkables as passing acts, rather than lasting aspects of a child's personhood or way of being in the world. We found this discursive move to be particularly compelling, as it constructed the child as: (1) capable of changing their behaviour, and (2) behaviour as not static or even unitary.

Delicately accounting for 'good' or 'bad' behaviours

Across the data, the therapists and children went about accounting for the behaviours marked as good or bad. Quite often, Superflex and the unthinkables were used when offering explanations for why a bad and/or good behaviour occurred in a given context and how a bad behaviour might be changed. We viewed this finding as pertinent, as the very act of teaching someone what counts as normative or non-normative is a delicate task, particularly in that how one behaves is tightly bound to identity. More specifically, marking a child's behaviour as appropriate and/or inappropriate is potentially face-threatening, as such a move may threaten one's positive self-image (Goffman, 1967). Further, if one's behaviour is presumed to be a part of one's identity, pointing out inappropriate aspects of one's way of being may threaten the very core of who they are. As such, when dealing with delicate matters, speakers often work to discursively save face and avoid threatening the positive image of themselves and/or others. This is something that has been identified as relevant in therapeutic talk between therapists and children with ASD (Lester & Paulus, 2014). We, therefore, were interested in further examining how lexical choices functioned in the participants' accounting for good and bad behaviours, which we positioned as a delicate interactional task.

In Extract 3, a therapist, Bria, invited TR, one of the participating children to evaluate his performance during the therapy session. Just prior to this invitation, TR had completed a task (i.e. handwriting) that typically resulted in him being reminded to 'defeat Rock Brain... with Superflex'. Rock Brain was described by the therapists and participating children as 'only doing what he wants to do' and often including behaviours like 'screaming, punching, pulling hair, and kicking'.

Extract 3

```
1   Bria:  Did you do a good job↓ (1)
2   TR:    Yeah (.)
3   Bria:  Did Glassman come out↑ =
4   TR:    =No=
5   Bria:  =Did Rock Brain come out=
6   TR:    =Yes↑
7   Bria:  Rock Brain came out↑
8   TR:    No::o kidding (laughs) (.)
9   Bria:  Kay were you Superflex while we were working↑ (.)
10  TR:    Yes (.)=
11  Bria:  =O::kay↑ hop on the swing good choice man↑
```

Bria began by inviting TR to evaluate whether he did a 'good job' (line 1), which was a typical way that the therapists concluded each session or transitioned between activities. Immediately, TR took up Bria's invitation to evaluate his performance, responding with 'Yeah'. Bria prompted further, listing some of the unthinkables that may have 'come out' during the session. In this way, the therapists invited the child to account for his performance in relation to the language of the Superflex curriculum. In contrast, the child could have simply been asked, 'Were you stubborn?' (aka Rock Brain) or 'Were you angry today?' (aka Glassman). Instead, Bria couched the evaluation within the language of Superflex, producing a more indirect and perhaps delicate means by which to evaluate the session. Further, there was evidence that the child oriented to this approach as perhaps something to be dealt with playfully, as he joked that Rock Brain came out (line 7). Laughter is often used in talk to deal with sensitive or uncomfortable interactions. In fact, speakers often manage 'trouble lightly' by infusing laughter within a topic that might otherwise be a bit delicate (Jefferson, 1984, p. 351). Understandably, being asked to evaluate whether one's performance was good or bad is a sensitive interactional task. Yet, here, like the majority of the 175 hours of data we analysed, the child moved to evaluate his behaviour within the context of the Superflex curriculum, with a bit of laughter and little resistance.

While Extract 3 illustrates how children went about evaluating and accounting for their behaviours after a therapy session, Extract 4 highlights how the children were invited to provide an account of their behaviours outside of The Green Room.

Extract 4

```
1   Michelle:  Oh DW did you have any Superflex moments that you
2              wanted to put up on the wall↑ or have [I forgot to
3              ask you↑
4   DW:                                                        I
```

```
5                    don't kno::w] (7)
6    Michelle:       Not that you can recall this week (2)
7    DW:             Um well I think there was one I can't remember what
8                    it was now though=
9    Michelle:       Okay if you think of it let me know okay=
10   DW:             Oh yeah there was (.) a field trip to the bowling
11                   alley at first yeah the first day I didn't wanna go
12                   but I went to [it
13   Michelle:                                 hh uh↑] you did↑=
14   DW:             =Yeah=
15   Michelle:       =Good for you that is definitely↑ defeating who↑
16   DW:             Probably Rock Brain (.)
17   Michelle:       And↑ (.)
18   DW:             Probably Glassman or Worry Wall (.)
19   Michelle:       I think all three (.) I love it good for you↑
```

Here, Michelle began by asking DW whether he had any 'Superflex moments' (line 1). Superflex, characterised as being flexible and adaptable to whatever life brings, was constructed as something that might be experienced in a series of moments across contexts. When DW 'recalled' a Superflex moment, he shared of a field trip/outing that he did not want to go on, but 'went to' despite his hesitations (lines 10–12). In a previous therapy session, DW had been working on developing strategies (e.g. social stories, deep breathing) to assist him with participating in social outings, as he described these types of events as being 'just horrible'. In fact, coming to The Green Room was his primary activity outside of his home. Michelle's response to his description of going on 'a field trip' highlighted her surprise ('hh uh') at him leaving his house. In talk, surprise displays typically serve to establish a norm around what is expected and what falls outside of the ordinary (Wilkinson & Kitzinger, 2006). Perhaps, then, in this instance DW's delivery of the field trip news was unexpected and fell outside of the ordinary. Nonetheless, Michelle moved to affirm ('Good for you') his choice and invited him to account for his 'Superflex moment' by naming the unthinkables he defeated (line 15). DW named three unthinkables, with Michelle again affirming him again (line 19).

While Extract 4 illustrates how a therapist invited a child to give an account about an event happening outside of The Green Room, Extract 5 illustrates how the therapists invited children to explain and deal with 'bad' behaviours displayed within The Green Room. Here, TR and Bria engaged in an interaction after TR had transitioned from a social group to an individual occupational therapy session. During this transition, TR began screaming, fell to the floor, and refused to move. After he calmed down, he came to see Bria, who had witnessed the event from a distance. Bria wove the 'screaming event' into her therapy session with TR, using the event as a means to work on one of his therapy goals: developing calming strategies. Extract 5 shows some of the initial moments of the interaction.

Extract 5

```
1   Bria:  =Kay Glassman has a huge reaction to a small problem (1)
2          you have a small problem↑ and your mouth wants to yell↑
3          look at my face (points to her face with her right
4          pointer finger) what can you do
5   TR:    Calm
6   Bria:  (Starts breathing in and out)]
7   TR:    Calm calm down=
8   Bria:  =What is this called (points to her chest moving up and
9          down as she inhales and exhales)
10  TR:    Deep breaths=
11  Bria:  =Yeah you may choose deep breaths with your mouth (1) you
12         may not choose screaming with your mouth (.) it is scary
13         for your friends (.) how do you think your friends at The
14         Green Room felt↑ (1)
15  TR:    S::cared=
16  Bria:  =Probably scared (upward) I bet they were scared you were
17         <so loud> (2) we have to do a better job of calming down
18         okay↓=
19  TR:    =°°Okay°° (.)
20  Bria:  You have to have a Superflex moment (.)
```

Bria began by defining what Glassman 'is' and the impact he has (lines 1–2), while inviting TR to describe what he 'can…do' when Glassman comes. TR responded by repeating the word 'calm', while Bria simultaneously began breathing in and out. The emphasis on breathing (lines 5–9) tightly coupled Bria's non-verbal actions with 'what can you (TR) do', highlighting how to 'defeat' the unthinkable. Bria emphasised the notion of choice in relation to this strategy (line 10), positioning TR as capable of *not* screaming. Further, she offered another reason why 'screaming with your mouth' may be detrimental, locating TR's friends as being 'scared'. The therapist stated that 'we have to a do a better job', with the pronoun 'we' functioning to index some level of solidarity between Bria and TR (Ostermann, 2003). Across the data, therapists often described sensitive matters as something 'we need to deal with', rather than locating the problem as solely being the child's to solve. We oriented to these pronoun displays as functioning to construct some level of equal footing, even though institutionally Bria was 'in charge'. Similar to Michelle in Extract 4, Bria invoked the language of 'Superflex moments', making explicit the role of Superflex in changing bad behaviour and accounting for good behaviour.

Clinical relevance summary

While it is recognised that these social actions are bound within a cultural discourse of normal or appropriate behaviour, it is important to consider how this can impact the therapeutic environment. The social distance created between the 'bad' behaviour and the behaviour linked to the identity of the child is in some ways successfully created through the distant characters classed as the 'unthinkables'. This allowed the child to have some distance from their own behaviour with less threat to face, which is particularly pertinent given that children can be positioned by other family members in negative ways (Parker & O'Reilly, 2012). Nonetheless, the 'good' and 'bad' behaviours were still marked interactionally and thus brought to the attention of the child. However, by using a fictional character and maintaining social distance, the notion that the 'bad' behaviours were flexible was left open, suggesting that with therapeutic assistance the child was open to change the way in which they behaved.

Arguably therapists face a challenging task in shaping behaviours and teaching children with ASD appropriate social skills, so that they may interact more successfully outside of the therapeutic arena. This is a delicate interactional achievement and our analysis illustrates that therapists work in ways that orient to the identity and face of the children in their care. The lexical choices have a face-saving function, as marking a child's behaviour as directly inappropriate has the potential to threaten the child's sense of identity. By inviting children to provide an account of their behaviours through the use of characters, the children were able to identify their own behaviours as appropriate or inappropriate, which have potential to lead to a better understanding of social rules. For a simple summary of the implications for practice, see Table 18.3.

Summary

In this chapter, we highlighted how lexical items used in therapy marked behaviours as good or bad. These lexical choices were evoked when children with ASD accounted for their behaviours. Such lexical items functioned to create distance between the child's identity and the inappropriate behaviour in particular.

Table 18.3   Clinical practice highlights

1. Therapists need to be mindful of the child's sense of self and personal identity.
2. Therapists can create social distance between the 'bad' behaviour of the child's identity through the use of characters.
3. Creating distance between the child and their behaviour is less face-threatening.
4. Maintaining the child's positive identity in light of 'bad' or 'inappropriate' behaviour is a delicate discursive achievement.

References

American Psychiatric Association (2014). *Diagnostic and statistical manual of mental disorders* (5th edition). US: APA.

Asperger, H. (1944). 'Die "Autistischen Psychopathen" im Kindesalter [Autistic psychopaths in childhood]' (in German). *Archiv für Psychiatrie und Nervenkrankheiten, 117,* 76–136.

Attwood, T. (2000). Strategies for improving the social integration of children with Asperger Syndrome. *Autism, 4,* 85–100.

Bagatell, N. (2007). Orchestrating voices: Autism, identity and the power of discourse. *Disability & Society, 22*(4), 413–426.

Baker, D. (2011). *The politics of neurodiversity: Why public policy matters.* Boulder, CO: Lynne Rienner.

Becker, H. (1963). *Outsiders: Studies in the sociology of deviance.* New York: Free Press.

Beidel, D., Turner, S., & Morris, T. (2000). Behavioral treatment of childhood social phobia. *Journal of Consulting and Clinical Psychology, 68*(6), 1072–1080.

Bettelheim, B. (1967). *The empty fortress.* New York: Free Press.

Biklen, D., Attfield, R., Bissonnette, L., Blackman, L., Burke, J., Frugone, A., Mukhopadhyay, R. R., & Rubin, S. (2005). *Autism and the myth of the person alone.* New York: New York University Press.

Blum, L. (2007). Mother-blame in the prozac nation: Raising kids with invisible disabilities. *Gender and Society, 21*(2), 202–226.

Borthwick, C., & Crossley, R. (1999). Language and retardation: Target article on language retardation. *Psychology, 10*(38), Article 1.

Brownlow, C. (2010). Presenting the self: Negotiating a label of autism. *Journal of Intellectual & Developmental Disability, 35*(1), 14–21.

Canguilhem, G. (1989). *The normal and the pathological.* Brooklyn, NY: Zone Books.

Chamak, B. (2008). Autism and social movements: French parents' associations and international autistic individuals' organizations. *Sociology of Health and Illness, 30*(1), 76–96.

Donnellan, A. M., Hill, D. A., & Leary, M. R. (2010). Rethinking autism: Implications of sensory and movement differences. *Disability Studies Quarterly, 30*(1), 1–23.

Edwards, D., & Potter, J. (1992). *Discursive psychology.* London: Sage.

Foucault, M. (1965). *Madness and civilization: A history of insanity in the age of reason.* New York: Random House.

Gilson, S., & DePoy, E. (2000). Multiculturalism and disability: A critical perspective. *Disability and Society, 15*(2), 207–218.

Goffman, E. (1967). *Interaction ritual: Essays on face-to-face behavior.* New York: Anchor Books.

Hartup, W. (1989). Social relationships and their developmental significance. *American Psychologist, 44*(2), 120–126.

Jefferson, G. (1984). On the organisation of laughter in talk about troubles. In J. M. Atkinson & J. Heritage (Eds.), *Structures of social action: Studies in conversation analysis* (pp. 346–369). Cambridge: Cambridge University Press.

Jefferson, G. (2004). Glossary of transcript symbols with an introduction. In G. H. Lerner (Ed.), *Conversation analysis: Studies from the first generation* (pp. 13–31). Amsterdam: John Benjamins.

Kanner, L. (1943). Autistic disturbances of affective contact. *Nervous Child, 2,* 217–250.

Karim, K., Ali, A., & O'Reilly, M. (2014). *A practical guide to mental health problems in children with autistic spectrum disorder: It's not just their autism!* London: Jessica Kingsley Publishers.

Karim, K., Cook, L., & O'Reilly, M. (2012). Diagnosing autistic spectrum disorder in the age of austerity. *Child: Care, Health and Development,* doi:10.1111/j.1365-2214.2012.01410.x

Kenworthy, L., Anthony, L. G., Naiman, D. Q., Cannon, L., Wills, M. C., Luong-Tran, C., ... & Wallace, G. L. (2014). Randomized controlled effectiveness trial of executive function intervention for children on the autism spectrum. *The Journal of Child Psychology and Psychiatry, 55*(4), 374–383.

Lester, J. N. (2012). A discourse analysis of parents' talk around their children's autism labels. *Disability Studies Quarterly, 32*(4), Article 1.

Lester, J. N., & Paulus, T. M. (2012). Performative acts of autism. *Discourse & Society, 12*(3), 259–273.

Lester, J. N., & Paulus, T. M. (2014). 'That teacher takes everything badly': Discursively reframing non-normative behaviors in therapy sessions. *International Journal of Qualitative Studies in Education, 27*(5), 641–666.

Madrigal, S., & Winner, M. G. (2008). *Superflex: A superhero social thinking curriculum.* San Jose, CA: Think Social Publishing, Inc.

Matson, J., Matson, M., & Rivet, T. (2007). Social-skills treatments for children with autism spectrum disorders: An overview. *Behavior Modification, 31*(5), 682–707.

Muskett, T., Body, R., & Perkins, M. (2013). A discursive psychology critique of semantic verbal fluency assessment and its interpretation. *Theory and Psychology, 23*(2), 205–226.Nadesan, M. H. (2005). *Constructing autism: Unraveling the 'truth' and understanding the social.* New York: Routledge.

Oliver, M. (1996). A sociology of disability or a disablist sociology? In L. Barton (Ed.), *Disability and society* (pp. 18–42). London: Longman.

O'Reilly, M., Karim, K., & Lester, J. N. (in press). Separating 'emotion' from 'the science': Exploring the perceived value of information for parents and carers of children with ASD. *Clinical Child Psychology and Psychiatry,*doi:10.1177/1359104514530735.

Ostermann, A. C. (2003). Localizing power and solidarity: Pronoun alternation at an all-female police station and a feminist crisis intervention center in Brazil. *Language in Society, 32,* 251–281.

Paul, R. (2003). Promoting social communication in high functioning individuals with autism spectrum disorders. *Child and Adolescent Psychiatric Clinics of North America, 12,* 87–106.

Parker, N., & O'Reilly, M. (2012). 'Gossiping' as a social action in family therapy: Thepseudo-absence and pseudo-presence of children. *Discourse Studies, 14*(4), 1–19.

Phelan, J. (2005). Geneticization of deviant behavior and consequences for stigma: The case of mental illness. *Journal of Health and Social Behavior, 46*(4), 307–322.

Rao, P., Beidel, D., & Murray, M. (2008). Social skills interventions for children with Asperger's Syndrome or high-functioning autism: A review and recommendations. *Journal of Autism and Developmental Disorders, 38,* 353–361.

Rosqvist, H. B. (2012). Practice, practice: Notions of adaptation and normality among adults with Asperger Syndrome. *Disability Studies Quarterly, 32*(2), Article 4.

Sacks, H. (1992). *Lectures on conversation.* Oxford: Blackwell.

Scattone, D. (2007). Social skills interventions for children with autism. *Psychology in the Schools, 44*(7), 717–726.

Tantam, D. (2003). The challenge of adolescents and adults with Asperger Syndrome. *Child and Adolescent Psychiatric Clinics of North America, 12*(1), 143–163.

Williams-White, S., Keonig, K., & Scahill, L. (2007). Social skills development in children with autism spectrum disorders: A review of the intervention research. *Journal of Autism Developmental Disorders, 37*(10), 1858–1868.

Wing, L. (1981). Language, social and cognitive impairments in autism and severe mental retardation. *Journal of Autism and Developmental Disorders, 11*(1), 31–44.

Wood, L. A., & Kroger, R. O. (2000). *Doing discourse analysis: Methods for studying action in talk and text.* Thousand Oaks, CA: Sage.

Recommended reading

- Brownlow, C. (2010). Presenting the self: Negotiating a label of autism. *Journal of Intellectual & Developmental Disability, 35*(1), 14–21.
- Karim, K., Ali, A., & O'Reilly, M. (2014). *A practical guide to mental health problems in children with autistic spectrum disorder: It's not just their autism!* London: Jessica Kingsley Publishers.
- Lester, J. N. (2012). A discourse analysis of parents' talk around their children's autism labels. *Disability Studies Quarterly, 32*(4), Article 1.

19
Name-Calling by a Child with Asperger's Syndrome

Johanna Rendle-Short, Ray Wilkinson, and Susan Danby

Introduction

This chapter focuses on one aspect of social interaction that is directly relevant to maintaining friendship, mental health and well-being, and supportive peer relations. The single case study is of a 10-year-old child diagnosed with Asperger's syndrome and her use of derogatory address terms, part of a wider pattern of behaviour evident in this child's interaction that resulted in behaviour that might be thought of as impolite or lacking in restraint. Analysis of these derogatory naming practices throws light on how conversational participants pursue affiliation and intimacy from a perspective of *language as action*. The chapter contributes to understandings of the difficulty in pinpointing, with precision and with clear evidence, what counts as a 'social interaction difficulty' due the context-specific nature of interaction. This chapter responds to the call in the literature on children with Asperger's syndrome for finer micro-level analysis of social behaviours in naturalistic settings with familiar peers (Macintosh & Dissanayake, 2006; Sterponi, de Kirby, & Shankey, Chapter 15, this volume).

Asperger's syndrome

It is well-documented that children diagnosed with Asperger's syndrome find it difficult to interact with adults and other children (e.g. Attwood, 2000; Fine, Bartolucci, Szatmari, & Ginsberg, 1994; Gillberg & Gillberg, 1989; Minshew, Goldstein, & Siegel, 1995; Tager-Flusberg & Anderson, 1991; Wing, 1981) and that they find it difficult to make and keep friends (e.g. Bauminger & Kasari, 2000; Bauminger, Shulman, & Agam, 2003; Humphrey & Symes, 2011). Yet social interaction is crucial for developing peer relationships and forming friendships (Corsaro, 1985; Danby & Baker, 2000; Erwin, 1993; Evaldsson, 2007; Humphrey & Symes, 2011; Margalit, 1994).

Although children with Asperger's syndrome might *want* to make friends, they tend to approach other children in a clumsy and not very successful way (Prior et al., 1998); they have more limited social networks (Chamberlain, Kasari, & Rotheram-Fuller, 2007); and they are not skilled at talking about things in common (Bauminger & Kasari, 2000). Koning and Magill-Evans (2001) showed that nearly half a group of adolescent boys diagnosed with Asperger's syndrome reported having no friends and a study by Macintosh and Dissanayake (2006) reported that, in comparison to typically developing children, children diagnosed with high functioning autism and Asperger's disorder were significantly less likely to interact socially with their peers. Yet, friendship is identified as having immediate benefit for school adjustment, improving self-esteem and personal well-being, reducing loneliness and the possibility of depression (Diehl, Lemerise, Caverly, Ramsayn, & Roberts, 1998; Dunn, 2004; Erwin, 1993; Margalit, 1994).

To date, there has been little detailed analysis of the exact nature of the social interaction difficulties experienced by children with Asperger's syndrome (see however, Kremer-Sadlik, 2004; Ochs & Solomon, 2004; Ochs, Kremer-Sadlik, Gainer Sirota, & Solomon, 2004; Rendle-Short, 2003, 2014; Sterponi & Fasulo, 2010; Stribling, Rae, & Dickerson, 2007; Volden, 2004; Wootton, 2003), partly due to the context-specific nature of interaction. As argued by Muskett, Perkins, Clegg, and Body (2010), when analysing inflexibility in the case of an eight-year-old child with a diagnosis of Autism Spectrum Disorder, what can be interpreted as inflexible behaviour in one context may not appear inflexible or may pass without comment in another context. Behaviours are not inherently problematic; they are only interpreted as such by the participants themselves (Rendle-Short, Cobb-Moore, & Danby, 2014).

Address terms and name-calling

Address terms are a useful way of indicating who the current turn at talk is being addressed to, as well as having the possible effect of selecting next speaker (Lerner, 2003; Sacks, Schegloff, & Jefferson, 1974). In addition, use of address terms is a way of showing personal concern or stance (either positive or negative) towards the addressee (Butler, Danby, & Emmison, 2010; Lerner, 2003; Wootton, 1981). Butler et al. (2010) have emphasised the interconnectedness of address terms through their analysis of names within the context of telephone counselling. They demonstrated how personal names can be used to build and maintain rapport and trust at precisely the moment at which the counsellor might be challenging the client's version of events. Through the use of personal names, the counsellor is simultaneously able to highlight the relevance of what is about to be said while also mitigating the disagreeing or challenging action. As these studies show, address terms interconnect at a number of levels.

Naming practices require that the speaker selects an appropriate name (or other formulation of the addressee) and decides when and where that particular name should be interactionally positioned, both within the talk as a whole and within a particular turn at talk. Choice of what to say and where to put the address term provides information concerning the social relations between the two interactants, particularly about the relative status of the speaker with respect to the person being addressed (Lerner, 2003; Schegloff, 2004, p. 64). As Poynton (1985, p. 80) stated, the issue of address 'undoubtedly provides the most elaborate resource for linguistic realisation of social relations'.

Sequentially, address terms can occur in any position within the talk – at the beginning of a turn, in the middle of a turn, or at the end of a turn. Taking account of where names are sequentially positioned within turns at talk provides information as to how they should be heard and responded to. For example, address terms at the beginning of a turn at talk or TCU (Turn Construction Unit) can either function as means for getting the attention of the recipient, particularly in multi-party interactions (Lerner, 2003), or as a means for demonstrating a disjunct between the current and prior talk (Clayman, 2010; Rendle-Short, 2011). In contexts where address terms are not directly doing recipiency work, such as in dyadic interactional contexts, address terms tend to occur at the end of a turn or TCU (Lerner, 2003; Rendle-Short, 2007). In this position, they can be used to demonstrate a particular stance or relationship towards the other person (Lerner, 2003). In the extracts under discussion in this chapter, the address terms tend to occur at the end of the TCU.

Project overview

This chapter explores selected extracts from collected video-recorded data of 81 minutes of Sarah (pseudonym), a 10-year-old child diagnosed with Asperger's syndrome, interacting with a friend and with family members, including her mother and brother. The video-recording was collected as part of a larger study analysing talk-in-interaction of children diagnosed with Asperger's syndrome aged 8–12 years. Sarah had been diagnosed with Asperger's syndrome, in accordance with the *Diagnostic and Statistical Manual of Mental Disorders* (American Psychiatric Association, 2000, *DSM-IV*, 4th edition), two years prior to this recording (and we acknowledge that since that diagnosis a more recent edition of *DSM-5* has been produced, which has implications for the use of the term 'Asperger's syndrome'). Her mother arranged the video-recording between Sarah and a similar-aged friend, Ellie (pseudonym). Sarah's mother reported that Sarah found it difficult to make and keep friends, and that these two girls no longer were friends a week following the afternoon tea visit. All participants gave consent and were aware of the camera.

The analysis is grounded in the fundamental conversation analytic notion that people exhibit, in the design and timing of their own talk and conduct, their understanding and treatment of the others' prior talk and conduct (Schegloff, 1992). Conversation analysis is a qualitative micro-interactional methodology that focuses on talk-in-interaction (Schegloff, 1988), and specifically on what the talk is *doing*. It analyses the manner in which participants design and understand talk in terms of the question, 'Why that now?' or 'Why did that speaker say that particular thing in that way now?'

Analysis

During the recorded interactions, Sarah repeatedly engaged in name-calling. These instances included category terms such as 'bilkhead' used in relation to her younger brother ('Jack, you're a bilkhead'). Here we focus on one subset of name-calling behaviours by Sarah: the use of address terms. The address terms we discuss can be thought of broadly as derogatory since they are either clearly hearable that way in terms of their semantics (e.g. the use of 'poohface' in Extract 4) or because they are treated in that manner by the addressed recipient (e.g. Sarah's mother's reaction to Sarah addressing her as 'mimmy' instead of 'mummy' in Extract 6). While Sarah's use of an addressee's first name (e.g. 'Jack') or kinship name (e.g. 'mum') is the unmarked usage, these derogatory address terms are marked forms, making them noticeable to other participants in terms of what interactional work Sarah is doing by using this form of address. As such these address forms stand out from the rest of the TCU in a way in which an unmarked form would not and can lead to other participants orienting to them in ways that are either affiliative (such as laughing) or disaffiliative (such as rebuking Sarah for using them).

Sarah's name-calling is part of a wider pattern of behaviour evident in these interactions whereby she engages in behaviour that might be thought of as impolite or lacking in restraint. For example, she regularly shouts or uses a very loud voice (e.g. to call her mother or reprimand her younger brother) and also on several occasions she burps loudly, apparently on purpose and for effect. These behaviours are evident in some of the examples below.

We cannot show from this single case study that the pattern of name-calling behaviour by Sarah described here is directly linked to her diagnosis as someone with Asperger's syndrome. This pattern of behaviour by Sarah, however, does tie in with a well-reported feature of Asperger's syndrome and autism, namely difficulties with social interaction and the interpersonal relationships and friendships that are created, maintained, and renewed through the medium of social interaction and conversation (e.g. Attwood, 2000; Gillberg & Gillberg, 1989; Tager-Flusberg & Anderson, 1991; Wing, 1981). It may be that Sarah uses these particular types of address term in an attempt to joke, and affiliate, with those around her in these interactions, in particular Ellie who is the overhearing audience for Sarah's name-calling in relation to her brother and her mother. If this is the case, it does not appear to be successful. As we display in the subsequent analysis, in a number of instances Sarah's use of address terms is not affiliated with by the addressed recipient or by other participants (including, on some occasions, Ellie) who are not the addressee but who instead constitute an 'overhearing audience'. In some cases, Sarah's use of derogatory address terms even leads to an explicit rebuke. Also, if the use of these address terms was intended to increase affiliation and friendship with Ellie, this did not appear to be successful as Ellie did not return to visit Sarah after this first visit.

In the analysis of the examples below, we discuss what types of action these address terms occur as part of, and where within turns the address terms typically occur. The examples are grouped broadly into three categories to reflect how other participants respond to Sarah's use of derogatory address terms and the actions of which they are part. In the first group of instances seen in Extract 1, the derogatory address term 'dum dum' is used by both girls as part of a game and Sarah's use of the address term is responded to in a generally affiliative manner by Ellie. In the second group of examples (Extracts 2–4), Sarah's use of the address term is not acknowledged by the other participants. In the final set of examples (Extracts 5 and 6), Sarah's use of a derogatory address term is explicitly challenged by the addressee (Sarah's mother).

Affiliating with the use of the address term

The derogatory address terms used by Sarah (and by Ellie on the occasions she uses them) occur most typically as 'post-positioned' address terms (Lerner, 2003); they are produced at the end of the TCU in which they occur. As such, they are not here primarily doing work of identifying who the addressee is but rather carrying out other interactional work, such as adding a certain stance to the action being produced or, as in Extract 1, contributing to actions that constitute the activity underway as a game.

In Extract 1, Sarah uses the derogatory address term 'dum dum' as part of a game that involves the girls acting stupid (or 'dumb'). Ellie responds affiliatively to the use of this term, in particular by adopting the address term and reciprocally using it back to Sarah (lines 20, 26, 33–34).

Extract 1

```
1    Sarah        HEY LOOK it's a WALL to bang
2         →       in to <dum dum>¿ ((gazes at Ellie))
3    Ellie        m(h)eh-
4    Sarah        heh
5    Ellie        ↑hya,
6                 (0.5) moonwalk. ((Ellie moonwalks))
```

```
7                    (0.1) ((Ellie bumps into the wall))
8                    AU::GH:::::::::: ((Ellie falls back))
9     Sarah          ↑huh huh [huh
10    Ellie                  [huhuhu[h
11    Sarah                        [heheh
12    Ellie          u::::::::gh, ((Ellie falls back onto the mat))
13    Sarah  →       <Was that f::un, dum dum?>=
14    Ellie          =↑fyooo
15    Sarah          he he [he
16    Ellie                [he he hh
17    Sarah          ↑wo::w, do you want to do it again?
18                   ((Ellie walks into wall again, laughter))
19    Sarah          <no::w I'll sho:w you how to do [it-
20    Ellie  →                                      [HELLO DUM DUM.
21                   0.5)
22    Sarah  →       YEAH I'M DUM DUM.
23                   (2.0) ((mutual gaze, S. patting her chest))
24    Ellie          he he he
25    Sarah          (1.0) ((acting 'dumb' with finger on lip))
26    Ellie  →       you look really dum dum.
27    Sarah          o:h (that's not) very nice.
28                   ((lifts dress, dances))
29                   ((runs into wall, laughter, E claps flippers))
30                   that ws f:::un. (pe:ace.)
31                   ((more joint laughter))
32                   a:::::ow. ((mock crying))
33    Ellie  →       why do you keep banging on- (.) into
34                   the wall dum dum.
35    Sarah  →       because I'm the <dum dum>.
36                   ((play a clapping game together))
37                   ((Sarah walks into wall again))
38    Sarah          heheh. now I'm-=
39    Ellie  →       =du:m du:m here you go.
40    Sarah          eh eh e:::h
41                   (1.0)
42    Ellie          ↑ooh look I've got some flippers on
```

This extract shows both girls displaying enjoyment of the non-proper 'dum dum' name through laughter (see Jefferson, 1979; Jefferson, Sacks, & Schegloff, 1987) and reciprocal use (see Goodwin, 1990 and 2007, on how children use format tying to build playful exchanges). Although 'dum dum' is hearable, at face value, as an insult or possibly a mock-insult, it elicits affiliative responses from Ellie. There is nothing in the interaction to demonstrate that the name is treated negatively. Ellie provides an affiliative response to the name and the actions produced by the turns-at-talk of which the name is a part.

Drew (1987, p. 232) argues that teasing is built in such a way that it is recognisable as an exaggerated version of an action or in direct contrast to something both participants know. In this example, the girls demonstrate their understanding that the proposition is not to be taken seriously as a true statement and reflection of reality. Exaggeration is used to demonstrate contrastiveness with the surrounding talk. Sarah says 'dum dum' (line 1) in a clear well-enunciated voice. Even though the 'dum dum' form is not totally new, because Ellie had previously said 'me dumb' (not shown on transcript) about 40 seconds earlier introducing into the talk the negative assessment that the address term picks up, they both treat this exaggerated address term as funny, and it achieves the desired effect of getting Ellie to bang into the wall. This playing with the address term leads to joint laughter, creating intimacy, and affiliation.

Not responding to the use of the address term

While it is evident from Extract 1 that a derogatory address term can be a part of the turn that is oriented to by another participant within their response, in Extracts 2, 3, and 4, Sarah's use of a derogatory address term as part of her action is not responded to by others. In Extracts 2 and 3, this lack of affiliation is notable in that Sarah's use of the address term is an integral part of a turn in which she is continuing the 'dum dum' game, but Ellie neither affiliates with the address term (e.g. by using it in her response) nor appears to go along with the playing of the 'dum dum' game.

Extract 2

```
55      Ellie       ♫ uh (.) uh (.) uh uh- (whoa,)
56      Ellie       urrgh.
57                  (1.0)
58                  °that hurts° ((falls down, rolls towards S))
59                  (5.0)
60      Sarah       ((burps))
61          →       hello prize dum dum¿ eheh,
62      Ellie       ee:::: [::::::ew]
63      Sarah              [eheheh ]
64      Ellie       °how cn you keep on doing the same thing.°
65                  I want to dance with the teddy bear.
66                  (0.5)
67                  °↑ooh. he's got big eyebrows, ↑yip.°
68      Sarah       ((loud burp)) uh huh[ huh, eh-     ]
69      Ellie →                         [↓oh dum: dum:.]
70                  ♫ tuh duh du:: tu du du::
71                  ((Ellie dances with teddy while Sarah claps
                    flippers in time))
```

After laughter, Ellie says 'whoa', 'urrgh', '°that hurts°' (lines 55–58), clutches her stomach and rolls towards Sarah. But Sarah does not demonstrate any understanding that Ellie may have hurt herself, possibly requiring consolation, or that the game is over. Instead, Sarah continues clapping with her flippers.

During the five-second pause (line 59), they maintain mutual gaze, with frozen movements. The gaze continues as Sarah burps and says, 'h<u>e</u>llo prize dum dum¿' (line 61) followed by laughter. But Ellie does not respond affiliatively to this invitation to laugh (Jefferson, 1979) or to play along with the 'dum dum' game. Instead, she groans (line 62), presumably in response to Sarah's burp. The disaffiliation continues as Ellie complains, 'how cn you keep on doing (sm) same thing' (line 64). Sarah burps a second time (line 68), to which Ellie responds with 'oh dum dum' (line 69). This turn, in response to Sarah's burp, again shows clear lack of affiliation. The address term is no longer a fun playful name; instead, it is said in response to the burp and a misunderstanding that the current activity is over. The intended negative characterisation of Sarah by Ellie is evident. Ellie immediately starts a new game with teddy in line 70.

A few lines later in the interaction Sarah uses a final 'dum dum' address term. This time, the address term is not acknowledged by Ellie.

Extract 3

```
83    Ellie       what's th<u>a</u>t.
84                (1.5)
85    Ellie       (I wanna              )
86    Sarah →     that's part of a sc<u>ra</u>tching post du[m dum.]
87    Ellie                                        [YE::::]AH.
88                (3.0)
89    Sarah       ↑HOW DO YA PAUSE THIS THING MUM?
```

The video shows that Ellie's 'what's that' (line 83) is a 'real' question, asking for information. Sarah, however, attaches a post-positioned 'dum dum' to her answer as a way of continuing the name-calling game. That Sarah may have got the tone wrong and not picked up on the shift in activity is evident in Ellie's loud elongated 'YE::::::AH'. This non-aligned single-word response (line 87) closes the sequence and the game. A minute later, Sarah asks if she can turn the recording off.

Extract 4 occurs in the kitchen with Sarah and Ellie having afternoon tea; Sarah's mother and younger brother are nearby. Sarah calls out to her brother, using the derogatory address term 'poohface' as part of a directive/rebuke to him. While it is not possible to know how Jack, the addressee of Sarah's turn, reacts, since he is off-camera, there is no verbal response by either him or Ellie to the action and the address term.

Extract 4

```
1     Sarah       we're n<u>o</u>t laughing.
2                 (0.9)
3     Sarah       we're <u>vo</u>miting.
4                 (2.5) ((Ellie mimes 'vomiting' very quietly))
```

```
5     Jack         eeyy:::::uh, ↑yuk, banana.
6                  (1.5)
7     Jack         [↑yeoo:::::::oh, ]
8     Sarah →      [↑DON'T TOUCH IT] POOHFACE.
9     Ellie →      what the hell is that,
10    Sarah →      fig or something.
11                 (1.5) ((E. gazes at melon in hand, then eats it))
12    Sarah        ↑NO::: JAA::CK,
13                 (1.0)
14    Sarah        PUT IT BACK. °hm°
15                 (0.8)
16    Jack         why.
```

Although the name-calling (line 8) immediately follows an affiliative moment with Sarah and Ellie laughing about sharing a seat (not shown on the transcript) and Ellie pretending to vomit (lines 1–3), there is no response to Sarah's '↑DON'T TOUCH IT POOHFACE' (line 8). Instead, Ellie asks a new question about a fig (line 9), before continuing to eat her afternoon tea. Even though Ellie could have responded affiliatively, by for example laughing, this does not occur. Thus, the extreme name-calling (both in terms of volume and its negative connotations) passes without comment or laughter by either Ellie or the intended recipient, Jack. When Sarah repeats the shouted instruction, she uses his real name, Jack (line 12), rather than 'poohface'.

Disaffiliation through challenging the use of the address term

In Extracts 2–4, Ellie's lack of response to Sarah's use of a derogatory address term as part of her turn is potentially hearable as an implicit lack of affiliation with Sarah. In Extracts 5 and 6, however, disaffiliation is made explicit in that Sarah's use of a derogatory address terms is challenged. In both cases, it is Sarah's mother who is the addressee and who then explicitly challenges Sarah's use of the address term. In both cases, the mother's first response to Sarah's turn is an other-initiation of repair ('I beg your pardon' in Extract 5, and 'what'd you call me?' in Extract 6). However, each clearly act as a harbinger of disagreement or disaffiliation with what Sarah has just said (cf. Schegloff, Jefferson, & Sacks, 1977) and indeed this is what follows in each case.

Sarah's action in line 2 of Extract 5 is hearable as a (mock) complaint to her mother. Just prior to this, Sarah had said that she is hungry and her turn in line 2 addresses the fact that this implicit request for food has not yet been acted on by mother, with her mother suggesting instead that Sarah have a big glass of water. This complaint also contains the derogatory address term 'doofus'.

Extract 5[1]

```
1   Mother       have a big glass of water.
2   Sarah   →    I said I'm hungry. doofus-,
```

```
3    Mother →    I beg your pardon?
4           →    there y' stopped yourself halfway there.
5           →    cos you realised what you're saying. didn't you.
6    Sarah       (I          ).
7                I said doofus.
8    Mother →    yeah. that's not very nice is it.
```

The mother responds to the derogatory 'doofus' by saying 'I beg your pardon?' (line 2) before providing a candidate explanation as to why Sarah did not continue talking (lines 4 and 5). Sarah restates the derogatory address term (line 7) either as a literal response to non-understanding or as a challenge, to ensure that the name is heard in the clear. Either way, her mother responds disaffiliatively, treating the name as 'not very nice' rather than funny (line 8).

The following extract is again a multiparty interaction with multiple possible audiences. Sarah addresses her mother with 'Mimmy'. Her mother, as the intended recipient, topicalises the address term (as was also seen in Extract 5); her friend, Ellie, as part of the overhearing audience, laughs.

Extract 6

```
1    Mother   Sarah? ((Mother holds knife out to Sarah))
2             (0.5) ((Sarah takes knife))
3    Sarah    thank you Mimmy. ((off camera))
4    Ellie    $°fhfh°$ ((Ellie brief gaze toward Sarah))
5    Mother   what'd you call me?
6    Sarah    [↑MIMMY.] Huh huh huh .hh=
7    Ellie    [°mimmy° ]
8    Sarah    ↑huh huh [huh huh,
9    Mother            [you called me Mimmy,
10            (1.5) ((Ellie gazes toward Mother))
11   Mother   [well maybe you shou-]
12   Sarah    [↑↑(thank God) ]I call you Mimmy [you don't listen,]
13   Mother                           [Jack. [scuse me.  ]
14   Jack                              [ can we      ]
15            ↑open [that.
```

'What'd you call me' (line 5) topicalises the address term used in Sarah's turn and is hearable as a pre-disagreement or pre-complaint (Schegloff, 2007) and is oriented to as such by Sarah as shown through her post-positioned laughter. Although the laughter (lines 6 and 8) opens up a space for a display of affiliation by her mother (see Walker, 2013), neither her mother nor Ellie laugh in response. Instead, her mother demonstrates non-alignment by overlapping Sarah's laughter with 'you called me Mimmy' (line 9) with stress on the name 'Mimmy' and slight rising intonation. By speaking after the onset of laughter, mother declines a post completion invitation to laugh affiliatively (see Jefferson, 1979, p. 93). Her statement 'you called me Mimmy' (line 9) is complaint-implicative, but the explicit directive/rebuke which she starts to produce in line 11 is cut off before completion as she turns to deal with her other child, Jack (line 13).

Discussion

We have examined the use of derogatory address terms as part of actions produced by a child with Asperger's syndrome. While we cannot show that this behaviour is linked to the child's diagnosis as someone with Asperger's syndrome, the pattern of behaviour analysed here does fit with other research that shows children with Asperger's syndrome regularly produce behaviour that is poorly designed for the particular recipients they are addressing and which leads these children to have difficulties overall with social interaction and the relationships constituted through social interaction, as outlined above.

The present study highlights some of the unpredictable and precarious aspects of spontaneous interaction for children diagnosed with Asperger's syndrome. Derogatory address terms can be an extreme form of behaviour, linguistically marked within interaction possibly leading to displays of 'heightened affect' (Goodwin & Goodwin, 2000, p. 7) such as (joint) laughter. They are interactionally risky both in terms of how they are set up and in terms of their potential affiliative or disaffiliative responses. An interactant, for example, has to consider where to utilise derogatory names; how to position them sequentially within the talk; how turn design, laughter or gaze might affect the likelihood of an affiliative or disaffiliative response; how to balance the tension between play acting, having fun and real life. These actions are an example of the link between communication practices that support affiliative social relations.

The analysis is not prescriptively arguing that children diagnosed with Asperger's syndrome should not use derogatory address terms. As Extract 1 demonstrates, they can, within certain contexts such as a game, be very effective on occasion in terms of achieving affiliation. Use of the 'dum dum' name opened up opportunities for collaborative laughter, fun and intimacy, even though the address terms were potentially derogatory. In this instance, it enabled Sarah to build rapport and affiliation with her friend. The challenge for Sarah was in recognising when the 'dum dum' game had run its course.

Addressing her brother and mother (Extracts 4–6) was less effective in terms of encouraging rapport and affiliation, either with them or with Ellie. The addressees either ignored her turn with its post-positioned derogatory address term or produced disaffiliative responses (Extracts 5 and 6). It has already been shown that children diagnosed with high functioning autism are less likely to initiate first turns compared to neuro-typical children of a similar age (Jones &

Schwartz, 2008). This may be due in part to the unpredictability of the response due to lack of interactional structure and scaffolding (Rendle-Short, 2014). So not being quite sure of whether an address term will be treated as serious or funny, whether it will result in laughter or not, makes it interactionally difficult for children who may not fully draw on the nuances of interaction.

An additional difficulty in understanding how children diagnosed with Asperger's syndrome develop friendships and social relationships is the context-specific nature of interaction. What counts as 'social interaction difficulties' in one context may not appear as a difficulty in another context. Sarah's strategy of name-calling was varyingly successful, depending on how the talk was packaged, where the talk occurred within the larger interaction, whether it was in the context of multiparty interaction. This means that a deficit model in which behaviours that directly arise from the specific condition are interpreted as problematic should be treated with caution. Behaviours are not inherently problematic; they are only interpreted as such by the participants themselves (Rendle-Short, Cobb-Moore, & Danby, 2014).

Educational relevance summary

The analysis highlights three key points. First, there is an urgent need for better understanding of what occurs socially when children interact with each other, in this instance those diagnosed with Asperger's syndrome. As shown in the analysis of the context-specific interactions, a particular behaviour (such as use of derogatory address terms) may sometimes result in an affiliative response but at other times may result in a disaffiliative response. As such, it does not seem possible to say unequivocally that the use of such a term by Sarah is interactionally problematic in general.

The second point is that, due to the link between social skills deficits and higher anxiety (see Bellini, 2004), there is an urgent need for empirical research to inform social skills intervention programmes. Meta-analytic reviews by Rao, Beidel, and Murray (2008) and Williams-White, Keonig, and Scahill (2007) demonstrate limited empirical support for social skills intervention programmes, despite being widely used, and despite research evidence showing that children diagnosed with High Functioning Autism and Asperger's syndrome are receptive to social skills training (Owens, Granader, Humphrey, & Baron-Cohen, 2008). Evidence-based research is necessary to inform professional practice and training for social skill interventions for children diagnosed with Asperger's syndrome.

These two points lead to the third point. Any social skills intervention programme requires a focus on the local management of interaction rather than on learning macro-skills, such as how to ask questions or what counts as an appropriate response to a question. One possible innovative approach is to encourage professionals, parents, and children to take account of the context-specific nature of interaction. Such an approach would involve teaching children how conversations work, particularly in terms of enabling them to understand how the prior talk informs subsequent talk. In this way, they would become their own mini-analyst with tools for understanding how an address term, for example, can result in either an affiliative or disaffiliative response. Such improved understanding of the local management of talk would have the capacity to empower children by encouraging them to look for structure in interaction.

Such a social skills intervention approach would rely on previous talk-in-interaction research that has identified the structure and organization of conversation (Psathas, 1995; Sacks, 1995; Schegloff, 1988). Conversation is like a game, with rules, that is played between two or more people. However, the rules of the game can seem invisible. The methodology of conversation analysis can make the rules of the 'game' clear to participants in the interaction. It is well-known that children with Asperger's syndrome respond well to rules and structure (Attwood, 2000), thus encouraging them to focus on the structural aspects of interaction, by, for example, monitoring what others say and do, gives children greater control over their conversations. Greater social interactional awareness and control for children diagnosed with Asperger's syndrome may enable them to develop good social relationships, an essential component for good mental health and well-being. For a simple summary of the implications for practice see Table 19.1.

Summary

Social interaction is directly relevant to maintaining friendship, mental health, and well-being. Through social interaction, children initiate ideas, make and respond to play suggestions, as well as demonstrate their understanding of what has just occurred in the prior talk. Social interaction through laughter and spontaneous play displays feelings of belonging based on shared concerns, interests or other activities, showing mutual respect for each other (Margalit,

Table 19.1 Educational practice highlights

1. Having strong social support and good social relations contribute to an overall sense of well-being and health.
2. Children diagnosed with Asperger's syndrome find social interaction difficult and, as a result, they find it difficult to make and maintain friendships.
3. Understanding what 'counts' as social difficulty can only be understood in context.
4. Learning how to monitor interaction and become their own mini-analysts can give children the necessary skills to improve their conversations with peers.

1994). According to the World Health Organization (Wilkinson & Marmot, 2003), social support and good social relations with friends are closely linked to an overall sense of well-being and are protective factors for mental and physical health.

One difficulty for children diagnosed with Asperger's syndrome is that, although they might *want* to make friends with other children, they tend to approach the other child in a clumsy and not very successful way (Prior et al., 1998) and they have more limited social networks (Chamberlain et al., 2007). They also find it difficult to maintain the quality of the friendship as they are not skilled in interacting with each other and talking about things in common (Bauminger & Kasari, 2000). Yet, friendship offers long-term benefits for school adjustment, improves self-esteem and personal well-being, reduces loneliness and the possibility of depression (Diehl et al., 1998; Erwin, 1993; Margalit, 1994).

Social interaction can only be fully understood in situ and potential difficulties can only be understood by taking the context into account. A single case analysis (Schegloff, 1987), such as that presented here, can draw attention to the context-specific nature of interaction when working with children with Asperger's syndrome and with children more generally.

Note

1. The camera does not capture this interaction and so there is only an audio-recording.

References

American Psychiatric Association (2000). *Diagnostic and statistical manual of mental disorders (DSM-IV-TR)* (4th edition). Washington, DC: American Psychiatric Association.

Attwood, T. (2000). Strategies for improving the social integration of children with Asperger Syndrome. *Autism, 4*, 85–100.

Bauminger, N., & Kasari, C. (2000). Loneliness and friendship in high-functioning children with autism. *Child Development, 71*, 447–456.

Bauminger, N., Shulman, C., & Agam, C. (2003). Peer interaction and loneliness in high-functioning children with autism. *Journal of Autism and Developmental Disorders, 33*(5), 489–507.

Bellini, S. (2004). Social skill deficits and anxiety in high-functioning adolescents with autism spectrum disorders. *Focus on Autism and Other Developmental Disabilities, 19*(2), 78–85.

Butler, C., Danby, S., & Emmison, E. (2010). Address terms in turn beginnings: Managing disalignment and disaffiliation in telephone counselling. *Research on Language and Social Interaction, 44*(4), 338–358.

Chamberlain, B., Kasari, C., & Rotheram-Fuller, E. (2007). Involvement or isolation? The social networks of children with autism in regular classrooms. *Journal of Autism and Developmental Disorders, 37*(2), 230–242.

Clayman, S. (2010). Address terms in the service of other actions: The case of news interview discourse. *Discourse and Communication, 4*(2), 1–22.

Corsaro, W. A. (1985). *Friendship and peer culture in the early years*. Norwood, NJ: Ablex.

Danby, S., & Baker, C. D. (2000). Unravelling the fabric of social order in block area. In S. Hester & D. Francis (Eds.), *Local educational order: Ethnomethodological studies of knowledge in action* (pp. 91–140). Amsterdam: John Benjamins.

Diehl, D. S., Lemerise, E. A., Caverly, S. L., Ramsay, S., & Roberts, J. (1998). Peer relations and school adjustment in ungraded primary children. *Journal of Educational Psychology, 90*, 506–515.

Drew, P. (1987). Po-faced receipts of teases. *Linguistics, 25*, 219–253.

Dunn, J. (2004). *Children's friendships: The beginnings of intimacy*. Malden, MA: Blackwell.

Erwin, P. (1993). *Friendship and peer relations in children*. New York: John Wiley & Sons.

Evaldsson, A. C. (2007). Accounting for friendship: Moral ordering and category membership in preadolescent girls' relational talk. *Research on Language and Social Interaction, 40*(4), 377–404.

Fine, J., Bartolucci, G., Szatmari, P., & Ginsberg, G. (1994). Cohesive discourse in pervasive developmental disorders. *Journal of Autism and Developmental Disorders, 24*, 315–329.

Gillberg, C., & Gillberg, C. (1989). Asperger syndrome–some epidemiological considerations: A research note. *Journal of Child Psychology and Psychiatry, 30*, 631–638.

Goodwin, M. H. (1990). *He-said-she-said: Talk as social organization among black children*. Bloomington: Indiana University Press.

Goodwin, M. H. (2007). Occasioned knowledge exploration in family interaction. *Discourse and Society, 18*(1), 93–110.

Goodwin, M. H., & Goodwin, C. (2000). Emotion within situated activity. In A. Duranti & M. Malden (Eds.), *Linguistic anthropology: A reader* (pp. 239–257). Oxford: Blackwell.

Humphrey, N., & Symes, W. (2011). Peer interaction patterns among adolescents with autism spectrum disorders (ASDs) in mainstream school settings. *Autism, 14*(4), 397–419.

Jefferson, G. (1979). A technique for inviting laughter and its subsequent acceptance/declination. In G. Psathas (Ed.), *Everyday language: Studies in ethnomethodology* (pp. 79–96). New York: Irvington Publishers.

Jefferson, G., Sacks, H., & Schegloff, E. (1987). Notes on laughter in the pursuit of intimacy. In G. Button & J. R. E. Lee (Eds.), *Talk and social organisation* (pp. 152–205). Clevedon: Multilingual Matters Ltd.

Jones, C. D., & Schwartz, I. S. (2008). When asking questions is not enough: An observational study of social communication differences in high functioning children with autism. *Journal of Autism and Developmental Disorders, 39*, 432–443.

Koning, C., & Magill-Evans, J. (2001). Social and language skills in adolescent boys with Asperger syndrome. *Autism, 5*, 23–36.

Kremer-Sadlik, T. (2004). How children with autism and Asperger Syndrome respond to questions: A 'naturalistic' theory of mind task. *Discourse Studies, 6*, 185–206.

Lerner, G. H. (2003). Selecting next speaker: The context-sensitive operation of a context-free organisation. *Language in Society, 32*, 177–201.

Macintosh, K., & Dissanayake, C. (2006). A comparative study of the spontaneous social interactions of children with high functioning autism and children with Asperger's disorder. *Autism, 10*(2), 199–220.

Margalit, M. (1994). *Loneliness among children with special needs*. New York: Springer Verlag.

Minshew, N., Goldstein, G., & Siegel, D. (1995). Speech and language in high-functioning autistic individuals. *Neuropsychology, 9*, 255–261.

Muskett, T., Perkins, M., Clegg, J., & Body, R. (2010). Inflexibility as an interactional phenomenon: Using conversation analysis to re-examine a symptom of autism. *Clinical Linguistics & Phonetics, 24*(1), 1–16.

Ochs, E., Kremer-Sadlik, T., Gainer Sirota, K., & Solomon, O. (2004). Autism and the social world: An anthropological perspective. *Discourse Studies, 6,* 147–183.

Ochs, E., & Solomon, O. (2004). Introduction: Discourse and autism. *Discourse Studies, 6,* 139–146.

Owens, G., Granader, Y., Humphrey, A., & Baron-Cohen, S. (2008). LEGO therapy and the social use of language programme: An evaluation of two social skills interventions for children with high functioning autism and Asperger Syndrome. *Journal of Autism and Developmental Disorders, 38,* 1944–1957.

Poynton, C. (1985). *Language and gender: Making the difference.* Burwood, VIC: Deakin University Press.

Prior, M., Leekman, S., Ong, B., Eisenmajer, R., Wing, L., Gould, J., & Dowe, D. (1998). Are there sub-groups within the Autistic Spectrum? A cluster analysis of a group of children with Autistic Spectrum Disorder. *Journal of Child Psychology and Psychiatry, 39,* 893–902.

Psathas, G. (1995). *Conversation analysis: The study of talk-in-interaction* (Vol. 35). Thousand Oaks, CA: Sage.

Rao, P., Beidel, D., & Murray, M. (2008). Social skills interventions for children with Asperger's syndrome or high-functioning autism: A review and recommendations. *Journal of Autism and Developmental Disorders, 38,* 353–361.

Rendle-Short, J. (2003). Managing interaction: A conversation analytic approach to the management of interaction by an 8 year-old girl with Asperger's syndrome. *Issues in Applied Linguistics, 13,* 161–186.

Rendle-Short, J. (2007). Catherine, you're wasting your time: Address terms within the Australian political interview. *Journal of Pragmatics, 39,* 1503–1525.

Rendle-Short, J. (2011). Address terms in the Australian political news interview. In M. Ekström & M. Patrona (Eds.), *Talking politics in broadcast media: Cross-cultural perspectives on political interviewing, journalism and accountability* (pp. 93–111). Amsterdam/Philadelphia: John Benjamin Publishing Company.

Rendle-Short, J. (2014). Using conversational structure as an interactional resource: Children with Aspergers Syndrome and their conversational partners. In J. Arciuli & J. Brock (Eds.), *Communication in autism: Trends in language acquisition research series* (pp. 212–238). Amsterdam/Philadelphia: John Benjamin Publishing Company.

Rendle-Short, J., Cobb-Moore, C., & Danby, S. (2014). Aligning in and through interaction: Children getting in and out of spontaneous activity. *Discourse Studies, 16*(6), 792–815.

Sacks, H. (1995). *Lectures on conversation* (G. Jefferson, Trans., Vol. I and II). Oxford, UK: Blackwell.

Sacks, H., Schegloff, E. A., & Jefferson, G. (1974). A simplest systematics for the organization of turn-taking for conversation. *Language, 50,* 696–735.

Schegloff, E. A. (1987). Analyzing single episodes of interaction: An exercise in conversation analysis. *Social Psychology Quarterly, 50*(2), 101–114.

Schegloff, E. A. (1988). Description in the social sciences I: Talk-in-interaction. *IPrA Papers in Pragmatics, 2*(1/2), 1–24.

Schegloff, E. A. (1992). Repair after next turn: The last structurally provided defense of intersubjectivity in conversation. *American Journal of Sociology, 97*(5), 1295–1345.

Schegloff, E. A. (2004). Answering the phone. In G. H. Lerner (Ed.), *Conversation analysis: Studies from the first generation* (pp. 63–107). Amsterdam/Philadelphia: John Benjamins Publishing Company.

Schegloff, E. A. (2007). *Sequence organization in interaction*. Cambridge: Cambridge University Press.

Schegloff, E. A., Jefferson, G., & Sacks, H. (1977). The Preference for self-correction in the organisation of repair in conversation. *Language, 53*, 361–382.

Sterponi, L., & Fasulo, A. (2010). 'How to go on': Intersubjectivity and progressivity in the communication of a child with autism. *Ethos, 38*, 116–142.

Stribling, P., Rae, J., & Dickerson, P. (2007). Two forms of spoken repetition in a girl with autism. *International Journal of Language and Communication Disorders, 42*, 427–444.

Tager-Flusberg, H., & Anderson, M. (1991). The development of contingent discourse ability in autistic children. *Journal of Child Psychology and Psychiatry, 32*, 1123–1134.

Volden, J. (2004). Conversational repair in speakers with autism spectrum disorder. *International Journal of Language and Communication Disorders, 39*, 171–189.

Walker, G. (2013). Young children's use of laughter after transgressions. *Research on Language and Social Interaction, 46*(4), 363–382.

Wilkinson, R., & Marmot, M. (Eds.) (2003). *Social determinants of health: The solid facts* (2nd edition). Copenhagen: World Health Organization.

Williams-White, S., Keonig, K., & Scahill, L. (2007). Social skills development in children with Autism spectrum disorders: A review of the intervention research. *Journal of Autism and Developmental Disorders, 37*, 1858–1868.

Wing, L. (1981). Asperger's syndrome: A clinical account. *Psychological Medicine, 11*, 115–129.

Wootton, A. J. (1981). Children's use of address terms. In P. French & M. MacLure (Eds.), *Adult-child conversation* (pp. 142–158). London: Croom Helm.

Wootton, A. J. (2003). Interactional contrasts between typically developing children and those with autism, Asperger's syndrome, and pragmatic impairment. *Issues in Applied Linguistics, 13*, 133–160.

Recommended reading

- Gardner, H., & Forrester, M. (Eds.) (2010). *Analysing interactions in childhood: Insights from conversation analysis*. Oxford: Wiley.
- Rendle-Short, J. (2003). Managing interaction: A conversation analytic approach to the management of interaction by an 8 year-old girl with Asperger's syndrome. *Issues in Applied Linguistics, 13*, 161–186.
- Schegloff, E. (1987). Analyzing single episodes of interaction: An exercise in conversation analysis. *Social Psychology Quarterly, 50*(2), 101–114.
- Wootton, A. J. (2003). Interactional contrasts between typically developing children and those with autism, Asperger's syndrome, and pragmatic impairment. *Issues in Applied Linguistics, 13*, 133–160.

20
Attachment Processes and Eating Disorders in Families: Research and Clinical Implications

Rudi Dallos and Sarah Pitt

Introduction

Given the essentially interpersonal nature of eating and food, it is perhaps interesting to note that eating disorders such as anorexia nervosa (henceforth anorexia) have typically been considered an intra-psychic phenomenon. This has featured conceptualisations, such as distorted body image, internalisation of gendered expectations of slimness in women, and biological explanations of hereditary predispositions (O'Shaugnessy & Dallos, 2009). In contrast, there have also been explanations which suggest interpersonal processes; for example, Bruch (1973) conceptualised anorexia in terms of a conflictual and enmeshed mother–daughter relationship. This came to be regarded as 'mother-blaming', and consequently there has been little research which has attempted to explore the role of family dynamics in the causation of eating disorders. However, systemic family therapy developed a number of approaches to working with eating disorders which assumed family dynamics as causal (Minuchin, Rosman, & Baker, 1978; Palazzoli, 1974). Contemporary practice in the United Kingdom (NICE, 2004) recommends systemic family therapy as the treatment of choice for children under the age of 16 (Dare, Eisler, Russell, & Smukler, 1990).

This chapter aims to offer an explanatory framework of the developmental process of eating disorders in families and an overview of the clinical implications and applications of this approach, drawing examples from a case study (see also Woolhouse & Day, Chapter 23, this volume). It introduces existing systemic and attachment theoretical ideas about anorexia and provides an integrative framework for conceptualising anorexia. A research case study offers practical examples of this framework and the clinical implications are discussed.

Systemic processes and anorexia

Systemic therapies have suggested a number of common family processes in anorexia. These feature conflict avoidance and enmeshment (Minuchin et al., 1978) and patterns of 'covert coalitions' where a child can become 'triangulated' between their parents' marital conflicts. Research has shown the presence of triangulation processes in families where a young person is experiencing mental health difficulties (Beuler & Welsh, 2009; Dallos & Smart, 2011). There is also an indication that the problems with eating have a trans-generational pattern; for example, mothers of young women with anorexia may have also suffered from eating problems themselves (O'Shaugnessy & Dallos, 2009). However, the question remains as to why the patterns in families develop? Arguably, systemic theories offer a description of the family processes but do not attempt to explain their causal development.

In attempting to consider family dynamics as causal in the development of eating disorders, the intention is not to assign blame to families. Instead, it is suggested that a conceptualisation of the development of eating disorders may assist in guiding treatment to be more effective. This is important given that eating disorders, and especially anorexia, are one of the most dangerous psychiatric conditions with treatments still only being moderately effective (Fairburn & Brownell, 2002). Most importantly, the emphasis of our proposed model is that parents are almost invariably attempting to do the best they can for their children (Byng-Hall, 1995a; Tomm, 1984).

A guiding concept we employ in our writing is Byng-Hall's (1985) idea of corrective and replicative scripts. Most parents develop ideas about how they want to either correct or repeat aspects of their own childhood experience of being parented. However, the positive intentions behind their approaches do not always produce the desired effects. In fact, in many cases our 'attempted solutions' (De Shazer, 1982; Watzlawick, Weakland, & Fisch, 1974) can produce quite the opposite effects. To put this rather cryptically, 'the path to Hell is paved with good intentions'. These processes are discussed in more detail throughout the chapter and examples are provided.

Attachment theory: early attachment and eating patterns

Food is one of the earliest and most fundamental forms of pleasure and also distress experienced by an infant (Bowlby, 1979; Friedman, 1996). As adults, we can only imagine what it must feel like for an infant to experience the craving for food and drink without conscious awareness that this desperate need will be relieved. At the same time, it can constitute an extremely sensitive and sensual experience of contact with the mother's body and the comforting taste and smell of milk. Bowlby (1979) suggested that the child internalises this early relationship with the mother in terms of whether it offers a comforting, speedy release from the distress of hunger. In contrast, this feeding could also be an unpredictable, frustrating experience which could maintain an unpleasant state of distress and anxiety for the baby.

In these early interactions, mothers vary in how much they are able to manage the intimacy and sensuality of feelings relating to feeding. Some mothers appear to be comfortable in their interaction with their baby, others become very anxious and concerned to offer physical contact and others find contact uncomfortable. Bowlby (1979) argued that food and the resolution of the anxiety generated by hunger provide an important template for early relationships which are shaped and in turn shape patterns of eating. Importantly, accounts from people with symptoms of anorexia suggest that they do not transcend hunger but come to tolerate the suffering of intense discomfort (Orbach, 2008). By analogy, they can be seen as if trapped in this early distress concerning food and appear to resist being comforted.

Integrating systemic and attachment theory: trans-generational processes of attachment

Systemic conceptualisations resonate with perspectives from attachment theory (Dallos, 2003; Ringer & Crittenden, 2007), which suggest that the experiences that parents have had in their own childhood shape the emotional context that they offer for their own children. Systemic family therapy and attachment theory have an interest in understanding trans-generational patterns (Bowlby, 1973), whereby family culture and processes can be passed from one generation to the next, including conflicts, and ways of managing distress (Lieberman, 1979). Attachment patterns have been found to be transmitted across the generations in families with anorexia (Ward, Ramsay & Treasure, 2000; Zachrison, & Skårderud, 2010).

Parents bring to their relationship their own repertoire of attachment experiences and in turn children develop their own attachment strategies based on their parents' responses to their attachment needs (Ainsworth, Blehar, Waters, & Wall, 1978; Crittenden, 2006). Where parents are able to respond consistently to provide care and comfort to a child when he or she is distressed, then a secure pattern is seen to develop. In cases where the child develops an avoidant pattern, parents are seen to respond consistently, but with little care and affection and with a message that encourages the child to become overly self-reliant. In anxious–ambivalent patterns, parents have been found to respond inconsistently, at times not being available and at other times as excessively anxious themselves, or as intruding into the child's activities. Finally, some children develop 'disorganised' or extreme strategies typically in contexts where the parents have responded in frightening or abusive ways. In these situations, the child experiences a severe dilemma in that the person who is meant to provide attachment security may at the same time be a source of threat or distress.

Parents' emotional, romantic relationship also evolves alongside the child's development and constitutes a central attachment relationship and emotional context for the child (Hazan & Shaver, 1987). On the flip side, in many relationships there is also a continual sense of anxiety or threat that the relationship may deteriorate or end.

The experiences that we have had through our childhood come to be held as 'internal working' models (Ainsworth et al., 1978) or DRs – 'dispositional representations' (Crittenden, 2006). The latter term captures the idea that DRs shape our anticipations of the future and guide how we will approach relationships with others and the seeking and giving of support and comfort. These representations are both implicit – procedural and sensory memories – and also explicit – semantic, episodic and integrative/reflective memory processes. Crittenden emphasises that the decisions that we make may be influenced by the implicit representations largely outside of conscious awareness.

Corrective scripts and representational systems

It is important to consider the concept of representational systems both in terms of the nature and experience of eating disorders and also in terms of parents' actions and responses to their child. Ringer and Crittenden (2007) employed an analysis of Adult Attachment Interviews (AAIs) in their research sample and suggested that hidden family conflicts in parents' relationships and histories of trauma contributed to the development of insecure attachment patterns and eating disorders. Mothers wanted to defend their daughters from threats and disappointments that they faced in their own lives and sought closeness with their daughters they had not received. Dallos and Denford (2008) interviewed families where a young person had a diagnosis of anorexia and found that across the generations parents had memories of not receiving comfort and described how they came not to expect comfort and warmth when upset or ill. For some parents, this lack of expectation of caring had a long-lasting effect and they developed a wish for this to be different for their own children. Byng–Hall (1991, 1995a) has termed this a 'corrective script'. He developed the idea that family scripts may be founded on family belief systems based on attachment experiences in their family of origin. Expectations are developed from these experiences about family rules, roles and routines. Family members may attempt in their current families to preserve what they had experienced as 'good' (replicative scripts) or improve what they had experienced as 'bad' (corrective scripts).

The young person with anorexia in this study seemed to have a critical role in the parents' attempts to correct their own negative experiences from childhood.

The concept of corrective and replicative scripts contains the idea of parents making conscious 'choices' about how they will attempt to respond to and act as parents with their children. However, our research indicates that a very important dilemma may arise for parents in this process. For example, a corrective script can be seen as essentially an explicit semantic intention about how to act. But many parents appear to have experienced a lack of care, affection and comforting in their childhood. Hence, they do not have procedural and sensory memories of receiving comfort and affection, and without these representations to draw on, they may find it difficult to offer comfort and affection for their own children. For example, we can read books or even take courses on how to parent our children, but arguably the response to our child's distress needs to be a rapid and automatic one. Parents who have experienced secure attachment experiences in their childhood are able to access their representations of comfort and affection rapidly and automatically. In contrast, it may be a much more confusing process for parents who have to try and piece together a response from their conscious intentions.

In our view this offers a key answer to the puzzle of how anorexia and other conditions may evolve. Parents are generally trying their very best to do things better than they had experienced in their childhood. Yet, without the procedural and sensory memories to draw on, they lack an embodied map and may inadvertently respond in ways that do not constitute a sense of comfort and affection for their child. Furthermore, as children develop more complex and challenging demands, parents may come to feel increasingly confused and anxious about how best to act towards their child. This corresponds, for example, with the frequent observation that parents with an eating disorder child try to avoid conflict and find it hard to develop consistent strategies to manage them.

Our research also indicates that parents with a young person with an eating disorder have frequently experienced emotionally barren, neglectful or dangerous childhoods. This may mean that they have unresolved or traumatic memories from their childhoods. For example, one mother described that her own mother had been depressed, emotionally unavailable and had attempted suicide when she was a child (Dallos, 2003). She then commented that she had wanted a much warmer and closer relationship with her own daughter (corrective script). However, not having experienced this herself, she was unsure about how to achieve this and an attempt was to buy her daughter nice things and hope this would gain her love and affection. When this did not appear to work she became sad and then highly distressed as the memories of her own childhood intruded, which led to her feeling a failure and desperately trying even harder to buy her daughter's affection. Perhaps, the most ironic and painful part of this process is that parents experience the bitter sting of a double failure: their relationship with their child appears to be becoming difficult and this is in the context of them wanting so hard for it to be better than their childhoods had been. In fact, they may feel that despite their best intentions their child is developing more, not less, serious problems than had been their own experience. The taste of this can be very bitter indeed and may lead parents to both sadness and anger, to blame themselves, to blame their child and perhaps lead to a very tempting resolution – that is, simply to frame the eating problems as an 'illness'.

Project overview: a case study from our research

The family consisted of Kate, her mother Fran, her father Bill and her three siblings Keith, Lucy and Paul. Kate had begun to experience difficulties with eating while at college, and this coincided with Bill becoming unwell with a heart condition and being admitted to hospital. Kate was admitted to hospital two weeks later due to her dangerously low weight and three weeks later moved to an adult Eating Disorders Unit. At the time of the case study her eating difficulties were stable and she was living in stepped-down accommodation and was attending the unit as a day patient. The family was receiving family therapy on the unit. AAIs were completed with the parents and the Transition to Adulthood Interview with Kate (Crittenden, 2006). A family interview was completed with Kate and her parents and contained open-ended questions related to the family background, relationships within the family and to transgenerational processes of comfort, relationships with food, emotional closeness and conflicts.

Attachment strategies

Analysis was based on Crittenden's (2006) extension of attachment theory.

Kate: Overall Kate predominantly demonstrated a dismissive strategy featuring a pattern of inhibiting expression of emotions and compliance. She minimised her emotional needs and glossed over emotional repercussions; for example, the end of her first romantic attachment.

Fran: Traumatic memories of neglect and being frightened by her parents' conflicts were indicated. She had a complex mixed strategy: Dismissive – compliant/self-reliant and pre-occupied – and threatening/aggressive, in which she demonstrated attempts to inhibit her feelings but was also pre-occupied by unresolved feelings of anger towards her mother.

Bill: He predominantly demonstrated a dismissive strategy – inhibited/ compliant – such that any hint of complaint towards his parents would be moderated or dismissed. He described a lack of comfort in his childhood with an emphasis on food as a substitute for affection – 'being stuffed rotten with food' to such an extent that he had to have special socks knitted for him since both of his legs were so huge.

Family interview

We explored the interview in terms of dominant themes that appeared to shape the family's thinking and interactions. We also employed a form of conversational analysis to consider how the themes were employed in the family conversational processes. Throughout the interview the parents also talked about ways they had wanted to do things better or different from their own experiences during childhood. Hence, 'corrective scripts' can be seen as a meta-theme running through all discourse themes described below:

- *Relationship with emotions*
- *Negative relationships with food*
- *The central role of illness*
- *Feeling powerless/not knowing what to do*

The following notations from conversational analysis are employed in the extracts below (but see Table 1 in the Preface of the handbook for a full description of the Jefferson symbols):

Emphasis
LOUD
(2.00) pauses
u::m hesitations

Relationship with emotions

The family talked about difficulties in sharing emotions and demonstrated self-protective strategies for avoiding difficult emotions; for example, steering towards safer conversations and softening critical comments. In the interview, the main focus was on the family's experience of conflict:

Kate: When I was growing up we were very different from every other
        family I think. Every other family I knew I would get into
        a lot of conflict and I KNOW WE HAD our conflict but it
        wasn't like everyone else's I don't think, it wasn't to the
        stage of screaming the house down.

Conflict here is exaggerated and possibly signalled as dangerous and out of control – 'screaming the house down'. Kate later acknowledged some conflict within the family but Bill remained silent, while Fran acknowledged the conflict but steered the conversation away. There was a strong ethos of self-reliance within the interview and a family discourse of 'keeping to ourselves', which strongly links with an avoidant self-protective strategy of relating. These discourses appeared to function to keep emotions hidden. There was a discourse of a lack of emotional comfort across the generations as well as within the current family. Bill and Fran talked about how they have tried to provide Kate with comfort they had not received themselves; for example, when she had the flu. Importantly, comfort was described in relation to physical illness rather than emotional distress. They may have been replicating some of their own childhood experience of lack of direct comfort in relation to emotional distress. Both in this interview and in the AAIs they all found it difficult to recall any specific examples of emotional comfort. Bill made justifications for the lack of comfort they received during his childhood in terms of it being a different era. Fran exonerated her parents in terms of them having to work a lot and therefore being on her own growing up. But both of them articulated a corrective script of wanting to be more emotionally available for their own children:

```
FRAN:    u::m (2.0) I just felt very strongly that I just wanted
         someone to be there for the children when they were
         growing up,
[Bill: yep me too]
FRAN     to be one of us, we decided that there would be one of us at
         home one of us would be at work.
```

Kate's Transition to Adulthood Attachment Interview (TAAI) revealed that, despite their corrective intentions, Bill and Fran may not have been as emotionally present in Kate's childhood as they had hoped – she could not recall many memories of her mother from childhood and instead remembered them both working a lot. Although Fran and Bill had corrective intentions of being more physically present as parents, they may have replicated some of their own experiences and Kate was unable to explicitly say this in front of them.

Negative relationships with food

Bill and Fran described negative, but opposing, experiences of mealtimes during childhood: excess versus restriction. Fran and Bill described wanting to do things differently with their children and developed a shared attachment framework where 'sitting at the table' and hearing about their children's day is highly valued by Bill, Fran and Kate.

```
Fran:   I wanted them all to talk say what they wanted to:: if they'd
        had a good day bad day what was good [Bill: Yes] what the
        teacher had don::e, or said or not, or how things got on in
        the playground, and in college really we tried to? you know?
        [quieter]
Kate:   To be honest I don't really remember a lot from when I was at
        (h) college'
```

The family later had a discussion about when Keith stopped sitting with the family to eat and would not speak to them for days at a time. Kate became critical towards Bill in relation to how he responded to this and was aroused in this passage:

```
Kate:   '...he wouldn't be around and you
        would always always AL:::WAYS do him a packed lunch to take.
        and you used to put SIX sandwiches in there, three packets of
        crisps, loads of chocolate bars, loads of fruit, and you know
        he's a bean pole anyway so he puts it where nobody knows
        where. But you'd always make sure he had food and never
        went without. (2.0) You wouldn't let him fend for himself.
        And we always used to tell you "dad don't do it" and you were
        like "NO NO IT'S FI::::NE I'LL DO IT."
```

Bill offered justification for feeding his son this way due to his role as a parent but may not have had experiences of how to emotionally respond differently as a parent. A trans-generational lack of boundaries around food and meal times is evident when Bill talked about his experiences of excessive food during childhood:

```
Bill:   '...I was stuffed ROTTEN Um:::...I had food > coming out of my
        ears literally.'
Bill:   'And I could have whatever I wanted to snack between and I
        wasn't stopped'.
```

Bill may be indicating that he needed boundaries to be provided around his food intake. Some of the lack of boundaries that Bill experienced around food may have inadvertently been re-experienced in their current family experience of meal times.

The central role of illness

Anorexia was described as an illness throughout the interview. In his AAI, Bill described illness as having an important role in his childhood due to his brother's epilepsy. In the interview, the family shared their experience of Bill also being ill at the same time as the onset of Kate's anorexia. The family talked about their fears of 'fragmenting' throughout the interview and placed importance on the family being kept together. They discussed the role illness had in preventing fragmentation:

```
Bill:   because of her illness has seemed to come back into the
        picture
[Fran and Kate: U::m]
Bill:   because losing your friends for one thing and another and you
        were ill and I was ill at the same time so we seemed to
        gravitate to each other
```

It was later acknowledged that although they physically came together, emotionally there was still distance:

```
Kate:   we came together but I think we were still very isolated in
        our own worlds
(Bill:  yes::::)
Kate:   together
```

Feeling powerless/not knowing what to do

Throughout the interview Bill and Fran constructed a powerless position of not knowing what to do to help Kate. They positioned their intentions to be 'problem solvers', but Fran acknowledged that although she tried to solve problems, she was often not able to:

```
Fran:   'I don't feel that I've done much hhh to solve problems I've
        worried about them but not been able to hhh (4.0) physically
        do anything about it.
Kate:   I think going to the doctors didn't really help because
        going to Dr Ross like said to me it's anorexia but then as
        soon as I said no it isn't he just went "^alright then".
        And that was it he didn't really argue back
Fran:   no
Bill:   we didn't sort of follow it up did we really? It was a case
        of you're going to the doctors they'll know best. Whether or
        not we were sort of passing the buck over to the doctors
        thinking he'll be able to take control of the situation and
        we didn't have to do anything about it we could say he'll be
        able to come up with this wo::nderful answer. I I I I I felt
        powerless.'
```

This powerlessness may reflect the dominant medical discourse that professionals hold the answers and doctors 'know best' and position anorexia as an illness. Through passing power to the doctors, Bill is positioned as both wanting someone else to take control of a situation he felt helpless in and in doing so enhancing the feeling of powerlessness. His arousal when talking about feeling powerless emphasises a connection with this feeling.

An attachment/systemic formulation

This research case study allowed us to consider the family both from a systemic lens in terms of their conversation and also from the 'inside' in terms of their attachment interviews. Their conversation revolved around four relational topics or themes which are consistent with some of the observations from systemic family therapy. The first theme, *relationship with emotions*, indicated that both in the family's shared beliefs and in their interactions there was an indication of anxiety about conflict and negative emotions. In their conversation, they repeatedly detoured away from addressing difficult emotions, attempted to soften any critical comments, and avoided conflict by talking about family members not present. When emotions were expressed, this felt dangerous, with a risk that the family might be 'torn apart' or 'fragmented' and conflict may remain unresolved. This also supports research indicating that individual attachment patterns influence the way family members communicate with one another (Crittenden & Kulbotten, 2007; Dallos & Smart, 2011). Specifically, it connects with previous research indicating conflict avoidance within families experiencing anorexia (Dallos & Denford, 2008; Minuchin et al., 1978; Palazzoli, 1974).

This 'conflict avoidance' was constructed around a replicative script from Bill's childhood, where his father used silence as a way of avoiding emotions and a corrective script from Fran whereby she felt triangulated between parents in severe conflict with each other. In their attempts to correct their own childhood experiences, the parents in this family may have moved to the opposite extreme and unintentionally created new problems (Byng-Hall, 1991, 1995a, 1995b; Crittenden, 2006). Parents may also be influenced by the culturally shared, but unrealistic, notion that they should be self-sacrificing and show consistent positive emotion to their children (Dallos, 2003). This links with the family discourses of 'keeping to ourselves' and being 'self-reliant' and that family members should not unduly worry each other with problems and should individually manage difficult emotions. Other families may share an opposing attachment framework whereby all feelings and concerns should be immediately shared due to the centrality of having a close emotional connection (Crittenden & Kulbotten, 2007; Hill et al., 2003, 2011). This shared attachment framework may have been influenced by parents' own experiences during childhood, where there was also a strong framework of self-reliance and keeping feelings in. The family in this research were trying to change their attachment framework to become more open with one another with regard to their feelings and concerns. A risk is that the family moved to the other extreme and created a new problem. Clinical interventions could help the family strike a balance between these opposing frameworks.

The second theme, *negative relationships with food*, identified opposing negative discourses around mealtimes during Bill and Fran's own childhood: excess versus restriction. With opposing difficult experiences during childhood, and a strong desire from both parents to do things differently, it would be difficult to develop a shared framework for their current family. However, the family did develop a shared attachment framework of 'sitting at the table' and talking about the day. This was highly valued as a way of providing a positive, corrective experience of mealtimes and preventing family 'fragmentation'. This may reflect a cultural value in society that 'sitting at the table' projects a happy, well-functioning family. However, mealtimes were experienced as 'spontaneous' and not important by Kate and it was acknowledged that, although unintentional, mealtimes may have become difficult. A secure attachment response requires the parent to help the child to be 'safe' not only through responding to their emotional 'attachment' needs but also through offering discipline, guidance, and boundaries (Crittenden, 1999, 2002; Hill et al., 2011). Through attempting to meet their children's attachment needs and correct their own childhood experiences, mealtimes may have become un-boundaried and therefore 'unsafe'. Negative, painful experiences with food in families where a young person has anorexia are supported by Dallos and Denford's (2008) study and broader practice-based clinical evidence.

The third discourse theme, *the central role of illness*, suggested that Kate's and Bill's illnesses may have served to prevent family fragmentation. Paradoxically, anorexia may also have been a way for Kate to gain independence and leave the family home. This family pattern which emerged in the family's discourse around the centrality of illness is similar to the systemic concept of triangulation reported by other researchers and clinicians (Crix, Stedmon, Smart, & Dallos, 2012; Dallos & Denford, 2008; Dallos & Smart, 2011). The original conceptualisation of triangulation was that through focusing on an ill child marital conflicts may be prevented (Dallos, 2003; Minuchin, 1985; Minuchin et al., 1978). There was some indication of Kate being in a triangulated position between her parents' conflicts during the interview. Minuchin (1985) also suggests that, once established, a child's triangulated position also in turn maintains the distance between the couple. For example, Kate's 'illness' may have made it harder for the parents to develop their relationship. It also seemed that in Kate's family, illness held a central role in holding the family together more generally and preventing 'fragmentation'. What is interesting in this family is that there is a double illness serving this purpose: illness of father and of child. Here both illnesses seemed central in physically keeping the family together. The family was able to reflect that although they were brought physically together, emotionally they remained isolated.

Related to this discourse of 'illness', a fourth discourse theme was identified: *feelings of powerlessness/not knowing what to do*. A medical discourse emerged in this theme and not only served the intention of relieving the family of their responsibility in helping Kate but also enhanced their feeling of powerlessness. It also placed anorexia as an 'illness'. Arguably, the medicalised constructions of anorexia have emerged as the dominant discourse, placing women experiencing this 'illness' in a passive, helpless position (Hepworth & Griffin, 1995; Rich, 2006). This case study suggests that this helpless position can be experienced more broadly by the family of the young person with anorexia.

These themes are inter-linked: The sense of powerlessness impacts on family relationships in general; for example, Bill's and Fran's confidence as parents and as a couple may have been disrupted and in turn their feeling less confident with each other aggravated the sense of powerlessness. Parents may feel at fault and in part blame their own relationship for the anorexia. The discourse of anorexia as an 'illness' can help soften a sense of self-blame. Although the medical discourse of anorexia as an 'illness' emerged in this research, the family's construction of anorexia was not focused.

Clinical relevance summary: treatment strategies – attachment narrative therapy

The research confirms the authors' previous research and clinical experience to suggest a number of core orientations for working with families with anorexia. These are based on the four stage models of attachment narrative therapy (ANT – Dallos, 2003):

1. *Establishing a safe therapeutic base* – The parents in this study and other families had clearly experienced difficult and emotionally unsupported childhoods. They were unsure about their own abilities as parents and could easily feel blamed and at fault for their child's problems. Hence, it is essential to recognise this and to offer a new experience of not blaming them, listening to their concerns, and validating their attempts to cope. This involves a recognition that this reassurance has to take place at multiple representational levels; for example, in the pace and emotional tone of the therapy (procedural and sensory level), and not simply in semantic terms focused on understanding and explanations. Bowlby described therapy as the therapist becoming a transitional attachment figure for the family.

2. *Exploration* – Working alongside the family in exploring their understandings and explanations has to be done carefully and with respect to the family's feelings. In particular, the focus of this exploration should include the following:

    a. Their relationships with food as a source of comfort.

    b. Their corrective and replicative scripts regarding what they feel they have learnt from their childhood and how they may wish to change how they were parented.

Typically, families' explanations may have coalesced to a focus on illness as a way of avoiding the negative impact of a sense of failure as parents.

It is important not only to be sympathetic to this sense of failure but also to recognise that it may be part of a wider pattern of the parents not having experienced emotional care and support through their childhoods. The therapist offering a sense of safety, consideration of their feelings and sympathetic emotional responding may be a new and potentially anxiety-provoking experience for them. 'Can I take the risk of trusting the therapist to look after my/our feelings?'

3. *Attempting change* – Parents may have the intention of correcting their own childhood attachment experiences. In this family, the parents' corrective intentions at more positive parenting and provision of comfort were hindered by their own lack of experience of positive attachments and comfort in childhood. It is also important to explore whether the parents' corrective scripts may have become overly rigid as a result of traumatic memories that mobilise the corrective script in the first place and their sense of failure and desperation at their child's condition. Both of these may result in a corrective script becoming dysfunctional, such that the parents are unable to utilise feedback about whether it is working or not and adapt it as necessary. The therapeutic process of change may hence focus not on what they are wrongly doing or not doing but how their attempts to do things better are currently not working. Instead of a deficit approach, the focus can be on assisting them to add to their explicit semantic intentions of caring for their child at the procedural and sensory levels. This involves exploring positive and more difficult and conflictual experiences with them in therapy in a safe and slow way so that they start to feel at an embodied level that conflict and more negative feelings can be managed. In part the relationship of a secure attachment from the therapist regarding the parents can become transferred by the parents towards their child.

4. *Maintaining the relationship* – since the problems with eating disorders can be seen as essentially about comfort and safety, it is important to maintain a sense of connection with the parents and the young person. This can involve discussions about how the therapist and family will keep each other in mind and more specifically offer a phased transition of the relationship. Rather than a focus on ending, the therapy can be seen as a shift so that the family comes to rely less on the presence of the therapist but knows that they remain as a potential source of support.

For a simple summary of the implications for practice, see Table 20.1.

Summary

This chapter has attempted to provide a conceptualisation of anorexia from an integrated systemic and attachment perspective. The chapter aimed to

Table 20.1   Clinical practice highlights

---

1. Evidence from attachment theory suggests that despite painful attachment histories, people can transcend these and move on to relate to each other emotionally in different ways than their traditional attachment frameworks (Dallos, 2003). Exploring parents' own childhood experiences during therapy can help a family understand their intentions as parents, where these intentions may have come from, and how their actual parenting behaviour may not be based on their intentions. This can help build empathy between family members (Lieberman, 1979). The family therapist on the unit working with this family reported that Fran acknowledged wanting an absence of conflict in her current family due to her childhood experiences of being caught between significant conflicts. This may fuel a pattern of helplessness where any normal conflict appears dangerous and both attachment and necessary discipline/guidance responses are inhibited. This may invite 'illness' as a way of managing family issues.
2. A possible clinical step is for parents to feel safe enough about their child's physical safety in relation to anorexia as a way of negotiating their marital issues. This may then release them and the child to move on from being locked in an illness framework. Likewise, many young people like Kate need to feel safe enough about their parents' well-being to move on with their own lives. Here the illness appeared to play both these contradictory roles: Kate stayed dependent while ill but also independent to some extent in having to move away from home. In the longer term, it is important that family issues are managed and independence achieved without needing 'illness'.
3. A consideration of attachment styles during therapeutic conversations can help guide therapists in thinking about the conversations they invite families into. For a family showing dismissive attachment patterns, such as the family in this research, the therapist may encourage the expression of feelings. The family therapist on the unit suggested this was one of the guiding therapeutic approaches with this family. Attachment patterns are not exclusive and should not promote a prescriptive approach but be used as a tentative guidance.

---

demonstrate through a research case study how trans-generational processes of attachment can influence the development of anorexia. A particular guiding principle has been the idea of 'corrective' and 'replicative scripts' whereby parents' attempt to 'correct' their own childhood attachment experiences and often inadvertently replicate their own experiences. Practical implications have been discussed and include discussing parents' own childhood experiences within family therapy sessions. ANT is identified as a useful treatment strategy.

References

Ainsworth, M. D., Blehar, M. C., Waters, E., & Wall, S. (1978). *Patterns of attachment: A psychological study of the strange situation*. Hillside, NK: Lawrence Erlbaum.

Beuler, C., & Welsh, B. (2009). A process model of adolescents' triangulation into parents' marital conflict: The role of emotional reactivity. *Journal of Family Psychology, 23*(2), 167–180.

Bowlby, J. (1973). *Attachment and loss, vol. II: Separation: Anxiety and anger.* London: Pimlico.

Bowlby J. (1979). *The making and breaking of affectional bonds.* London: Tavistock.

Bruch, H. (1973). *Eating disorders: Obesity and anorexia and the person within.* New York: Basic Books.

Byng-Hall, J. (1985). The family script: A useful bridge between theory and practice. *Journal of Family Therapy, 7,* 301–305.

Byng-Hall, J. (1991). The application of attachment theory to understanding and treatment in family therapy. In C. M. Parkes, J. Stevenson-Hinde, & P. Marris (Eds.), *Attachment across the life cycle* (pp. 199–215). London: Tavistock/Routledge.

Byng-Hall, J. (1995a). *Re-writing family scripts: Improvisations and family change.* New York: Guildford Press.

Byng-Hall, J. (1995b). Creating a secure base: Some implications of attachment theory for family therapy. *Family Process, 34,* 45–58.

Crittenden, P. M. (1999). Danger and development: The organisation of self-protective strategies. *Monographs of the Society for Research in Child Development, 64*(3), 145–171.

Crittenden, P. M. (2002). Attachment theory, information processing, and psychiatric disorder. *World Journal of Psychiatry, 1,* 72–75.

Crittenden, P. M. (2006). A dynamic maturational model of attachment. *Australian and New Zealand Journal of Family Therapy, 27*(2), 105–115.

Crittenden, P. M., & Kulbotten, G. R. (2007). Familial contributions to ADHD: An attachment perspective. *Tidsskrift for Norsk Psykologforening, 44*(10), 1220–1229.

Crix, D., Stedmon, J., Smart, C., & Dallos, R. (2012). Knowing 'ME' knowing you: The discursive negotiation of contested illness within a family. *Humans Systems: The Journal of Therapy, Consultation & Training, 23*(1), 27–49.

Dallos, R. (2003). Using narrative and attachment theory in systemic family therapy with eating disorders. *Clinical Child Psychology and Psychiatry, 8*(4), 521–535.

Dallos, R., & Denford, S. (2008). A qualitative exploration of relationship and attachment themes in families with an eating disorder. *Clinical Child Psychology and Psychiatry, 13*(2), 305–322.

Dallos, R., & Smart, C. (2011). An exploration of family dynamics and attachment strategies in a family with ADHD/conduct problems. *Clinical Child Psychology and Psychiatry, 16*(3), 535–555.

Dare, C., Eisler, I., Russell, G., & Smukler, G. (1990). The clinical and theoretical impact of a controlled trial of family therapy in anorexia nervosa. *Journal of Marital and Family Therapy, 16*(1), 39–57.

De Shazer, S. (1982). *Patterns of brief therapy: An ecosystemic approach.* New York: Guilford Press.

Fairburn, C. G., & Brownell, K. D. (2002). *Eating disorders and obesity.* London: Guilford Press.

Friedman, M. E. (1996). Mother's milk: A psychoanalyst looks at breastfeeding. *The Psychoanalytic Study of the Child, 51,* 473–490.

Hazan C., & Shaver P. (1987). Conceptualising romantic love as an attachment process. *Journal of Personality and Social Psychology, 52,* 511–524.

Hepworth, J., & Griffin, C. (1995). Conflicting opinions? 'Anorexia Nervosa', medicine and feminism. In C. Kitzinger & S. Wilkinson (Eds.), *Feminism and discourse: Psychological perspectives* (pp. 68–85). London: Sage.

Hill, J., Fonagy, P., Safier, E., & Sargent, J. (2003). The ecology of attachment in the family. *Family Process, 42(2),* 205–221.

Hill, J., Wren, B., Alderton, J., Burck, C., Kennedy, E., Senior, R., Aslam, N., & Broydon, N. (2011). The application of a domains-based analysis to family processes: Implications for assessment and therapy. *Journal of Family Therapy*, doi:10.1111/j.1467-6427.2011.00568.x.

Lieberman, S. (1979). Transgenerational analysis: The genogram as a technique in family therapy. *Journal of Family Therapy, 1*, 51–64.

Minuchin, P. (1985). Families and individual development: Provocations from the field of family therapy. *Child Development, 56*, 289–302.

Minuchin, S., Rosman, D. L., & Baker, L. (1978). *Psychosomatic families: Anorexia nervosa in context.* Cambridge, MA: Harvard University Press.

National Institute for Clinical Excellence (2004). *Guidelines for eating disorders CG9.*

Orbach, S. (2008). *Fat is a feminist issue.* London: Arrow Books.

O'Shaugnessy, R., & Dallos, R. (2009). Attachment research and eating disorders: A review of the literature. *Clinical Child Psychology and Psychiatry, 14*(4), 549–574.

Palazzoli, M. S. (1974). *Self-starvation: From the intra-psychic to transpersonal approach to anorexia.* London: Chaucer.

Rich, E. (2006). Anorexic dis(connection): Managing anorexia as an illness and an identity. *Sociology of Health and Illness, 28*(3), 284–305.

Ringer, F., & Crittenden, P. (2007). Eating disorders and attachment: The effects of hidden family processes on eating disorders. *European Eating Disorders Review, 15*, 119–130.

Tomm, K. (1984). One perspective on the Milan systemic approach: Part 1. Overview of development, theory and practice. *Journal of Marital and Family Therapy, 2*, 113–125.

Ward, A., Ramsay, R., & Treasure, J. (2000). Attachment in eating disorders. *British Journal of Medical Psychology, 73*, 35–51.

Watzlawick, P., Weakland, J., & Fisch, R. (1974). *Change: Principles of problem formation and problem resolution.* New York: Norton.

Zachrison, H. D., & Skårderud, F. (2010). Feelings of insecurity: Review of attachment and eating disorders. *European Eating Disorders Review, 18*, 97–106.

Recommended reading

- Dallos, R. (2003). Using narrative and attachment theory in systemic family therapy with eating disorders. *Clinical Child Psychology and Psychiatry, 8*(4), 521–535.
- Hepworth, J., & Griffin, C. (1995). Conflicting opinions? 'Anorexia Nervosa', medicine and feminism. In C. Kitzinger & S. Wilkinson (Eds.), *Feminism and discourse: Psychological perspectives* (pp. 68–85). London: Sage.
- Zachrison, H. D., & Skårderud, F. (2010). Feelings of insecurity: Review of attachment and eating disorders. *European Eating Disorders Review, 18*, 97–106.

21

Using Discourse Analysis to Study Online Forums for Young People Who Self-Harm

Janet Smithson

Introduction: young people talking about self-harm online

In this chapter, I consider the use of a discourse analytic approach to studying online forums for young people who self-harm. Diagnosis of self-harm is increasing substantially, especially among children and young people – a recent report states that there has been a threefold increase in UK teenagers who self-harm in the last decade (World Health Organization (WHO), 2014). Young people do not readily consult health professionals and often rely heavily on peer advice and support, including, increasingly, a range of online discussion forums. With increasing numbers of young people accessing the Internet for support for self-harming behaviour, there is a need to understand how such sites might benefit or harm their users and to consider the implications for moderation of online forums.

A number of recent studies have demonstrated how the study of online forum talk can highlight people's understanding of their own mental health issues and of the services which attempt to support them. Recent studies suggest that online forums can provide not only information (Vayreda & Antaki, 2009) but also mutual support (Stommel & Koole, 2010), even a 'sanctuary' (Gavin, Rodham, & Poyer, 2008), although these researchers also raised concerns that support sites could also normalise eating disorders, possibly encouraging users not to seek external help. Whitlock, Powers, and Eckenrode (2006) found a similar tension in their study of online forums for young people who self-harm; young people found these sites to be a valuable source of support, advice, and community, but the authors raised concerns over the possibilities of reinforcing self-harming behaviour as normal or even desirable. Giles and Newbold (2011) compared moderated and unmoderated forums and suggested that unmoderated online communities offer 'largely peer support rather than trusted advice and information' (p. 427). The unease from mental health professionals about vulnerable young people's use of online forums contrasts with the enthusiasm from young adults for communicating on social media.

A discursive approach moves away from the assumption of mental illness as an objective fact, towards a focus on the importance of language in the understanding of people's experiences of mental health conditions (see also Karim, Chapter 2, this volume). How are mental health problems identified, labelled, and interpreted by children and young people who post on online forums? What are the categories and practices associated with self-harm for young people who perform these activities?

Methodological approach: online discourse analysis

An understanding of how young people build up an identity in their talk as 'a self-harmer', how they make sense of this activity and identity construction, and how this is oriented towards and treated as authentic or not by others is crucial in developing ways to support young people clinically or therapeutically. In common with many of the book's authors, I argue that a discourse analysis approach has important uses for the study of mental health, and in particular, I suggest that this approach can help in moving away from the construction of mental illness as an objective truth, focusing instead on the importance of language in constructing and deconstructing the assumptions that surround the diagnosis and treatment of mental health disorders. I am taking a broad discursive approach, based on the version of discourse analysis (DA) developed by Potter and Wetherell (1987), but also drawing on Critical Discourse Analysis' concern with power in terms of how discourses make available certain versions of reality and identity, while marginalising alternative accounts (Parker, 2005). Discursive approaches have no recognisable 'method' in terms of formalised procedures, but they share a set of assumptions about language practices – an 'analytic mentality' (Schenkein, 1978). The focus is on examining discursive practices: how are topics and concerns constructed in people's accounts? Attention is paid to variability of these accounts, rhetorical aspects, and the functions of talk as a social action or behaviour. Language is understood to be a resource through which all sorts of interactional work can be accomplished (Potter, 2004), and analysis begins from what participants themselves are making relevant in their talk.

Appropriateness of online forum data for Discourse Analysis

Discourse analysts have a strong preference for naturally occurring data. Lamerichs and te Molder (2003) defined naturally occurring talk as comprising 'interactions that would have occurred regardless of whether a researcher was involved' (p. 458) and argued that such talk can be found in computer-mediated communication. Meredith and Potter (2014) argued that online talk is inherently interactional, in that 'it is designed for a particular recipient or recipients; it unfolds sequentially responding to what has come before and building a context for what comes next; and its intelligibility is centrally related to its role in building and responding to particular actions' (p. 370) and therefore is best analysed by approaches which foreground interaction. Studies which consider forum talk as a social practice in its own right include studies of forum membership (Smithson et al., 2011b), how new members are integrated (Stommel & Koole, 2010), and how appropriate ways of talking for the topic are maintained (Vayreda & Antaki, 2009).

Discourse analysts have demonstrated the role that hegemonic discourses play in shaping mental health problems and solutions. Online forums are a space in which participants – who may or may not have an official diagnosis of a mental health problem – choose to post about their experience and understanding of mental health issues. This provides the possibility of a changed power differential to that found in much research or in official health service discourses. On these sites, participants choose to join in and self-identify as having a problem or illness, even while they may be contesting the physical or psychological category of this experience or behaviour (Horton-Salway, 2007). Horton-Salway argued that online support groups are engaged in the 'collaborative production of identities' (p. 896) and members are not permitted to veer from the institutionally preferred themes. Site members orient to accepted ways of discussing a problem – in this case self-harming behaviour. Forum talk therefore is a way of understanding the shared nature of an experience or illness.

Some areas in which DA can be particularly relevant for understanding child mental health issues as presented on online forums include the following:

1. Constructions of health, mental health, and illness. What are the impacts of the categories and practices associated with a specific mental health issue?
2. Positioning. How do participants position themselves, the health services, and their parents?
3. Talk as action. What do young people use online forums for? Which activities do they perform, and with what results?
4. Power. What power dynamics are exhibited, prioritised, and marginalised in the forum interactions and in young people's descriptions of mental health issues?

People posting on health or mental health support forums are often seeking a place to be 'legitimised' or listened to, especially if they have a contested condition that is not always acknowledged by the health professionals or by wider society. With children and young people there are additional power differentials to consider. Children's and young people's health and mental health problems are regularly presented through their parents, and it is much easier to talk about valuing 'the voice of the child' than it is to include it. Hutchby and O'Reilly (2010) demonstrated how children are routinely awarded 'half-membership' status by turn-taking and sequence organisation patterns in family interactions with professionals (in their study, family therapists). In contrast, internet forums are a space in which children and young people often engage with great enthusiasm, and usually away from the 'familial moral order' (Hutchby & O'Reilly, 2010).

Ethical considerations in online research on child mental health

There are specific ethical concerns inherent in online research, those specific to mental health online forums, and for children and self-harm in particular.

Private and public space

Public forums are legally in the public domain; however, many health researchers are uncomfortable with just using data without asking posters' permission, especially as posters often view 'their' forum as a private chat space.

Many researchers aim for informed consent from site users and or owners, even when this is not strictly insisted on (British Psychological Society, 2013; Sharkey et al., 2011). McKee and Porter (2009) recommended that the researcher makes a 'case-based approach' to online data; Whiteman (2010) similarly argued that doing ethical research (in general, but particularly concerning internet sites) requires a situated approach to ethics rather than decontextualised 'solutions'.

Valid consent for children

In online research, children are often online without parental knowledge, and this is particularly likely for young people accessing self-harm forums.

Anonymity and confidentiality

Open forums are public spaces and so are accessible by search engines, which means that even if a researcher changes names and identifying features in a publication, it is possible to search the text cited and find the original thread. This may lead to identifying features or more details about the person posting.

Participant safety

Participant safety is clearly of crucial concern in child mental health research. If research uses existing sites, there is no chance of providing safeguards, but the researcher is not actually carrying out any activity in data collection. For online interventions, the researchers can offer safeguards and helplines, but with the caveat that online participants can retain anonymity/privacy whether or not this is in their best interest. This can lead to questioning whether this whole area is too risky to conduct research in. A United Nations Children's Fund (UNICEF) (2011) report on child safety online stressed the right of children 'to information, freedom of expression and association, privacy and non-discrimination' (2011, p. vii) and recommended that focus be on 'building a safer internet', educating children in risk, rather than trying to stop children using it (p. vii). The report noted that children are often active in protecting other children who they view as more vulnerable and suggests 'a potential role for children as peer educators, mentors and advisers' (2011, p. 7). Livingstone (2003) considered the risks for children on the Internet and put adult anxieties about risk for children in the context of earlier, similar worries about the risks of television. She advocated for a 'child-centred' approach which avoids constructing children as just vulnerable but acknowledges their agency in their own right, including (similarly to the UNICEF report) their right to communication and information.

Duty of care

How is appropriate care to be offered to young people in distress, perhaps suicidal? For existing forums, it can be argued that the research activity does not increase risky behaviour, but this leaves researchers, especially those who are trained health professionals and clinicians, in a difficult situation, feeling a 'duty of care' which they cannot act on (Giles & Newbold, 2011; Sharkey et al., 2011).

I will return to some of these ethical concerns in the discussion section, after the analysis of forum talk. In general, ethical guidelines for online research (e.g. British Psychological Society, 2013) suggest that the general ethical principles of research (respect for individuals, social responsibility, scientific value, maximising benefits, and minimising harm) should be applied to online research but cannot always be applied in the traditional ways.

There are few specific ethical issues for a DA approach, but one difference is how participant involvement and feedback are viewed. Georgaca and Avdi (2012) pointed out that DA relies on an assumption that participants are positioned by and using discourses of which they may not be fully aware, so it is inappropriate to ask participants to validate analysis. However, a DA can be used as a way of empowering mental health service users (Georgaca & Avdi, 2012), and one benefit of online forums is that they provide an obvious place to post analysis and generate discussion.

Project overview: data and participants

To understand how young people construct self-harm behaviour, I use extracts from three types of sites:

1. A longstanding, moderated support forum ('Recover Your Life'). They welcome 'an open and non-judgemental atmosphere' but stress that 'We do not condone any pro-activity or encourage anyone to hurt themselves in any way, and there are a few rules to hold that in place'. Posts viewed as encouraging self-harm are removed or amended, and site users who flout the rules are banned. Posts are from 2014.
2. A research-based site ('SharpTalk') set up over three months to provide discussion and support about self-harm for young people and health service professionals. The site was moderated by the research team, but aimed for a 'light touch' moderation compared to the established support sites, to encourage debate (for more on the SharpTalk project, see Jones et al., 2011; Owens et al., 2012). Posts are taken from the site during its existence in 2009.
3. A participant-led unmoderated forum, anonymised here as 'ProSite'. The site introduction stresses that 'it is not a place to encourage people to further their self-harm. Please be aware of this and make "ProSite" the support community it needs to be'. However, posts are not moderated, and posting 'pro' (in favour of) self-harm is permitted. Posts are from 2014.

In all three forums, participants identified themselves as self-harming. They used self-chosen pseudonyms or usernames. The extracts here are from threads started by those who present themselves as under 18 years of age. Interaction was asynchronous; online discussion forums allow members to log in when and wherever they choose, read what others have posted, and respond by typing their own message. Posts were arranged in 'threads' (separate discussions). Ethically, the 'Recover Your Life' participants are aware that they are posting on a large, well-known public forum. The ProSite participants are aware they are posting on a public forum but that it is less well known. Pro self-harm sites tend to change name and move server regularly, as they often get closed down when server owners discover the nature of the site. It was not feasible to gain informed consent for these public forums. The SharpTalk participants were fully aware and had given consent for their posts to be analysed and included in publications. There were threads discussing their feelings about this. Six months after the forum closed, it was reopened for a week, during which participants viewed and commented on initial research findings.

The extracts in this chapter were selected as examples of discussions about self-harming behaviour and some problems it brings, from the younger posters on the forums – they either identified their age as under 18 or talked about the role of school, worries about school exams and bullying, relationships with parents, and involvement with the UK Child and Adolescent Mental Health Services (CAMHS). Posts are shown in original language, punctuation, and structure, including abbreviations and forum or text norms, but with posting names (already pseudonyms), exact dates, and other possible identifiers changed.

Analysis: young people's construction of self-harm, health services, and parental support

Extract 1. Recover Your Life (mid thread)[1]

1.   **Mon 22.10. Sal:** I wish parents could be more understanding about things like SH.[2] If
2.   they were things would be so much easier. So, so, so much easier.
3.   **Mon 23.01. Tig:** Yeah it would be nice ... But think about it from their viewpoint.
4.   They've lived probably 30, 40, maybe 50 years. Pain was always a BAD thing ...
5.   They've probably never heard of self-harm. People were suicidal, but that was horrible
6.   too, cause no one understood that either. Self-harm sounds to them like something
7.   only someone crazy would do. I mean cutting yourself just sounds so foreign if you
8.   don't understand it. So then they find out their child is doing it ... they're bound to be
9.   upset, when all the know about self-harm is that people in mental hospitals do it. Not
10. that this is true, it's just all they know. That's why we need to somehow educate
11. people that self-harm is just a bad coping mechanism, just like alcohol, drugs, sex
12. addictions, and so forth.

In Extract 1, two young people are discussing the problem of parents not understanding about self-harm (SH). Tig's response to Sal includes a number of assumptions about self-harm. First, Tig describes it as a generational behaviour. Parents, who have 'lived probably 30, 40, maybe 50 years', are unlikely to understand this new form of behaviour. Tig contrasts the parental understanding – 'something only someone crazy would do', 'foreign', 'people in mental hospitals do it' with their understanding 'just a bad coping mechanism, just like alcohol, drugs, sex addictions, and so forth'. Tig here places self-harm as a normal coping mechanism for their generation, comparable to, and no worse than, the normal coping mechanisms for their parents' generation. From this perspective, the parental generation needs educating, so that they can be more understanding of the resources their children draw on.

Extract 2. Recover Your Life (thread opening)

1. **Fri 18.35. Tinder: Damn glass "trigger" help**
2. We went on a field trip first period.... I found a piece of glass in the road. I broke it,
3. and took a shard with me.
4. I have second period off.... I'm sitting in the library...I've cut myself with it. Damn it.
5. I need this...why?! ****
6. **Fri 18:45. Oddball: Hi,**
7. Really sorry to hear you're struggling. Have you sought medical attention? With
8. something like dirty glass you probably should have it looked at due to infection risk
9. Is there anything in particular that made you want to cut?
10. **Fri 21:14. Tinder:** I'm still at school, so no medical attention from the nurse. It's
11. glass from a beer bottle. They're minor cuts, just a lot of them.
12. I wanted to see the blood, I felt really excited that I found the glass. I just wanted to
13. see blood.
14. What's wrong with me?
15. **Fri 21:49. Oddball:** Nothing's wrong with you. I know a lot of people who SH just
16. to see blood, I know I used to do it a lot just to prove to myself that I was real/still
17. alive.
18. What time do you finish school? I'd be exhausted if I was there this late lol. Could
19. you go to the nurse, or someone confidential. Glass from a used beer bottle could
20. carry infection which you don't want.
21. Is there anyone at school or at home you could talk to about this?
22. **Sat 00:25. Petal:** I agree there is nothing wrong with you, alot of people also cut
23. because they want to see the blood so your not alone in this
24. Also I will attach a list of alternative things you could try that you might find helpful
25. [Attaches long list titled "SI urges: what to do instead"]

In Extract 2, Tinder is posting online while at school (in a North American time-zone). Tinder describes his/her[3] self-harming behaviour, and asks 'why?! ***'. Her request, therefore, is for an answer to why they performed this activity, rather than, say, a direct request for health advice. Ten minutes later, Oddball replies. It is relevant to note the order and content of her response. First, she empathises 'sorry to hear you're struggling'. Second, she offers health advice 'Have you sought medical attention?' and reminds Tinder about the infection risk. Third, she orients to Tinder's question about 'why?', asking about possible triggers for cutting. Two and a half hours later Tinder replies with details about why she isn't taking the advice on seeking health attention, details about the injury, and a response to Oddball's question about the reason for cutting 'I just wanted to see blood'. She ends with another question about the practice of self-harming 'What's wrong with me?' The responses to this last question are illuminating. 'Nothing's wrong with you', says Oddball, who elaborates that s/he knows a lot of people who do this, and s/he used to do it herself, 'just to prove to myself that I was real/still alive'. Petal similarly states, 'I agree there is nothing wrong with you', a lot of people do this, 'your not alone'. This normalisation of self-harming behaviour, and the reason for doing it (to see blood) as common, understandable, and not 'wrong', is common on the self-harm forums but in marked contrast to most standard health messages on this topic.

Aside from the normalisation of self-harming activity, however, the health advice offered is notably standard, or 'mundane' (Smithson et al., 2011a). There are many concerns from health professionals about the nature of advice given on online forums. In this extract, the advice given – to seek medical attention, to check for infection, to talk to someone at home or school – is very similar to professional advice. In her post, Petal, after reinforcing the message that there is nothing wrong with the poster, adds 'a list of alternative things' – a list of suggested distraction activities is very standard in self-harm support, and many forums provide these lists.

Extract 3. SharpTalk (mid thread)

1. **Thu 17:37. Poppy:** im sorry. i only just read that. i just got let out hospital. xxx
2. **Thu 18:01. Lex:** Are you OK hun?
3. **Thu 18:10: Poppy** no
4. im worse than i was yday but . . . i didnt get the f*ckin 'magic number' so i live to tell
5. the tale once more
6. people dont care. in the past wen i done it they dont care so they wouldnt now.
7. but im to pathetic to even get death right

8.   **Thu. 18:27: Starr** im sorry you felt that you had to do that.have they offered you any
9.   support now? were they ok with you?
10. **Thu 18:32. Poppy:** my mom refused it, said here nothing wrong, that i attention
11. seeking. thats why i couldnt have her go with me,
12. im a useless, pathetic, fat waste of a life!
13. and i got forced to eat 3 rolls!
14. **Thu 18:41. Starr:** When i went to see someone at CAMHS[4] they didnt tell my
15. parents becuase i told them that i didnt want them to. it was surprisingly easy to hide
16. it as it was at times that mostly i could choose. if you want someone to help you with
17. MH[5] stuff IRL[6] then i recon the first place to start is geting a decent GP that isnt a
18. to$$er like the one you went to see about the cut. then asking them to sort out an
19. assessment – most will suggest it before you do. one argument that may or may not
20. work with your parents is that if after the assessment they dont think there is a
21. problem then at least you know. (i dont think they are likly to do that)
22. i guess you gotta remember its your life…if you feel you need that support its your
23. decision now not your parents

In this thread previously, Poppy has been struggling with self-harming (cutting), and anorexia, and has not been impressed with her treatment by her GP (General Practitioner). Here she posts after a recent overdose leading to a hospital stay. She posts about people not caring (line 6) when she has overdosed before, which elicits a response from Starr: 'have they offered you any support now? were they ok with you?' The focus here for both Poppy and Starr is the perceived lack of care and support by unspecified 'people'. From her next post, it appears that Starr was offered some additional support but 'my mom refused it, said here nothing wrong, that i attention seeking'. Moreover, the hospital stay led to enforced eating: 'and i got forced to eat 3 rolls!' This leads Starr to write a long post about her experiences of negotiating the health services, in particular CAHMS (Child and Adolescent Mental Health Services), and the GP. Starr offers advice from her experience on how to manage the CAHMS if you do not want your parents involved. Of keeping the CAHMS appointments without her parents knowing, she says that 'it was surprisingly easy to hide it as it was at times that mostly i could choose'. A previous GP who Poppy visited about a cut is designated 'a to$$er' for their treatment of Poppy, and Starr recommends that Poppy instead gets a 'decent' one. Starr ends with an exhortation to autonomy, in the face of problems with negotiating the CAHMS service, the GP, and parents: 'you gotta remember its your life…if you feel you need that support its your decision now not your parents'.

Of interest here is the construction of the various people trying to help Poppy. The CAHMS, the GP, and parents are all portrayed by Poppy and upheld by Starr as potentially ranged against seeing the child/young person as an autonomous person. Within this, Starr aims to position them both as independent individuals who can choose their involvement with the services. In this extract, CAHMS are positioned as potentially enabling the young person in their desired autonomy, though this requires careful negotiation. GPs are positioned as variable in their support for young people who self-harm, best dealt with by changing GP if possible. Parents are positioned by Poppy as intrusive and unsupportive.

Extract 4. SharpTalk (mid thread)

1. **Tue 16:41: Mouse** i hate this.
2. Can't believe I did that.
3. She took my control from me.
4. Want to cut.
5. I want out.
6. Hate myself.
7. Why? Why? Why?.
8. **Tue 16:51: Sherbertpop** what did you do hun? what control has been taken from
9. you? you've always got some control remember.
10. **Tue 17:04. Mouse:** I've been tryng so hard to eat. But within my limits. Cause I don't
11. wanna be like this forever. We were out. Went to McDonalds. Not allowed to eat
12. salad. Had burger.
13. I feel disgusting. Didn't want it. My mum took my control away from me.
14. I felt like being sick.
15. I want to hurt myself so bad.
16. I hate me. More waah.
17. **Tue 17:09: Sherbertpop** have you explained to your mum how it made you feel? try
18. not to hurt, you're stronger than this. was this a kind of one-off meal? because maybe

19. you could regain control over your other meals, if your mum isn't there to make you

20. eat things you don't want to. what distractions could you use? maybe go for a run or

21. stomp about a bit if you're feeling angry and hating, it might take away some of your

22. negative energy xx

In this extract, Mouse posts about a problem with a specific person which is making her want to cut herself: 'she took my control away'. This elicits a response from Sherbertpop about the nature of control: 'you've always got some control remember'. The 'remember' here suggests a shared body of knowledge about the psychology of self-harm and how to deal with it. Mouse then elaborates. Her family went to McDonalds, where her mother insisted she ate a burger, not a salad. This has triggered a compulsion to cut in Mouse. Sherbertpop's advice is oriented to three things. She suggests that Mouse try and explain how she feels to her mother. She reminds Mouse that she can regain control over other meals. She ends with suggesting a distraction activity. We can note here, as with previous extracts, firstly, the relative mundanity of advice. Sherbertpop's three suggestions are drawn on regular therapeutic norms. The activity of self-harm is not seen as shocking but is seen as something to be dealt with in a variety of possible ways.

Extract 5. ProSite (thread opening)

1. **Mon 18:55. Symbolic:** I'm freaking out about whether or not to tell my therapist

2. about SH. She's new and this will be my first appointment with her but I can't let my

3. parents find out and I'm freaking out oh my godd. Do you think I'll get in trouble for

4. cutting and do they have to tell my parents?

5. What has your experience been with therapists/psychologists?

6. All help is very much appreciated.

7. Thank you! ☺

8. **Mon 19:03. Bigears:** dooooon't do it, friend. i think therapy can be super helpful and

9. i'm all for honesty and openness with your doctors but every time i have told a

10. therapist about a vice or self-destructive habit before i've felt ready to, i've regretted it

11. big time. once you open that door, she'll ask you about it every session and monitor it
12. and try to get you to talk about it. this can be super helpful if you're really impaired by
13. your self-harm and emphatically want to stop, but if you're still ambivalent or feeling
14. unsure or secretive about it then letting your therapist in could be something you'll
15. really regret. if you're a minor, i'm not sure but i don't think your therapist has to tell
16. your parents... *unless* she's acutely worried about your safety. unfortunately, in my
17. experience therapists aren't great at understanding that self-harm is usually (1) not
18. inherently dangerous (as long as you stay away from major blood vessels and keep
19. your cuts clean) and (2) NOT an indicator of suicidality. so if you tell her you self-
20. harm, even if you feel perfectly safe, she might overrule you and go over your head to
21. talk to your parents.

This final extract is from the unmoderated forum. Symbolic is concerned about what to tell her new therapist, and particularly about what the therapist might tell her parents about her self-harming. Bigears provides an extended response to this post, starting with an exhortation to not tell the therapist about self-harming. She argues that she is 'all for honesty and openness with your doctors but' she has regretted the occasions when she has opened up to a therapist. She makes a distinction between people who are definitely wanting to stop self-harming, those who are 'really impaired', and those who are ambivalent, or not really impaired by this activity. As with previous posters, she is unclear about what therapists have to disclose but does not trust therapists to keep confidential informations of this nature. The opening up to the therapist carries the risk of not being able to control the content of future discussions. We can see more normalisation of self-harming behaviour in this post. Bigears argued that self-harm is 'not inherently dangerous' if practiced safely and feels that therapists cannot be relied on to understand that it is 'NOT an indicator of suicidality'.

The tone of advice is different from the two moderated sites, where most posters are already in the healthcare system. Here there appears to be more negotiation about whether to avoid telling professionals about behaviour and also about the possibility of not wanting to be treated, or cured. On this ProSite in general, posters tended to be encouraged towards using health services and therapists, but the readiness to engage, and wanting recovery, was not taken for granted.

Discussion: online forums as a safety net?

I have aimed to demonstrate how a DA can provide an understanding of how young people use online forums to talk about self-harm. In particular, this analysis highlights the ways in which constructions of self-harm differ for young people, parents, and professionals. The young posters demonstrate the gaps in their understanding of their behaviour, and treatment options, and what they perceive as their parents' and professionals' perspectives and wishes. It is of course not the case that all professionals have the same view of self-harm, and research suggests that different professionals (nurses and psychologists, for instance) may conceptualise it differently – the former as an illness, the latter as a coping mechanism. The young people on these forums tend to characterise mainstream health system talk of self-harming behaviour as problematic, and most commonly as an illness to be cured, while on the forums self-harm is mutually agreed to be normal – not unproblematic, but a typical coping strategy for this age group, comparable to other more socially acceptable coping strategies. Self-harming, especially cutting, is talked about as a normal form of stress release, something you try to limit for safety reasons, but also accept. The triggers for the self-harm activity are viewed as the problem that needs addressing, as is the negotiation with parents and professional services. In choosing to join self-harm sites, these young people identify themselves as engaging in behaviour which is widely disapproved of, but not necessarily having an illness that needs to be cured (Stommel & Koole, 2010). Mudry and Strong (2013) described how users of a gambling support site, by the act of joining, present themselves through discursive performances, which are modified to fit the forum norms. Similarly, posters on self-harm sites modify their talk to fit site norms. On the two moderated sites in particular, they drew on acceptable normative healthcare discourses (Smithson et al., 2011a). While self-harm is constructed as a coping strategy, it is acknowledged to have problematic aspects, or to need boundaries. In particular, posters were encouraged to consider distractions, talk to people, and attempt to regain control.

The temporal aspect of the forum talk is noteworthy. Posters often post at times when standard healthcare options are not available – late at night, or at school, and receive attention and advice from peers. Often, posters are trying to distract themselves, by posting, with varying degrees of success. Stommel and Lamerichs (2014) showed how online requests for support (on a Veganism forum) can take place over multiple posts.

The construction of mental health services is markedly different from official descriptions. Young people often post with concerns about how to negotiate the support services available, and a recurrent issue is how to negotiate these as a child, or minor, while attempting to retain autonomy and control. The loss of control is viewed by posters as a trigger to further self-harm, and anxieties about therapists reporting back to parents are reinforced by other young people's experiences of this happening.

Another regular topic is the difficult relationships many posters have with parents. In these extracts we could note the tension, with parents positioned here as triggering self-harming behaviour by their concern. Parents, in general and specifically concerning how to talk to them about self-harm, are agreed on these forums to be a problem, with a variety of solutions offered.

Clinical relevance summary

One advantage, arguably, of online support sites is often perceived as the main disadvantage. They are lightly regulated, or not at all. Participants are anonymous and may not be of the age, gender, or any other characteristics they say. All this, however, can lead to more freedom to express. From a DA perspective, the accuracy of accounts is not necessarily problematic. The talk illustrates how young people choose to position themselves and how they form and negotiate their identities. The difficulties in talking about self-harm with parents and services, the different ways in which young people feel able to talk online, and in other contexts, and the importance of a sense of control over their own behaviour in terms of not triggering self-harming behaviour, all provide insight into young people's understanding of their own mental health problems and the implications of these. Another advantage of online forums which this analysis highlights is the possibility of being there in the moment of need. Participants post during the process of feeling a trigger, trying not to self-harm, just having self-harmed, or trying to stop. Temporally, the forums are open and accessible at times when conventional services are not, and the young people who self-harm often post and respond to each other late at night. Hillier, Mitchell, and Ybarra (2012) argued that lesbian, gay, and bisexual young people use the Internet as a safe place, a way of accessing health information and support not easily available to them offline, and that their 'risky' online behaviour needs to be understood in this context. Similarly, the young people on the self-harm sites appear to be accessing peer-delivered information and support, often of a sensible nature, certainly empathetic and aiming to be supportive. They contrast this acceptance of self-harming activity, and advice given within this framework, with their difficulties in negotiating their (often extensive) offline support systems. There are important implications for clinical practice from this study, especially concerning the importance of autonomy and control for these young people, the need to be able to position themselves in a discussion

Table 21.1 Clinical practice highlights

---

1. Forums as a safety net. Young people use these forums to post at times of crisis and receive support, empathy, and advice. As such they may provide a crucial safety net for some vulnerable young people, including those who are self-harming in secret.
2. Being there in the moment of crisis. The temporality and accessibility of the online forums can complement more traditional services. Many young people posting on these forums are already receiving real life support, but this provides help at exactly the time it is needed.
3. Anonymity and virtual intervention. A major drawback for clinicians is the impossibility of providing real-life intervention if a young person posts about suicidal behaviour. Only virtual support can be offered. However, this is often precisely why young people feel able to describe their activities in such detail.
4. Reconstruction of self-harm and implications for health services and parents. It is salutary to note the priorities and meanings which young people ascribe to self-harm, and to the sources of support offered. While young people realise that parents and health professionals are generally trying to help them, their accounts are fraught with negotiating the complexities of these relationships, and especially in regard to trying to regain autonomy and control of their lives and healthcare behaviour.

---

rather than be positioned through adult talk, and the impact of loss of control on triggering behaviour. For a simple summary of the implications for practice, see Table 21.1.

Summary

Overall, the forum talk gives a picture of young people struggling to cope with their behaviour and the triggers, while performing a complex set of interrelated negotiations with school counsellors, therapists, GPs, CAHMS, parents, and peers. With all these support systems, young people may be children legally, but they are attempting to take control of their own situation, including whether to disclose their behaviour, who to, and whether to aim for treatment or a cure or just empathy. The potential for online forums dominated by young people in increasing a sense of agency and control over the characterisation of their experiences is noticeable in this study and contrasts with the less agentic positioning of children in many other healthcare and therapeutic environments (Hutchby & O'Reilly, 2010).

Notes

1. Original spellings and punctuation retained. All usernames changed.
2. Self-harm.
3. The gender of the person posted is not always specified. The majority of the posters are however female and thus I use the generic 'she' throughout the chapter.
4. Child and Adolescent Mental Health Services.

5. Mental health.
6. In real life.

References

British Psychological Society (2013). *Ethics guidelines for internet-mediated research*. INF206/1.2013. Leicester. Retrieved from www.bps.org.uk/publications/policy-andguidelines/research-guidelines-policydocuments/research-guidelines-poli. Easier to find here: http://www.bps.org.uk/system/files/Public%20files/inf206-guidelines-for-internet-mediated-research.pdf.

Gavin, J., Rodham, K., & Poyer, H. (2008). The presentation of 'pro-anorexia' in online group interactions. *Qualitative Health Research, 18*(3), 325–333.

Georgaca, E., & Avdi, E. (2012). Discourse analysis. In D. Harper & A. R. Thompson (Eds.), *Qualitative health research methods in mental health and psychotherapy: A guide for students and practitioners* (pp. 147–162). Chichester: Wiley-Blackwell.

Giles, D. C., & Newbold, J. (2011). Self- and other-diagnosis in user-led mental health online communities. *Qualitative Health Research, 21*(3), 419–428.

Hillier, L., Mitchell, K. J., & Ybarra, M. L. (2012). The Internet as a safety net: Findings from a series of online focus groups with LGB and non-LGB young people in the United States. *Journal of LGBT Youth, 9*(3), 225–246.

Horton-Salway, M. (2007). The 'ME bandwagon' and other labels: Constructing the genuine case in talk about a controversial illness. *British Journal of Social Psychology, 46*(4), 895–914.

Hutchby, I., & O'Reilly, M. (2010). Children's participation and the familial moral order in family therapy. *Discourse Studies, 12*(1), 49–64.

Jones, R., Sharkey, S., Smithson, J., Ford, T., Emmens, T., Hewis, E., Sheaves, B., & Owens, C. (2011). Using metrics to describe the characteristics of, and the participative stances of members within, small discussion forums on self-harm. *Journal of Medical Internet Research, 13*(1), e3.

Lamerichs, J., & te Molder, H. F. M. (2003). Computer-mediated communication. From a cognitive to a discursive model. *New Media and Society, 5*(4), 451–473.

Livingstone, S. (2003). Children's use of the internet. Reflections on the emerging research agenda. *New Media and Society, 5*(2), 147–166.

McKee, H. A., & Porter, J. E. (2009). *The ethics of internet research. A rhetorical, case-based process*. New York: Peter Lang.

Meredith, J., & Potter, J. (2014). Conversation analysis and electronic interactions: Methodological, analytic and technical considerations. In H. L. Lim & F. Sudweeks (Eds.), *Innovative methods and technologies for electronic discourse analysis* (pp. 370–393). Hershey: IGI Global.

Mudry, T. E., & Strong, T. (2013). Doing recovery online. *Qualitative Health Research, 23*(3), 313–325.

Owens, C., Sharkey, S. J., Jones, R., Ford, T., Emmens, T., Smithson, J., & Hewis, E. (2012). Building an online community to promote communication and collaborative learning between health professionals and young people who self-harm: An exploratory study. *Health Expectations*, doi:10.1111/hex.12011.

Parker, I. (2005). *Qualitative Psychology: Introducing radical research*. Milton Keynes: Open University Press.

Potter, J. (2004). Discourse Analysis as a way of analysing naturally occurring talk. In D. Silverman (Ed.), *Qualitative research: Theory, method and practice* (pp. 200–221). Thousand Oaks, CA: Sage.

Potter, J., & Wetherell, M. (1987). *Discourse and social psychology: Beyond attitudes and behaviour*. London: Sage.

Schenkein, J. (1978). Sketch of an analytic mentality for the study of conversational interaction. In J. Schenkein (Ed.), *Studies in the organization of conversational interaction* (pp. 1–6). New York: Academic Press.

Sharkey, S., Jones, R., Smithson, J., Ford, T., Owens, C., Emmens, T., & Hewis, E. (2011). Ethical practice in internet research involving vulnerable people: Lessons from a self-harm discussion forum study (SharpTalk). *Journal of Medical Ethics*, doi: 10.1136/medethics-2011-100080.

Smithson, J., Sharkey, S. J., Jones, R., Ford, T., Emmens, T., Hewis, E., & Owens, C. (2011a). Advice, support and troubles telling on a young people's self-harm forum. *Discourse Studies, 13*(4), 487–502.

Smithson, J., Sharkey, S. J., Jones, R., Ford, T., Emmens, T., Hewis, E., & Owens, C. (2011b). Membership and boundary maintenance in an online self-harm forum. *Qualitative Health Research, 21*(11), 1567–1575.

Stommel, W., & Koole, T. (2010). The online support group as a community: A micro-analysis of the interaction with a new member. *Discourse Studies, 12*(3), 357–378.

Stommel, W., & Lamerichs, J. (2014). Interaction in online support groups. In H. Hamilton & W. S. Chou (Eds.), *The Routledge handbook of language and health communication* (pp. 198–211). London: Routledge.

UNICEF (2011). *Child safety online: Global challenges and strategies*. Florence, Italy: The UNICEF Office of Research, Innocenti.

Vayreda, A., & Antaki, C. (2009). Social support and unsolicited advice in a bipolar disorder online forum. *Qualitative Health Research, 19*(7), 931–942.

Whiteman, N. (2010). Control and contingency. Maintaining ethical stances in research. *International Journal of Internet Research Ethics, 3*(12), 6–22.

Whitlock, J., Powers, J., & Eckenrode, J. (2006). The virtual cutting edge: The Internet and adolescent self-injury. *Developmental Psychology, 42*(3), 407–417.

World Health Organisation (2014). *Health behaviour in school aged children*. Geneva: World Health Organisation Collaborative Cross-National Survey. HSBC Network.

Recommended reading

- McKee, H. A., & Porter, J. E. (2009). *The ethics of internet research. A rhetorical, case-based process*. New York: Peter Lang.
- Smithson, J., Sharkey, S. J., Jones, R., Ford, T., Emmens, T., Hewis, E., & Owens, C. (2011b). Membership and boundary maintenance in an online self-harm forum. *Qualitative Health Research, 21*(11), 1567–1575.
- UNICEF (2011). *Child safety online: Global challenges and strategies*. Florence, Italy: UNICEF Office of Research, Innocenti.

22
Using Conversation Analysis for Understanding Children's Talk about Traumatic Events

Amanda Bateman, Susan Danby, and Justine Howard

Introduction

There is a wealth of research and literature investigating children's traumatic experiences from a psychological perspective, whereas there is relatively less literature using a sociological approach that includes discourse, narrative, and conversation. This chapter aims to demonstrate the importance of investigating children's trauma talk through this latter approach by providing a theoretical overview of literature that uses conversation analysis (CA) to explore children's interactions related to trauma and associated mental health matters.

In relation to understanding trauma talk, CA provides a detailed insight into the systematic ways in which people discuss experiences of trauma through their talk-in-interaction. Sacks, Schegloff, and Jefferson's (1974) development of CA afforded a closer look at the turn-taking evident in conversation, so as to reveal the orderly and systematic ways in which people co-constructed every-day interactions with one another. CA has helped counsellors to become aware of how their turn at talk prompts a desired response from people experiencing trauma and of turns at talk that may hinder the progression of disclosing talk about trauma. The relatively new approach of using CA to understand trauma in childhood reveals the importance of talk in the process of recovery, as well as *how* the participants co-construct talk about traumatic experiences.

The first part of the chapter explores literature specifically attending to children's trauma talk with people who have been professionally trained in psychological techniques from a CA approach (e.g. Butler, Potter, Danby, Emmison, & Hepburn, 2010; Danby, Baker, & Emmison, 2005; Hutchby, 2007) where 'the discourse is in a significant sense task-oriented' (Hutchby & O'Reilly, 2010, p. 50). The second section of the chapter discusses how children talk about their traumatic experiences with people who are not qualified therapists or psychiatrists, people such as their peers and teachers (e.g. Bateman & Danby, 2013; Bateman, Danby, & Howard, 2013a, 2013b; O'Connor, 2013; Osvaldsson, 2011). The final section of the chapter summarises how children's trauma talk compares and contrasts when co-constructed with either professionally trained psychotherapists or teachers and peers. Need for more research using a CA approach for investigating children's traumatic experiences is called for due to the significance it holds in providing insight into each child's personal sense-making of the traumatic event, and with a range of people (see also Fasulo, Chapter 1, this volume).

Introducing trauma

Trauma can be defined as any deeply disturbing or distressing experience. While some experiences are clearly traumatic, it is difficult to identify an exhaustive list of events. This is because the ability to manage the challenge of trauma is dependent on numerous factors – for example, the complexity of the trauma, the individuals' coping abilities, their subjective experience of the trauma itself, and also the level of external resources and support mechanisms available to them at the time (Hendry & Kloep, 2002). It is vital, however, that children who have experienced events that are likely to have been traumatic receive specialist support in order to reduce the potential impact of the trauma as well as the possibility of their experiencing post-traumatic stress disorder (PTSD) and any other secondary problems (National Institute of Clinical Excellence (NICE), 2005). PTSD is diagnosed in those who have experienced a highly stressful 'event or situation ... of an exceptionally threatening or catastrophic nature, which is likely to cause pervasive distress in almost anyone' (World Health Organization, 1992, p. 147).

Specialist support in the form of therapy or counselling is available for children and young people with symptoms of trauma. The majority of child counsellors, psychotherapists, and play therapists tend to adopt an integrative approach to their work, which utilises a variety of techniques to best support the child in moving from an emotionally disturbed and dysfunctional state towards resolution and adaptive functioning (Prendiville, 2014). The child largely determines the pace of the therapeutic process, maximising the potential for them to reach timeous resolution while minimising the potential for further sensitisation (Kaminer, Seedat, & Stein, 2005). Therapeutic intervention is designed to enable the child to disclose and describe the traumatic event, to deal with difficult and destructive emotions, and to develop new ways of thinking and feeling. Each stage utilises different counselling styles, and the nature of the interaction and conversational techniques is important in facilitating the therapeutic process. One example of an integrative approach is Sequentially Planned Integrative Counselling for Children (SPICC). In this model, Geldard, Geldard, and Foo (2013) propose five stages of the therapeutic process, each

Table 22.1 Interactional and conversational orientation during each stage of the SPICC process

Phase	Aim	Nature of interaction and conversation
1	To facilitate the therapeutic relationship between child and counsellor and encourage the child to begin telling their story	Based on child-centred psychotherapy, to facilitate an open and authentic relationship, skills which permit the child to tell the story in whichever way they wish will be adopted. These might include reflection, summarising, feeding back what has been said, and using open questions.
2	To encourage more storytelling and an awareness of emotions	Based on Gestalt therapy, following disclosure, issues revealed in the story might be returned to by the counsellor for exploration and the child will be encouraged to talk about their emotions.
3	To develop a new perspective of themselves or the issue, enhancing self-perception	Based on narrative therapy, in this phase the counsellor will encourage the child to tell and retell their story in numerous ways to begin the process of altering their perspective. As a new or preferred story emerges, the counsellor will support the child in retaining this to enhance sense of self.
4	To challenge unhelpful emotions or beliefs and consider alternatives	Based on cognitive behavioural therapy, here, talk is directly focused on challenging thoughts and behaviours and learning new ways of thinking.
5	To try out alternatives and experience consequences to reinforce adaptive change	Based on behavioural therapy, the consequences of different behaviours and ways of thinking are directly experienced and evaluated through reward and incentives. Positive and negative outcomes might be recorded.

with a specific aim and related interactional and conversational techniques (see Table 22.1).

NICE (2005) proposes that one of the main symptoms of PSTD is re-experiencing aspects of the traumatic event. For children, the re-experiencing is likely to manifest in repetitive re-enactment through play, drawing, or dreams, in more or less recognisable forms. Specialist developmentally appropriate support enables the child to benefit from both the process of play as abreaction in itself and the exploration of emotions through talk in the therapeutic context. Joining with the child during their play within the context of the therapeutic relationship provides opportunities for the therapist to facilitate the child's understanding of their experience and the reframing of their trauma responses

(Prendiville, 2014). Of importance, however, is the observation that children spontaneously re-enact and talk about trauma in their everyday play (Bateman et al., 2013a, 2013b). Indeed, Terr (2003) contended that post-traumatic play is more likely to be seen in these spontaneous play situations than in the therapy room itself, as it is often played out in a particular location with the same materials each time. Understanding the nature of interactions and talk during this everyday play alongside work inside of the therapeutic relationship is therefore important.

Children can be supported in resolving their experiences of trauma using a variety of therapeutic techniques. O'Connor (2012) demonstrated the benefit of theatrical creative-arts-based therapy with children following the Christchurch earthquakes, reframing children as actors in the event. The emphasis on theatre and performance created dramatical distance, which enabled children to talk about the earthquake very soon after the event, enabling healing to proceed through framing and reframing. Ruf et al. (2010) documented the benefits of a narrative exposure therapy for children and adolescents (KidNet) for supporting refugees traumatised by organised violence. Children were supported by a therapist in constructing a narrative of their experiences and were encouraged to relive and become more habituated to the associated emotional responses. After only eight sessions, the authors report a significant reduction in PTSD symptoms and a return to normative functioning. Both these approaches demonstrate the therapeutic value of children talking about their experiences.

However, as argued by Gibbs, Mutch, O'Connor, and MacDougall (2013), there is a need for a wider range of critical informed studies about children's experiences of disaster. Of particular note is that, currently, a majority of studies tend to focus on children's talk within the context of the therapeutic relationship or therapeutic process. While children benefit from this specialist intervention, their coping mechanisms are extensively supported through everyday interactions with parents, teachers, and friends (Prinstein, La Greca, Vernberg, & Silverman, 1996). Understanding the nature of children's spontaneous talk about disaster using CA has powerful implications for the development of children's services; for example, in training teachers and healthcare staff, and in developing training courses and support groups for parents and families.

Trauma talk in therapy settings

In therapy situations, psychiatrists and therapists are classed as professionals as they are observable as such through their co-production of 'institutional talk' between themselves and their patients. Within the domain of CA, Sacks suggested that everyday interactions could occur among people who either are 'professionals' in the know or those who are less knowledgeable, 'laypeople' (Sacks, 1992). Referencing membership categories he described collections of people, including therapists, as members of the knowledge (K) collection category (Sacks, 1972). In relation to the work of therapists, collection K is 'a collection constructed by reference to special distributions of knowledge existing about how to deal with some trouble' (Sacks, 1972, p. 37). The therapist's professional relationship with their patient is made evident as a standard relational pair (SRP) where the therapist asks the patient questions about their trouble, and a preferred answer from the patient is to provide some information for discussion (Stivers & Robinson, 2006). A patient–therapist SRP involves knowledge distribution between the members, where it is the role of the therapist to open the meeting, ask questions, offer advice, and close the meeting:

> A 'beginning of a therapy session' is an attended part of the therapy session, as compared to just drifting into therapy talk, or 'therapy talk' being undifferentiated from any other talk that gets done while these persons are more or less together.
>
> (Sacks, 1992, p. 106)

Sacks' statement here highlights the difference between the systematic ways that people use talk to co-produce a therapy session, as well as the talk that they use to co-produce everyday interactions where they 'drift' between topics of conversation. The observable way that the features of talk are systematically different in various situations suggests that conversational exchanges will look distinctive in different institutions; for example, talk in school classrooms will look different to talk in therapy sessions. With regard to children's talk about traumatic events, children may have the opportunity to talk to adult specialists with professional training to deal with issues surrounding people's traumatic experiences in a therapeutic way, such as psychiatrists, psychologists, and therapists. It is also recognised that children do discuss traumatic events that they have experienced with adults who have not received such medical training, namely teachers and parents who engage with traumatised children on a regular basis.

Children's trauma talk: Insights from CA

In the case of the youngest members of society, CA has been used to investigate children's talk to therapists (e.g. Hutchby, 2007; Hutchby & Moran-Ellis, 1998; Hutchby & O'Reilly, 2010; O'Reilly, 2008) and children's talk with teachers (e.g. Bateman, 2013; Bateman & Waters, 2013; Bateman et al., 2013a; Gardner, 2013; Gardner & Mushin, 2013). Some instances of teacher–child interactions in early childhood education in New Zealand reveal that there are often similarities between exploring an educational 'task' problem and the features of 'emotional' problem talk in therapy sessions (Bateman, 2013). Likewise, when children discuss a lived traumatic event with their teachers in an early childhood setting, as was documented following the Christchurch earthquakes in New Zealand, there are elements of teaching and learning evident in the conversation exchanges about the traumatic event (Bateman & Danby, 2013; Bateman et al., 2013a, 2013b).

Although therapy sessions and early childhood settings can both be classed as 'institutions', the main focus of a therapy session is to talk about emotional troubles whereas early childhood education settings provide a range of educational experiences where the topic of a traumatic event could be drifted in and out of. The use of CA is paramount in identifying the differences and similarities of conversational features during these different situations as it affords the opportunity to reveal how such talk about traumatic experiences is achieved by the participants in situ. These findings have implications for practice where the importance of children sharing their traumatic experiences in order to avoid future complications such as PTSD is noted (NICE, 2005).

Children's trauma talk with therapists

During professional, clinical-type settings such as therapy sessions, talk that stimulates disclosure about a traumatic event is purposeful where the professionally trained adult steers the conversation towards a discussion of the child's traumatic experience, making this the central conversation item during the interaction where there is a specific agenda for the topic of the talk (Hutchby & O'Reilly, 2010). Through engaging in talk in such a way the turn-by-turn verbal actions of the interlocutors work to co-produce the context of the interaction as institutional talk, as articulated by Hutchby (2007):

> [A]t the same time as acknowledging that institutional settings clearly involve participants adopting particular roles and engaging in relatively specialized speaking practices, CA emphasises the active work of participants in rendering these roles and speaking practices into a lived reality. In other words, the observably specialized nature of institutional discourse must be seen as actively produced by the participants. (p. 32)

Hutchby goes on to demonstrate the ways in which counsellors prompt children to disclose information about the traumatic event that led to them needing professional help. These conversations revealed the frequent use of questions by the counsellor about the child's feelings; however, Hutchby points out that the surrounding literature discourages the use of questions by counsellors as they may be perceived by the children as intrusive. Furthermore, Hutchby discusses the delicate way that counsellors work towards identifying the child's trouble and perspective of events in order to provide some future guidance in dealing with it. The CA approach to the analysis revealed that the counsellors 'picked up on aspects of the discourse environment – not just the words that were spoken, but also objects or circumstances that were implicated in what was said – in such a way as to translate them into therapy objects' (Hutchby, 2007, p. 60). This was accomplished through the counsellor attending to the child's talk through 'active listening' by repeating or summarising what the child has disclosed in their turn at talk with a formulation. These findings are important for practice as they not only inform of how conversational strategies work to elicit information about a child's trauma where the child subsequently offers some information about the trouble, but can also be met with the utterance 'I don't know' from the child as they resist the opportunity to elaborate.

Other institutional contexts in which children may be involved in disclosing to professionally trained adults their traumatic experiences involve police interviews. Children's talks with police officers about instances of sexual abuse have been investigated using CA (Fogarty, 2010; Fogarty et al., 2013). Within this study, police officers who were members of the Sexual Crime Investigation Branch (SCIB) of South Australian Police (SAPOL) were observed to use specific language that progressed the talk about the topic of the abuse in order to prompt further information from the teller. Within this research, the CA approach revealed how the police officer who interviewed the abused child about their traumatic experience demonstrates specific conversational strategies that enable the child's disclosure about the event through such features as continuers, non-verbal actions and '*progressivity* – the smooth, collaborative completion of actions within a given sequence of interaction' (Fogarty et al., 2013, p. 414, emphasis in original).

There is increasing attention being directed to the use of CA as an intervention tool for self-reflection on talk and mental health. The Discursive Action Model (DAM), introduced by Edwards and Potter (1993), shifted an agenda from focusing on cognition to focusing on how discursive actions are part of everyday life. The DAM model called for a re-examination of language-based approaches to expand theoretical work, with a shift from perceptually focused approaches to one that investigated how participants managed issues such as blame, compliments, invitations, and so on. Topicalising the discursive actions within activity sequences, to examine talk as social action, makes possible an investigation of the interactional work that happens in descriptions and practices of everyday life. Building on this early work by Edwards and Potter (1993), is a specific intervention approach for young people called the 'discursive action method', which was developed by Lamerichs and te Molder (2011). This approach has been used to encourage adolescents to think about how they talk about health and to develop interventions with a health focus designed to support their peers (Lamerichs, Koelen, & te Molder, 2009; Lamerichs & te Molder, 2011).

Through the use of CA, these studies demonstrate the importance of using specific conversational strategies for eliciting information from children in professional settings, revealing the formal, frequently adult-governed interactions co-produced between professionally trained counsellors and SCIB officers and traumatised children as institutional talk. However, adults must be aware that the sensitive timing of therapeutic intervention is of paramount importance as children must be ready to work with their trauma (Prendiville, 2014). As such, adults can take the child's lead through providing helplines and online forums that are available for children whenever they feel ready to communicate about their trauma.

Children's use of helplines, online forums, and emergency call centres

The early work of Sacks and his investigation of calls to a suicide prevention hotline (Silverman, 1998) was the beginning of a body of research investigating calls to helplines, warm lines, and emergency call centres (Baker, Emmison, & Firth, 2001; Butler, Danby, Emmison, & Thorpe, 2009; Edwards & Stokoe, 2007; Hepburn, 2005; Landqvist, 2005; Leppanen, 2005; Potter & Hepburn, 2003; Pudlinski, 2002; Raymond & Zimmerman, 2007; Shaw & Kitzinger, 2007; Stokoe & Hepburn, 2005; te Molder, 2005; Wakin & Zimmerman, 1999, Watson, 1981, 1986; Whalen & Zimmerman, 1998; Zimmerman, 1992). It has only been within the last 15 years that research has begun to focus more specifically on children and young people contacting organisations for help.

Children and young people are documented to communicate their troubles to counsellors. Kids Helpline is an Australian helpline staffed by professional counsellors operating 24 hours a day with more than 500,000 contacts from children and young people per year. Their use of different technological modalities (telephone, email, online chats) has been investigated to show their impact on the patterns of communicative interaction; that is, to identify the differences that technologies make to young peoples' reporting of troubles. A comparison of telephone and online counselling in opening sequences found differing modalities shape what is possible in terms of counsellors displaying 'active listening' (Danby, Butler, & Emmison, 2009). The usual sorts of signals (e.g. continuers) counsellors use to indicate they are listening in phone calls are not possible with web chat, so counsellors have designed new ways of displaying active listening, such as commenting on the length of time to type messages.

Modality impacts on the sequential organisation of counsellor–client interactions: web counselling involves a quasi-synchronous interaction where turns can overlap and a response to one turn does not always follow the turn it is responding to, which has implications for how the interaction is understood. As Harris, Danby, Butler, and Emmison (2012) found in their investigation of counsellor strategies as they shift the client from email to phone/web chat counselling, there can be significant periods between email messages. Counsellors used multi-layered approaches to request and persuade clients to change their counselling relationship from text to talk without placing that relationship in jeopardy. For example, counsellors asked questions to encourage certain responses from clients to maintain counselling aspects of the interaction, helping progress the chain of emails from troubles telling to helping clients identify and carry out potential problem solutions.

The Kids Helpline research has focused on how young clients tell their troubles and how counsellors respond to young callers and provide counselling support. Findings include showing how counsellors provide interactional space for young callers to design their own entry into the counselling sessions, dimensions of interactional sensitivity, and skill of callers and counsellors (Danby et al., 2005). Such narratives result from the sequential environment and opening turns by counsellors. Specific constructions provided by clients in formulating reasons for contacting were identified (Emmison & Danby, 2007a). Counsellors use specific practices to avoid advice giving in line with the institutional remit of providing support, and not advice (Butler et al., 2010), so that strategies include questioning to help clients solve problems without being directive, while empowering clients and promoting self-directiveness. Counsellors use address terms to manage structural and interpersonal aspects of counselling, client rapport, and trust, particularly when counsellors disagree or challenge callers (Butler, Danby, & Emmison, 2011). Another strategy used by counsellors is the proffering of script proposals to give specific examples of what clients could say to a third party (Emmison, Butler, & Danby, 2011), and praise and compliments help callers identify strengths and resources, such as dealing with bullying at school (Danby, Butler, & Emmison, 2011). Moral versions of callers are collaboratively constructed to display moral reasoning, making visible counsellors' interactional work to support and empower clients (Danby & Emmison, 2014). Analysis of prank/testing calls show the collaborative and artful collaborations by caller and counsellor working towards the success of the call, showing how counsellors use professional skills effectively (Emmison & Danby, 2007b).

The emerging body of work involving children calling for help is now allowing for comparative analysis of how different service organisations orient specifically to young callers. For instance, comparing calls to the helpline designed specifically for children and calls made by young people to a Swedish emergency call centre (Cromdal, Persson-Thunqvist, & Osvaldsson, 2012) shows how different services orient to their institutional remit, with implications for call organisation, showing not just differences in how call-takers interact with clients and control decision-making, but also similar practices in terms of fact-finding, orientation to background noise and events, call back invitations, and comfort/emotion work.

Online forums are increasingly interactional spaces where young people experiencing mental health problems participate in Internet-enabled communities. Osvaldsson (2011), for example, examined how young people describe on an online forum their experiences of being bullied to show how young people use the telling of their experiences to construct their identities as victims or to seek help and advice from the online (non-professionally trained) peer community to deal with the ongoing bullying. In addition to this, Smithson (Chapter 21, this volume) has explored how young people identify, label, and interpret their problems of self-harming in an online space, and she has also examined how the young people provided peer support in this environment.

Children's everyday trauma talk with teachers

Although literature discussed so far has a focus on empirical research that has used CA to reveal the structures of talk evident between counsellors and children regarding traumatic experiences, policy documents and guidance accessible to parents and families that offer advice on how to help children cope with traumatic experiences are often produced from psychological research. For example, Goodman (2012) offered advice to both parents and teachers with reference to the American Psychological Association, where children's reaction to trauma in relation to their developmental stages is outlined. Advice is provided regarding what to say to children who have experienced a traumatic event, such as engaging in talk with children about their experiences that includes discussing 'unpleasant, confusing feelings' (Goodman, 2012, p. 20). Such documents are often published in response to some traumatic event, such as that of the Christchurch earthquake in New Zealand where 'parents and school/early childhood centre staff were concerned for the physical and psychological well-being of children, and were uncertain whether or not they should be seeking professional help for them' (Brown, 2012, p. 88). Furthermore, teachers stated that they wanted 'information on children's reactions and recovery, behaviour, self-care and positive recovery stories and supports available to families' (Dean, 2012, p. 95). Such reactive government documents often recommend talking to children about their traumatic experiences in order to stimulate recovery as people come to terms with events through sharing their stories with supportive others in a timely manner.

Adults can take the child's lead in determining their readiness to discuss traumatic experiences through the signs and signals they demonstrate in their talk and play, or through changes to their other day-to-day behaviours (e.g. toileting, sleeping, or eating patterns). Therefore, increased understanding about how children play out and talk about their trauma experiences in everyday activities is particularly important, as it is during this everyday play where symptoms of trauma or PTSD will first manifest. Parents and educators can be alert to the signs of trauma in children's play and conversation, and be ready to respond sensitively, ensuring that their reactions to this early re-enactment do not lead to increased emotional sensitivity. Equipping parents and educators with non-directive play and child-centred counselling skills (the interactional and conversational techniques which are associated with the early stages of the SPICC process) may include ensuring that no judgements about the play are made (whether these be positive or negative) so that children feel safe and permitted to play as they wish. Reflection and summarising techniques can demonstrate their witnessing and acceptance of the child's play (Howard, 2010). In addition, psycho-education to normalise the trauma, along with being mindful as to how their own emotional reactions to the traumatic event or the content of the child's play and talk could influence their responses and inhibit the child's healing process, is important (Kaminer et al., 2005).

Insights can be gained by investigating conversations about traumatic events in the everyday interactions between children and their teachers who are not clinically trained in trauma counselling, but who recognise the importance of considering the child's readiness to disclose information and feelings about their traumatic event. Like in clinical settings, schools are institutional settings where members' identification of specific roles and identities are demonstrated by their orientation to these observable identities as they co-produce the context (Hutchby, 2007). In primary schools and early years settings, the social organisation within is specific to each individual institution as it is co-constructed through the participants' talk and actions (Butler, 2008; Danby & Baker, 2000). Although this is apparent, there remain some similarities in each institution such as the teacher instigating the allotment of turns between themselves and the children, usually through question and answer scenarios (Bateman, 2013). Likewise, similarities between teacher–child and therapist–child conversations are evident when the topic of talk is having experienced a traumatic incident, observable through the conversational features of adults prompting a telling about the event (Bateman & Danby, 2013; Bateman et al., 2013a) and demonstrating active listening (Bateman et al., 2013a). However, there are also conversational differences evident in trauma talk between teachers and children compared to children and therapists, where the teachers have been found to co-produce affiliation with the child through their collaborative storytelling when discussing the event, as both members were part of the same traumatic experience (Bateman & Danby, 2013).

When examining children's talk with teachers about traumatic experiences, the usefulness of a CA approach is evident in an emerging body of work. For example, investigating teachers' communication with preschool-aged children in relation to the severity and unexpectedness of natural disasters in a post-earthquake situation has found that teachers have particular roles in supporting families to come to terms with what has happened. Bateman and colleagues (Bateman & Danby, 2013; Bateman et al., 2013a, 2013b) show how preschool teachers worked with the children and families in preschool classrooms following the Christchurch earthquakes where, months after the earthquake, families and communities were still dealing with the aftermath of these traumatic events. Children often initiated talk about their memories of the earthquake, and what happened to them and their families, prompting other children and teachers to tell their second stories of what happened. The teachers skilfully shifted the talk and conversation from children's memories of the earthquake to encourage stories of how their communities were managing through helping each other and working together. Talk about the traumatic earthquake experience was also evident when teachers and children went on outdoor excursions where observing broken buildings and roads led to talk, through pivotal utterances, about the damage to the environment during the actual earthquake event. The use of CA was found to be key to revealing how children's traumatic experience of a natural disaster were managed on a daily basis between children and their teachers, and this afforded an understanding about the ongoing processes of normalising such events in a timely manner, when the children were ready to talk about them.

Up until now there has been resistance towards the integration of ideas from the field of counselling and therapy into education (Mayes, 2009). Given that we are now far more aware of the qualities inherent in everyday interactions that render it powerful from a developmental and therapeutic perspective, perhaps we must reconsider this reluctance. When children are in a playful state, they are free from the fear of failure, and while in control of their activities, they are able to try out and try on emotions and ideas (Howard & McInnes, 2013). Within a true play-based curriculum, where children are afforded autonomy and control, teachers arguably function as therapists to some extent 'not in the formal sense of conducting therapy ... but in the practical sense of being alert and responsive to psychological needs' (Basch, 1989, p. 772).

While there is some caution associated with the rise of therapeutic education (Eccleston & Hayes, 2008), with increased emphasis on emotional health and well-being within play-based curricula, we are beginning to recognise the shift away from the *teacher–master* conception to the *teacher–therapist* conception (Shalem & Bensusan, 1999), and educators are requiring professional development necessary to prepare them for this role. Soles, Bloom, Health, and Kargiannakis (2009) argued that teachers have such a fundamental role in supporting children's mental health and identifying when intervention might be required, and that their professional and academic development must be reviewed to ensure adequate emphasis on typical and atypical development.

Clinical/educational relevance: The co-construction of trauma talk with children

Hutchby (2007) documents the usefulness of investigating child trauma through CA:

> Studies of counseling and psychotherapy using CA show how a close examination of the ways people talk in such settings can provide valuable insights into the kinds of strategies used by counselors to help make sense of and find ways of dealing with their situations. (p. 125)

The analysis of how children's disclosure of talk about trauma is co-produced in clinical settings offers insight into how participants take on roles in the co-production of institutional talk (Hutchby, 2007). The features of such clinical talk are asserted through the counsellor demonstrating active listening and formulations or repeats of children's utterances (Hutchby, 2007). The disclosure about traumatic events is also investigated through the use of CA where it is documented as being achieved with progressivity and continuers in special police interviews (Fogarty, 2013), offering implications for clinical practice through recognising the importance of the police interviewer conversation techniques when eliciting sensitive information from children about their traumatic experience. It is important to

> focus on a prospective interviewer's skill in restoring progressivity to interactions, since this is likely to be a marker of their capacity to move the interaction back and forth between less delicate and more delicate topics, so as to maintain the child's engagement in the interview until the required information is gathered.
>
> (Fogarty, 2013, p. 417)

Guidance for helping aid children's recovery following a traumatic event is often aimed at people who are not trained in professional counselling and highlights the significance of talking about traumatic experiences to help children to come to terms with the event and so help minimise PTSD (Brown, 2012). The need for this further understanding about how to help support children who have experienced a traumatic event, and so also their families, is a significant matter to be addressed (Petriwskyj, 2013). CA is a useful way to explore how children talk about trauma (see Lamerichs, Alisic, & Schasfoort, Chapter 33, this volume). CA demonstrates the ways in which talking about the traumatic event is achieved as a joint project, informing of the ways that adults untrained in therapeutic techniques do manage such talk about trauma when children approach them during everyday activity, when they are ready (Bateman & Danby, 2013; Bateman et al., 2013a, 2013b).

The literature discussed in this chapter sheds light on how interactions with children are co-produced in practice with adults who have been professionally trained in children's mental health issues, as well as with those who interact with children on a daily basis but may not have had professional psychological training. Therapy talk is unique in that both child and counsellor are present at particular and usually pre-specified times to focus on discussing the child's trauma, whereas children may disclose to a teacher their perspectives of a traumatic event by drifting into the topic during everyday interactions. These various studies reveal that CA offers findings that inform practice about the ways in which trauma talk is attended to by participants in situ; adults can provide help and support for children in structured, formal settings or be sensitive to their orientation to trauma talk during everyday interactions. For a simple summary of the implications for practice, see Table 22.2.

Summary

There have been three core issues that have been raised in this chapter. First is that an approach that draws on sociological understandings highlights the

Table 22.2 Clinical/educational practice highlights

1. Therapy talk about a traumatic event between therapist and child is deliberate as both members are present in that place at that time to focus on discussing the child's trauma.
2. Trauma talk with teachers is more spontaneous where children may disclose information about the traumatic event to a teacher or parent by drifting into the topic during everyday interactions.
3. There are similarities between teacher–child and therapist–child conversations about traumatic experiences as, in both situations, the adult can prompt a telling about the event and demonstrate active listening. However, there are also differences in the talk between teacher–child and therapist–child in their conversations about traumatic experiences where teacher–child interactions can display affiliation through their collaborative storytelling when discussing a traumatic event that both members experienced.
4. New knowledge regarding the conversational features involved in the co-production of trauma talk between teachers and children could inform therapeutic talk in counselling sessions as it reveals how children turn to talk about their traumatic experience in a timely manner when they are ready.

value of investigating trauma talk as a therapeutic intervention. Further, CA as a method offers an analytic tool to study, in fine detail, how children and young people initiate, participate, and engage in talk about traumatic events, and how adults including professional counsellors and therapists, and teachers, can engage in talk that offers therapeutic intervention. A close examination of what and how matters are raised shows what is important for the participants, both children and adults.

Second, we have noted that there are some similarities between children's talk with counsellors and teachers in that there were interactional opportunities for children to initiate topics of concern to them. One advantage of educators engaging in this very specific kind of talk is that these moments may happen in everyday situations where contextual events may provide a natural prompt for a child or children to talk about what has happened. As well, teachers may know the children well, and children have the opportunity to talk about the traumatic event with peers and not only in a one-one relationship with an adult. The prompting to talk about a therapeutic event may be more difficult within a therapeutic professional context, although counsellors and therapists may make use of artefacts to support this process. As well, therapists and children may need time to build trusting relationships.

Finally, it is evidence that to date, there are relatively few studies using CA to investigate young children's conversations about traumatic events. In our search, we did not find studies that investigated, for example, how parents and other family members might participate in family talk around traumatic events involving young children. Additionally, we searched for studies of how educators of children in elementary and secondary schooling might engage in trauma talk with students.

Acknowledgements

The authors thank Joyce Lamerichs for her insightful feedback on an earlier draft of this chapter.

References

Baker, C. D., Emmison, M., & Firth, A. (2001). Discovering order in opening sequences: Calls to a software helpline. In A. McHoul & M. Rapley (Eds.), *How to analyse talk in institutional settings: A casebook of methods* (pp. 41–56). London/New York: Continuum.

Basch, M. (1989). The teacher the transference and development. In K. Field, B. Cohler, & G. Wood (Eds.), *Learning and teaching in education: Psychoanalytic perspectives*. Madison, CT: International University Press.

Bateman, A. (2013). Responding to children's answers: Questions embedded in the social context of early childhood education. *Early years: An International Research Journal, 33*(3), 275–289.

Bateman, A., & Danby, S. (2013). Recovering from the earthquake: Early childhood teachers and children collaboratively telling stories about their experiences. *Disaster Management and Prevention Journal – Special Issue: Lessons from Christchurch, 22*(5), 467–479.

Bateman, A., Danby, S., & Howard, J. (2013a). Everyday preschool talk about Christchurch earthquakes. *Australia Journal of Communication – Special Issue: Disaster Talk, 40*(1), 103–123.

Bateman, A., Danby, S., & Howard, J. (2013b). Living in a broken world: How young children's well-being is supported through playing out their earthquake experiences. *International Journal of Play – Special Issue: Play and Well Being, 2*(3), 202–219.

Bateman, A., & Waters, J. (2013). Asymmetries of knowledge between children and teachers on a New Zealand bush walk. *Australian Journal of Communication – Special Issue: Asymmetries of Knowledge, 40*(2), 19–32.

Brown, R. (2012). Principles guiding practice and responses to recent community disasters in New Zealand. *New Zealand Journal of Psychology, 40*(4), 86–89.

Butler, C. W. (2008). *Talk and social interaction in the playground.* Aldershot: Ashgate.

Butler, C. W., Danby, S., & Emmison, M. (2011). Address terms in turn beginnings: Managing disalignment and disaffiliation in telephone counselling. *Research on Language and Social Interaction, 44*(4), 338–358.

Butler, C. W., Danby, S., Emmison, M., & Thorpe, K. (2009). Managing medical advice seeking in calls to Child Health Line. *Sociology of Health and Illness, 31*(6), 817–834.

Butler, C. W., Potter, J., Danby, S., Emmison, M., & Hepburn, A. (2010). Advice implicative interrogatives: Building 'client centred' support in a children's helpline. *Social Psychology Quarterly, 73*(3), 265–287.

Cromdal, J., Persson-Thunqvist, D., & Osvaldsson, K. (2012). 'SOS 112 what has occurred?' Managing openings in children's emergency calls. *Discourse, Context & Media, 1*(4), 183–202, doi:10.1016/j.dcm.2012.10.002.

Danby, S., & Baker, C. (2000). Unravelling the fabric of social order in block area. In S. Hester & D. Francis (Eds.), *Local educational order: Ethnomethodological studies of knowledge in action* (pp. 91–140). Amsterdam: John Benjamins.

Danby, S., Baker, C., & Emmison, M. (2005). Four observations on opening calls to Kids Help Line. In C. D. Baker, M. Emmison, & A. Firth (Eds.), *Calling for help: Language and social interaction in telephone helplines* (pp. 133–151). Amsterdam/Philadelphia: John Benjamins.

Danby, S., Butler, C. W., & Emmison, M. (2009). When 'listeners can't talk': Comparing active listening in opening sequences of telephone and online counselling. *Australian Journal of Communication, 36*(3), 91–113.

Danby, S., Butler, C. W., & Emmison, M. (2011). 'Have you talked with a teacher yet?': How helpline counsellors support young callers being bullied at school. *Children & Society, 25*(4), 328–339.

Danby, S. & Emmison, M. (2014). Kids, counsellors and troubles-telling: Morality-in-action on an Australian children's helpline. *Journal of Applied Linguistics and Professional Practice, 9*(2), 263–285.

Dean, S. (2012). Long term support in schools and early childhood services after February 2011. *New Zealand Journal of Psychology, 40*(4), 95–97.

Eccleston, K., & Hayes, D. (2008). *The dangerous rise of therapeutic education.* London: Routledge.

Edwards, D., & Potter, J. (1993). Language and causation: A discursive action model of description and attribution. *Psychological Review, 100*(1), 23–41.

Edwards, D., & Stokoe, E. (2007). Self-help in calls for help with problem neighbors. *Research on Language and Social Interaction, 40*(1), 9–32.

Emmison, M., Butler, C., & Danby, S. (2011). Script proposals: A device for empowering clients in counselling. *Discourse Studies, 13*(1), 3–26.

Emmison, M., & Danby, S. (2007a). Troubles announcements and reasons for calling: Initial actions in opening sequences in calls to a national children's helpline. *Research on Language and Social Interaction, 40*(1), 63–87.

Emmison, M., & Danby, S. (2007b). Who's the friend in the background? Interactional strategies in determining authenticity in calls to a national children's helpline. *Australian Review of Applied Linguistics, 30*(3), 31.31–31.17, doi:10.2104/aral0731.

Fogarty, K. (2010). 'Just say it in your own words': The social interactional nature of investigative interviews into child sexual abuse. Unpublished doctoral dissertation, The University of Adelaide.

Fogarty, K., Augoustinos, M., & Kettler, L. (2013). Re-thinking rapport through the lens of progressivity in talk to develop peer-based health activities. *Qualitative Health Research, 19*(8), 1162–1175.

Gardner, R. (2013). Conversation analysis in the classroom. In J. Sidnell & T. Stivers (Eds.), *The handbook of conversation analysis* (pp. 593–612). Oxford: Wiley-Blackwell.

Gardner, R., & Mushin, I. (2013). Teachers telling: Informings in an early years classroom. *Australian Journal of Communication, 40*(2), 63–81.

Geldard, K., Geldard, D., & Foo, R. (2013). *Counselling children: A practical introduction* (4th edition). London: Sage.

Gibbs, L., Mutch, C., O'Connor, P., & MacDougall, C. (2013). Research with, by, for and about children: Lessons from disaster contexts. *Global Studies of Childhood, 3*(2), 129–141.

Goodman, R. F. (2012). *Caring for kids after trauma and death: A guide for parents and professionals.* New York: The Institute for Trauma and Stress.

Harris, J., Danby, S., Butler, C. W., & Emmison, M. (2012). Extending client-centered Support: Counselors' proposals to shift from email to telephone counseling. *Text and Talk, 32*(1), 21–37.

Hendry, L. B., & Kloep, M. (2002). *Lifespan development: Resources, challenge and risk.* London: Centage, Thomson Learning.

Hepburn, A. (2005). 'You're not takin' me seriously': Ethics and asymmetry in calls to a child protection helpline. *Journal of Constructivist Psychology, 18*(3), 253–274.

Howard, J. (2010). The developmental and therapeutic value of children's play: Re-establishing teachers as play professionals. In J. Moyles (Ed.), *The excellence of play* (pp. 201–215). Buckingham, England: Open University Press.

Howard, J., & McInnis, K. (2013). *The essence of play: A practice companion for professionals working with children and young people.* London, England: Routledge.

Hutchby, I. (2007). *The discourse of child counselling.* Amsterdam/Philadelphia: John Benjamins.

Hutchby, I., & Moran-Ellis, J. (Eds.) (1998). *Children and social competence: Arenas of social action.* London: Falmer Press.

Hutchby, I., & O'Reilly, M. (2010). Children's participation and familial moral order in family therapy. *Discourse Studies, 12*(1), 49–64.

Kaminer, D., Seedat, S., & Stein, D. J. (2005). Post-traumatic stress disorder in children. *World Psychiatry, 4*(2), 121–125.

Lamerichs, J., Koelen, M., & te Molder, H. (2009). Turning adolescents into analysts of their own discourse: Raising reflexive awareness of everyday talk to develop peer-based health activities. *Qualitative Health Research, 19*(8), 1162–1175.

Lamerichs, J., & te Molder, H. (2011). Reflecting on your own talk: The discursive action method at work. In C. Antaki (Ed.), *Applied conversation analysis: Intervention and change in institutional talk* (pp. 184–206). Basingstoke: Palgrave Macmillan.

Landqvist, H. (2005). Constructing and negotiating advice in calls to a poison information center. In C. D. Baker, M. Emmison, & A. Firth (Eds.), *Calling for help: Language and social interaction in telephone helplines* (pp. 207–234). Amsterdam/Philadelphia: John Benjamins.

Leppanen, V. (2005). Caller's presentations of problems in telephone calls to Swedish primary care. In C. D. Baker, M. Emmison, & A. Firth (Eds.), *Calling for help: Language and social interaction in telephone helplines* (pp. 177–205). Amsterdam/Philadelphia: John Benjamins.

Mayes, C. (2009). 'The psychoanalytic view of teaching and learning, 1922–2002'. *Journal of Curriculum Studies, 41*(4), 539–567.

National Institute of Clinical Excellence (NICE) (2005). *Post-traumatic stress disorder. The management of PTSD in adults and children in primary and secondary care*. National Institute for Clinical Excellence, London: BPS Publications.

O'Connor, P. (2012). *Applied theatre: Aesthetic pedagogies in a crumbling world*. London: Routledge.

O'Connor, P. (2013). A teaspoon of light: Expressions of light and understanding through the voices of children in Christchurch. In B. Clark, A. Grey, & L. Terreni (Eds.), *Kia Tipu Te Wairua Toi – Fostering the creative spirit: Arts in early childhood education* (pp. 79–86). Wellington: Person.

O'Reilly, M. (2008). What value is there in children's talk? Investigating family therapists' interruptions of parents and children during the therapeutic process. *Journal of Pragmatics, 40*, 507–524.

Osvaldsson, K. (2011). Bullying in context: Stories of bullying on an internet discussion board. *Children & Society, 25*, 317–327.

Petriwskyj, A. (2013). Reflections on talk about natural disasters by early childhood educators and directors. *Australian Journal of Communication, 40*(1), 87–101.

Potter, J., & Hepburn, A. (2003). 'I'm a bit concerned' – Early actions and psychological constructions in a child protection helpline. *Research on Language and Social Interaction, 36*(3), 197–240.

Prendiville, E. (2014). Abreaction. In C. Schaefer & A. Drewes (Eds.), *The therapeutic powers of play: 20 Core agents of change*. Hoboken, NJ: John Wiley and Sons.

Prinstein, M. J., La Greca, A. M., Vernberg, E. M., & Silverman, W. K. (1996). Children's coping assistance: How parents, teachers, and friends help children cope after a natural disaster. *Journal of Clinical Child Psychology, 25*, 463–475.

Pudlinski, C. (2002). Accepting and rejecting advice as competent peers: Caller dilemmas on a warm line. *Discourse Studies, 4*(4), 481–500.

Raymond, G., & Zimmerman, D. H. (2007). Rights and responsibilities in calls for help: The case of the mountain glade fire. *Research on Language & Social Interaction, 40*(1), 33–61.

Ruf, M., Schauer, M., Neuner, F., Catani, C., Schauer, E., & Elbert, T. (2010). Narrative exposure therapy for 7- to 16-year-olds: A randomized controlled trial with traumatized refugee children. *Journal of Traumatic Stress, 23*(4), 437–445.

Sacks, H. (1972). An initial investigation of the usability of conversational data for doing sociology. In D. Sudnow (Ed.), *Studies in social interaction* (pp. 31–74). New York: The Free Press.

Sacks, H. (1992). *Lectures on conversation*. Oxford: Blackwell.

Sacks, H., Schegloff, E. A., & Jefferson, G. (1974). A simplest systematics for the organization of turn-taking in conversation. *Language, 50*(4), 696–735.

Shalem, Y., & Bensusan, D. (1999). Why we can't stop believing? In S. Appel (Ed.), *Psychoanalysis and pedagogy* (pp. 27–43). Westport: Bergin and Garvey.

Shaw, R., & Kitzinger, C. (2007). Memory in interaction: An analysis of repeat calls to a home birth helpline. *Research on Language and Social Interaction, 40*(1), 117–144.

Silverman, D. (1998). *Harvey Sacks: Social science and conversation analysis.* Cambridge: Polity Press.

Soles, T., Bloom, E., Health, N., & Kargiannakis, A. (2009). An exploration of teachers' current perceptions of children with emotional and behavioural difficulties. *Emotional and Behavioural Difficulties, 13*(4), 275–290.

Stivers, T., & Robinson, J. (2006). A preference for progressivity in interaction. *Language in Society, 35*(3), 367–392.

Stokoe, E., & Hepburn, A. (2005). 'You can hear a lot through the walls': Noise formulations in neighbour complaints. *Discourse & Society, 16*(5), 647–673.

te Molder, H. (2005). 'I just want to hear somebody right now': Managing identities on a telephone helpline. In C. D. Baker, M. Emmison, & A. Firth (Eds.), *Calling for help: language and social interaction in telephone helplines* (pp. 153–173). Amsterdam/Philadelphia: John Benjamins.

Terr, L. (2003). Wild child: How three principles of healing organised 12 years of psychotherapy. *American Academy of Child and Adolescent Psychiatry, 42*(12), 1401–1909.

Wakin, M. A., & Zimmerman, D. H. (1999). Reducation and specialization in emergency and directory assistance calls. *Research on Language and Social Interaction, 32*(4), 409–437.

Watson, D. R. (1981). Conversational and organisational uses of proper names: An aspect of counsellor-client interaction. In P. Atkinson & C. Heath (Eds.), *Medical work: realities and routines* (pp. 91–106). Farnborough: Gower.

Watson, D. R. (1986). Doing the organization's work: An examination of aspects of the operation of a crisis intervention centre. In S. Fisher & A. D. Todd (Eds.), *Discourse and institutional authority: Medicine, education and law.* Norwood, NJ: Ablex.

Whalen, J., & Zimmerman, D. H. (1998). Observations on the display and management of emotion in naturally occurring activities: The case of 'hysteria' in calls to 9-1-1. *Social Psychology Quarterly, 61*(2), 141–159.

World Health Organization (1992). *The ICD–10 classification of mental and behavioural disorders: Clinical descriptions and diagnostic guidelines.* Geneva: World Health Organization.

Zimmerman, D. H. (1992). Achieving context: Openings in emergency calls. In G. Watson & R. M. Seiler (Eds.), *Text in context: Contributions to ethnomethodology* (pp. 35–51). Newbury Park: Sage.

Recommended reading

- Bateman, A., Danby, S., & Howard, J. (2013a). Everyday preschool talk about Christchurch earthquakes. *Australia Journal of Communication – Special Issue: Disaster Talk, 40*(1), 103–123.
- Danby, S., Butler, C. W., & Emmison, M. (2011). 'Have you talked with a teacher yet?': How helpline counsellors support young callers being bullied at school. *Children & Society, 25*(4), 328–339.

- Edwards, D., & Potter, J. (1993). Language and causation: A discursive action model of description and attribution. *Psychological Review, 100*(1), 23–41, doi:10.1037/0033-295X.100.1.23.
- Hutchby, I. (2007). *The discourse of child counselling.* Amsterdam/Philadelphia: John Benjamins.
- Lamerichs, J., Koelen, M., & te Molder, H. (2009). Turning adolescents into analysts of their own discourse: Raising reflexive awareness of everyday talk to develop peer-based health activities. *Qualitative Health Research, 19*(8), 1162–1175.

23

Food, Eating, and 'Eating Disorders': Analysing Adolescents' Discourse

Maxine Woolhouse and Katy Day

Introduction

Female adolescents have long been identified as a group 'at risk' of developing an eating disorder (Fairburn & Harrison, 2003). Moreover, figures from the Health and Social Care Information Centre (HSCIC, 2014) indicate that the number of hospital admissions for treatment of an eating disorder had risen by 8% in the preceding 12 months. In terms of girls who were admitted, the most common age was 15 (300 out of 2,320), whereas for boys, this was 13 (50 out of 240). Published statistics must be treated with caution, however. First, as these tend to be based on those receiving treatment, they provide only a partial account as many 'cases' remain unidentified. Second, as we shall argue, treating anorexia nervosa and bulimia nervosa (and other so-called eating disorders such as 'binge eating disorder') as identifiable conditions is fraught with problems. Nevertheless, there appears to be a consensus that eating practices that are a cause for concern are on the rise, and that young people are particularly vulnerable.

Although attention has been paid to those eating disorders associated more with weight reduction and restricted eating (e.g. anorexia and bulimia), recent years have also witnessed an intense focus on the so-called global epidemic of obesity. Furthermore, within this rhetoric it is *children*'s bodies that have come under particular scrutiny fuelled by concerns over their health on reaching adulthood (Evans & Colls, 2011). Despite the medicalised language of disease and illness pervading 'obesity discourse', there still remains a pervasive understanding of obesity as a result of bad lifestyle choices made by individuals or, in the case of young people, those of their carers (but commonly mothers) (Rich, Monaghan & Aphramor, 2011). Although discourses around eating disorders such as anorexia and bulimia differ in many ways to those around childhood obesity (e.g. *every* child is deemed at risk of becoming obese – see Evans & Colls, 2011), critical and discursive analysts have questioned the usefulness of biomedicalised, individualistic understandings of fatness; highlighted the moralising and stigmatising effects of obesity discourse; and cast doubt on this rhetoric as facilitating individual and political responses that are actually conducive to health (Rich et al., 2011). As such, we wish to highlight the inextricable links between discourses around obesity, healthism (Crawford, 1980), anorexia, bulimia, and so forth and acknowledge how these intersect in multiple and complex ways (e.g. see Rich & Evans, 2008). However, we argue that young people labelled as 'having anorexia' or 'having bulimia' are more powerfully associated with being 'psychologically disturbed' than those labelled as 'obese' and are therefore more likely to come to the attention of mental health practitioners.

In this chapter, then, we review and critique psychological theories that seek to explain eating disorders such as anorexia and bulimia in children and adolescents (see also Dallos & Pitt, Chapter 20, this volume). We then move on to discuss how feminist discursive work which has drawn upon social constructionist and post-structuralist theories of language has addressed many of the limitations associated with these analyses (see Graham, Chapter 28, this volume, for further discussion of post-structuralism). Finally, we conclude by summarising contributions of our own and others' research in this area and examine how the findings might be put into practice to prevent dangerous body modification practices such as self-starvation and help young people experiencing problems with food and their bodies.

We acknowledge that some boys and young men also experience body distress and that rates of eating disorders are said to be on the rise among these groups (e.g. Maine & Bunnell, 2008). For example, recent research has drawn attention to how the Western 'culture of appearance' is now targeted at boys/men as well as girls/women (e.g. Magallares, 2013) and the increased sexualisation of the male body in media texts from the mid-1980s onwards (e.g. Gill, 2008). Gill (2008) discussed how increasingly within popular culture, 'overweight' male bodies are pathologised in such a way as to cast doubt upon the masculinity of boys and men who are deemed to be fat. Such cultural shifts have, in turn, been associated with an increase in body dissatisfaction, weight concerns and in some cases, various 'body image disorders' among young men (e.g. Gill, 2008; Grogan, 1999). Further, the pursuit of male body ideals has been identified among school-aged boys (e.g. Martin & Govender, 2011). However, there is still a notable gender imbalance where eating disorders are concerned, at least in relation to those cases that are officially identified and recorded (Beat, 2014). The chapter will, therefore, focus mostly on work that seeks to explain body modification practices such as voluntary self-starvation among girls and young women.

Explaining 'eating disorders': Mainstream psychological work

Within psychology, there is no shortage of literature on eating disorders and the bulk of theories and empirical studies are set within a positivist-empiricist framework (e.g. Malson, 1998). It is beyond the scope of the chapter to provide a full review of this work. Here, we shall review and critique, first, theories and models that focus on the development of problems associated with food and eating during childhood and adolescence and, second, those that enjoy particular dominance within the discipline.

Social cognitive accounts and family influences: Blaming mother

Social cognition has become a dominant framework within psychology for explaining a host of phenomena (Gough & McFadden, 2001). In relation to eating disorders, social cognitive accounts focus on patterns of maladaptive thinking related specifically to body size, shape, and weight (Jones, Leung, & Harris, 2007) as being the core aetiology of eating disorder symptoms (Fairburn & Harrison, 2003). An example would be a tendency to judge self-worth almost exclusively in relation to eating, shape, or weight and the ability to control these (Fairburn, Cooper, & Shafran, 2003). Some more recent work has examined the role of negative 'core beliefs' or 'early maladaptive schemas' (i.e. stable and generalised beliefs about oneself such as 'I am a failure' – see Jones et al., 2007) and the role that these play in the onset and maintenance of disordered eating. Investigations into the origins of such cognitions and core beliefs have frequently pointed to the role of patterns of parent–child interaction and childhood experiences as a primary cause (e.g. Ward Ramsey, & Treasure, 2000 cited in Jones, Harris, & Leung, 2005), in particular, the role of the mother and the level and type of maternal 'care'.

Other research, regarding 'family influences' on eating behaviours and attitudes has examined the effects of appearance-focused family culture and parental commentary about body weight, shape, and size. For example, mothers' direct negative feedback to daughters regarding weight/size and encouragement to diet have been found to be associated with increased weight concerns in daughters (e.g. Baker, Whisman, & Brownell, 2000) which in turn can lead to restrained eating (Francis & Birch, 2005). Several studies have suggested that direct comments may not be necessary to incite body modification practices; rather, a mere 'modelling' effect may occur whereby simple exposure to a parent's 'problematic' eating attitudes and behaviours can influence those of children (Cooley, Toray, Wang, & Valdez, 2008).

Although there is acknowledgement within this body of literature that young people do not exist in a vacuum, cognitive-behavioural accounts ultimately provide individualised understandings of eating disorders by locating the source of problems within the individual in the form of dysfunctional cognitions (e.g. Malson, 1998). The 'social' is largely conceived of as the (nuclear) family environment, which in turn is often reduced to mother–infant or mother–adolescent interactions (Hepworth, 1999). Further, this overemphasis on the mother–daughter relationship is problematic through the way in which it is conceived, this being as pathologic and faulty. How constructs such as 'maternal overprotection' (e.g. Turner, Rose, & Cooper, 2005) are understood is often left unexplained and suggests deviation from an ideal and normal way of mothering which is implied rather than made explicit. Rather, constructs such as 'the family' and a mother's rightful role within it are taken for granted (Hepworth, 1999). Similarly, cognitions associated with disordered eating are often described as *faulty* and/or *distorted* (Fairburn & Harrison, 2003), the implication being that there exists a more accurate way of perceiving and thinking about food, eating, weight, and shape.

What this amounts to is a failure to locate practices associated with eating disorders within wider socio-cultural and political conditions that make possible the many ways in which mothers and young people make sense of food, eating, weight, shape, and so forth. Further, not only is there a failure to examine constructions of, for example, ideal bodies and ideal mothering, but such accounts arguably contribute to motherhood as a site of regulation and surveillance by blaming mothers when 'things go wrong' (O'Reilly, 2010). It is acknowledged by many that the mother–daughter relationship is important because of shared gender experiences (Hepworth, 1999; Malson, 1998) and to deny this importance is perhaps naïve. However, analyses are limited when these relationships are examined in isolation from the socio-cultural conditions through which they are constructed. Similarly, treating those 'with' and those 'without' an eating disorder as clearly identifiable groups implies that these operate in separate discursive realms and arguably reproduces the pathologisation of young people who are experiencing problems with food and eating. Therapeutic interventions then become targeted at the individual by aiming to alter their 'faulty' cognitions and eating behaviour while leaving gendered power relations and social structures unexamined (Malson, 1998).

Socio-cultural analyses: Fashion victims and mean girls

Failure to understand body distress within a wider socio-cultural context has to some extent been addressed by a body of literature that has sought to examine the role of socio-cultural conditions in the development of eating disorders. A key focus has been cultural messages promoting the 'thin ideal' (Stice, Schupak-Neuberg, Shaw, & Stein, 1994), which are believed to be transmitted not just through family members, but also through the young person's peers and the mass media (Stice, 2002). These ideals then become internalised by the individual (Polivy & Herman, 2002). Various factors have been identified as increasing the likelihood that this will result in eating disorder symptomology, including existing body dissatisfaction (Stice, 2002), heightened thin-ideal internalisation (Stice, 2002), greater gender-role endorsement (Stice et al., 1994), and maternal pressure to lose weight (Pike & Rodin, 1991).

During the last 50 years (in developed, industrialised societies at least) the ideal body for girls and women has decreased in size to the extent that some women featured in magazines have weights that satisfy the diagnostic criteria for anorexia nervosa (Wiseman, Gray, Mosimann, & Ahrens, 1992). It is argued to be no coincidence that there has been a parallel trend in the rise of eating disorders (e.g. Stice, 2002). A vast number of studies have been conducted to examine the possible relationship between various forms of media consumption and 'body image and eating disturbance' (Stice, 2002, p. 104). Groesz, Levine, and Murnen (2002) conducted a meta-analysis of data from 25 experimental studies on the effect of media portrayals of the thin ideal on female body image. It was found that girls' and womens' body images were more negative after viewing slender 'idealised' media images than after viewing 'non-idealised' or control images. These results were argued to offer support for the socio-cultural perspective that links media images with how girls and women feel about their weight, shape, and size.

Socio-cultural analyses can be credited with considering wider social forces such as the mass media that may be implicated in eating disorders, and it is difficult to disregard this body of research which points to the negative effects of media propagated images of idealised bodies (Malson, 2009). Nevertheless, it can be argued that research within this domain is problematic on several levels. First, such research is firmly rooted within mainstream understandings of social psychology as a scientific enterprise, which seeks to investigate the effects of social factors on the individual, assessed mostly through the employment of quantitative research methods (Gough & McFadden, 2001). The 'social' within this tradition is reduced to a number of relatively discrete, and supposedly measurable factors such as 'media images' (e.g. Groesz et al., 2002), 'parental comments' (e.g. Fairburn, Cooper, & Doll, 1999), and so forth, all of which are examined for their purported effects on an individual's 'body image' (e.g. Groesz et al., 2002) and 'eating behaviours' (e.g. Pike & Rodin, 1991). As Hepworth (1999) argued, ultimately, these explanations have not reconceptualised the aetiology of 'eating disorders', such as anorexia nervosa; social dimensions have become accepted as 'influences' or 'factors', but these have become subsumed within psychiatric discourse. Notions of psychopathology and individualism have therefore largely remained.

A further consequence of this positivist empiricist framework is the scant attention paid to the multiplicity of meanings attached to food, consumption, slenderness, gendered bodies, and so forth. The various ways in which girls and young women are positioned, and position themselves within gendered, classed, and racialised relations of power are also overlooked. Girls and young

women, it seems, are generally presented as passively absorbing cultural messages transmitted via family, peers, and the mass media and 'those diagnosed
as "anorexic" or "bulimic" are made to appear as the ultimate fashion victims'
(Malson, 2009, p. 136). It seems that within this literature the only (implicit)
explanation for the gendered distributional patterns of 'eating disorders' is that
girls and women are subject to greater socio-cultural pressure to be slim and
to engage in activities to achieve this, compared with their male counterparts
(Andersen, 2002). However, during the last few decades, feminist scholars have
recognised the need to examine girls' and women's experiences of eating and
body management practices 'within the context of the oppressive gender ideologies and inequalities in gender power-relations operating in (Western/ised)
patriarchal cultures' (Malson & Burns, 2009, p. 1).

We now turn to examine recent discursive feminist work which has drawn
upon social constructionist and post-structuralist theories of language in order
to develop 'new *critical* feminist approaches to understanding "eating disorders"' (Malson & Burns, 2009, p. 2, emphasis in original). Within this
literature, ways of eating, embodied subjectivities, and body management practices are regarded as historically and culturally contingent, produced within
and through the prevailing discursive landscape. Rather than searching for the
'causes' of eating disorders, the concern lies with how different practices are
constructed, as well as consideration of the various cultural meanings attached
to food, eating, the female body, and so forth (e.g. see Bordo, 2003).

Feminist discursive research on eating and 'body distress'

There is, of course, a broad landscape of physical cultures that may be considered as 'sites of learning' through which young people's understandings
of body management and food practices are shaped. Among these are the
mass media (e.g. magazines, television, film, music videos, advertisements,
and the Internet), health promotion materials (e.g. the UK-government-backed
'Change4Life' campaign), fashion and cosmetic industries, schools, and clini-
cal/healthcare settings. Indeed, critical (and often feminist-informed) discursive
work has engaged with and examined a variety of such discursive sites in order
to interrogate the multiple ways through which food, eating, and body management practices are constructed (e.g. as ideal/undesirable, healthy/unhealthy,
normal/abnormal, good/bad) and the possible implications of these for subjectivities and practices (for examples of this work see Aphramor & Gingras, 2008;
Burns & Gavey, 2004; Kokkonen, 2009; Madden & Chamberlain, 2004; Piggin &
Lee, 2011; Rich & Evans, 2008).

In attempting to illustrate some of the discursive work undertaken in this
domain, we provide an overview of studies focusing on three areas. These are
the everyday talk of those not identified as eating disordered; constructions of

eating disorders and anorexic identities in schools and within online communities; and finally, constructions of bulimia within treatment settings. We selected this work to look at in some detail. First, it specifically engages with *young* people and illuminates how they make sense of prevailing discourses around food, bodies, 'eating disorders', and so forth; and second, this selection allowed for a consideration of the 'sense-making' of those identified and *not* identified as 'eating disordered', something which we consider important given questions raised earlier in the chapter about the usefulness of this dichotomy.

In an attempt to challenge the pervasive notion of 'normal' and 'disordered' eating as two distinct categories, Woolhouse, Day, Rickett, and Milnes (2012) conducted focus group discussions with primarily white, working-class adolescent girls (from the United Kingdom) around the themes of food, eating, femininities, and body management practices. None of the participants were formally or self-identified as 'eating disordered'. A key aim was to examine ways in which classed and gendered discourses were drawn upon in their sense-making of various eating and body management practices. Employing post-structuralist discourse analysis to examine the data, a central finding was that the girls constructed eating as an 'unfeminine' activity associated with greed, animality, and involving expressions of desire and appetite. For example, whereas eating generous quantities and displaying enthusiasm for food were deemed as normative and socially sanctioned for boys, girls were construed as needing to restrict food intake, show a lack of interest in food, exhibit 'good manners' and be weight-conscious. In accounting for this gender disparity, one of the girls commented, "'cos girls aren't supposed to eat like pigs are they?' to which another responded 'like [girls should be] ladylike'.

To make sense of this talk, the authors considered long-standing constructions of women 'as body' (Bordo, 2003), and ruled by their bodies, which are regarded as unstable, out of control, and inherently weak (Ussher, 1989), yet simultaneously voracious, threatening, and in need of control (Bordo, 2003). Therefore, a woman must exercise control over her appetites in order to signify moral and sexual virtue and constitute her as 'properly' feminine (e.g. controlled, delicate, dainty, and passive). Yet, as implied by the participants, this idealised version of femininity is very much classed (i.e. 'ladylike'), built upon bourgeois feminine characteristics (Walkerdine, 1996).

In addition, it was noted how discourses around hegemonic femininity and social class intersected with those around 'healthism' (Crawford, 1980) and neoliberalism to construct the ideal girl/woman. For example, taking personal responsibility for eating and one's health through adopting a rational approach was often implied or directly advocated by the girls:

> Lisa: Or just yeah, that's what I always do, just eat whatever I want like, but just eat sensibly like if you wanna crisp, don't buy three crisps just buy one [packet of] crisps.

Here, Lisa deployed a 'discourse of moderation' which involves eating 'whatever I want' but not to excess ('don't buy three crisps'). In doing so, she could position herself as 'sensible', balancing entitlement with self-restraint. Lisa thus construes 'sensible' eating as simply a matter of personal responsibility and individual choice, achievable through rational decision-making; the ideal feminine neoliberal subject is controlled, sensible, health-conscious, and personally accountable for her actions.

Clearly then, culturally available discourses are constantly reproduced, challenged, negotiated, and reworked as we attempt to make sense of the world around us. Research such as that discussed above underscores the argument that eating and body management practices are informed and shaped by the discursive landscape, and therefore – attempting to understand girls' ways of eating – their relationships with food, embodied experiences, and so forth in isolation from socio-political, cultural, and discursive contexts produces limited and impoverished accounts. Furthermore, analyses of the everyday talk of adolescents who are not considered (or consider themselves) eating disordered draws our attention to the discursive similarities between the supposed 'disordered' and 'non-disordered' girls. To illustrate these arguments, we now turn our attention to some fascinating discursive research on adolescents who are formally identified as eating disordered.

In order to examine intersecting discourses around schooling and anorexia, Halse, Honey, and Boughtwood (2007) interviewed Australian adolescent school girls who had all been diagnosed with anorexia. The authors identified three overarching discourses around notions of virtue pertaining to self-discipline, achievement, and healthism. For example, in relation to one of the discourses ('Discipline as a virtue and the self-disciplined girl' – Halse et al., 2007, p. 224), it was noted how discipline as a virtue was woven into the structural fabric and everyday practices of schools and, importantly, how this was taken up by the girls and reproduced through a variety of practices in their daily lives. For instance, girls talked about carefully planning their day allocating precise time slots for various self-monitored activities such as homework, exercise, eating, and relaxation. Similarly, they imposed strict regulations on their food consumption which dictated what, where, and when they ate.

Within and outside of school, the authors also noted how 'the high achiever' was a desirable and rewarded subject position and conversely, non-achievement attracted 'punishment' through the withholding of rewards which were provided to high achievers. Interestingly, the girls in the study not only deployed this discourse of 'achievement as a virtue' in their talk about school work, but also as a way of accounting for their eating practices; weight loss itself was constructed as an achievement and as something that required personal expertise (Halse et al., 2007).

Finally, schools as a site for the reproduction of a discourse of healthism was also noted (Halse et al., 2007). Unsurprisingly, some of the girls articulated their weight-loss practices through this discourse, for example by suggesting that they started dieting in order to be healthier. However, becoming 'healthier' was not just about physical health; avoidance of 'bad' foods and controlling their eating was construed as a means of avoiding all the negative moral attributes associated with fatness. The authors argued that 'virtue' discourses operating within the school environment around achievement, discipline, and healthism offer the highly desirable subject position of the self-disciplined, healthy/thin/beautiful high achiever and yet, virtue is configured as a state without boundaries; in other words, 'it is not possible to be too disciplined, too high achieving or too healthy. Consequently, the project of becoming a "virtuous" girl is a never-ending project of self-perfection and self-production' (Halse et al., 2007, p. 230). As such, the authors argue that rather than anorexia being regarded as an individual psychopathology or deviance from a generalised norm, it can be read as produced in and through the discourses of virtue operating within schools and throughout wider society.

In recent years with the proliferation of Internet 'blogs' and web-based discussion boards, interest has grown into the analysis of 'body talk' material posted on relevant sites. For example, using feminist-informed post-structuralist discourse analysis, Day and Keys (2008) analysed the posts of contributors to 'pro-eating disorder' websites and found that bulimia and anorexia were constructed not only as means to achieve a culturally sanctioned 'look', but also as an admirable, rule-bound way of living, the rules of which can offer a route to salvation. Constructed in this way, anorexia and bulimia made available the identity of the 'super-compliant servant' (Day & Keys, 2008, p. 8), as one who complies with the rules of the 'eating disorder religion' (Day & Keys, 2008, p. 8). However, in contrast to such notions of compliance and conformity, the authors also point to alternative material on the sites that suggested resistance to biomedical and psychiatric discourses around eating disorders which serve to position women who engage in 'eating disordered practices' as victims of disease/mental disorder. Day and Keys argued that by constructing self-starvation and other related practices as 'ways of living' (Day & Keys, 2008, p. 10), the contributors were able to position themselves as agentic subjects who make active lifestyle choices and strive for independence and control over and through, for example, their bodies. This discursive strategy may represent attempts to defy the many practices which subject women/female bodies to surveillance and control. In addition, framing their activities as 'freely chosen' may be a means of negotiating the tension between discourses which promote female passivity and conformity and the highly prized (neoliberal/male/bourgeois) values of autonomy and independence (Day & Keys, 2008).

Within the work reviewed so far, a number of themes can be detected. These include eating as incompatible with constructions of heteronormative femininity; ways of eating as understood through discourses around gender, social class, health(ism), and neoliberalism; the construction of gendered identities through intersecting yet conflicting discourses around food, bodies, virtue, compliance, and autonomy, and resistance to dominant biomedical and psychological understandings of eating disorders. This latter theme, in particular, has been noted within research conducted in treatment and therapeutic settings, to which we now turn.

Within the therapeutic setting, Guilfoyle (2001) examined the ways in which bulimia is discursively produced within 'expert' discourse and considered the implications such constructions have for those diagnosed with bulimia. Psychological discursive constructions of bulimia depend upon the modern Western concept of the individual as a self-contained entity (Sampson, 1989, cited in Guilfoyle, 2001), and, as such, the bulimic individual not only displays bulimic behaviours, but also is believed to possess an array of psycho(patho)logical characteristics, such as low self-esteem, impulsivity, body image problems, depression, and so forth. Such characteristics, Guilfoyle argues, 'allow for the construction of a complex and psychologized bulimic' (Guilfoyle, 2001, p. 160). In this case study, drawn from his therapeutic work in South Africa with a 19-year-old woman, Guilfoyle demonstrates how the client is discursively manoeuvred from an initial position of refusing to accept a psychological account of her bulimic practices to her eventual acceptance of a thoroughly psychologised account of her 'self' and her body management practices. Through this, Guilfoyle reveals how power invested in 'psy' discourse and the therapist renders possible the construction of the bulimic subject while simultaneously foreclosing alternative accounts of eating practices.

What such work brings to the fore is how psychomedical discourse brings into being the bulimic and anorexic subject (Guilfoyle, 2001), constituting subjectivity, and precluding alternative understandings of women's eating and related practices. Despite this, as noted in the work hitherto discussed, it appears that resistance to hegemonic constructions is always possible. This underscores some important ongoing issues and debates within feminist poststructuralist work around agency and resistance. It has been contended that girls and women have the ability to 'rewrite' dominant ideologies, often in ways that may attribute them with more power (e.g. Eckermann, 1997). One advantage of this is that it avoids deterministic analysis, which positions girls and women as being passively inscribed by oppressive discourses (Day, 2010). For example, Guilfoyle (2001) notes how the client initially rejected being positioned as psychologically defective/ill and instead construed her 'bulimic' activities as a reasonable strategy for weight loss. Likewise, as outlined previously, those posting on pro-eating disorders websites often construed their practices in terms of 'lifestyle choice' as opposed to 'sickness' or 'disease'. However, we would caution against humanistic readings of agency which undermine the constraints placed upon subjects and the power of social institutions. For example, Guilfoyle (2001) argues that such points of resistance to 'psy' accounts of eating practices might be difficult to sustain, particularly within contexts such as that of therapy. Such difficulties, he argues, are related to the privileged position that the institution of 'psy' occupies and the operations of power implicated within it. Part of this position serves to dismiss alternative accounts of subjectivity and practices. Further, psy discourse has entered into common parlance, offering 'techniques and a language for individual self-knowledge and self-government' (Guilfoyle, 2001, p. 156) and serves to produce psychologised subjectivities. In addition, what could be read as 'resistance' occurring on the Internet could also be read as the 'buying into' of a different set of social norms and discourses such as those that promote individualism and autonomy (see also Burns & Gavey, 2004). We therefore agree with Weedon (1987) that girls and women have agency in so far as they will variably take up, negotiate, and resist available discourses and the identities that these constitute. Yet, people do not 'make' discourses in conditions of their own choosing (see Collier, 1994); different discourses are variably available to different groups (depending upon their relative access to power and the contexts that they are immersed in) and the options for construing are not limitless (for a fuller discussion, see Day, 2010).

Clinical relevance summary: Implications for practitioners

The literature reviewed here has illuminated a number of key points which have implications for those who seek to help alleviate eating and body distress experienced by many adolescents. First of all, discursive research with girls who *are not* identified as 'eating disordered' (Woolhouse et al., 2012) powerfully draws attention to the common discursive ground shared by those 'with' and 'without' formal labels. For instance, ideas such as girls should eat with restraint, show little enthusiasm for food, and be 'ladylike', sensible, moderate, and health-conscious. Considering girls' 'everyday' talk forces us to re-evaluate dominant understandings of eating practices as comprising distinct categories (e.g. 'normal' and 'disordered'), and therefore we may begin to view those practices that are commonly considered 'pathological' as an understandable response to particular structural and ideological conditions.

Schools are a powerful social institution in the lives of young people and are often a vehicle for delivering health-promoting initiatives such as those aimed at tackling obesity (Rich & Evans, 2008). However, messages around health, weight, and exercise are intricately bound with discourses around gender, class, morality, neoliberalism, and so forth and therefore may be read and taken up in complex and problematic ways (see Rich & Evans, 2008). Given this, we argue that school-based initiatives which aim to work in partnership with young people to explore and deconstruct taken-for-granted assumptions embedded within, for example, gender ideologies and health messages is a useful starting point. Similarly, Day and Keys' (2008) research draws attention to the potential of online communities in offering young people a space to challenge and resist oppressive and harmful discourses around eating and body management practices.

The discursive research discussed here also highlights resistance from girls diagnosed with eating disorders towards dominant biomedical and psychological accounts of their experiences. Further, work such as that of Guilfoyle (2001) (see also Boughtwood & Halse, 2010), which indicated that the girls' own understandings are sometimes dismissed by clinicians, has been taken as further evidence of the girls' faulty thinking or the girl is encouraged to refashion her account to fit a psychological framework. However, such strategies can be counter-productive leaving girls feeling undermined and encouraging them to engage in practices which may be unhelpful to their short- or longer-term well-being. As such, Boughtwood and Halse (2010) recommended that clinicians respect the fact that girls often do not share medical and 'expert' understandings of their experiences. This, they argue, is a necessary starting point in order to have a more open and honest relationship with those they are seeking to help rather than one which girls sometimes construe as 'patronizing and insincere' (Boughtwood & Halse, 2010, p. 92). Part of that relationship involves acknowledging the limitations of current knowledge about eating distress and appropriate interventions (Boughtwood & Halse, 2010).

Finally, research reviewed here draws attention to the operations of power implicated in the therapeutic relationship and, as Guilfoyle (2001) argued, how these may be obscured by the therapist's insistence on investing a client's resistance with psy discursive meaning. However, alternatively a client's refusal to accept psychological interpretations of her practices could be treated by the therapist as a political struggle, and as such Guilfoyle (2001) suggested that a non-psychologised feminist dialogue with girls could offer a more helpful form of intervention (see Guilfoyle, 2001, p. 170 for examples of questions that might form the basis of group discussion about women's eating and bodily distress). Some of the advantages to such an approach are that operations of power are highlighted rather than concealed, the individual becomes a 'politically located subject' (Guilfoyle, 2001, p. 171) whose practices take on a different meaning, and emphasis is placed on the need for political and social change rather than at the level of the individual. For a simple summary of the implications for practice, see Table 23.1.

Summary

Feminist discursive research on girls' eating and bodily distress draws attention to the wider cultural and discursive conditions that inform and make possible various forms of embodiment, eating and body management practices, and subjectivities. As such, this research underscores how, in contrast to individualist

Table 23.1 Clinical practice highlights

---

1. Prevailing cultural discourses form the 'conditions of possibility' for eating and bodily distress. As such, encouraging adolescents to question the assumptions embedded within – for example, gender ideologies, beauty ideals, and so forth – helps to locate problems in the political realm rather than at the level of the pathologised individual (see Guilfoyle, 2001).
2. Adolescents' own understandings of their distress may differ significantly from 'expert' theories. In addition, there are varied and different ways among adolescents for making sense of their practices and experiences. This underscores the importance of acknowledging and respecting the adolescent's own understandings without assuming that bio-medical/psychological accounts represent a form of superior knowledge.
3. Rather than interpreting an adolescent's refusal to accept psychological accounts of their practices as a (further) sign of illness, this can be understood as a political struggle (resistance to powerful 'psy' and expert discourse) and therefore harnessed as an opportunity to explore alternative (non-psy) understandings of eating and bodily distress.

---

biomedical/psychological understandings of eating and weight management, pathologised forms of embodiment and practices do not exist outside of the discursive realm (Malson & Burns, 2009).

For example, research discussed in this chapter highlights how girls may draw on a range of intersecting discourses (e.g. around gender and bodies, class, health, neoliberalism, and so on) to make sense of eating and body management practices. Yet, it was also noted that women do not passively absorb such discourses, but rather, may adopt various discursive strategies in order to challenge and resist cultural diktats. However, it was also indicated that the prevailing socio-political climate of neoliberalism, healthism, and the vilification of fatness make available alternative accounts and subject positions that might also be deemed problematic, placing a strong emphasis on the individual and thus obscuring socio-political explanations and discouraging collective forms of action (Day & Keys, 2008).

In arguing for feminist discursive approaches to bodily distress, which also entails questioning the conceptual distinctions between 'normal' and 'disordered', we do not wish to downplay the severity of distress and often life-threatening complications encountered by many girls and women. However, we strongly believe that current dominant modes of understanding are limited at best and perhaps even counter-productive to the promotion of well-being in girls and women. As such, we agree with Guilfoyle (2001) in his call for interventions which aim to engage girls in feminist political discourse as a way of shifting the focus of their distress away from individual psychology to the wider discursive landscape and socio-political structures which form the conditions of possibility for adolescent bodily distress.

References

Andersen, A. (2002). Eating disorders in males. In C. Fairburn & K. Brownell (Eds.), *Eating disorders and obesity: A comprehensive handbook* (2nd edition) (pp. 188–192). New York: The Guildford Press.

Aphramor, L., & Gingras, J. (2008). Sustaining imbalance – Evidence of neglect in the pursuit of nutritional health. In S. Riley, M. Burns, H. Frith, S. Wiggins, & P. Markula (Eds.), *Critical bodies: Representations, identities and practices of weight and body management* (pp. 155–174). Basingstoke: Palgrave Macmillan.

Baker, C., Whisman, M., & Brownell, K. (2000). Studying intergenerational transmission of eating attitudes and behaviours: Methodological and conceptual questions. *Health Psychology, 19*(4), 376–381.

Beat (2014). Eating disorder statistics. Retrieved June 4, 2014 from http://www.b-eat.co.uk/about-beat/media-centre/facts-and-figures/.

Bordo, S. (2003). *Unbearable weight* (10th anniversary edition). Berkeley, CA: University of California Press.

Boughtwood, D., & Halse, C. (2010). Other than obedient: Girls' constructions of doctors and treatment regimes for anorexia nervosa. *Journal of Community & Applied Social Psychology, 20*, 83–94.

Burns, M., & Gavey, N. (2004). 'Healthy weight' at what cost? 'Bulimia' and a discourse of weight control. *Journal of Health Psychology, 9*(4), 549–565.

Collier, A. (1994). *Critical realism: An introduction to Roy Bhaskar's philosophy.* London: Verso.

Cooley, E., Toray, T., Wang, M. C., & Valdez, N. N. (2008). Maternal effects on daughters' eating pathology and body image. *Eating Behaviors, 9*(1), 52–61.

Crawford, R. (1980). Healthism and the medicalization of everyday life. *International Journal of Health Services, 10*(3), 365–388.

Day, K. (2010). 'Pro-anorexia' and 'binge-drinking': Conformity to damaging ideals or new, resistant femininities? *Feminism & Psychology, 20*(2), 242–248.

Day, K., & Keys, T. (2008). Starving in cyberspace: A discourse analysis of pro-eating-disorder websites. *Journal of Gender Studies, 17*(1), 1–15.

Eckermann, E. (1997). Foucault, embodiment and gendered subjectivities: The case of voluntary self-starvation. In A. Peterson & R. Bunton (Eds.), *Foucault, health and medicine* (pp. 151–169). London: Routledge.

Evans, B., & Colls, R. (2011). Doing more good than harm? The absent presence of children's bodies in (anti-)obesity policy. In E. Rich, E. Monaghan, & L. Aphramor (Eds.), *Debating obesity: Critical perspectives* (pp. 115–138). Basingstoke: Palgrave Macmillan.

Fairburn, C. G., Cooper, Z., & Doll, H. A. (1999). Risk factors for anorexia nervosa: Three integrated case-control comparisons. *Archives of General Psychiatry, 56*, 468–476.

Fairburn, C. G., Cooper, Z., & Shafran, R. (2003). Cognitive behaviour therapy for eating disorders: A 'transdiagnostic' theory and treatment. *Behaviour Research and Therapy, 41*, 509–528.

Fairburn, C. G., & Harrison, P. (2003). Eating disorders. *The Lancet, 361*(1), 407–416.

Francis, L., & Birch, L. (2005). Maternal influences on daughters' restrained eating behaviour. *Health Psychology, 24*(6), 548–554.

Gill, R. (2008). Body talk: Negotiating body image and masculinity. In S. Riley, M. Burns, H. Frith, S. Wiggins, & P. Markula (Eds.), *Critical bodies: Representations, identities and practices of weight and body management* (pp. 101–116). Basingstoke: Palgrave Macmillan.

Gough, B., & McFadden, M. (2001). *Critical social psychology: An introduction.* Basingstoke: Palgrave Macmillan.

Grogan, S. (1999). *Body image: Understanding body dissatisfaction in men, women, and children.* London: Routledge.

Groesz, L., Levine, M., & Murnen, S. (2002). The effect of experimental presentation of thin media images on body satisfaction: A meta-analytic review. *International Journal of Eating Disorders, 31,* 1–16.

Guilfoyle, M. (2001). Problematizing psychotherapy: The discursive production of a bulimic. *Culture & Psychology, 7*(2), 151–179.

Halse, C., Honey, A., & Boughtwood, D. (2007). The paradox of virtue: (Re)thinking deviance, anorexia and schooling. *Gender and Education, 19*(2), 219–235.

Health and Social Care Information Centre (HSCIC) (2014). Eating disorders: Hospital admissions up by 8 per cent in a year. Retrieved June 3, 2014 from http://www.hscic. gov.uk/article/3880/Eating-disorders-Hospital-admissions-up-by-8-per-cent-in-a-year.

Hepworth, J. (1999). *The social construction of Anorexia Nervosa.* London: Sage.

Jones, C., Harris, G., & Leung, N. (2005). Parental rearing behaviours and eating disorders: The moderating role of core beliefs. *Eating Behaviors, 6,* 355–364.

Jones, C., Leung, N., & Harris, G. (2007). Dysfunctional core beliefs in eating disorders: A review. *Journal of Cognitive Psychotherapy: An International Quarterly, 21*(2), 156–171.

Kokkonen, R. (2009). The fat child – a sign of 'bad' motherhood? An analysis of explanations for children's fatness on a Finnish website. *Journal of Community & Applied Social Psychology, 19*(5), 336–347.

Madden, H., & Chamberlain, K. (2004). Nutritional health messages in women's magazines: A conflicted space for women readers. *Journal of Health Psychology, 9*(4), 583–597.

Magallares, A. (2013). Masculinity, drive for muscularity and eating concerns in men. *Suma Psicológica, 20*(1), 83–88.

Maine, M., & Bunnell, D. (2008). How do the principles of the feminist, relational model apply to treatment of men with eating disorders and related issues? *Eating Disorders: The Journal of Treatment & Prevention, 16*(2), 187–192.

Malson, H. (1998). *The thin woman: Feminism, post-structuralism, and the social psychology of Anorexia Nervosa.* London: Routledge.

Malson, H. (2009). Appearing to disappear: Postmodern femininities and self-starved subjectivities. In H. Malson & M. Burns (Eds.), *Critical feminist approaches to eating dis/orders* (pp. 135–145). Hove: Psychology Press.

Malson, H., & Burns, M. (2009). Re-theorising the slash of dis/order: An introduction to critical feminist approaches to eating dis/orders. In H. Malson & M. Burns (Eds.), *Critical feminist approaches to eating dis/orders* (pp. 1–6). Hove: Psychology Press.

Martin, J., & Govender, K. (2011). 'Making muscle junkies': Investigating traditional masculine ideology, body image discrepancy, and the pursuit of muscularity in adolescent males. *International Journal of Men's Health, 10,* 220–239.

O'Reilly, A. (2010). (Out)lawing motherhood: A theory and politic of maternal empowerment for the twenty-first century. *Hecate, 36*(1/2), 17–29.

Piggin, J., & Lee, J. (2011). 'Don't mention obesity': Contradictions and tensions in the UK Change4Life health promotion campaign. *Journal of Health Psychology, 16,* 1151–1164.

Pike, K., & Rodin, J. (1991). Mothers, daughters and disordered eating. *Journal of Abnormal Psychology, 100*(2), 198–204.

Polivy, J., & Herman, C. P. (2002). Causes of eating disorders. *Annual Review of Psychology, 53,* 187–213.

Rich, E., & Evans, J. (2008). Learning to be healthy, dying to be thin: The representation of weight via body perfection codes in schools. In S. Riley, M. Burns, H. Frith, S. Wiggins, & P. Markula (Eds.), *Critical bodies: Representations, identities and practices of weight and body management* (pp. 60–76). New York: Palgrave Macmillan.

Rich, E., Monaghan, L. F., & Aphramor, L. (2011). Introduction: Contesting obesity discourse and presenting an alternative. In E. Rich, L. Monaghan, & L. Aphramor (Eds.), *Debating obesity: Critical perspectives* (pp. 1–35). Basingstoke: Palgrave Macmillan.

Sampson, E. (1989). The challenge of social change for psychology: Globalization and psychology's theory of the person. *American Psychologist, 44*(6), 914–921.

Stice, E. (2002). Sociocultural influences on body image and eating disturbance. In C. Fairburn & K. Brownell (Eds.), *Eating disorders and obesity: A comprehensive handbook* (2nd edition) (pp. 103–107). New York: The Guilford Press.

Stice, E., Schupak-Neuberg, E., Shaw, H. E., & Stein, R. I. (1994). Relation of media exposure to eating disorder symptomology: An examination of mediating mechanisms. *Journal of Abnormal Psychology, 103*, 836–840.

Turner, H., Rose, K., & Cooper, M. (2005). Parental bonding and eating disorder symptoms in adolescents: The mediating role of core beliefs. *Eating Behaviors, 6*, 113–118.

Ussher, J. M. (1989). *The psychology of the female body.* Taylor & Frances/Routledge.

Walkerdine, V. (1996). Working class women: Psychological and social aspects of survival. In S. Wilkinson (Ed.), *Feminist social psychologies* (pp. 145–164). Buckingham: Open University Press.

Ward, A., Ramsey, R., & Treasure, J. (2000). Attachment patterns in eating disorders: Past in the present. *International Journal of Eating Disorders, 28*, 370–376. Cited in Jones, C., Harris, G., & Leung, N. (2005) Parental rearing behaviours and eating disorders: The moderating role of core beliefs. *Eating Behaviors, 6*, 355–364.

Weedon, C. (1987). *Feminist practice and poststructuralist theory.* Oxford: Basil Blackwell.

Wiseman, C. V., Gray, J. J., Mosimann, J. E., & Ahrens, A. H. (1992). Cultural expectations of thinness in women: An update. *International Journal of Eating Disorders, 11*(1), 85–89.

Woolhouse, M., Day, K., Rickett, B., & Milnes, K. (2012). 'Cos girls aren't supposed to eat like pigs are they?' Young women negotiating gendered discursive constructions of food and eating. *Journal of Health Psychology, 17*(1), 46–56.

Recommended reading

- Hepworth, J. (1999). *The social construction of anorexia nervosa.* London: Sage.
- Malson, H. (1998). *The thin woman: Feminism, post-structuralism, and the social psychology of anorexia nervosa.* London: Routledge.
- Malson, H., & Burns, M. (2009). *Critical feminist approaches to eating dis-orders.* London: Routledge.
- Riley, S., Burns, M., Frith, H., Wiggins, S., & Markula, P. (2008). *Critical bodies: Representations, identities and practices of weight and body management.* Basingstoke: Palgrave Macmillan.
- Saukko, P. (2008). *The anorexic self: A personal, political analysis of diagnostic discourse.* Albany: State University of New York Press.

Part V
Managing Problem Behaviour

24

Presuming Communicative Competence with Children with Autism: A Discourse Analysis of the Rhetoric of Communication Privilege

Jessica Nina Lester

Introduction

In this chapter, I explore how parents and therapists negotiate the meaning of communication with minimally and non-verbal children with autism. Specifically, I share findings from a study of the interactions of children with autism and their parents and therapists. This study was situated within a discursive psychology framework (Edwards & Potter, 1992; Potter & Wetherell, 1987) and informed by critical notions of disability (Oliver, 1996; Thomas, 1999), critical perspectives on communication (Biklen et al., 2005; Biklen & Burke, 2006), and certain aspects of conversation analysis (Sacks, 1992).

To set the context, the chapter begins with an overview of the history of autism. I then discuss how verbalness and the notion of competence have been historically coupled and privileged. To illustrate the relevance of this to autism, I share a detailed analysis of three representative extracts from interviews with parents and therapists of children with autism and two representative extracts from audio and video data of therapy sessions with children with autism. Specifically, I attend to the talk about non-verbalness and the function of bodily movements and other modes of communication. I conclude by discussing how privileging non-normative ways of communicating makes evident the tensions between the goal of assisting children to become more functional outside of the therapy setting and the goal of and commitment to self-expression.

The discursive making of autism

Autism has most often been defined in and through biomedical discourses, and typically described as a neurodevelopmental condition 'characteristized by persistent deficits in social communication and social interaction across multiple contexts' (American Psychiatric Association, 2014, p. 31). At present, it remains the most commonly studied childhood mental health disorder (Wolff, 2004), with the majority of this research focused on studying aetiologies and interventions. Popularised and academic understandings of autism have most often been framed in pathological terms (Nadesan, 2005).

Embedded within a matrix of institutional and professional discourses, autism was constructed in the 1940s as a diagnostic category. The two most prominent figures in the history of autism are Leo Kanner (1894–1981) and Hans Asperger (1906–1980). Although Kanner is credited for coining the term 'autism' as a diagnostic category, it was Bleuler (1857–1939) who first used the term in 1908 and again in 1911 to describe what he viewed as one of the primary psychological symptoms of schizophrenia. Bleuler used the term 'autistic' to explain his 'psychotic' patients' proclivity to withdraw into a world of fantasy (Alexander & Selesnick, 1966). With much writing and debating about schizophrenic autism occurring within psychiatry circles during the 1920s and 1930s, it is not surprising that both Kanner and Asperger elected to use the term 'autism' to describe the children they viewed as affectively withdrawn. In addition to the influences of the psychiatric discourses specific to schizophrenic autism, the institutionalised segregation of people deemed 'mentally ill' (Foucault, 1965), as well as the increase in the social surveillance of children with the advent of compulsory education engendered the necessary social conditions for the construction of autism as a diagnostic category (Nadesan, 2008; Strong & Sesma-Vazquez, Chapter 6, this volume).

Despite the prolific clinical work around autism since Kanner's 1943 publication, it was not until 1980 that the American Psychiatric Association incorporated the criteria for the diagnosis of autism within the *Diagnostic Statistical Manual of Mental Disorders* (DSM) (American Psychiatric Association, 1980). Since that time, autism has remained within the diagnostic manual, with expansive changes to the criteria being made since its inclusion. With the reality of what counts as autism extending its boundaries to include more and more children over the last 60 years, the prevalence of autism cases has not surprisingly escalated. Nonetheless, with the recent publication of the *DSM-5* (American Psychiatric Association, 2014), a new categorical structure for the diagnosis of autism has emerged, further highlighting its contingent and shifting meaning (Lester & Paulus, 2012).

While biomedical discourses have pervaded the discussions surrounding autism, in this chapter, I take a social constructionist orientation to autism. In doing so, I attempt to move away from purely medicalised perspectives of autism – which frame autism as being a biological 'truth' comprised of a triad of deficits, often including: (1) impaired social interaction, (2) lack of or limited imagination, and (3) delayed and/or limited communication (Frith, 1989).

Rather, I suggest that the emergence, labelling, and treatment of autism is embedded within institutionalised histories and cultural and discursive practices (Giles, Chapter 12, this volume; Nadesan, 2005). From this perspective, disability more generally is always already considered at the intersection of culture and biology. As such, throughout this chapter, I consider the function of communicative behaviours, as they are often marked in society as evidence of competence and/or incompetence – something that I explore in greater detail next.

The construction of communicative competence

Historically, verbalness has been tightly coupled with one's 'humanness' (Lewiecki-Wilson, 2003, p. 157). As Wickenden (2011) noted, 'being able to talk is seen as part of being human, and this ability is often privileged over other aspects of humanness' (p. 2). The *audibility* of one's voice has been positioned as a window into one's 'mind' and thereby evidence that one is thinking and ultimately part of the human race (Fry, 1977). With verbalness so tightly connected to popularised notions of intelligence, being non-verbal in a verbal world frequently results in being cast as incompetent (Biklen et al., 2005), particularly in a society structured around norms of communication defined by 'vocal bodies' (Paterson & Hughes, 1999, p. 604). As such, individuals who are not verbal are often presumed to be 'mentally retarded', and subsequently consigned to the position of 'judged' (Bogdan & Taylor, 1976, p. 47) (see Chapter 32).

More particularly, since its inception, autism has been described as being evidenced by communication challenges. For instance, the *DSM-5* outlined three levels of severity for autism spectrum disorders. The third level, which requires the most substantial support, presented social communication deficits as being 'severe' and exhibited by 'very limited initiation of social interactions'. The example provided within the *DSM-5* described an individual 'with few words of intelligible speech who rarely initiates interaction' (American Psychiatric Association, 2014). Not surprisingly, then, many autism interventions are centred on providing communication supports to individuals diagnosed with autism. While such interventions are central to the long-term success of individuals with autism, there remains a need to explore how non-intelligible speech are oriented to by therapists, parents, and society, as communication patterns are frequently linked to one's identity as an (in)competent social actor.

Ashby and Causton-Theoharis (2009) conducted a narrative study of seven autobiographical accounts of adolescents and adults with autism, exploring how the texts addressed issues of competence and noting how the participants made sense of the constructs of intelligence and mental retardation in relationship to broader institutional practices, such as intelligence testing (see Fatigante, Bafaro, & Orsolini, Chapter 32, this volume, for further discussion of intellectual disability). Across the texts, the researchers noted that what was often assessed by psychologists and other experts was not the individual's level of understanding but their ability to perform; that is to do something in a particular way, with the 'correct' way being specific to a given test and assessor's perspective. For instance, many of the participants described experiences where they failed to comply with a professional assessor who then named them incompetent. For them, it was not that they did not understand; they simply did not communicate as they were asked. Additionally, many of the authors of the analysed texts described an awareness of how the speech of people with autism was not conventional, as they did not always use words to convey their ideas. Consequently, the inability 'to convey' ideas 'through speech' resulted in constructions of incompetence (p. 506). The authors of this study pointed to how communication and competence function as intersecting concepts, often resulting in the exclusion of people who communicate in non-verbal or simply non-normative ways. They suggested that educators and researchers alike expand their definitions of normative performance, with an intent to 'no longer disqualify people from this thing we call the human race' simply because of communicative differences (p. 514).

In theoretically similar work, Rossetti, Ashby, Arndt, Chadwick, and Kasahara (2008) carried out a qualitative study exploring the performances of individuals with autism while being trained to use a communication device. They specifically explored the idea of competency and agency alongside 'behaviours and actions traditionally linked with incompetence' (p. 364), such as laughing loudly or producing presumably non-meaningful sounds. The researchers suggested that the three participating teenagers and five participating adults with autism displayed their agency through non-verbal movements that would often be interpreted by neuro-typical people as evidence of incompetence. Their findings align closely with the autobiographies of adults with autism who have emphasised their awareness of being perceived by others as incompetent due to their delayed or absent verbal response (Mukhopadhyay, 2011; Rentenbach, 2009; Sinclair, 1992).

To date, however, little research has examined how the notion of communicative competence is negotiated in the talk of children with autism and their parents and therapists. Thus, this chapter focuses specifically on this issue.

Project overview

The data included within this chapter were part of a larger ethnography focused on the everyday practices of children with autism and their parents and therapists (see Lester, 2011). Specifically, this study was informed by discursive psychology (Potter & Wetherell, 1987), with a discourse analysis being carried out.

Table 24.1 Participating therapists' demographic information

Pseudonym	Professional Title	Total Years at the Site
Bria	Occupational Therapist	4
Drew	Speech Pathologist/Clinical Director	4
Jennifer	Speech Pathologist	2
Megan	Speech Pathologist/Clinical Director	4
Michelle	Teacher/Autism Specialist	4
Patricia	Physical Therapist	1
Samantha	Medical Secretary	½
Seth	Occupational Therapist	½

The setting and participants

The study occurred at The Green Room – a paediatric clinic located in the Midwestern region of the United States. This clinic served children with disabilities who lived in a bi-state area. At the time of this study, approximately 80 families were served at The Green Room. A total of eight therapists agreed to participate in this study, with relevant demographic information listed in Table 24.1.

Twelve families with children clinically diagnosed with autism agreed to participate in this study. All of these families received therapeutic services of some kind at The Green Room. In total, 12 children with autism labels (see Table 24.2), aged 3 to 11 years, 6 fathers, and 11 mothers participated. Four of the children were considered functionally non-verbal, with all of the participating children being described as 'communicatively impaired'.

Data sources

The data drawn upon for this chapter were: (1) 175 hours of audio and video-recordings of the therapy sessions with the participating children and their therapists; (2) eight interviews with the therapists, ranging from 10 minutes to 42 minutes; (3) 14 interviews from parents, ranging from 22 minutes to 84 minutes; (4) 654 pages of handwritten field notes primarily focused on the therapy sessions; and (5) audio-recordings of and 20 pages of observational/field notes from the face-to-face meetings with therapists focused on my initial interpretations of the data.

Data analysis

I carried out an inductive analysis of the data, with four overlapping steps completed. First, I created modified Jeffersonian transcripts (Jefferson, 2004) of the therapy and interview data, focusing specifically on those segments of the 175 hours of therapy data and interviews that centred on those children who were described by parents and therapists as functionally non-verbal. I used Transana for transcription and the early stages of the analysis (Lester, 2015).

Table 24.2   Relevant information for participating children (*indicates those children considered functionally non-verbal)

Child's Pseudonym	Age	Gender	Race	Diagnostic Label
Billy	6	Male	Caucasian	Asperger's
Chance*	3	Male	Caucasian	Autism
George	4	Male	Caucasian	PDD-nos[a]
Noodle/Nancy*	7	Female	Caucasian	Chromosomal Deletion/Autism
Picasso	9	Male	Caucasian	Autism
Saturn	7	Male	Native American	Autism
The Emperor/TE*	7	Male	Caucasian	Autism
Thomas	5	Male	Caucasian	Autism
Tommy	7	Male	Caucasian	PDD-nos/ADHD[b]
T-Rex/TR	7	Male	Caucasian	Autism/ADHD
Diesel Weasel/DW	11	Male	Caucasian	PDD-nos/Bipolar/OCD[c]/ ADHD/Tourette's
Will*	8	Male	Caucasian	Mental Retardation/Autism

[a] Pervasive Developmental Disorder, not otherwise specified.
[b] Attention Deficit Hyperactivity Disorder.
[c] Obsessive Compulsive Disorder.

Second, I completed a 'more intensive' study of the emergent discourse patterns (Potter & Wetherell, 1987, p. 167). I considered how both non-verbal and verbal communication were positioned and constructed as being functional and legitimate. Third, I met with the participating therapists and shared the emergent findings to generate new analytical perspectives. Finally, I produced the two overarching discourse patterns along with line-by-line analyses of representative extracts. I used ATLAS.ti to support both the management of the data and the analysis process, primarily using the memoing and coding features.

Ethics

Prior to beginning the study, I acquired approval from my institutional review board. Further, because research with children with communicative differences often entails unique ethical challenges (Lester & Barouch, 2013), I invited each participant's ongoing consent. Throughout, I applied self-selected pseudonyms to maintain anonymity.

Findings

I situate these findings in the larger body of research that highlights how caretakers collaborate with young children with minimal speech (Ochs, &

Schieffelin, & Platt, 1979) and people with aphasia (Goodwin, 1995) to make sense of their communication. More particularly, I discuss two discursive patterns: (1) positioning of non-verbal behaviours as explainable, and (2) co-constructing legitimate communication.

Positioning non-verbal behaviours as explainable

Across the data, the parents of the children reframed their child's silence or non-responses as a communicative difference rather than a deficit or evidence of incompetence. Responding to the question, 'what things might you want someone who has just met your child to know about him,' Amelia, a participating parent, emphasised the need to 'be patient' and give her son, The Emperor, the time he needs to respond.

Extract 1

```
1    Amelia:    a lot of people ask him questions and then (1) it
2               takes him it will take The Emperor a good thirty seconds
3               sometimes to think of the answer but it's in↑ there=
4    Jessica:   =mm hm=
5    Amelia:    =I mean he knows what you're saying to him and he
6               knows the answer to what you're asking him↑=
7    Jessica:   =mm hm (1)
8    Amelia:    sometimes it comes out and sometimes it doesn't
9               so most of it is patience most of it is you know (1)
10              don't expect >even though he looks< like he should
11              react↑=
12   Jessica:   =mm hm=
13   Amelia:    =like everybody else's kid (1) he's not gunna
14              react [like
15   Jessica:   mm hm]
16   Amelia:    everybody else's kid um (2) but he's (.) he's
17              smart I mean he knows (1) he knows what's going on he
18              knows (2) everything anybody else's child that age knows
19              he [just can't
20   Jessica:   mm hm]
21   Amelia:    just can't get it out so
```

Amelia first made relevant her son's need for time to respond to a question ('a good thirty seconds'), while also making evident that indeed he does have a response. Jefferson (1989) noted that comfortable silences often last no more than 1.2 seconds. Such a finding perhaps explains why The Emperor's 30-second response time is worthy of comment by his mother. Further, in talk, non-responses often result in moral conclusions being drawn about people (Drew & Toerien, 2011) and may also signal trouble in talk (Pomerantz, 1984).

More generally, pauses might be viewed as evidence of a speaker being disengaged (Sacks, Schegloff, & Jefferson, 1974) or that non-conversational activities are occurring, such as writing something down (Goodwin, 1981). Yet, here, Amelia reframed the meaning of her son's silence, highlighting that he 'knows what you're saying' and 'the answer'. In this way, her son was positioned as knowledgeable, which stands in contrast to how many non-verbal people with autism have been described (e.g. Rentenbach, 2009). This is therefore a significant discursive move on the part of Amelia, as if she had suggested that her son did *not* have the answer or was anything less than knowledgeable, his competence would be questioned.

Amelia also acknowledged that her son is not 'like everybody's kid' – something that many of the participating parents articulated. Comparative frameworks are often evoked when describing that which is 'normal' and that which is 'abnormal', as 'normal' is only possible in comparison to that which is 'abnormal' or 'pathological' (Canguilhem, 1989). Yet, even while acknowledging that The Emperor was not like other children his age, he was positioned as 'smart', with his non-responses explained as the challenge of getting 'it out'. In this way, Amelia's talk functioned to at once locate her son as a competent, non-normative communicator who does encounter communicative challenges.

Similar to the participating parents, the therapists talked often about the importance of orienting to the children as competent and capable, regardless of their mode of communication. In Extract 2, Bria, an occupational therapist, described what she shares with new staff about the children who receive services at The Green Room.

Extract 2

```
1    Bria:      Um::: the biggest thing that I tell new staff is that
2               probably↓ in their life they've met a lot of people who
3               have had (.) low expectations or expectations that <weren't
4               high> enough=
5    Jessica:   =h[m:::
6    Bria:      for] what they're you know capability probably is so
7               I think the biggest thing that I would want other people
8               to know↑ is that these children (.) probably >hear
                understand
9               and see things< more than we give them credit for <even if
10              it's differently than we see them> it's probably at a
11              greater capacity or a greater extent than what we realize↑
                and
12              tuh (1) play into that into a positive way but also have
13              enough expectations that (.) we can meet a maximum potential
```

Bria began by making relevant the 'biggest thing' that she tells 'new staff', with her use of the word 'biggest' underlining the importance of her subsequent claims. Throughout, she contrasted the 'capability' of the children with the 'low expectations' that the children encounter from 'a lot of people'. These 'low expectations' were implicitly located in society at large, rather than within the child. In this way, the embodied differences of the children were positioned as being unfairly 'read' by others as evidence of incompetence. Similar to Rossetti et al.'s (2008) study in which outsiders interpreted the non-verbal performances of people with autism through a pathological lens and thereby constructed them as incompetent (p. 506), Bria positioned other's interpretation of children who 'hear, understand and see... differently' as resulting in 'low expectations'.

With the communicative differences of people with autism frequently viewed as odd, inappropriate, or even threatening, a primary challenge for children with autism is navigating how other's interpret their non-normative communication patterns (Lester & Paulus, 2014). In Extract 2, Bria highlighted that these societal misunderstandings should be made explicit to new staff at The Green Room, thereby also making visible one of the institutional goals. This institutional orientation was framed as one that results in children reaching a 'maximum potential'. Implicitly, then, a presumption of competence was constructed as central to (1) the therapeutic work that occurred at The Green Room, and (2) to whether a child made 'progress'.

While the participating therapists spoke often about the importance of 'self-expression' and the creation of a therapeutic space in which children felt free to communicate through a variety of modalities (e.g. invented sign language, communication devices, nonverbal body movements, writing), they also described the importance of 'teaching children' to navigate a highly verbal world. For the therapists, then, the notion of 'progress' was constructed always already in relationship to a society that has long coupled verbalness with competence. Nonetheless, the therapists spoke frequently about the importance of not simply interpreting nonverbal behaviours as being equivalent to low cognitive functioning, as illustrated in Extract 3.

Extract 3

```
1    Megan:     she's completely nonverbal um she's got some behaviours
2               and she looks extremely low [functioning
3    Jessica:   hmm (.)
4    Megan:     but] I think there's a lot of >little fantastically<
5               secret things that unless you know how to [elicit it
6    Jessica:   hmm mm
7    Megan:     you] never really [know↑
8    Jessica:   Hmm mm
9    Megan:     what] the potential is so (.) yeh I would say as a
10              whole there's there's more inside their their heads than
                (.)
11              meets the eye
```

Megan, a speech pathologist, described Nancy, a child with autism, as 'completely nonverbal'. She evoked an image of 'low functioning' as potentially including 'some behaviours' and being 'completely nonverbal'. Here, the non-disabled gaze was positioned as reading Nancy's body as 'low functioning', with this gaze being 'the product of a specific way of seeing which actually constructs the world it claims to have discovered' (Hughes, 1999, p. 155). In other words, there are certain bodily performances that are viewed as 'low functioning'. In this case, however, Megan contradicted this interpretation of 'low functioning' and claimed that there are also 'secret things' that can be known to those who know 'how to elicit' them. Here, Megan implicitly oriented to her role as a 'special' therapist, as she was presumably one of those who knows 'how to elicit' these 'fantastically secret things'. This further highlighted the central role that caretakers and trusted others play when interacting with children with little to no speech, as they are often the one's that generate co-constructed understandings of what comes to be counted as meaningful and functional communication (Goodwin, 1995). Like Bria in Extract 2, Megan concluded by highlighting how the way one 'reads' embodied differences shapes 'the potential' that will be known.

Co-constructing legitimate communication

Across the therapy sessions, a co-construction of what came to be counted as legitimate communication occurred, with the child's 'voice' not being limited to audible words (Wickenden, 2011). I share next two representative extracts drawn from the 175 hours of therapy data, recognising that what I share here is partial, as there is much left unsaid (Noblit, 1999). Nonetheless, for the purposes of this chapter, these two extracts provide key examples of how the very notion of communication was worked up within the therapy session.

In Extract 4, Nancy, a non-verbal child with autism, had just entered the room where she was scheduled to have a 30-minute speech therapy session with Megan, her therapist. One of the primary goals for her sessions was to be more skilled at making requests via her Dynavox – a communication device/box that allowed Nancy to touch a screen to identify symbols, pictures, letters, and words that were then audibly read in a voice of her choosing. During these sessions, her therapist always had her favourite foods and games on or near the table at which they worked. During this particular therapy session, Nancy's favourite game, the 'ball popper', was positioned on the centre of the table, with her Dynavox positioned to the right of it.

Extract 4

```
1  Megan:   =[(walks to chair and sits down behind table)
2  Nancy:   (turns toward table and gazes at ball popper)](.)
```

```
3    Megan:   (pulls Nancy's chair up to table) °okay you can
4             sit° (.)
5    Nancy:   (gazes at Dynavox)=
6    Megan:   =(places Dynavox in front of self and Nancy and looks
7             at Dynavox screen) Oh↑">you were doing< alphabet stuff
              huh↑=
8    Nancy:   =[((taps the Dynavox screen with left pointer
9             finger, moving the screen back to where request for
10            playing with toys are located)
11   Megan:   (gaze shifts to watch Nancy's hand and the
12            Dynavox screen)] Oh look at you go right back to where
13            you need to be↓=
14   Nancy:   =(gaze shifts to ball popper on table and she
15            points toward the ball popper with her left pointer
16            finger)
17   Megan:   you [wanna do the ball popper right away↑
18   Nancy:   (sits down in chair)]
19   Megan:   AWESOME can you tell me that↑=
20   Nancy:   =(clicks on Dynavox screen)(2)
21   Megan:   okay you might have to close that [one
22   Nancy:   eh:::m:::
23   Megan:   I don't think the ball popper's on that one]
```

In Extract 4, Megan began by making relevant the institutional norms, which frequently entailed the request for the child to 'sit' down, with the Dynavox or other communication device then being placed near the sitting child and therapist. In this case, Nancy took up her first conversational turn by turning her body to the table and shifting her gaze towards the ball popper, with her gaze making evident her request to play with the game. However, Megan responded with a request for her to 'sit', with Nancy then shifting her gaze to the Dynavox. Megan then immediately moved the Dynavox in front of herself and Nancy, orienting to Nancy's gaze shift as a request to communicate with the Dynavox. Nancy made an initial attempt to request the ball popper, beginning by moving the screen to where requests for toys and games were typically made.

Similar to Lewiecki-Wilson's (2003) description of the need for people who can 'carefully and ethically co-construct narratives and arguments from the perspective of the disabled person' (pp. 161–162), Megan and Nancy co-crafted an account of what Nancy wanted to do. In this case, Nancy wanted to play with the 'ball popper', with her gaze shifts (line 2, line 14), pointing (lines 15–16), and initial attempts to make the request with the Dynavox (line 8, line 20 making this evident. However, the institutionality of this interaction was also made visible with Megan's insistence that a 'full' request with the Dynavox be made, as the aim of the session was to teach Nancy how to make a request with a complete sentence (e.g. 'I want to play ball popper'). As such, in lines 17–23, there was both an acknowledgement that Nancy wanted to play with the ball popper and a request to 'tell me that', with the 'telling' being presumably something that must be hearable; that is, an audible request made via the Dynavox. Thus, here, we see that the therapist and child negotiated the request to play with the ball popper, with gaze shifting and pointing oriented to as legitimate, but not sufficient communicative efforts. Sufficiency was perhaps based in the reality that few outsiders would be familiar or attentive enough to the 'embodied nonverbal performances and gestures' that made Nancy's interaction with Megan possible (Lewiecki-Wilson, 2003).

Yet, for many children learning to use a communication device, outsiders questioned the legitimacy of their 'voice' and their need to use communication devices. While the therapists and parents created opportunities for the children to use communication devices, the cultural and historical presumption that communication *is* verbal and *should* take place orally often pervaded, with Extract 5 illustrating this well.

Extract 5 is drawn from a therapy session in which Bria, an occupational therapist, attempted to work with Nancy, despite her Respite Care Worker failing to bring her communication device to the session. Prior to beginning the therapy session, the clinical directors asked the Respite Care Worker to return home and retrieve the communication device, as this is 'one way Nancy talks to us'. The extract begins just after the Respite Care Worker entered the therapy room with the communication device, nearly 30 minutes after the start of the session.

Extract 5

```
1  Nancy:          (Crying as she runs to the door as the respite care
2                  worker brings the communication device into the therapy
                   room)=
3  Respite Care    =Hi I thought your nap would have helped↑
   Worker:
4                  (.)(gives the communication device to Bria)
5  Bria:           Well I think she [is genuinely frustrated
6  Nancy:          (Crying and moves to stand beside Bria)]
7  Bria:           that she doesn't have her words↑ (sets up the
8                  communication device on the table directly in front of
                   where
9                  Nancy who is now standing) look now you can tell me
                   what you
10                 are mad about (3) come on let's talk about it(2)let's
                   talk
11                 about it tell me what's going on(4)
12 Nancy:          (Navigates to communication page in communication
13                 device and clicks the words/pictures)Please please
                   I want I
14                 want please please all done (.)
```

In Extract 5, Nancy's 'frustration' is evidenced through her crying and request to be 'done'. While the Respite Care Worker positioned Nancy's tears as being due to her being tired ('I thought your nap would have helped'), Bria countered this claim by coupling Nancy's frustration with not having 'her words'; that is, her communication device. 'Words', in this case, were positioned as the central explanation for the 'frustration'. Edwards (1997) has suggested that emotional states are often performed 'bodily or linguistically' and mark 'that something is wrong' (p. 170). Here, then, Nancy was not simply frustrated for the sake of being frustrated, but there was a cause, which Bria oriented to as not having access to her 'words'. Historically, many people with autism have not been given access to communication devices, resulting in 'behaviours' perceived to be inappropriate or even threatening (e.g. Rentenbach, 2009). Yet, here, we see the therapist orient to the child's frustration as explainable, and her non-verbal performance as a legitimate means by which to communicate her frustration. Bria concluded by inviting the child to 'talk about it', with Nancy requesting to be 'all done'.

For both parents and therapists, then, the children's non-normative ways of communicating were positioned as commonly misinterpreted by others, yet *not* evidence of 'true' incompetence. Rather, for many of the participating parents and therapists, the children's embodied differences called for not simply the child to learn new ways to communicate, but the 'community and society to also change', as one therapist shared in an interview (Lester, 2012).

Therapists' responses to the findings

When I met with the therapists to share the findings and invite their responses to my interpretations, they each emphasised the importance of reframing the very definition of communication. In response to Extracts 1 and 5, Jennifer stated:

> The ability to communicate right away doesn't directly correlate with level of intelligence. If you don't talk, people automatically assume that your cognitive level is lower than expected, and that's for any kid who doesn't speak as we think they should or speak like we do, and that's always parents' first concern. You know, that's always something that they express, I think, when their kid doesn't talk, 'I just don't want people to think that he doesn't know.'

Referring to Extract 5, Bria pointed to how being non-verbal or using a communication device to communicate functions to mark a child as incompetent. She continued by sharing that Nancy had recently learned how to type on a keyboard with minimal physical support. After we discussed the extensive body of literature that reports that people who use something other than words to communicate are presumed to be retarded, incompetent, and perhaps not even quite human (Lewiecki-Wilson, 2003), Bria proceeded to share that one of the school officials questioned whether Nancy is 'really typing, really communicating.' She stated:

> One of the ladies at school, who isn't like a super invested women as a side note, said essentially that Megan was hitting the keys for her and that Megan just didn't realize that . . . right away people were like, 'She is not typing. She's not the one who's doing it.' And I just thought how unfortunate right away to discount it because even if it was kind of a fluke [i.e., Nancy typing words and accurately hitting the keys], she's learned so much already that even if we were hitting all the keys for her, I still think she would learn from that. So even if she believes that, I felt like she should have a different response than like totally negating everything and saying, 'She doesn't need a keyboard. She doesn't need an iPad. She doesn't need any of that 'cuz she's not actually doing it.'

With the consequences of being viewed as not being a legitimate speaker or exhibiting 'odd' communication patterns often resulting in exclusion from *full* participation in society, it is paramount that clinicians, as well as society at large, consider interpreting the discourses of the bodies of those with autism from a perspective of competence.

Clinical relevance summary

I emphasize here three key points relevant to clinicians. First, these findings and the broader discussion make visible the felt tensions between creating opportunities for self-expression and supporting children as they learn to navigate a society that privileges particular ways of communicating. The clinician, then, must learn to at once address the needs of a child with autism and their family as they relate to learning to successfully navigate the world, while also valuing and building upon the unique communicative competencies that are produced.

Second, while not denying the challenge of an impairment effect (Thomas, 2004), what remains perhaps equally difficult is learning how to live in a society that 'reads' non-normative communication and embodied differences as 'abnormal' and something to be 'fixed' (Lester, 2012). It is paramount, then, for clinicians to create spaces for parents and children with autism to reframe non-normative communication patterns, moving away from presumptions that result in the child being cast as incompetent and incapable of participating in society. Rather, participation may be conceived of as occurring in non-normative ways.

Table 24.3   Clinical practice highlights

---

1. Therapists should presume communicative competence regardless of the mode (e.g. bodily movements, communication device, sign, verbal communication) by which a child communicates.
2. Therapists and parents of children with autism can participate in the co-construction of an identity of communicative competence, particularly as they orient to a child's language (in all its forms) as meaningful and functional.
3. A balance should be struck between creating opportunities for self-expression and teaching a child how to communicate in a way that is translatable across contexts.

Finally, it is the clinician who plays a central role in the positioning and co-constructing of communicative acts, with what comes to be counted as legitimate, meaningful, and functional often taking place through what Lewiecki-Wilson's (2003) called 'mediated rhetoricity' (p. 161). Lewiecki-Wilson (2003) described 'mediated rhetoricity' as the 'language used for the benefit of the disabled person, which is (co)constructed by parents, advocates and/or committed caregivers' (p. 161). Such language, she argued, requires 'thoughtful attention' on the part of the clinician, particularly as the child's 'embodied nonverbal performances and gestures' must be 'learned' and oriented to as meaningful (p. 161). Ultimately, as Osteen noted (2008), 'we must strive to speak not *for* but *with* those unable or unwilling to communicate through orthodox modes of communication' (p. 7). For a simple summary of the implications for practice, see Table 24.3.

Summary

In this chapter, I discussed the way in which verbalness has been privileged and coupled with the notion of human competence. By presenting a discourse analysis of data drawn from a larger study of the everyday practices of children with autism and their parents and therapists, I illustrated how meaningful and functional communication might be negotiated and positioned as something that is far more than verbal production.

References

American Psychiatric Association (1980). *Diagnostic and statistical manual of mental disorders* (3rd edition). Washington, DC: Author.

American Psychiatric Association (2014). *Diagnostic and statistical manual of mental disorders* (5th edition). Washington, DC: APA.

Alexander, F. G., & Selesnick, S. T. (1966). *The history of psychiatry: An evaluation of psychiatric thought and practice from prehistoric times to the present.* New York: Harper & Row.

Ashby, C. E., & Causton-Theoharis, J. N. (2009). Disqualified in the human race: A close reading of the autobiographies of individuals identified as autistic. *International Journal of Inclusive Education, 13*(5), 501–516.

Biklen, D., Attfield, R., Bissonnette, L., Blackman, L., Burke, J., Frugone, A., Mukhopadhyay, R. R., & Rubin, S. (2005). *Autism and the myth of the person alone.* New York: New York University Press.

Biklen, D., & Burke, J. (2006). Presuming competence. *Equity & Excellence in Education, 39*(2), 166–175.

Bogdan, R., & Taylor, S. (1976). The judged, not the judges: An insider's view of mental retardation. *American Psychologist, 31,* 47–52.

Canguilhem, G. (1989). *The normal and the pathological.* Brooklyn, NY: Zone Books.

Drew, P., & Toerien, M. (2011). *An introduction to the methods of conversation analysis (short course on conversation analysis).* York, UK: University of York.

Edwards, D. (1997). *Discourse and cognition.* London: Sage.

Edwards, D., & Potter, J. (1992). *Discursive psychology.* London: Sage.

Foucault, M. (1965). *Madness and civilization: A history of insanity in the age of reason.* New York, NY: Random House.

Frith, U. (1989). *Autism: Explaining the enigma.* Cambridge, MA: Blackwell Publishers.

Fry, D. B. (1977). *Homo loquens: Man as a talking animal.* Cambridge: Cambridge University Press.

Goodwin, C. (1981). *Conversational organization.* New York: Academic Press.

Goodwin, C. (1995). Co-constructing meaning in conversations with an aphasic man. *Research on Language and Social Interaction, 28*(3), 233–260.

Hughes, B. (1999). The constitution of impairment: Modernity and the aesthetic of oppression. *Disability & Society, 14*(2), 155–172.

Jefferson, G. (1989). Preliminary notes on a possible metric which provides for a 'standard maximum' silence of approximately one second in conversation. In P. Bull & R. Derek (Eds.), *Conversation: An interdisciplinary approach* (pp. 166–196). Clevedon: Multilingual Matters.

Jefferson, G. (2004). Glossary of transcript symbols with an introduction. In G. H. Lerner (Ed.), *Conversation analysis: Studies from the first generation* (pp. 13–31). Amsterdam: John Benjamins.

Kanner, L. (1943/1985). Autistic disturbances of affective contact. In A. M. Donnellan (Ed.), *Classic readings in autism* (pp. 11–50). New York: Teachers College Press.

Lester, J. N. (2011). *The discursive construction of autism: Contingent meanings of autism and therapeutic talk.* Unpublished dissertation: University of Tennessee.

Lester, J. N. (2012). Researching the discursive function of silence: A reconsideration of the normative communication patterns in the talk of children with autism labels. In G. S. Cannella & S. R. Steinberg (Eds.), *Critical qualitative research reader* (pp. 329–340). New York: Peter Lang.

Lester, J. N. (2014). Negotiating the abnormality/normality binary: A discursive psychological approach to the study of therapeutic interactions and children with autism. *Qualitative Psychology, 1*(2), 178–193.

Lester, J. N. (2015). Leveraging two computer-assisted qualitative data analysis software packages to support discourse analysis. In S. Hai-Jew (Ed.), *Enhancing qualitative and mixed methods research with technology* (pp. 194–209). Hershey, PA: IGI Global.

Lester, J. N., & Barouch, A. (2013). Inviting the assent of children described as functionally nonverbal. In I. Paoletti, A. Duarte, I. Tomas, & F. Menéndez (Eds.), *Practices of ethics:*

An empirical approach to ethics in social science research (pp. 65–84). Cambridge, UK: Cambridge Scholars Press.

Lester, J. N., & Paulus, T. M. (2014). 'That teacher takes everything badly': Discursively reframing non-normative behaviors in therapy sessions. *International Journal of Qualitative Studies in Education, 27*(5), 641–666.

Lewiecki-Wilson, C. (2003). Rethinking rhetoric through mental disabilities. *Rhetoric Review, 22*(2), 156–167.

Mukhopadhyay, T. R. (2011). *How can I talk if my lips don't move? Inside my autistic mind.* New York: Arcade Publishing.

Nadesan, M. H. (2005). *Constructing autism: Unraveling the 'truth' and understanding the social.* New York, NY: Routledge.

Nadesan, M. H. (2008). Constructing autism: A brief genealogy. In M. Osteen (Ed.), *Autism and representation* (pp. 78–95). New York: Routledge.

Noblit, G. W. (1999). *Particularities: Collected essays on ethnography and education.* New York: Peter Lang Publishing.

Ochs, E., Schieffelin, B., & Platt, M. (1979). Propositions across utterances and speakers. In E. Ochs & B. Schieffelin (Eds.), *Developmental pragmatics* (pp. 251–268). New York: Academic Press.

Oliver, M. (1996). A sociology of disability or a disablist sociology? In L. Barton (Ed.), *Disability and society* (pp. 18–42). London: Longman.

Osteen, M. (2008). Autism and representation: A comprehensive introduction. In M. Osteen (Ed.), *Autism and representation* (pp. 1–47). New York: Routledge.

Patersen, K., & Hughes, B. (1999). Disability studies and phenomenology: The carnal politics of everyday life. *Disability & Society, 14*(5), 597–610.

Pomerantz, A. (1984). Agreeing and disagreeing with assessment: Some features of preferred/dispreferred turn shapes. In J. M. Atkinson & J. Heritage (Eds.), *Structure of social action: Studies in conversation analysis* (pp. 57–101). Cambridge, UK: Cambridge University Press.

Potter, J., & Wetherell, M. (1987). *Discourse and social psychology.* London: Sage.

Rentenbach, B. (2009). *Synergy.* Bloomington, IN: AuthorHouse.

Rossetti, Z., Ashby, C., Arndt, K., Chadwick, M., & Kasahara, M. (2008). 'I like others to not try to fix me:' Agency, independence, and autism. *Intellectual and Developmental Disabilities, 46*(5), 364–375.

Sacks, H. (1992). *Lectures on conversation.* Oxford: Blackwell.

Sacks, H., Schegloff, E. A., & Jefferson, G. (1974). A simplest systematics for the organization of turn-taking for conversation. *Language, 50*(4), 696–735.

Sinclair, J. (1992). Bridging the gaps: An inside-out view of autism (or, do you know what I don't know?). In E. Schopler & G. B. Mesibov (Eds.), *High-functioning individuals with autism* (pp. 294–302). New York: Plenum Press.

Thomas, C. (1999). *Female forms: Experiencing and understanding disability.* Buckingham, UK: Open University Press.

Wickenden, M. (2011). Whose voice is that?: Issues of identity, voice and representation arising in an ethnographic study of the lives of disabled teenagers who use Augmentative and Alternative Communication (AAC). *Disability Studies Quarterly, 31*(4), Retrieved from: http://dsq-sds.org/article/view/1724/1772.

Wolff, S. (2004). The history of autism. *European Child & Adolescent Psychiatry, 13*(4), 201–208.

Wood, L. A., & Kroger, R. O. (2000). *Doing discourse analysis: Methods for studying action in talk and text.* Thousand Oaks, CA: Sage.

Recommended reading

- Biklen, D., Attfield, R., Bissonnette, L., Blackman, L., Burke, J., Frugone, A., Mukhopadhyay, R. R., & Rubin, S. (2005). *Autism and the myth of the person alone.* New York, NY: New York University Press.
- Lester, J. N. (2012). Researching the discursive function of silence: A reconsideration of the normative communication patterns in the talk of children with autism labels. In G. S. Cannella & S. R. Steinberg (Eds.), *Critical qualitative research reader* (pp. 329–340). New York: Peter Lang.
- Rossetti, Z., Ashby, C., Arndt, K., Chadwick, M., & Kasahara, M. (2008). 'I like others to not try to fix me:' Agency, independence, and autism. *Intellectual and Developmental Disabilities, 46*(5), 364–375.

25
Parents' Resources for Facilitating the Activities of Children with Autism at Home

Monica Ramey and John Rae

Introduction

In the United Kingdom, and in many societies, whatever level of support might be available outside the child's home, the primary carers for children with autism are often their own family members, usually their parents. Consequently, parents' resources for supporting their children's capacity to engage in activities are of considerable practical importance and merit careful analysis. Indeed, there is evidence from a randomised control study that a joint engagement programme can have a positive long-term effect on children's abilities (Kasari, Gulsrud, Wong, Kwon, & Locke, 2010). A general aim of parents' participation in children's activities (including neurotypical children) is to enable children to achieve more than they could unaided. The importance of such co-participation is captured by Vygotsky's influential idea of the zone of proximal development (ZPD). This refers to a region which is just beyond a child's current sphere of competence but into which the child can move through the involvement of a more experienced peer or carer. For Vygotsky's socio-cultural theory of learning, it is not merely that the child can achieve more with support but rather that such supportive encounters are fundamental for development. He states that the ZPD 'awakens a variety of internal developmental processes that are able to operate only when the child is interacting with people in his environment' (Vygotsky, 1978, p. 90). The term 'scaffolding' (Wood, Bruner, & Ross, 1976) is a related metaphor that is widely used in learning environments to describe the supportive conduct. While Wood et al. (1976) do not cite Vygotsky, the intimate connection between scaffolding and Vygotsky's work is pointed out by Cazden (1979). It can be said that scaffolding is the support that is provided by a 'more knowledgeable other' who can assist the learner in completing a task or solving a problem by assisting the learner into the ZPD. In scaffolding, language and interaction are important as the child internalises routines and procedures from the social and cultural context where he or she learns (Applebee & Langer, 1983). Tasks should build on what the child already knows and ongoing assessment of the child is important in order to tailor scaffolding as required (Langer & Applebee, 1986). Additionally, dialogue in interaction keeps the recipient actively engaged in the activity and through monitoring both parties can coordinate the direction of the activity (Brown & Palincsar, 1989).

Parents can take on the role of the facilitator providing strategies for completing a task. Additionally, parents are able to expand the child's learning by using modelling, perspective taking, clues, and pointing out important information that can help learners think (Wood et al., 1976). In Wood (1980, cited in Carr & Pike, 2012), *contingent shifting* is used to describe change in the level of support offered depending on the child's performance, such as a reduction in the level of support in response to the child making progress. However, increasingly the responsibility for task completion is given to the child, thereby expanding his or her knowledge. Carr and Pike (2012) suggest that effective *contingent shifting* in scaffolding by the parent may be indirectly influenced by the *parenting style* in managing challenging behaviour and developing a positive relationship with the child. Although previous research into parents scaffolding their children's activities has used observational measures and has focused on specific areas of interactions that may affect educational and developmental outcomes (e.g. Carr & Pike, 2012; Salonen, Lepola, & Vauras, 2007), a study by Freeman and Kasari (2013) observed the scaffolding styles of parents of children with ASD during a ten-minute play session and found that parents of children with ASD created more play scenarios and offered and directed play more than parents of typical children. Parents of children with ASD also showed high interest in their child's play attempts, responding with high levels of play, while parents of typical children would expand their children's play. In conclusion, parents who were less directive and played just above the child's level, combined with imitating the child's actions, resulted in longer interactions.

These observational studies involve the careful coding of specific practices of parental scaffolding in order to examine scaffolding in a systematic way. Conversation Analysis (CA) offers a complementary research approach in which a more exploratory stance is taken to interaction and in which the *organisation of interactional* phenomena is the focus. A particular focus of CA research involves the delineation of the sequential relationship between one participant's action and another participant's action-in-response. A previous CA-based analysis of scaffolding examined the nature of a learning support assistant's (LSA) supportive conduct in a classroom setting (Stribling & Rae, 2010). This case study of a single extended episode examined how the LSA co-participated in a number of sequences initiated by a class teacher in order to facilitate a response from a girl with autism. The interactional organisation of instructional settings has been examined quite extensively through CA methods;

for example, there are multimodal analyses of object-related interactions in adult learning situations such as manual crafts (Ekström, Lindwall, & Säljö, 2009; Lindwall & Ekström, 2012) and surgery (e.g. Mondada, 2014; Zemel & Koschmann, 2014).

For Vygotsky, the concept of the ZPD was fundamentally concerned with *development;* it provided a way of understanding how support from others can contribute to development. In a related way, the concept of *scaffolding* suggests a temporary supportive structure, one that can be removed in due course. In the case of atypical development, it is not necessarily the case that supportive behaviour can be withdrawn (Stone, 1998). It can be briefly noted that a related conceptualisation occurs within the Applied Behavior Analysis (ABA) or the Lovaas approach to teaching children with autism. Here, a trainer's or parent's support behaviours are construed as *prompts.* Based on the principles of operant conditioning, a common concern is the *fading* of these prompts such that the child can perform the socially relevant target behaviour independently (e.g. Krantz & McClannahan, 1998; MacDuff, Krantz, & McClannahan, 2001).

Given the importance of parental scaffolding and the need to understand what practices it consists in, this chapter uses CA to identify moment-to-moment actions in episodes on interaction involving children with autism and their parents carrying out spontaneous activities together at home.

Project overview

In order to examine the organisation of talk-in-interaction and scaffolding procedures provided by parents and children with an Autism Spectrum Disorder (ASD) taking part in naturally occurring interactions at home, families with children with ASD were emailed a letter of invitation to take part in the study. Parents who agreed and gave full informed consent were additionally provided with a debrief form with contact details if further information is needed or if they wish to withdraw. Our analysis is based on video-recordings of four children with an ASD to whom we shall refer using the following pseudonyms: Mary (aged 14) interacting with her Mum and doing pottery and poetry classes; Ben (aged 12) interacting with his Mum, Dad, and also with a sibling and an ABA tutor carrying out money-related activities, playing with Lego and playing with a video game; and Will and Anna (both aged 13), baking with Will's Mum. A total of two hours of video-recordings were made.

Different analytical approaches can be brought to bear on video data from learning settings (Barron, 2006). CA is applied for this study due to its focus on social interaction in its own terms. A notable early formulation of CA refers to the aim of developing '... a naturalistic observation discipline that could deal with the details of social action(s) rigorously, empirically and formally' (Schegloff & Sacks, 1973, p. 233). This paper further states, 'Our analysis has sought to explicate the ways in which our materials are produced in orderly ways that exhibit orderliness and have orderliness appreciated and used...' (p. 290). This orderliness is frequently, but not exclusively, exhibited in the analysis of sequentially unfolding moment-by-moment interaction (Goodwin & Heritage, 1990; Schegloff, 2007). Fundamentally then, CA takes stretches of interactional data and asks, first, what is the social organisation? Secondly, how is this organisation produced? It should be noted that CA originated within sociology and that often the basic sociological question of *who is doing what?* It is this that is relevant in the analysis. Indeed, although our focus is on domestic activities, in our analysis the classical sociological issues of the division of labour, who controls the means of production, and matters of identity will be relevant. While CA can be used to underpin quantitative Content Analytic studies which incorporate inter-rater agreement measures, it is, in the first instance, a qualitative discovery-oriented methodology.

Video-recordings were inspected in terms of what activities are underway and how they are organised. In addition to attending to the participants' talk and other vocalisations, their gaze, gesture, and handling of objects were frequently the focus of analysis. Video-recorded data can be especially useful for the analysis of interactions involving children with autism, as gaze and body language can be particularly pertinent (see Dickerson & Robins, Chapter 4, this volume, for a good example). In this report then, we are primarily concerned with the parents' resources but in order to get them, the actions of both parents and children need to be examined. Episodes of interest were transcribed using the Jefferson transcription system for CA extended with annotations for visible action. While CA analysis commonly, and perhaps preferably, selects a specific interactional phenomenon and examines it in depth and detail, the present treatment locates a number of phenomena that are constitutive of the setting that is under analysis. The analysis here is geared to identifying the facilitating practices that occur, and how they are implemented, in a small opportunity sample.

Analysis

The analysis will first focus on a single extend episode, a pottery-making session, and identify a range of supportive actions within this. We will then consider the generality of some of the phenomena observed through comparisons with episodes involving other families in other settings. In the pottery session (which is of about 12-minutes duration), Mary fashions a clay pot, with help from her Mum, on a small battery-powered portable potter's wheel. The analysis will draw on a number of transcribed episodes within this activity. Readers should note that these extracts seek to *represent* the activity that occurred rather than to *describe* it. As such, readers are encouraged to try to *visualise* the unfolding actions.

Extract 1 [MR2012 MM Clay] simplified

Mary is seated at a table with a potter's wheel in front of her with a tool in her right hand. Her mother is approaching the table and is beside Mary, placing a small water bowl on the table. Mary is turned to her right, towards her mother.

```
00        Mar:    nh::::::.
01        Mum:    Righ'!
02                (-[-)
                  [X Mum: water bowl audibly placed on table
                  [Mar: turns towards potter's wheel
03        Mum:    [Water
                  [Mar: brings tool to the potter's wheel and
                  [begins to clean the wheel
04        Mum:    Move the phaper °(huh)°
05                (----------1---------2---------3)
06                (----------4---------5---------6)
07        Mum:    °°( ) ( ) you°°
08        Mar:    nhah::uh::
                               [Mum: points towards clay
09   ➔    Mum:    Are you goina put [the ^clay on.
10                (---------1)
11   ➔    Mum:    Is it plugged in (°just a second)°
12        Mar:    nnnn
13        Mar:    (   )  (   )  (   )  ( .) dah duh nun °huh °huh
14        Mar:    °(just a second)°
15                (12) ((Mum gets pedal))
     ➔                ((Mar gets clay))
16        Mar:    (nothing)
17                (5.0)    (Mum plugs pedal in)
18        Mar:    (baddu)
19                (4.0)
20      (Mum):    gnnn
21        Mum:    °That's for your° foo:t.
22                (---------1---[- - - - 2 - - -[- - - -[-3)
                              [turns        [turns [turns
                              [towards      [back  [towards
                              [mum                 [mum
23   ➔    Mar:    [Mama help
24        Mum:    [YeAH
25   ➔    Mum:    [Yeah¿ >Of course? I'm going to help¿< (1.0)
                  [Mum sits
```

In Extract 1, Mum assists with some material preparations, commenting on them as she performs them: setting a bowl of water next to Mary (lines 2–3) and removing some irrelevant pieces of paper (line 4). When Mum sets the water down, Mary begins to scrape the potter's wheel with a clay-working tool. Mum produces an initiating action, 'Are you goina put the ^clay on.', pointing at the clay, which is apparently designed to progress the pottery activity, encouraging

Mary to move from cleaning the wheel to getting the material to be worked on into position. Mary continues cleaning the wheel, but at this point the mother engages in another preparatory matter, namely the proper positioning of the pedal which controls the rotation of the potter's wheel. She launches this through a question, 'Is it plugged in (°just a second)°' and moves to get the pedal, to plug into the wheel, and she positions it on the floor. As Mum reaches for the pedal, which is on the table to Mary's left, Mary reaches for the clay, which is also to her left. As Mum unwraps the wire from the pedal, Mary unwraps the clay. As Mum becomes available after placing the pedal on the floor under the table, Mary says, 'Mama he'p' (line 23), to which Mum replies, 'Yeah¿ >Of course? I'm going to help'.

The tape only begins with Mary already sitting at the table, so we do not have access, on video, to the background to this episode of activity. We can see, however, that Mum is engaged in a number of actions that are accountably designed to prepare for, and to progress, the activity of pottery. Some of these actions appear to be fundamentally unilateral, for example setting down the bowl of water next to Mary; however, it is notable that this is done in a conspicuous and accountable way – the arrival of the bowl to the table being accompanied by 'right', the occurrence of a loud clink as it meets the table, and the provision of verbal commentary on this, 'Water' (line 3) By contrast, the instruction to put the clay onto the wheel 'Are you goina put the ^clay on', implemented through a question, is an initiating action which implicates that Mary should transfer the clay onto the wheel. Previous work on classroom settings has focused on how the child's response to a teacher's initiating action was occasionally delayed and was produced through supportive conduct from a learning support assistant (Stribling & Rae, 2010). In the present setting, Mary's response does not come about straight away – but neither is it pursued or awaited by Mum. Rather, in this case, Mum launches an alternative activity (getting, and setting up, the pedal). Mary's response occurs while Mum is doing this, indeed it occurs in parallel with it. Although there is no evidence here that launching a side activity (getting the pedal) is a specific strategy to allow a child a space to respond to a just prior instruction, or that engagement in a side action might have a modelling function, it is notable that Mum refrains from pursuing a response from Mary. Routinely in social interaction, participants have a range of resources for pursuing a response from co-participants who have failed to provide one (Pomerantz, 1984). It is a feature of informal spontaneous activities that they involve parents having to address contingencies, such as dealing with an incomplete practical matter (here setting up the foot pedal); here it can be seen that addressing such contingencies can provide an opportunity space for a child to respond to an initiating action.

Throughout this extract then, Mary engages in task-relevant activities. For example, she is appropriately oriented to the potter's wheel and engages with it; she gets the clay following her Mum's instruction to do so. It is notable that she produces a task-related initiating action, instructing her mother to help with the activity, 'Mama help' (line 23). This does not appear to implicate assistance with an immediate difficulty but seems to work prospectively to indicate a requirement to assist with the later stages of the forthcoming activity. Nevertheless, Mum responds immediately with a reassuring reply, 'Yeah of course. I'm going to help'. The production of parental instructions is further illustrated in Extract 2 (this episode follows immediately from the stretch of activity presented in Extract 1).

Extract 2 [MR2012 MM Clay] simplified
Continues immediately from Extract 1

```
26          Mar:   n[nuh
27    ➜     Mum:   [Put it in [the    [mid:dle
      ➜                       [point [pressing]
28    ➜     Mum:   If you ca:n.
29                 (--[-------1--------2
                   [Gaze at Mary
                   [Gaze at Mum
                   [(hih)(hih)(hih) (hee)
30          Mum:   [That's the [hard part
                   [           [smack
                   [Mary and Mum gaze at clay in Mary's hands
31          Mum:   [To [>put it in the middle<
                   [   [Mum: gaze at Mary
                   [Mar: gaze at mum eyes closed then opens eyes
32          Mar:   [Hee[ hee   hee   hee    [hee hee he¿
                   [   [Mum: gaze at wheel [
                       [Mar: gaze at mum [
33    ➜     Mum:                          [Are you looking.
                   [Mum: gaze at potter's wheel
34          Mar:   [^hee ^hee?
      ➜     Mar:   [Places clay on wheel
35          Mar:   [hih
                   [Pats clay several times
36          Mum:   Then you wet your ^ha:and:.
37                 [(2.0)
                   [Mar: pats clay looks up
38                 [(^hyi!) (0.5) (no)
                   [claps LH against raised R forearm
39          Mum:   (Qu- You-) [Find the little] foo- (.)[ pedal.
            Mar:              [ sliding chair ]
            Mar:                                        [(^nyi!)
41:         Mum:   ( ) ( ) ( ) ( ) ((animated vocalisation))
42          Mum:   ^Yeah.
43          Mar:   ( ) [( ) ( ) ( ) ((animated))
                       [wheel motor turns
```

```
44     Mum:   (Yah) Wet your [^hand].
                             [point]
45:    Mar:   ( ) [( ) ( ) ((singing)
                   [wets hand
46:    Mar:   ((singing))
```

In Extract 2, both Mum and Mary are now seated and Mary has a ball of clay in her hands. Mum produces another action that is geared to progressing the task, the verbal instruction 'Put it' (i.e. the ball of clay) 'in the mid:dle' (i.e. in the centre of the wheel). As she says this, she points at the wheel bringing her pointing finger into a prominent contact-gesture which involves pressing the centre of the wheel. She extends her turn with the clause 'If you can', thereby extending the response space and perhaps indicating a lowering of her expectations that Mary can accomplish this. A spate of joint engagement follows which seems to be a site of positive affect with smiling and laughter. Mum again uses a question to progress the activity, gazing at the wheel she says 'Are you looking' (line 33), where upon Mary places the clay on the centre of wheel and pats it. It is notable that attentional gaze, together with this instruction to look, appears sufficient to direct Mary's focus of attention, and action, to the potter's wheel. Mum then produces an instruction in the form of a report 'Then you wet your hand' and then a direct instruction 'Find the little foo- (.) pedal' and 'Wet your hand' (pointing).

So far, the parental resources that we have examined involve the use of talk and gesture, in coordination with objects, to guide and direct the child's conduction. However, in Extract 3, Mum uses tactile resources to guide Mary's hands. First, she uses her left hand to guide Mary's right hand by holding the base of her hand (line 54), then she brings in her right hand to control Mary's fingers (line 55). She subsequently prompts Mary by saying 'And again on the si:ide' (line 59) simultaneously miming the required hand position.

Extract 3 [MR2012 MM Clay] simplified
Continues immediately from Extract 2

```
47        Mar:   ((starts wheel))
48        Mar:   Mumma he|p
49        Mum:   Ye:ah that's alright (yeah)
50        Mar:   Difficu' part
51        Mum:   Difficult yeah
52        Mum:   But we try:y.
53               (10)
54   →    Mum:   ((places LH over base of Mar's right hand))
55   →    Mum:   ((brings RH into to hold & guide Mar's fingers))
56        Mar:   ((Takes hands off clay and examines fingers
                     of RH))
57        Mar:   ((Presses R thumb into centre of clay))
```

```
58        Mar:   (mmm)
59    ➔   Mum:   And [again                    ] on the si:ide.
      ➔          [Mum: cupping gesture]
60        Mar:   ((stops wheel glances at Mum))
61        Mum:   ((glance and smile at Mar))
62        Mum:   It's fun! isn't it.
63        Mum:   Yeah but it is a bit [difficult
64    ➔   Mar:                        [More water
65        Mum:   Yeah we'll put more water.
66        Mum:   Smoo:th. ^it
67        Mum:   And you see there there's a little bit on
68               the wro:ng side
69        Mum:   Put that there where there's a little hole
70        Mar:   ((Soft singing))
71        Mar:   erghhh: ((creaky))
72        Mum:          [Go around ] ^it
                 [Mum: cupping gesture]
73        Mum:   (Look) Got a little bit on my finger
74        Mum:   Look you're making a hole there you need to
75               put your finger there
76        Mum:   You know you can jus' hold
77    ➔   Mum:   ((Gestures then takes and guides RH with
                 both hands))
78        Mum:   ((Releases hands))
79        Mar:   ((Gazes at Mum))
80        Mum:   ((Gazes closely at clay)
```

In Extract 4, which occurs towards the end of the activity, Mum engages in direct work on the clay pot herself; however, this involvement is highly collaborative in two respects. First, it comes about as a result of a request from Mary, Mary says 'Help' (line 2), pushing the wheel towards Mum (line 2) and pursues this request for help (line 5) with a prominent gesture. Mum seeks clarification 'Help doing hwha:t:' (line 7) and Mary provides it using talk and by gesturing towards a diagram. Secondly, Mum's direct involvement is collaborative in the sense that Mary controls the rotation of the wheel while her mother shapes the clay on the wheel; the work on the pot thereby being distributed between them.

Extract 4 [MR2012 MM Clay] simplified
8:55 into recording Mary has been smoothing the clay pot that she has formed, sometimes freehand, sometimes turning it on the wheel.

```
01    Mar:   °(ooo) (mee)° ((soft vocalisation))
02    Mar:   (uh) [He'p! (hee) (hih)(hih) (hee)
                  [pats wheel with both hands thereby
                  [pushing it slightly towards Mum
03    Mar:   ugh? ugh. [((Clap))    [((Clap))
                       [gaze at Mum [eyes closed
```

```
04       Mum:   Is [it finished?
                   [gaze at Mar
05       Mar:   [HE P! (huh (huh)
                   [rapid hands together upward gesture
07       Mum:   [Help doing hwha:t:
                   [Mum: cupping gesture
08       Mum:   Help do=
09       Mar:   =Pa ern! ((points at diagram))
10       Mum:   What (.) [doing [that shape.=
                        [       [Mum: points at diagram
                        [Mar: hold hands apart
11       Mar:   =>Yeah<
12       Mum:   Well [we can try:
                     [Mum: wets RH
13       Mum:   Need to make the hole bigger
14       Mum:   [We try:
  →              [Brings RH to clay
15  →    Mum:   (We/you) press the ^button
16       Mar:   ((looks at the pedal))
17       Mum:   ((looks at the pedal))
18       Mum:   Find the ^pedal.
19  →           (10.0) ((Mar presses the pedal making the wheel turn))
    →                  ((Mum shapes the clay with her won RH))
20       Mum:   (You take a turn)/(You like to try)
21       Mar:   ((Wets fingers and brings to the clay))
22       Mum:   [(            [             )
                [gestures [guides hand
```

Across Extracts 1–4, and across the rest of pottery-making episode, several different classes of supportive conduct can be identified. Table 25.1 lists them in order of apparently increasing involvement.

The following section will consider further examples of (a) parents' resources for directing their children's attention; (b) parents' resources for prompting and pursuing responsive actions; and (c) parents' resources that are concerned with task-related contingencies and opportunities.

Directing the child's attention

In the pottery session, it is notable that the child has primary access to the objects through which the task is carried out. That is, the potter's wheel is placed directly in front of Mary and she apparently controls the foot pedal. As such, she is largely responsible for the progress of the task. By constrast in a kit construction task in which a boy, Benjamin, works with his Dad, the father places the box of components in front of himself and invites Benjamin to locate items from it and to fit them together. Despite the differences in this arrangement, and the difference in the nature of task itself (shaping clay as opposed

Table 25.1   Classes of parent's supportive conduct

Action	Level of parental involvement
1 Present, but otherwise engaged	Low
2 Observing	
3 Responding to questions or requests	
4 Reassuring	
5 Commenting	
6 Instructing (verbally)	
7 Instructing (using talk and gesture)	
8 Guiding child's hand	
9 Performing the task by guide the child's hand(s)	
10 Performing the task by physically directing the child's hand(s)	
11 Performing part of the task directly	high

to assembling components), certain of the supportive practices located in the pottery episode occur here too. For example, the management of the child's attention can become an interactional concern.

Extract 5 is taken from an episode in which Benjamin ('Ben') and his father ('Dad') are seated at a table with a Lego calendar box. Dad has the Lego box in front of him, Benjamin is holding an electronic soundbox toy which is emitting a clapping sound effect. Dad produces an orienting remark and a summons (line 1) to which Benjamin responds with a vocalisation (lines 2–3); however, Benjamin remains engaged with the soundbox, pressing a button which produces a laughter and applause sound effect (shown as 'xxx . . . '); as the soundbox emits this, Ben waves his hands (line 4).

Extract 5 [MR2012 Benjamin & Dad Table]
Ben and Dad are seated side by side at a table, Dad is to Ben's right and has a Lego box open towards his right-hand side. Ben is handling a sound effects toy.

```
01   Dad:   Now where are we. (0.3) Benjamin?
02   Ben:   (uhn uhh)
03          (uhh uh[n)
                   [((soundbox stops clapping))
04   Dad:        [Um:[:
                     [((Ben presses button))
                     [((Ben waves hands))
05                   [xxxxxxxxx1xxxxxxxxx2xxxxxxxxx3
                     [((soundbox laughter/applause))

             [((Ben: glance at Lego box))
             [                        [((Ben: Gaze
                                      [at soundbox))
```

```
06                  xxxxxxxxx[4xxxx          [x x [xx[ x 5)
   →                      [((Dad: Gaze    [((Dad: grasps
   →                      [ at soundbox))[ soundbox))
07 →     Dad:  Shall we put that away for a moment
   →                      [ ((Father removes soundbox)) ]
08       Ben:  [It should be [ ( W a s h i ng t ] on )
                             [((Ben: Gaze at Lego box))
                             [((Dad: Turns Lego box
                             [ Towards Ben))
09       Dad:  Can you find number thirteen
```

The father's practical problem here then is that Ben is not properly aligned with the proposed activity of playing with the Lego box. Dad's strategy is to remove the soundbox from Ben. First, he does not remove the box at once but allows for some time (about 4 seconds) to elapse before doing so (lines 5–6). It is notable that during this period Ben apparently shows some attention to the Lego box and glances at it briefly (line 6). Dad then brings his gaze to the soundbox, he takes it in his hand, and as he says 'Shall we put that away for a moment', he removes it from Ben. Ben appears to respond by producing an apparently unrelated vocalisation 'It should be Washington' (line 8) but he nevertheless brings his gaze to the Lego box such that Dad now considers it relevant to produce an activity-related instruction, 'Can you find number thirteen' (line 9). Consequently, we can see here that the father accomplishes the removal of the soundbox, and thereby accomplishes Ben's attending to the construction kit, in a progressive and accountable way.

Prompting and pursuing responsive actions

The issue of pursuing a response also occurs in the episode involving Ben and his Dad. Having produced the instruction 'Can you find number thirteen' (Extract 6, line 9), Dad waits for two seconds and then produces a simplified and emphasised repeat (Extract 6, line 11). Benjamin says, 'Thirteen wait' (line 13), thereby using vocal resources to show that although he has yet to produce the responsive action that has been implicated, he is nonetheless attending to it. In response to a further pursuit from Dad, 'Where's thirteen' (line 14), he produces an upscaled action to show that he is attending to the matter 'Way weh (.) wait a minute' (line 15). On the one hand, Ben's uses of vocal resources to display attention to the task of responding contrast with the use of tapping gestures that have been described in this sequential position (Dickerson, Stribling, & Rae, 2007); however, his rapid co-occurring shake of his hand (line 15) bears some comparison with such tapping gestures. Thus here, Dad displays resources for pursuing a response, in this case largely verbal resources – and Ben displays resources for showing that he is relevantly attending.

Extract 6 [MR2012 Benjamin & Dad Table]
Follows from Extract 5.

```
09  ➔  Dad:   Can you find number thirteen
10             (----------1---[------2)
                               [Ben reaches towards box
11  ➔  Dad:   [Thirtee:na:
12             (------[---)
                          [Dad glances at Ben))
13     Ben:   Thirteen [ wait ]
       Dad:            [Where's-]
                       [Dad glance at Ben
14  ➔  Dad:   Where's thir[teen
                           [Dad: gaze at box
15     Ben:   [        Way [ weh (.) wait ] a minute
              [    [Ben: rapidly shakes RH]
              [Dad: gaze at Ben
16  ➔  Dad:   Where's [thirt=
                      [Ben: reaches towards the box
17     Dad:   =Aht's right well done
18            (-----)
19     Dad:   'kay let's see what we have to do with this
```

While Dad in Extract 6 is chiefly using verbal resources to prompt Ben to produce a nonverbal response (locating an object), Extract 7 demonstrates how a
parent can use multimodal resources in order to elicit a verbal response. Extract
7 is drawn from an afternoon baking session organised by Mum for Will, her
son, and Anna, a friend; both Will and Anna are 13 years old and diagnosed
with autism. Here Will is getting the ingredients ready to make brownies with
Mum's assistance.

Extract 7 [Will & Mum & Anna Baking_I, We need flour]
Anna is putting an apron on, Mum is has a recipe book in front of her on a worktop;
Will is moving about.

```
02  ➔  Mum:   We need Fl[uh
03     Wil:             [Flour
                        [Mum: gaze at recipe
04     Mum:   And what  [else
                        [adjusts recipe book
05            (-------[--1)
                      [Mum: gaze at Will then recipe book
                      [Will walks back towards mum
                      [Will: gaze returns to recipe book
06     Ann:   [((yoo      [mee mah))
              [Mum: points [
              [ to page    [
```

```
07  →   Mum:              [We <Nee:D
                          [Will and mum look at recipe book
                          [mum moves finger on recipe book
08      Will:   °chocolate ba [r
09      Mum:                  [chocolate
10  →   Mum:    =and we some shu
                ((mum pointing at recipe book))
12      Wil:    sugar
13              (-------[--1--------2)
    →   Mum:           [Mum: Mouths an E sound,
14      Wil:    e- [e- eggs
                   [Mum: gaze at Will
15      Mum:    eggs [and
    →                [point at recipe gaze at Will
16      Wil:    Butter
17      Mum:    Ok:ay? so let's get them out!
18      Wil:    ((walks towards the fridge))
```

This episode takes place in the kitchen where Mum and Will are looking at the recipe book while Anna is putting on her apron. In line 1, Mum produces a model sentence by partially forming a sentence omitting the last utterance so that Will can complete it. This prompting technique has the same format as an *intraverbal* strategy used in behavioural-based approaches (Sundberg & Michael, 2001). Will is gazing at the recipe book with his hand on Mum's shoulder. Will responds with 'Flour' while simultaneously moving away with his hand sliding off Mum's arm (line 3). As the task at hand is to collect all the ingredients to make brownies, it may be that Will wants to get the flour straight away. However, Mum repositions herself in front of the recipe book and points to the list of the ingredients and summons Will with 'and what else'. At this point, Anna who is present but not immediately involved in the exchange between Mum and Will engages in an echolalic utterance (line 5). Will returns and his gaze returns to the recipe book. Having secured Will's attention to the recipe book, Mum produces an incomplete sentence 'we need' while pointing at the recipe book. Will responds with 'chocolate bar' and Mum says 'chocolate' thereby confirming but also correcting Will's response. Mum then continues with another incomplete sentence for Will to complete 'and we need some shu' to which Will responds 'sugar'. For the next item in the list of ingredients, Mum just shapes her mouth to the 'e' sound (for the response of *eggs*) and Will reads and responds 'e e eggs'. Mum confirms and continues with 'eggs and' whereupon Will reads and responds 'butter' without Mum prompting him. Mum confirms this item and indicates movement into a new activity with 'okay, so let's get them out'. Across this extract then, Mum uses a number of prompts to get Will to name the ingredients that they will need for the baking project. Consequently, his production of a list of the ingredients is scaffolded through Mum's prompts. Will responds quite readily to each prompt, so in this episode the pursuit of responses is limited. In most cases, these prompts are built out of a grammatically incomplete turn-at-talk coupled with gestures involving the recipe book; however, in the case of the item 'eggs', the prompt consists of a point at the recipe book. In this case, apparently in response to a latency in Will's responding, the mother pursues a response by mouthing the start of the required word. (As it happens, Will is not monitoring her and appears to produce his response independently of this pursuit.) While it is necessary to identify and retrieve the ingredients for the baking task, it is not necessary to enumerate a list of the ingredients together. Here then, Mum is creating an opportunity for Will's co-participation in the larger baking project where the task of the moment becomes listing the ingredients. This establishment of a local task exploits two general but contrasting properties of prompts, namely that there is a high degree of freedom about when they can be produced yet the range of relevant responses can be highly constrained. Furthermore, an interactionally important feature of prompts as supportive conduct is that they make the progress of the activity conditional upon the child's response. These features make prompts a potentially powerful and ubiquitous resource in supporting children's involvement in practical activities and in structuring those activities.

Task-related contingencies and opportunities

The issue of the parent being required to attend to a task-related contingency is also evident in the episode involving Ben and his Dad, as shown in Extract 8. It transpires that Dad's nomination of a particular component Part 13 is a mistake (lines 21, 23–25), and it becomes relevant to get this piece back from Ben. Having conveyed that they have already done this component, Dad reaches his left hand over to Ben's right hand in which he is holding Part 13 and then brings his right over and slides the component into it, producing an account as he does so 'Looks as though we've done it' (lines 26–27).

Extract 8 [MR2012 Benjamin & Dad Table]
Follows from Extract 6

```
20              (-----)
21  ➜   Dad:    oh wa[i
22      Ben:         [can I have
23  ➜   Dad:    oh no we've done that one
24  ➜   Dad:    we've done hang on we've done that one
25  ➜   Dad:    [sorry
    ➜           [Dad: LH to Ben's RH
    ➜           [Ben: Rotates RH
26      Ben:    [(I'm sorry)/(I saw it)
        Dad:    [(looks like) looks
27      Dad:    [as though we've done it        ]
    ➜           [Dad slides part 13 into his RH]
28      Ben:    ((Rapidly reaches towards a component in the box))
```

In Extract 1, the parent's engagement in a practical contingency provided an opportunity for action, on that occasion it provided a space for Mary to carry out a responsive action. In Extract 8, the nature of the contingency disrupts the progress of the activity – it involves having to hand back a part which it had taken some effort to locate. At just this point, Ben darts to locate another component in the box; thus it appears that addressing the contingency that arose disrupts the joint progress on the task that had recently been established. Extract 9, drawn from the baking session, further suggests that breaks in the progress of an activity can lead to momentary disruptions. Here William has retrieved a box of eggs and has been waiting with them while Mum had to attend to something else. It occurs to Mum that it is desirable to use up some older eggs first. She swiftly brings out these newer eggs and puts away the ones that William had retrieved. At just this moment, William covers his ears and then begins engaging in self-stimulatory behaviour, rubbing his cheeks, and jumping up and down. Although we cannot be clear about the causal link on this single case, it appears that the change to the trajectory of this task (i.e. the substitution of the eggs) has created some disruption that has occasioned this self-stimulatory behaviour.

Extract 9 [Will and Anna baking] <1:35>
William is standing at the kitchen counter with the egg box that got out a minute ago. Mum comes back from having helped a younger sibling with another project.

```
01   Mum:    So William (.) eggs (.) have you got the eggs=
02   Mum:    [=Let's have a look (.) is that today's    ]
             [Mum comes round and examines the egg box1]
03   Mum:     [Let's get these ones                   ]
             [Mum: opens draw and gets out egg box2]
04   Wil:    [(            )]
             [Wil: opens and closes egg box]
05   Mum:     [Those ones are              ] older ]
             [Mum    puts    egg box1    away]
             [Will Steps away from the    ]
             [worktop, brings hand to head ]
06   Wil:    (    )=
07   Mum:    =okay
08   Mum:    [So what do we have to do William
             [Will rubs cheeks and jumps up and down
```

Domestic activities then, in contrast to more routine educational tasks, can involve contingencies where objects need to be located or substituted. A key difference can be seen within this class of occurrences. On the one hand, there are cases where a parent's addressing a contingency occurs at a moment where a child is occupied in something else and the parent's engagement with another matter does not disrupt the child. On the other hand, where the parent needs to remedy a state of affairs that the child has brought about, remediation becomes the joint activity of the moment. A parallel is seen here with the talk-based phenomena of *exposed* and *embedded* correction (Jefferson, 1987). In exposed correction, one speaker corrects another speaker by producing a turn that is dedicated to carrying out correction, thus correction comes to the surface of the interaction. Indeed, the action of correcting becomes the activity of the moment, and the progression of the activity that was in progress is temporarily suspended. By contrast, in embedded correction, one speaker produces a turn that responds to another speaker's turn, but in the course of producing that turn embeds the correction of a component of the turn to which they are responding. The activity that was underway is thereby progressed, with correction being carried out in passing.

A further kind of contingency involves taking an opportunity to engage in a stretch of action which is relevant, but not immediately necessary for, a course of action that is in progress. For example, in the baking session, Extract 10, Mum takes the opportunity to appreciate the mixture that they have created. Within this bout of activity, she initiates a sequence by asking Will 'Is it nice' (line 5) to which he replies 'Delicious' (line 6), thereby achieving a moment of shared appreciation. Such sequences of actions are essentially *side sequences* (Jefferson, 1972).

Extract 10 [MR2012 WMA] (05:34) (Schematic)

```
01   Mum:   Smell ((proffers bowl to Anna))
02   Ann:   ((Smells mixture))
03   Mum:   ((Proffers bowl to William))
04   Wil:   ((Smells mixture))
05   Mum:   Is it nice
06   Wil:   (Delicious)
07   Mum:   Delicious
```

Discussion

Across the episodes examined, the parents show their use of a range of practices for supporting their children in domestic, practical activities. As with the practices identified in the work of a learning support assistant (a participant who has an institutional job to provide support) (Stribling & Rae, 2010), these practices are richly multimodal and commonly have a *contingent and progressive* character. That is, they are commonly carried when they become contingently relevant and their degree of directiveness can be scaled up, or down, as becomes as necessary. Some practices, such as Mary's Mum guiding Mary's hand on the potter's wheel, are highly directive but nonetheless require appropriate participation from Mary: both in terms of allowing her hand to be guided and also in terms of operating the pedal to control the wheel. Others pose questions for the child to solve. For example, when Mum gazes at the potter's wheel and asks 'Are you looking?', Mary has to work out the practical relevance of this gestalt, this particular configuration of utterance, gaze, and object. Similarly, when Will's Mum points at the recipe book and says 'what else?' this is not informative statement delivering concrete information; its import is to be found in the current course of action (Garfinkel, 2002). Will comes back towards his mother and gazes at the book thereby disentangling his mother's question, her pointing, and her eye-gaze.

Clinical relevance summary

This analysis has described a number of practices that are spontaneously used in a small opportunity sample from families who consented to take part in this research. Further research, with a larger sample, is necessary to establish the generality of the practices used and to assess their effectiveness. The study does, however, provide some examples of contingencies that are commonly encountered and of the practices used by parents in addressing them. Furthermore, there are some interesting implications for practice with children diagnosed with ASD and their families. Our analysis has demonstrated that parents use many different resources for supporting children with their everyday tasks and this could involve directing the child's attention or prompting responsive actions. Furthermore, there were different ways of dealing with task-related contingencies and opportunities. This shows how parents can create opportunities for facilitating and promoting participation within the interaction. For a simple summary of the implications for practice, see Table 25.2.

Summary

Fundamentally then, the parents' tasks in carrying out everyday practical activities with their children, as in much social interaction, consists in deciding how to co-participate with the unfolding events (Goodwin, 2007). In these settings,

Table 25.2   Clinical practice highlights

1. Parents' resources for supporting children carrying out everyday tasks involve (a) directing the child's attention, (b) prompting and pursuing responsive actions, and (c) ways of dealing with task-related contingencies and opportunities.
2. An important class of activities involves creating opportunities for various forms of participation (by the parent or child) which are not directly supportive of the child's conduct, but which allow for the progression of the task or which manage the child's involvement with it.

the parents draw on interactional resources in order to accomplish the practical management of their child's distinctive ways of seeing the world and acting within it, their *autistic intelligence* (Maynard, 2005). It seems likely that some of the practices used, such as Mary's Mum's use of manual guidance, or her question 'Are you looking?', and Will's Mum's vocalisations for the beginnings of words, are adaptations of ABA prompts, but it is beyond the scope of the current study to establish this. Certain practices however are neither *prompts* nor are they *scaffolding*. The term 'scaffolding' implies supportive conduct, that is actions that are done in order to support or assist the child. However, an important class of activities involve creating opportunities for various forms of participation (by the parent or child) which are not directly supportive but which allow for the progression of the task or which manage involvement with it. This study is clearly limited. Further work is needed to better understand parents' resources and their effectiveness.

Acknowledgements

We greatly appreciate the help of the children and parents who have contributed to this project, without whom it would not have been possible. An earlier version of this presentation was given at ICCA 2014 at UCLA; we're grateful for the comments received there. Thanks also to Mats Andrén for helpful comments.

References

Applebee, A. N., & Langer, J. A. (1983). Instructional scaffolding: Reading and writing as natural language activities. *Language Arts, 60*, 168–175.

Barron, B. (2006). Video as a tool to advance understanding of learning and development in peer, family and other informal learning contexts. In R. Goldman, R. Pea, B. Barron, & S. J. Derry (Eds.), *Video research in the learning sciences* (pp. 159–187). Routledge: Abingdon.

Brown, A. L., & Palincsar, A. S. (1989). Guided, cooperative learning and individual knowledge acquisition. In L. B. Resnick (Ed.), *Knowing, learning and instruction: Essays in honor of Robert Glaser* (pp. 393–451). Hillsdale, NJ: Lawrence Erlbaum Associates.

Carr, A., & Pike, A. (2012). Maternal scaffolding behaviour: Links with parenting style and maternal education. *Developmental Psychology, 48*(2), 543–551.

Dickerson, P., Stribling, P., & Rae, J. (2007). Tapping into interaction: How children with autistic spectrum disorders design and place tapping in relation to activities in progress. *Gesture, 7*, 271–303.

Ekström, A., Lindwall, O., & Säljö, R. (2009). Questions, instructions, and modes of listening in the joint production of guided action: A study of student–teacher collaboration in handicraft education. *Scandinavian Journal of Educational Research, 53*(5), 497–514.

Freeman, S., & Kasari, C. (2013). Parent-child interactions in autism: Characteristics of play. *Autism, 17*(2), 147–161.

Garfinkel, H. (2002). *Ethnomethodology's program: Working out Durkheim's aphorism*. Lanham, MD: Rowman & Littlefield.

Goodwin, C. (2007). Participation, stance and affect in the organization of activities. *Discourse Society, 18*(1), 53–73.

Goodwin, C., & Heritage, J. (1990). Conversation analysis. *Annual Review of Anthropology, 19,* 283–307

Jefferson, G. (1972). Side sequences. In D. N. Sudnow (Ed.), *Studies in social interaction* (pp. 294–33). New York: Free Press.

Jefferson, G. (1987). On exposed and embedded correction in conversation. In G. Button & J. R. E. Lee (Eds.), *Talk and social organization* (pp. 86–100). Clevedon, UK: Multilingual Matters.

Kasari, C., Gulsrud, A. C., Wong, C., Kwon, S., & Locke, J. (2010). Randomized controlled caregiver mediated joint engagement intervention for toddlers with autism. *Journal of Autism and Developmental Disorders, 40*(9), 1045–1056.

Krantz, P. J., & McClannahan, L. E. (1998). Social interaction skills for children with autism: A script-fading procedure for beginning readers. *Journal of Applied Behavior Analysis, 31*(2), 191–202.

Langer, J. A., & Applebee, A. N. (1986). Reading and writing instruction: Toward a theory of teaching and learning. In E. Z. Rothkopf (Ed.), *Review of research in education* (Vol. 13, pp. 171–194). Washington, DC: American Educational Research Association.

Lindwall, O., & Ekström, A. (2012). Instruction-in-interaction: The teaching and learning of a manual skill. *Human Studies, 35*(1), 27–49.

MacDuff, G. S., Krantz, P. J., & McClannahan, L. E. (2001). Prompts and prompt-fading strategies for people with autism. In C. Maurice, G. Green, & R. M. Foxx (Eds.), *Making a Difference: Behavioral Intervention for Autism* (pp. 37–50). Austin, TX: Pro-ed.

Maynard, D. W. (2005). Social actions, gestalt coherence, and designations of disability: Lessons from and about autism. *Social Problems, 5,* 499–529.

Mondada, L. (2014). Instructions in the operating room: How the surgeon directs their assistant's hands. *Discourse Studies, 16*(2), 131–161.

Pomerantz, A. (1984). Pursuing a response. In J. M. Atkinson & J. Heritage (Eds.), *Structures of social action: Studies in conversation analysis* (pp. 152–163). Cambridge: Cambridge University Press.

Salonen, P., Lepola, J., & Vauras, M. (2007). Scaffolding interaction in parent-child dyads: Multimodal analysis of parental scaffolding with task and non-task oriented children. *European Journal of Psychology of Education, 22*(1), 77–96.

Schegloff, E. A. (2007). *Sequence organization in interaction: A primer in conversation analysis* (Vol. 1). Cambridge: Cambridge University Press.

Schegloff, E., & Sacks, H. (1973). Opening up closings. *Semiotica, 8*(4), 289–327.

Stone, C. A. (1998). The metaphor of scaffolding its utility for the field of learning disabilities. *Journal of Learning Disabilities, 31*(4), 344–364.

Stribling, P., & Rae, J. (2010). Interactional analysis of scaffolding in a mathematical task in ASD. In H. Gardner & M. Forrester (Eds.), *Analysing interactions in childhood insights from conversation analysis* (pp. 185–208). New York: Wiley.

Sundberg, M. L., & Michael, M. (2001). The benefits of Skinner's analysis of verbal behaviour for children with autism. *Behaviour Modification, 25,* 678–724.

Vygotsky, L. S. (1978). *Mind in society: The development of higher psychological processes.* Cambridge, MA: Harvard University Press. (Original manuscripts [ca. 1930–1934].)

Wood, D. J. (1980). Teaching the young child: Some relationships between social interaction, language, and thought. In D. R. Olsen (Ed.) *The Social Foundations of Language and Thought* (pp. 280–296). New York, NY: Norton.

Wood, D., Bruner, J., & Ross, G. (1976). The role of tutoring in problem solving. *Journal of Child Psychology and Child Psychiatry, 17*, 89–100.

Zemel, A., & Koschmann, T. (2014). 'Put your fingers right in here': Learnability and instructed experience. *Discourse Studies, 16*(2), 163–183.

Recommended reading

- Stone, C. A. (1998). The metaphor of scaffolding and its utility for the field of learning disabilities. *Journal of Learning Disabilities, 31*(4), 344–364.
- Tarplee, C., & Barrow, E. (1999). Delayed echoing as an interactional resource: A case study of a 3-year-old child on the autistic spectrum. *Clinical Linguistics & Phonetics, 13*, 449–482.

Transcription codes

The notation scheme used in for these data was originally developed by Gail Jefferson. For further explanation of Jefferson's transcription conventions, see Jefferson, G. (1984) Transcription notation. In Atkinson, J. & Heritage, J. (Eds.) *Structures of Social Action* (pp. ix–xvi). New York: Cambridge University Press. Superscribed letters, as in Mary's rendering of 'help', shown as 'he^{l}p', are used to show sounds that are partially present, for example which condition the pronunciation of other sounds in the word but are not fully expressed.

26

Managing and Normalising Emotions and Behaviour: A Conversation Analytic Study of ADHD Coaching

Louise Bradley and Carly W. Butler

Introduction

Attention Deficit Hyperactivity Disorder (ADHD) is the most commonly diagnosed disorder in childhood with worldwide prevalence estimated around 5% (Polanczyk, de Lima, Horta, Biederman, & Rohde, 2007). Those that are given a diagnosis of ADHD often present with emotional and social difficulties, including poor emotional regulation and a greater excessive emotional expression, especially for anger and aggression (Wehmeier, Schacht, & Barkley, 2010). Such difficulties impact on self-esteem and self-concept, although this impact has rarely been addressed in research (Ryan & McDougall, 2009; Wehmeier et al., 2010). Instead, research has focused on assessment, diagnosis, and treatment (Barkley, 2006), or behaviour management for parents or carers to reduce and manage undesirable behaviour (Gavita & Joyce, 2008).

Discussions around ADHD are contentious and there has been much debate about its origin and validity that has divided professional and public opinion (Horton-Salway & Davies, Chapter 9, this volume; Timimi & Timimi, Chapter 8, this volume). The research behind this chapter does not set out to enter into this debate, but instead it seeks to consider the emotional and social problems experienced by children that have largely been ignored. The focus for the research project then is about the everyday lives and experiences of children who live with a diagnosis of ADHD and to understand more about the way they understand and define themselves, their emotions and behaviours, and descriptions of ADHD.

This chapter will look at the practices used to help children understand, share, and co-produce their knowledge of emotions during a cognitive-behavioural programme for children with a diagnosis of ADHD. Drawing on the ideas of Edwards and Mercer (1987), the concern will be about how 'common knowledge' becomes constructed, developed, and displayed in talk. As such the chapter will examine how 'knowledge is actually built and shared' (Edwards & Mercer, 1987, p. 156) within the interactions and to what ends. The analysis will show how shared knowledge and understanding of emotions are collaboratively constructed through a series of step-by-step questions and discussions aimed at helping children recognise emotions in themselves and others, for the purpose of assisting them to manage and control their emotions in the future.

The chapter will also draw on Harvey Sacks' work on 'doing being ordinary' which he suggests is done by 'spend[ing] your time in usual ways, having usual thoughts, usual interests' (Sacks, 1984, p. 415). Sacks' idea is that people use 'ordinariness' as a rhetorical alternative to 'extraordinariness', to normalise events and to counter any negative inferences that could otherwise be made (see also Burridge, 2008; Lawrence, 1996; Sneijder & te Molder, 2009). Within this chapter, we show how questions, descriptions, and collective person references are used to normalise emotions and counter any negative inferences that could be made about children and ADHD. The chapter will also consider the ways in which emotional knowledge is built, shared, and co-produced to claim 'ordinariness' as a device to 'unpathologise' emotion and offer children an alternative identity to the 'disordered' construct that is so often bound to ADHD (see Bradley & Butler, forthcoming; Brady, 2014; Danforth & Navarro, 2001; Horton-Salway, 2011). The effects of a disordered identity on a child's social, emotional, and behavioural well-being have been reported (see Houck, Kendall, Muller, Morrell, & Wiebe, 2011; Krueger & Kendall, 2001), but we know very little about the interactional practices that can help support, protect, or enhance a child's well-being. The specific interactional practices identified and discussed in this chapter will have practical implications for those working with vulnerable children.

Project overview

The research project from which this chapter derives was interested in identifying the supportive practices used by professionals to help children with emotional, social, and behavioural difficulties construct a more positive sense of self. Data for the project were collected from two settings, one of which was a charity organisation that supports families, children, and adults with ADHD through regularly run support groups and training programmes. One such programme is 'RAPID', a cognitive-behavioural psychoeducational programme aimed at helping children achieve future goals by helping them improve their ability to pay attention, reduce impulsive behaviours, problem-solve, and develop social skills (Young, 2009, 2013).

Over the course of a nine-week programme the sessions were video-recorded and then analysed to identify supportive practices that help children enhance their behaviour and social and emotional well-being. As such, the programme was a rich site for the supportive practices that were the focus of the project, and of particular interest here was the way in which the children's knowledge and skills were built week by week. A specialist ADHD coach delivered the RAPID programme using a manual based on standard cognitive-behavioural tools and techniques, to provide direct intervention to teach children cognitive, social, and emotional skills, and moral values (Young, 2013). The delivery involved teaching, coaching, and reinforcing positive behaviours through a range of cognitive tools, which formed a real-life 'toolbox' that the children could draw on to help manage their thoughts, emotions, and behaviours.

What is often missing from programmes, such as RAPID, and in research that develops and assesses them, is a focus on the interactional practices through which the concepts and tools are delivered. As such, we demonstrate how the co-production of emotional knowledge and the ordinariness of emotional states underpin the practical delivery of the tools in the RAPID programme, and how these aspects are presented through the micro-detail of talk-in-interaction. We begin by showing how emotional knowledge is co-produced as ordinary and then how these fundamental aspects of emotional management are drawn on in demonstrating the tools to the children. In so doing, our analysis demonstrates clinical relevance by bridging the gap between theory and practice, as delivery of the programme will be examined as a practical concern. This interactional approach identifies the practices being used by the coach to deliver the programme. Discussion of these practices will be valuable for practitioners as research often treats such methods as implicitly, common-sensibly, or intuitively known by those delivering intervention.

Co-producing emotional knowledge as ordinary

Throughout the nine-week programme, developing emotional knowledge was an ongoing concern, based on the understanding that if children can recognise their own as well as others' emotions, they can learn to control and adapt their behaviour. The following extract is from a session designed to help children develop their problem-solving skills, which has involved discussing how paying attention to the emotional states of others can help them 'spot and recognise problems' to 'avoid getting told off'. Here the focus is on 'listening' as a tool for emotion recognition. Our focus is on how understandings about emotional displays are co-produced between the coach and the children. Standard Jeffersonian conventions are used for the transcription (see Jefferson, 2004). A table (complete with symbols and the descriptions) is also provided in the Preface of this volume.

Extract 1

```
1   Maureen:   >You have to listen to what people are
2              sa:ying.
3              (0.6)
4   Maureen:   An' listen to what their to:ne of voice
5              is= >Does anyone know what I mean< by a
6              tone of voic:e.
7              (0.2)
8   Tristan:   When voi[ce sound [good.
9   Dan:              [Hand goes up
10  Mason:                      [°A wha:t?
11             (0.3)
12  Maureen:   Yes Dan.
13  Dan:       E:rm (.) it mea:ns er:m if they're gru:mp,
14             (0.2) e:rm:: [they could be lo:↑w (0.3) ↓er:=
15  Tristan:                [Yes Dan huh.
16  Dan:       =un' they could be high (.) ↑um:: li[ke that.
17  Maureen:                                       [And what
18             does that mea:n- w- if somebody's got ↓a low
19             grumpy voice what does that of[ten mean.
20  Tristan:                                 [.HH ((Hand up))
21  Dan:       I[t mea:ns they're not ha[ppy with it.
22  Tristan:    [°They're unhappy°      [That they're unhappy
23  Dan:       An:d >[they're not go-
24  Maureen:         [They're not happy= Well done D:an,
```

Maureen introduces the need to listen to people's 'tone of voice' (lines 4–5) but does not assume any shared knowledge as she initiates an understanding check, 'does anyone know what I mean by a tone of voice' (lines 5–6). The question starts off an Initiation–Response–Evaluation (IRE) sequence (Mehan, 1979), a classic pedagogic tool, which invites displays of existing knowledge to generate a shared understanding. Dan is selected to answer and offers, 'Erm it means erm if they're grump, (0.2) erm they could be low (0.3) ↓er: un' they can be high (.) ↑um like that' (lines 13–16), to demonstrate his understanding of different tones of voice that he embodies through his prosodic production. He also reveals that he knows someone's tone of voice can indicate mood, which occasions Maureen's question, 'And what does that mea:n- w- if somebody's got ↓a low grumpy voice what does that often mean' (lines 17–19), uttered using the same words (low and grumpy) and prosodic embodiment as Dan. The question invites Dan to elaborate, 'It mea:ns they're not happy with it' (line 21), to show he understands the relationship between tone of voice and emotional states. Maureen issues an 'affiliative repeat' (Margutti & Drew, 2014, p. 7) and positive assessment, 'They're not happy = Well done D:an' (line 24), to evaluate and close the sequence. The IRE has been used to elicit a display of recognition about tone of voice and how talk can indicate emotional states.

A second feature of the extract is the work accomplished by Maureen and Dan's use of unmarked person references: 'people' (line 1), 'their' (line 4), 'they're' (lines 13, 21, 22, 23, 24), 'they' (lines 14, 16), and 'somebody's' (line 18). These collective and non-specific references provide the speaker a useful ambiguity to talk about what people do in general (Sacks, 1984). For Maureen and Dan, the ambiguity of collective references has allowed them to speak about the indefinite 'everyone' without excluding themselves, in a way that suggests there is nothing extraordinary about using tone of voice to indicate that someone is 'happy' or 'unhappy'. Topicalising their tacit understanding in this way introduces tone of voice as an ordinary practice that people use to display and recognise others' emotional states, and as such tone of voice is made available for the children to use. This ordinary tool can be used by the children for future problem-solving to help them spot and recognise 'problems' based on other people's emotions in relation to their own behaviour or actions. This theme continues throughout the analysis; the tools introduced throughout the programme are offered as practical methods for the children to use to manage ordinary everyday problems.

The IRE sequence and usefully inclusive and ambiguous person references are used to construct emotional knowledge and behaviour as shared and ordinary. However, the basis for these sessions taking place is that the children's behaviour in reaction to their emotions is often extraordinary – hence the diagnosis of ADHD – and that Maureen, as coach, has expert knowledge on managing emotions. A key aspect of the work then is that both Maureen's authority and the extraordinary behaviour of the children are minimised throughout. The next extract begins to identify the boundaries of ordinary emotions using similar practices to those already identified (questions, descriptions, and collective person references) building a common knowledge of 'strong feelings' to, first, construct an ordinary alternative and, second, to introduce the need for control.

Extract 2

```
1   Maureen:   So:me feelings are easier to express than
2              others a:ren't they an::d,=
3   Tristan:   =Yeah=
4   Maureen:   =If thing:s ~are~ (.) t:oo strong >if our
5              feelings are too strong (0.6) the:n that's
6              not good for us is it.
7   Tristan:   N[o. ((Shakes head))
8   Maureen:    [.Tch and sometimes we express them in ways
9              that (.) we don't want t:o (.) and that
10             makes us feel silly.
11             (.)
12  Maureen:   .Tch (.) So. We nee:d to learn to control our
```

```
13              feelings >don't we.
14              (0.4)
15  Tristan:    Yeap.
16  Shay:       (Nods head)
17  Benji:      Ye[a-
18  Maureen:      [Yes:. Okay.
```

Maureen's opening turn, '<u>So:</u>me feelings are easier to express than others <u>a:</u>ren't they', is a declarative + tag question. While Tristan confirms this (line 3), the tag is not designed to elicit a response from the children as Maureen's elongated 'an::d' allows her to hold open her turn by projecting more to come (Schegloff, 2007). The mid-turn tag treats the statement 'some feelings are easier to express than others' as already known by the children (see Hepburn & Potter, 2010) and draws attention to this as part of the knowledge-building process. The dialogic format is preserved within a 'monologic' turn that allows for (as in this case), but does not require, responses by the children. The upshot of this is that the children are constructed as 'knowing' and as sharing access to the experience of emotional expression.

Maureen completes her turn with 'if thing:s ~are~ (.) t:oo strong >if our feelings are too <u>strong</u> (0.6) the:n that's not good for us is it' (lines 4–6). This starts to establish boundaries of normal emotion, with some feelings being 'too strong' and 'not good'. However, these 'strong feelings' are still produced as ordinary via Maureen's use of the possessive pronoun 'our' and collective pronoun 'us' to speak on behalf of her, Lucy, the children, and all possible members, as in Extract 1. The ambiguity manages the notion that it is not just the children who have feelings that are 'too strong' and 'not good'; the children are not being singled out as anything other than 'ordinary', normalising any potential inferences that could be made about children and ADHD (Sacks, 1984).

The ordinariness of strong emotions is also evident in Maureen's turn-final tag question, 'is it' (line 6), which mobilises support for an assertion made within the speaker's domain (Heritage, 2012), providing the children the opportunity to display their independent access to this emotional understanding through their own experience or general social knowledge. This works to position the children as collaborators in the knowledge and assessment being constructed as Maureen invites their participation, and Tristan's 'no' and headshake (line 7) confirm this is something he either knows already or can agree with.

Maureen continues with a further collective understanding, 'and sometimes we express them in ways that (.) we don't want t:o (.) and that makes us feel silly' (lines 8–10). This presents the circular nature of emotions – we behave in certain ways because of strong feelings, which results in further emotions. For Sacks, 'we' is a reference that 'may refer to all members of a category that have ever lived and may ever live' (Sacks, 1992, p. 335), so Maureen again pushes back against any inference that the children are being singled out. Instead, Maureen's use of collective person references normalises, contextualises, and formulates this generalised pattern (Edwards, 1994). This continues in Maureen's upshot, '<u>S</u>o. We <u>nee</u>:d to learn to control our <u>f</u>eelings >don't we' (lines 12–13), which hints at 'control' as a solution to inappropriate emotional expression, which we *all* 'need'. The turn-final tag, 'don't we', positions the children as participatory collaborators who share access to this solution.

Normalising anger

We have shown how understandings about emotions are co-produced as ordinary within the coaching sessions. This work forms a vital component of the steps taken to address the children's ability to recognise the emotions of themselves and others as both visible and normal. In this section we show how the generic 'everybody-ness' of emotions is recalibrated in a way that also opens up the specificity of emotional experiences in 'angry moments'. By bringing in the experiences of the individual children, the session moves on to co-construct a sense of what 'anger' is and where the boundaries of ordinary, valence-free anger lie. This forms part of a step-by-step progression towards the introduction of the 'self-talk thinking tool'.

While the collective proterms construct the universality of emotions and the tag questions highlight the shared understandings about them, as the session progresses Maureen begins to invoke specific emotions and the idea that it is *individual* experiences that are shared. This is evident in the following extract that follows a discussion after a vignette is shared to evoke feelings of anger for the children.

Extract 3

```
1    Maureen:   We a:ll feel angry >sometimes don't we<.
2    Shay:      ((Nods))
3    Chloe:     ((Nods))
4    Mason:     Ye[ah.
5    Tristan      [Hu:huh.
6               (1.0) ((Miles nods))
7    Maureen:   Pu[t your hand up if you felt angry this we:ek.
8    Tristan:     [Yeah.
9               (3.8) ((Everyone puts hand up, except Mason))
10   Maureen:   Yep. We've all felt angry >haven't we<. O:kay.
```

Maureen opens the discussion with the summative 'We <u>a</u>:ll feel <u>a</u>ngry >sometimes don't we<', using the by-now familiar collective proterm 'we', the inclusive 'all', and the turn-final tag. These work together to position the children within the 'normal, standard or expected' category of people who 'sometimes' feel angry, establishing this as common knowledge (Edwards, 2007). At this point, Maureen begins to 'zero in' (Schegloff, 2000, p. 715) by inviting the children to raise their hands 'if you felt <u>a</u>ngry this we:ek' (line 7). This initiates a shift from the general and inclusive towards the specific, in terms of both individual experience and time frame. Schegloff's (2000) notion of 'granularity' is important within interaction because it is at this level of detail that people gain access to experiences.

Therefore, the turn is designed to do two things. First, the hand raising seems symbolic of being 'counted' in both senses of the word. Maureen's claim that 'we all feel angry sometimes' is reinforced as everyone (except Mason) raised their hand, which is visible to the children and reinforces the 'normalness' of anger. In this sense it defines the importance of the message being conveyed in line 1 and the normalising action being performed by it. Second, the turn brings into play the children's everyday reality. It is asking the children to locate and bring to mind an actual experience within their week that they can then use to contextualise the feeling of anger in a meaningful way, to make sense of the unfolding discussion. This sharing of individual experience to produce a collective understanding is ratified as Maureen acknowledges the raised hands reaffirming, 'Yep. We've all felt <u>a</u>ngry >haven't we<. <u>O</u>:kay' (line 10). Having established the ordinariness of anger, Maureen invites the children to assess experiences of anger.

Extract 4

```
11  Maureen:  I̲s̲ it a g̲ood thing o̲r a bad thing.
12  Mason:    Ba:[d.
13  Tristan:     [Bad.
14            (1.3)
15  Tristan:  Bad.
16  Mason:    B̲ad.
17            (0.8)
18  Maureen:  ↑I̲t's not bad is it ↓i̲f- (.) feeling an̲gry,
19            (0.3)
20  Tristan:  >It's ↑good thing<.
21  Maureen:  Isn't b̲:ad.
22            (0.5)
23  Maureen:  I̲f you know how to control it.
24  Mason:    ~Y̲e:h~.
```

The fixed choice question requires the children to assess anger as either 'good' or 'bad'. Both Mason and Tristan respond 'bad', but with her evaluative third-position turn Maureen challenges this (line 18). The assertion 'it's not bad' counters the boys' answer, but this corrective work is minimised by the tagged 'is it'. The tag implies that the children 'already know' anger isn't bad, despite their answers, and as such Maureen supports the children's participation.

Tristan then asserts a new understanding, 'it's a good thing' (line 20), and Maureen repeats, 'isn't b:ad' (line 21), to reiterate her third-turn evaluation. She then introduces a contingency or conditionality for this not-negative assessment of anger, 'If you know how to control it' (line 23). There is a logical inference to be made here: uncontrolled anger is bad. However, this is not made explicit as Maureen takes a more positive valence to leave this message inferable. The children's involvement in this sequence, starting in Extract 3, shows how they have engaged as active participants in producing understandings about anger via a step-by-step progression, with increasing granularity and specificity.

In the extracts shown so far, the children's engagement in the discussion has been carefully scaffolded by Maureen in subtle but important ways. The collective person references are 'specifically vague' (Garfinkel, 1967, p. 41) in that they include anyone and everyone while they also directly specify the children. Throughout, the questions are designed not only to invite and support the engagement of the children but also to construct them as knowing, thereby orienting to their expertise and authority over their own experiences and lifeworlds (Butler et al., 2010). One upshot of these techniques is that in addition to being treated as 'ordinary experts', either implicitly through turn-medial tag questions or explicitly through invitations to participate, the methods support interactional spaces where the children can initiate their own contributions to the discussion. This is evident in the following example, which continues directly on from the previous extract.

Extract 5

```
25   Maureen:   Cos it's a na:[tural feeling i:sn't it.= Feeling=
26   Tristan:              [>I don't know how to control it.
27   Maureen:   =an:ger (0.3) is nat:ural. We all feel cross
28              and angry sometimes.
29              (0.4)
30   Maureen:   I[t mi:ght b:e,
31   Mason:      [And if you felt it- if you felt befo:re an:
32              if you do it again you know how contr- (.) how
33              contr- (.) to control it.
34   Maureen:   Exactly Mason. Ex:actly=
35   Lucy:      =°Erm°=
36   Maureen:   =Because you know so:meti:mes (1.0) .tch it
37              mi:ght be (0.5) that somebody's said something
38              to upset yo:u,= >Might'n it.
39              (0.3)
40   Maureen:   Might be that somebody's eaten that last
41              chocolate biscu↑it,
42   Lucy:      HuHuh.
43   Maureen:   It might be that (0.4) >somebody's: >if- if
44   Maureen:   it's Lucy or myse:lf somebody might have cut
```

```
45              us up when we're dri:ving.
46              (0.8)
47  Maureen:    It might be that somebody's not done so↑mething
48              you've asked them to do: ↓o:r (0.7) or you've
49              forgotton something and you feel angry at
50              yourself.
```

The extract begins with more of the normalising work (Edwards, 2000, 2007) that Maureen has been doing throughout, asserting the 'naturalness' of feeling 'cross and angry' (lines 27–28) as both inclusive ('we' and 'all') and known-in-common ('isn't it'). In overlap with this, Tristan responds to Maureen's earlier assertion (line 23, Extract 4) reporting a subjective experience, 'I don't know how to control it', as if verbalising a realisation or acknowledgement in light of Maureen's talk. This turn is neither oriented to by Maureen as a contribution to the ongoing talk nor is it oriented to as an intrusion. This demonstrates how in these sessions, non-interactionally relevant contributions are both possible and unsanctioned and allow for individual experiences to be heard, while a group-relevant focus is maintained.

Mason then initiates his contribution, 'And if you felt it- if you felt befo:re an' if you do it again you know how contr- (.) how contr- (.) to control it' (lines 31–33). The prefacing 'and' connects his reasoning to Maureen's prior turn, thereby actively constructing common knowledge. Mason's use of generic person references ('you') is aligned with Maureen's use and constructs his contribution as being true for all and as an ordinary understanding for recognising and controlling anger. He uses his social understanding and the 'collectivity's corpus of knowledge' (Sharrock, 1974, p. 45) to make sense of the experience by connecting what people know and what they do as a practice for managing anger. Maureen's high-grade assessment (Antaki, 2002) affirms Mason's reasoning, 'Exactly Mason. Ex:actly' (line 34) and ratifies his expressed understanding.

In the same way that Mason connected his talk by using the preface 'and', Maureen uses 'because' to include Mason's contribution before continuing her abandoned turn from line 30, 'Because you know so:meti:mes (1.0) . tch it mi:ght be (0.5) that somebody's said something to upset yo:u,=>Might'n it' (lines 36–38). She continues to list hypothetical things that 'might' make 'somebody' angry, descriptively working anger up as being directed *at* ordinary everyday things, therefore rational and spontaneous rather than dispositional and irrational (Edwards, 1997), to continue the 'unpathologising' work seen throughout.

Using metaphors to construct experiential aspects of anger

As the session progresses, Maureen increases the focus on the children's individual experiences. Although the generic ordinariness of emotions, and anger in particular, is a vital part of redressing the 'pathologisation' of ADHD behaviours, there is a need to return to the children's own individual experiences of anger which is problematic at times. She begins to address this in the following extract as she introduces the metaphor of anger as a volcano with the potential to erupt (see Lakoff (1987) on emotion metaphors).

Extract 6

```
1   Maureen:    We need to think about what those triggers ar:e.
2              (0.8)
3   Maureen:    Cos s:ometimes it can make you feel like a
4              volcano >can't it= And what do volcanoes do.
5   Benji:     Eru[pt.
6   Mason:        [Ex[plo::d[e,
7   Paige:        [((Gestures and mouths explosion
8   Maureen:            [Erupt.
9              (0.4)
10  Maureen:    And when is it you get into trouble.
11  Tristan:   When you fight,
12  Maddie:    Huh hh h h[ h
13  Maureen:             [>Yes Mason.
14  Mason:     When you fi:ght or say swe:ar words. >Becos
15             I- (.) usually do that often now.
16             (0.5)
17  Maureen:   Yep.
18             (0.3)
19  Maureen:   And i[t is that ((Points to Mason))
20  Tristan         [When you run out of school.
21             (0.6)
22  Maureen:   >When you [run out of school.
23  Lucy:                [Uh:m.
24  Tristan:   When yo[u climb over the school fence.
25  Maureen:          [So it's when you erupt i:sn't it.
26  Maureen:   It's when you're tha[t volcano and you erupt.
27  Tristan:                       [I don't do that.
28  Paige:     Ye[s:.
29  Maureen:     [That's: (.) when you get into trouble.
30  Maureen:   If you can ke:ep [it,
31  Maddie:                     [°Skive school sometimes°
32  Tristan:   Don['t bubbling.
33  Maureen:      [Bubbling along and manage i:t,(1.0) then:
34             you don't erupt do you.
35  Tristan:   >Then the volca:no won't n- never erupt<.
36             (0.5)
37  Maureen:   °°No. Okay°°.
```

Metaphors are a conceptual resource used to construct a narrative description of emotional expressions in real life. Metaphors of anger as bubbling or boiling are ways to construct anger as passive and experiential (Edwards, 1997). Examples of 'erupting emotions' are shared in this sequence as part of a shift from the more general and hypothetical to the children's own experiences, which is managed in part by Maureen's referential shift from 'we' to 'you', in a narrowing of recipiency. The children provide a selection of appropriate actions (lines 5–8) in response to Maureen's question about what volcanoes do. They then contribute examples of when they 'get into trouble' by offering, 'when you fight' (line 11), 'say swe:ar words' (line 14), 'run out of school' (line 20), or 'climb over the school fence' (line 24).

Interestingly, in proffering these examples, Tristan and Mason both use the generic 'you' to talk about what seem to be descriptions of personal experiences. This is a nice illustration of how the work of the session can help the children recognise their own experiences as ordinary, to unpathologise behaviours potentially seen as being bound to ADHD. Mason then shifts to the subjective 'I' to speak about his own experiences through his admission that he fights and says swear words when he is angry (lines 14–15). This disclosure seems to allude once more to the context in which this interaction takes place and the reasons why the children are attending.

Throughout, Maureen acknowledges and validates the children's experiences, then closes the 'sharing' by formulating these specific examples as 'eruptions' (lines 25–29). By using the volcano metaphor to contextualise anger's physical feeling, Maureen makes available another practical tool and ordinary method for the children to use, to recognise, and to manage their anger. The solution is 'if you can ke:ep it bubbling along and manage it then you don't erupt' (lines 30–34). Tristan formulates the upshot, '>Then the volca:no won't n- never erupt<' (line 35), prefacing 'then' to connect his talk to Maureen's in a collaborative building and discovery of knowledge and understanding. Tristan's use of 'the' and Maureen's use of 'it' construct the volcano as a separate entity that sits within the children but is not part of them. By separating the volcano from the person, an interesting psychological interpretation can be considered; the children are in charge of 'the' volcano and as such they can control it if they use the tools being taught to recognise their anger and externalise it in this way.

The extract comes at the end of a long stretch of talk that began in Extract 3 when Maureen made it known to the children that 'we all feel angry sometimes'. The discussion progressed to challenge the belief that anger 'is not a bad thing' in Extract 4, before moving to normalise anger and assert the need for control in Extract 5. The talk ended in Extract 6 above, when anger was described as a volcano, offering the children a practical tool for recognising and controlling their anger. The extended analysis of these extracts has shown how the children's knowledge and assessment of anger has been shaped and developed through a step-by-step progression, collaboratively constructed to provide a practical understanding and tools for the children to use to help them recognise and control their anger in the future. This collective, practical understanding is both generic and specific and serves as a basis for introducing a classic cognitive-behavioural method: self-talk.

Self-talk thinking tool

For Goffman (1978), self-talk is a ritualised behaviour, one that we practice for different reasons. Maureen is teaching this ordinary practice as a formalised method for controlling anger. In introducing self-talk into the children's toolkit, there is a marked shift in the participation structure, as Maureen positions herself and Lucy as instructors. In the moments leading up to the following extract the group have enacted, or watched, a role-play scenario in which one member of the group has acted being angry and another member has reacted to calm her down. The role-play was used to start a discussion that involved describing physical displays of anger in other people.

Extract 7

```
1    Maureen:   Okay. >So now we know (.) that you can
2               recognise when you're getting angry.
3               (1.1)
4    Maureen:   And no:w (.) we need to teach you a way (.)
5               to stop you from (0.2) le:tting ou:t (.)
6               those angry feelings.
7
8               (8 lines omitted)
9
10   Maureen:   The way not to get into trouble is to
11              manage, to control your anger. That is
12              so so important. And we do that
13              particularly (.) I am going to get
14              another tool out of my toolbox.
15
16              (3 lines omitted)
17
18   Maureen:   ((Holds self-talk sign up in air))
19   Maureen:   By s:elf-talk= Does anybody know what I
20              mean by s:[elf-talk.
21   Shay:               [Talk to your self.
22   Tristan:   >Talk to your self.
23   Maureen:   Talking to your self.
```

Maureen's summative turn (lines 1–2) assumes and attributes the children's competency in recognising 'when you're getting angry', on the basis that in their prior discussion the children did describe physical displays of anger. Maureen then introduces the self-talk tool as something that 'we need to teach you ... to stop you from letting out those angry feelings' (lines 4–6). Unlike previous instances, the all-inclusive 'we' refers only to Maureen and Lucy, as coach and assistant. This then begins to partition the group into a two-party organisation: the children and the instructors (Butler, 2008; Sacks, 1992). As such, there is a shift in the distribution of the rights, obligations, and expertise among the group. In lines 1–6, Maureen switches the tense from speaking in the present to the future, which lends itself to an instructional or advisory mode. The future-oriented footing in 'when you're getting' (line 2) and 'letting out' (line 5) evokes a time when the children will get angry and express that anger and highlights the shift from the earlier normalising work to the 'main business' of providing the children a specific anger management tool.

We also see a partitioning in terms of the ownership of angry emotions. Maureen's descriptions, 'those angry feelings' (line 6), separate the emotion from the children, similar to the distancing work seen in Extract 6. Anger is formulated as something 'other than' the individual, helpfully separating the person from the emotion to resist the idea that anger is dispositional (Edwards, 1997) and part of the children's identity. Having introduced 'self-talk' from her 'tool box' (line 14), Maureen invites the children to identify if they know what 'self-talk ' means (lines 19–20). The children's responses are affirmed as correct, and this then signals the introduction of self-talk phrases.

In data not shown, the children are invited to come up with things they could say to themselves to 'calm down'. Two phrases are nominated: 'tell yourself to calm down' and 'tell yourself not to get angry'. Using the generic person reference 'yourself', the children formulate these bits of self-talk as *instructions* to one-self, rather than *talk* to one-self. In the following sequence Maureen builds on their contributions to turn generic instructions into personalised self-talk.

Extract 8

```
1   Maureen:  So- there's lots of self-talk phrases
2             >that we can tell< ourselves aren't there.
3             (0.5)
4   Maureen:  Yeah. If I start w- a fight with that
5             bo:y I'll get into trouble.
6             (1.6)
7   Benji:    What.
8   Maureen:  >That's what you can say to yourself isn't
9             it< (.) if I start a fi:ght=
10  Benji:    =Yeah.
11  Maureen:  I'll get into trouble. ((Sing-song voice))
12  Benji:    Ye[ah.
13  Maureen:    [It's not worth arguing with him (.)
14             I need to find a teacher. ((Sing-song voice))
15             (0.8)
```

```
16  Maureen:  He a:lways does that kinda stuff (0.2)
17            that's / why / kids / don't / like / him.
18  Maureen:  It's about having a m:antra to put into your
19            head. >So (.) I need to calm down. Calm
20            down.
21            (0.5)
22  Maureen:  Walk awa:y. >Tell the teacher.= Whatever it
23            is to have that s:elf-talk. So instead of
24            (0.9) some of the things you want to
25            say (.) those <angry sweary things>.
26            (0.4)
27  Maureen:  We need to contro:l ya anger and ya
28            feelings. Okay and problem-solve
29            >so that we don't< get into that trouble.
```

Maureen begins this sequence with a return to the inclusive 'we' that flattens out the relationship between her, Lucy, and the children. The self-talk phrases are 'anyone's' and the children are positioned as equally able to recognise the truth in the statement, as shown with the tag. Maureen then offers a list of hypothetical phrases (lines 4–5) designed as 'script proposals' (see Emmison, Butler, & Danby, 2011), using direct reported speech to give the children access to what they could say in such situations. Maureen adopts the 'animator' role (Goffman, 1981) as if speaking on behalf of the children as authors of the self-talk. She widens the children's collective 'corpus of knowledge' (Sharrock, 1974) to the many possible ways of formulating such talk for themselves in practice: 'I need to find a teacher', 'I need to calm down'.

Stepping outside of the animator role briefly, Maureen explains it is about having a 'mantra to put into your head' (line 18). This explanation is a nice illustration of how the practical method (saying something to yourself) is treated as a cognitive practice. Putting a phrase into one's head makes it transportable and available to be drawn on in multiple situations. What is being done here could be described as the explicit seeding of a script formulation (Edwards, 1997) that can be made interactionally relevant and used in everyday situations. In closing this sequence, Maureen formulates 'that self-talk' (line 23) as a tool that the children can use to replace 'those angry sweary things' (line 25), 'to control ya anger and ya feelings' (lines 27–28), 'and problem solve >so that we don't< get into that trouble' (lines 28–29). Maureen returns to the inclusive collective 'we' (lines 27 and 29) to speak about self-talk as an ordinary practice that can be used by everyone to control anger and not get into trouble. This return to the ordinary is now done on the basis that a jointly produced understanding of self-talk has been formalised and scripted to be reflective of the children's everyday reality but constructed as an 'ordinary' practice that is used by everyone to manage anger. This continues the normalising work seen throughout to prevent the children from being singled out and dispositionally defined by their emotions and behaviours.

Clinical relevance summary

The RAPID programme has been found to be effective in helping children diagnosed with ADHD gain knowledge and skills to improve interpersonal relationships (Young, 2013). However, we believe that the therapeutic benefits of this programme are not merely the result of the cognitive-behavioural theories and principles being applied through the RAPID programme but also the less noticed and less understood interactional practices through which the programme is delivered. In this chapter, we have outlined how 'ways of speaking' in the delivery of the programme, which are often treated as implicit, common sense, or intuitive, have constructed a more positive version of ADHD; this in turn has implications for real-life constructions of ADHD identities and associated understandings.

There is vital clinical relevance to be found in the specific methods and practices used in the actual delivery of this programme and others like it. An interactional analysis has revealed the ways in which Maureen delivers the RAPID programme (1) to treat the children as *experts* about ADHD and their experiences of it, as their participation and agency are a vital part of the programme; (2) to introduce cognitive tools in a meaningful way that is inclusive of the children's knowledge and experiences; and (3) to normalise emotions and prevent excessive emotional expression from becoming a defining part of their identity. These findings bridge the gap between theory and practice because identification of the methods actually used in delivering the cognitive-behavioural programme makes these same methods available for practitioners and professionals to recognise and apply to their own practice. For a simple summary of the implications for practice, see Table 26.1.

Table 26.1 Clinical practice highlights

1. The asymmetrical relationship between coach or therapist and children can be minimised via questioning practices (tag questions position the children as already having knowledge) to invoke and promote the expertise and agency of the children.
2. The step-by-step introduction of cognitive-behavioural tools can be introduced through the collaborative production of shared knowledge and experience.
3. Normalising practices, such as collective and inclusive person references ('we'), can be used effectively to 'unpathologise' emotions, behaviours, and experiences that may otherwise be especially salient for children with a diagnosis of ADHD, which can help address issues of self-concept and identity.

Summary

This chapter has tracked the delivery of a small element of a cognitive-behavioural programme for children with a diagnosis of ADHD. The aim has been to highlight how much work is done not just through the content of the programme and the tools that are introduced to the children but in the ordinary interactional practices through which the programme is delivered. The skill of the coach does not rest so much in her use of cognitive-behavioural methods but in the specifics of how the material is packaged, organised, and delivered. These details are reliant on interactional communicative skills that are regularly 'seen but unnoticed' (Garfinkel, 1964). Simple methods, such as using the inclusive 'we', have immense power in constructing ADHD, its behaviours, emotional consequences, and the identities of children diagnosed with it more positively.

Through the implementation of cognitive-behavioural methods such as metaphors, tone of voice, and self-talk the sessions have focused on the behavioural, emotional, and social difficulties associated with a diagnosis of ADHD. The future-oriented focus of the coaching made the tools and strategies available to be drawn on and used by the children in real-life challenges within their everyday world. The analysis has identified how the teaching of emotional knowledge was co-produced between the coach and the children through the use of the IRE and tag questions; the ordinariness of emotional states was achieved through the use of person references, and both were fundamental to the delivery of the programme and the teaching of tools.

Knowledge and understanding was developed over a step-by-step sequence that invited increasing participation from the children and ultimately established understandings as shared. The IRE was used as part of this work to elicit understanding and to continually check and ensure that joint understanding was achieved (Edwards & Mercer, 1987). A further way this was done was via tag questions that invited displays of understanding by the children. Mid-turn tags were used to treat some bit of talk as already known by the children, or as something they could accept and agree with as part of the knowledge-building process, and to mark the importance or significance of an assertion. Slots for affirmation were not always left for the children, and as such Maureen assumed and attributed understanding. While turn-final tags invited agreement from the children, they also marked the talk as a kind of judgement or conclusion.

The practices used throughout effectively orient to the ordinariness of the children's experiences and behaviour. This was primarily done through the seemingly minor practice of using generic and inclusive person references. The collective person references used to normalise emotions seem to preclude (Extracts 1–5) and lead up to a shift from the collective to the subjective (Extracts 6–8), to speak about the children's own emotional experiences and enable instruction, before shifting back at the end of Extract 8 to speak once more in ordinary terms about self-talk as a practice available to all. Thus, the rather technical and theoretically driven notion of self-talk is sequentially packaged up and produced as an ordinary practice. More specifically, the 'unpathologising' of the children's behaviour was achieved through separating anger (Extracts 6 and 7) from the children; so this emotion did not become dispositional and bound to their identity as an 'extraordinary' description that could be used to categorise them. This 'unpathologising' is important given that previous research has rarely addressed the impact of ADHD emotions and behaviours on self-esteem and self-concept. Despite there being research that highlights how children tend to define themselves by their ADHD behaviours (Krueger & Kendall, 2001), and that experiences of ADHD often impact negatively on their self-esteem (Travell & Visser, 2006), studies have not focused on the ways in which children can be supported to counteract negative effects, but this has been a focus for this chapter.

References

Antaki, C. (2002). 'Lovely': Turn-initial high grade assessments in telephone closings. *Discourse Studies, 4*(1), 5–23.

Barkley, R. (2006). *Attention-deficit hyperactivity disorder: A handbook for diagnosis and treatment* (3rd edition). New York: Guildford Press.

Bradley, L., & Butler, C. W. (forthcoming). People with ADHD...: An interactional analysis of an ADHD support group.

Brady, G. (2014). Children and ADHD: Seeking control within the constraints of diagnosis. *Children & Society, 28*, 218–230.

Burridge, J. D. (2008). 'Hunting is not just for bloodthirsty toffs': The countryside alliance and the visual rhetoric of a poster campaign. *Text and Talk, 28*(1), 31–53.

Butler, C. W. (2008). *Talk and social interaction in the playground*. Aldershot: Ashgate.

Butler, C. W., Potter, J., Danby, S., Emmison, M., & Hepburn, A. (2010). Advice-implicative interrogatives: Building 'client-centred' support in a children's helpline. *Social Psychology Quarterly, 57*, 265–289.

Danforth, S., & Navarro, V. (2001). Hyper talk: Sampling the social construction of ADHD in everyday language. *Anthropology & Education Quarterly, 32*(2), 167–190.

Edwards, D. (1994). Script formulations: An analysis of event descriptions in conversation. *Journal of Language and Social Psychology, 13*, 211–247.

Edwards, D. (1997). *Discourse and cognition*. London: Sage.

Edwards, D. (2000). Extreme case formulations: Softeners, investment, and doing nonliteral. *Research on Language and Social Interaction, 33*(4), 347–373.

Edwards, D. (2007). Managing subjectivity in talk. In A. Hepburn & S. Wiggins (Eds.), *Discursive research in practice: New approach to psychology and interaction* (pp. 31–50). Cambridge: Cambridge University Press.

Edwards, D., & Mercer, N. (1987). *Common knowledge: The development of understanding in the classroom*. London: Methuen.

Emmison, M., Butler, C. W., & Danby, S. (2011). Script proposals: A device for empowering clients in counselling. *Discourse Studies, 13*(3), 3–26.

Garfinkel, H. (1964). Studies in the routine grounds of everyday activities. *Social Problems,* *11*(3), 225–250.

Garfinkel, H. (1967). *Studies in ethnomethodology.* Englewood Cliffs: Prentice-Hall.

Gavita, O., & Joyce, M. (2008). A review of the effectiveness of group cognitively enhanced behavioural based parent programs designed for reducing disruptive behaviour in children. *Journal of Cognitive and Behavioural Psychotherapies, 8*(2), 185–199.

Goffman, E. (1978). Response cries. *Language, 54*(4), 787–815.

Goffman, E. (1981). *Forms of talk.* Oxford: Oxford University Press.

Hepburn, A., & Potter, J. (2010). Interrogating tears: Some uses of 'tag questions' in a child protection helpline. In A. F. Freed & S. Ehrlich (Eds.), *'Why do you ask?': The function of questions in institutional discourse* (pp. 69–86). Oxford: Oxford University Press.

Heritage, J. (2012). Epistemics in conversation. In J. Sidnell & T. Stivers (Eds.), *The handbook of conversation analysis* (pp. 1–29). Chichester, UK: John Wiley & Sons, Ltd.

Horton-Salway, M. (2011). Repertoires of ADHD in newspaper media. *Health, 15*(5), 533–549.

Houck, G., Kendall, J., Muller, A., Morrell, P., & Wiebe, G. (2011). Self-concept in children and adolescents with attention deficit hyperactivity disorder. *Journal of Pediatric Nursing, 26*(3), 239–247.

Jefferson, G. (2004). Glossary of transcript symbols with an introduction. In G. H. Lerner (Ed.), *Conversation analysis: Studies from the first generation* (pp. 13–31). Amsterdam: John Benjamins.

Krueger, M., & Kendall, J. (2001). Descriptions of self: An exploratory study of adolescents with ADHD. *Journal of Child and Adolescent Psychiatric Nursing, 14*(2), 61–72.

Lakoff, G. (1987). *Women, fire and dangerous things: What categories reveal about the mind.* Chicago: University of Chicago Press.

Lawrence, S. G. (1996). Normalizing stigmatized practices: Achieving co-membership by 'Doing being ordinary'. *Research on Language and Social Interaction, 29*(3), 181–218.

Margutti, P., & Drew, P. (2014). Positive evaluations of student answers in classroom instruction. *Language and Education, 28*(5), 436–458.

Mehan, H. (1979). 'What time is it, Denise?': Asking known information questions in classroom discourse. *Theory into Practice, 18*(4), 285–294.

Polanczyk, G., de Lima, M., Horta, B., Biederman, J., & Rohde, L. (2007). The worldwide prevalence of ADHD: A systematic review and metaregression analysis. *American Journal of Psychiatry, 164*(6), 942–948.

Ryan, N., & McDougall, T. (2009). *Nursing children and young people with ADHD.* Oxon: Routledge.

Sacks, H. (1984). 'On doing 'being ordinary'. In J. M. Atkinson & J. Heritage (Eds.), *Structures of social action* (pp. 413–429). Cambridge: Cambridge University Press.

Sacks, H. (1992). *Lectures on conversation* (Vol. 1. G. Jefferson, Ed.) Padstow: T. J. Press Ltd.

Schegloff, E. A. (2000). On granularity. *Annual Review of Sociology, 26*, 715–720.

Schegloff, E. A. (2007). *Sequence organization interaction.* Cambridge: Cambridge University Press.

Sharrock, W. W. (1974). On owning knowledge. In R. Turner (Ed.), *Ethnomethodology* (pp. 45–53). Harmondsworth: Penguin.

Sneijder, P., & te Molder, H. (2009). Normalizing ideological food choice and eating practices. Identity work in online discussions on veganism. *Appetite, 52*, 621–630.

Travell, C., & Visser, J. (2006). 'ADHD does bad stuff to you': Young peoples' and parents' experiences and perceptions of attention deficit hyperactivity disorder (ADHD). *Emotional and Behavioural Difficulties, 11*(3), 205–216.

Wehmeier, P. M., Schacht, A., & Barkley, R. A. (2010). Social and emotional impairment in children and adolescents with ADHD and the impact on quality of life. *Journal of Adolescent Health, 46*(3), 209–217.

Young, S. (2009). *RAPID: Reasoning and problem solving for inattentive detectives.* London, England: Psychology Services Limited.

Young, S. (2013). The 'RAPID' cognitive-behavioural therapy program for inattentive children: Preliminary findings. *Journal of Attention Disorders, 17*(6), 519–526.

Recommended reading

- Mehan, H. (1979). 'What time is it, Denise?': Asking known information questions in classroom discourse. *Theory into Practice, 18*(4), 285–294.
- Sacks, H. (1984). Again, no 'a'? On doing 'being ordinary'. In J. M. Atkinson & J. Heritage (Eds.), *Structures of social action* (pp. 413–429). Cambridge: Cambridge University Press.

27

Interlocutor Influence on the Communication Behaviours Associated with Selective Mutism

Hanna Schäfer and Tom Muskett

Introduction

Selective mutism (SM) is a psychiatric diagnostic term used to describe children who pervasively do not speak in certain social situations where speaking would be expected. For the diagnosis to be made, this reluctant or absent speaking must (a) have been evident for more than one month, excluding the first month in school; (b) not be due to a communication, pervasive developmental or psychotic disorder, or insufficient knowledge of the language and social conventions associated with the context of non-speaking; and (c) be judged to cause functional difficulties in terms of educational attainment and social communication (American Psychiatric Association, 2000).

The reasons for a child behaving in a manner consistent with these criteria are inevitably complex, and there is much variation between these children's individual profiles. However, in spite of this complexity and diversity, there exists a general consensus in the child psychiatry literature that SM is a behavioural manifestation of some form of underlying anxiety, and hence it is classifiable as an anxiety disorder. This conjecture is based upon research findings demonstrating that children who are reluctant to speak in certain contexts have significantly higher rates of anxiety than their typically communicating peers (Bergman, Piacentini, & McCracken, 2002; Carbone et al., 2010; Elizur & Perednik, 2003; Sharp, Sherman, & Gross, 2007; Vecchio & Kearney, 2005; Yeganeh, Beidel, & Turner, 2006). In spite of this broad agreement, there remains considerable disagreement regarding the most appropriate diagnostic nosology of anxiety disorders for SM to be classified within. For instance, a social phobia is often diagnosed as co-morbid to SM, indicating that diagnostic criteria for these may partly overlap (Vecchio & Kearney, 2005). However, unlike those describable as exhibiting social phobia, children diagnosed with

SM often communicate comfortably through non-verbal means and may not actively avoid social situations (Omdal & Galloway, 2008).

Across the clinical literature, it is asserted that detailed and systematic observation of a child's communication across a variety of contexts is crucial for planning intervention (Omdal & Galloway, 2008). Such observation enables the identification of specific aspects of particular social situations that appear to evoke anxiety for the individual child. Importantly, they can also indicate extra-individual factors that may perpetuate the problematic communication behaviours in question. Indeed, in the SM literature, behaviourist theoretical frameworks have often been applied to argue that interlocutors may frequently inadvertently reinforce non-speaking behaviour, both verbally and non-verbally (Johnson & Wintgens, 2001; Omdal & Galloway, 2008; Scott & Beidel, 2011; Wong, 2010). Such behaviourist accounts imply that the nature of real-time *interactions* involving the child diagnosed with SM may be a factor in the perpetuation, or even exacerbation, of ostensibly problematic individual communication behaviours. However, there remains little research examining interactions involving such children, with most interactionally leaning accounts of SM having been extrapolated top-down from behaviourist theory rather than emerging bottom-up through empirical examination of interactions involving diagnosed individuals.

One exception is the work of Nowakowski et al. (2011) on joint attention behaviours by children diagnosed with SM and their parents. Here, recordings of parent–child dyads undertaking both structured and unstructured tasks were quantitatively analysed and compared against those involving both anxious and typically behaving children. Subtle between-group differences were identified: it was reported that the SM dyads established significantly less parent-initiated joint attention episodes during structured tasks than controls, although no such difference was identified during unstructured tasks and the frequency of child-initiated joint attention was comparable across conditions.

Nowakowski et al.'s (2011) findings provide some preliminary indications of potentially distinctive features of interactions involving children diagnosed with SM. However, the data analysed in this study are unlikely to be representative of the social encounters that constitute a childs everyday lives, as recordings were made in a laboratory setting where parents were provided with scripts for how to manage the different tasks, which were of a maximum of five minutes duration only. Furthermore, the quantitative coding approach adopted by the authors did not enable ideographic examination of individual interactional or relational differences across dyads. This is arguably problematic given the heterogeneity of profile associated with SM. Accordingly, Nowakowski et al. (2011) assert that a more fine-grained approach to analysis 'would provide further insight into the mechanisms through which parenting may be maintaining the disorder' (p. 89).

In this chapter, we present such a fine-grained, qualitative and ideographic account of an interaction involving a child diagnosed with SM. To do so, we apply the methodology of conversation analysis (CA). As outlined elsewhere in this volume (see Preface, and Fasulo, Chapter 1, this volume), CA is a qualitative methodology that takes naturally occurring interactions as its data and produces as findings detailed explication of the moment-by-moment structural organisation of these interactions. A key analytic premise in CA is a strict focus on so-called *participant orientation*. In other words, analytic arguments in CA are constrained as much as possible to being based upon the situated conduct of the participants in the recordings themselves, as they demonstrate through their real time understanding of one another's actions through their contributions to the unfolding interaction (cf. Schegloff & Sacks, 1973).

While the bulk of work in CA relates to interaction involving non-disordered individuals, the approach is increasingly used to analyse social encounters involving persons clinically considered to exhibit communication difficulties. This is in part because an analytic focus on participant orientation can offer unique insights on the nature and social significance of 'atypical' behaviour. For instance, CA enables identification of the real-time interactional consequences of ostensibly disordered communication across a vast range of social contexts, in turn providing insight into the extent to which apparent communication difficulties actually impede interactions as they happen (Wilkinson, 2008). CA also enables consideration of speakers' use of multiple semiotic and contextual resources to facilitate interaction in circumstances where one or more participants produce talk that is pervasively unintelligible or otherwise problematic (Garcia, 2012). Finally, as CA analysis involves scrutiny of the conduct of *all* parties in a given interaction, application to interactions involving participants categorised as disordered can demonstrate how emergent 'atypical' communication behaviours are, to a greater or lesser extent, jointly mediated, in that co-speakers are inevitably implicated in their moment-by-moment manifestation (see Muskett, Perkins, Clegg, & Body, 2010).

The established methodological utility of CA to examine ostensibly disordered interaction renders it well suited to application in this chapter. While it is accepted within the SM literature that the behaviours associated with the diagnosis 'disrupt the normal process of interaction' (Johnson & Wintgens, 2001, preface), the nature of this disruption is yet to be systematically identified. CA is apposite for such an investigation for three reasons. First, by drawing upon existing findings regarding, for instance, speakers' management of turn-taking (e.g. Sacks, Schegloff, & Jefferson, 1974) and apparently absent responses in mundane conversation, the course of encounters involving children who are reluctant to speak can be described in terms of the normative practices that pervade all social interactions. Second, CA's modelling of individuals' behaviour as *sequentially situated* necessitates consideration of the interactional trajectories of *all* participants preceding the manifestation of problematic behaviour. Therefore, local influences on the behaviours associated with SM can be explored, providing opportunity to identify interlocutor turns that appear facilitative or otherwise for a child's subsequent verbal and non-verbal participation. Finally, as children diagnosed with SM often communicate non-verbally (Omdal & Galloway, 2008), broadening analysis to capture participants' use of semiotic resources other than talk enables description of a child's interactional participation in spite of any differences in verbal conduct.

We now present our CA account of a clinical interaction involving a child diagnosed with SM. This account focuses on a specific phenomenon: that being how the *expectation* for the child to speak, as established by prior adult turns, appeared to influence significantly the nature of the child's immediately subsequent contribution. This phenomenon was identified bottom-up from close analysis of the recording of this interaction and provides a striking indication of the reciprocal and dynamic nature of the behaviours associated with SM. As we will go on to discuss, the analytic findings presented below generate intriguing implications relating to clinical practice with such children.

Project overview

Participants and data

The data below constitute a video-recording of one clinical speech and language therapy session of 45 minutes duration, involving 'Alexandra' (labelled A in the transcripts), a girl diagnosed with SM, and 'Caroline' (C), her speech and language therapist (SLT) (pseudonyms applied). Several professional disciplines may work with children with SM, with each drawing upon different theoretical frameworks for describing, conceptualising, assessing, and intervening upon the behaviours associated with the diagnosis. Unidisciplinary interventions delivered by SLTs are now relatively common, particularly in private practice. Here, SM is typically formulated in a manner that focuses solely on the presenting communication behaviours, rather than seeking to hypothesise what underlying thoughts, feelings, or past experiences may contribute to the child's present difficulties. Unsurprisingly therefore, SLT intervention typically aims to modify the child's communication behaviour through behaviourist desensitisation methods such as use of so-called speaking hierarchies, typically delivered through individual play-based clinical work in conjunction with coaching of significant adults around the child (Katz-Bernstein, 2011).

Alexandra, who was aged eight years, six months, when recorded, was completing such a desensitisation-based therapy. When first diagnosed with SM, Alexandra demonstrated considerable reluctance to speak and communicated verbally only with her parents. The analysed session, recorded at a juncture when she had already received weekly SLT for one year, is at an advanced point in treatment where a previously unknown person joins the session for approximately 20 minutes. At the time of recording, Alexandra's speaking habits had changed to the extent that she did speak to other children, the SLT, and sometimes relatives and teachers. In the analysed session, the unknown person is the first author of this chapter (labelled H in the transcripts). All participants are native German speakers and remain seated around a table facing the camera throughout.

Transcription and analysis approach

A rough German transcript of verbal contributions during the session, including overlaps and pauses, was first produced in ELAN (see Hellwig & Somasundaram, 2011) using standard CA conventions (e.g. Hepburn & Bolden, 2013; Jefferson, 2004). The transcript and recording were scrutinised in an unmotivated manner, enabling initial identification of noteworthy sequences and phenomena. Identified sequences were then re-transcribed in detail to include embodied action in addition to talk, using a compound partiture format (see Ten Have, 2007) consisting of Jeffersonian transcription alongside notations to record eye-gaze (based on Damico & Simmons-Mackie, 2002) (for details of gaze transcription, see Appendix at the end of the chapter). These detailed transcriptions were subjected to a further analytic pass in order to generate collections of comparable phenomena. All analysis was conducted in German; on the transcripts below, an idiomatic English translation has also been provided.

This chapter focuses on one identified phenomenon which relates to the influence of adult turns on Alexandra's subsequent responses. Our account is built around a collection of sequences involving adult interactional behaviours that, to a lesser or greater extent, somehow *require* Alexandra to contribute verbally subsequently. It may appear sensible that a child diagnosed with SM will find social situations where they are required to speak challenging. However, the turn-by-turn perspective offered by CA enables a more nuanced consideration of this point, as it enables examination of variation in child conduct at a turn-by-turn level. As we will demonstrate, this analytic focus indicates several important issues in relation to A's diagnosis, as well as enabling identification of facets of her communicative competence that may previously have been overlooked.

Analysis

It is well established in the CA literature that certain designs of turn project for certain kinds of response. For instance, as Stivers and Rossano (2010) describe, question forms strongly project for subsequent co-participant responses, as reflected by the sanctioning of any absence of answering. Conversely, turns such as assessments, announcements, or noticings do not necessarily require a response, and hence the absence of a subsequent co-participant turn may not necessarily be made relevant by speakers. However, response mobilising design features can be deployed by speakers to project more strongly for a particular form of next-action, thereby increasing the accountability of subsequent co-participant conduct. In this analysis, it will be demonstrated that such design forms have specific consequences for the nature of Alexandra's responses.

Extracts 1–3 illustrate instances where adult interlocutors either ask questions or use response mobilising features in their turns and thereby project strongly for Alexandra to speak next. Extract 1 is from the middle of the analysed session. Caroline and Alexandra have just finished their first therapy game, and Hanna has been present and participating for the previous ten minutes. Just prior to this extract, Caroline asks if Alexandra would like to choose the next activity. Alexandra nods. Caroline adds that Alexandra could show Hanna her favourite game. Alexandra shakes her head. Caroline says Alexandra could choose without showing her game to Hanna, but A does not respond to this suggestion. There is then a one-second pause, before Hanna starts speaking:

Extract 1 [34:29] c = cupboard

```
1      Cx_game_____,,,c__,,, game_____
       Ax_straight___,,,c___,,,H,,.straight_____
       Hx_cupboard_____,,,A_____
   H   sind das die Spiele hier¿
       are these the games over here¿
              ((H points at cupboard))
              ((A has hands in front of mouth))


2      Cx_A_____,,,game_____,,,A_____
       Ax_straight____,,,C_____,,,straight_____
       Hx_A_____,,,cupboard___
              ((C puts toys into box))
       (----------------------------7.6-----------------------------)


3      Cx_H_____
       Ax_straight_____
       Hx_cupboard_____,,,A_____
   H   ich kenn hier nämlich noch gar kein Spiel.
       I don't know any of the games here yet.
       ((A has hands in front of mouth...................))


4      Cx_A_____
       Ax_straight_____
       Hx_A_____
       ((A has index fingers in mouth.........))
       (-----------------------2.4------------------------)
```

5 Cx_A_____„,game_____
   Ax_straight_____
   Hx_A_____
  H m:h,(----------------2.4------------------)
   m:h,
   ((A has index fingers in mouth . . . .))

6 Cx_game_____„,H_____„,A_____
   Ax_straight_____
   Hx_cupboard_____„,A_____
  H oder sind <u>da</u> noch welche im Schrank,
   or are <u>there</u> more games in the cupboard,
      ((H points at cupboard))
   ((A has hands in front of mouth, bends body))

7 Cx_A_____
   Ax_straight_____
   Hx_A_____
  A ((nods . . . .))
   ((A has both index fingers in mouth))
    ((C nods . . .))

8 Cx_A_____
   Ax_straight_____„,upwards_____
   Hx_ A_____„,C____„,A_____
  H e:cht, ((flicks)) (0.6) <nicht schlecht,>
   rea:lly,   ***not bad,***
   ((H smiles . . . .))

In line 1, H's question form and accompanying gaze nominates A as the next speaker. However, A does not provide an answer, and there then follows a long pause of 7.6 seconds. Such latency is unusual: typically, a selected next speaker responds after a beat of silence, otherwise repair is initiated within one second (Hepburn & Bolden, 2013). H and C look at A for most of this period of silence, thereby jointly indicating that the floor remains hers. In line 3, H then accounts for her asking of the unanswered question, and again A does not react in spite of both adults' gazes indicating an ongoing expectation for her to speak. This continues until line 5, where H produces a gap-filling 'mh' before providing an alternative *oder* ('or')-prefaced candidate answer to her own first question. This finally occasions a minimal non-verbal response from A in line 7, to which C rapidly aligns. H accepts this minimal response as adequate by her desisting pursuit of a response and instead producing a positive third-position assessment (line 8).

 Sequences of this nature, in which an adult question precedes an adult-initiated repair sequence following a missing response by Alexandra, are common within these data. Similar consequences also arise following adult non-question turns where response mobilising features are deployed. An example of this phenomenon is demonstrated in Extract 2, which contains an exemplar of such a feature (use of the tag 'ne' (*'right'*) at turn-end). This extract is from a section of the session where the participants play a board game. This game involves players aiming to acquire as many fruit cards as possible and requires a special die on which are pictures of a basket and a raven.

Extract 2 [40:30]

```
1    Ax_game_____
     Cx_H_____,,,A____
     Hx_dice_____,,,A_____
H    vielleicht den Raben da abkleben¿
     perhaps glue it there on top of the raven¿
                    ((H points at die))

2    Ax_dice_____
     Cx_A__,,,dice_____,,,A____
     Hx_A_____,,,dice____,,,A_____
C    genau wir kleben einfach noch ein Körbchen drauf ne¿
     correct we just glue a basket on it right¿

3    Ax_,,,C_____,,,down_____
     Cx_A_____
     Hx_A_____
C    (.) ist eine gute Idee o:der (.) was denkst du,
     is a good idea o:r        what do you think,
           ((A grins))
               ((H laughs))

4    Ax_down____
     Cx_A_____
     Hx_A_____
C             ja¿
              yes¿
     ((A nods))

5    Ax_down_____,,,C_____
     Cx_A_____,,,dice_,,,A_____
     Hx_C_____,,,A____
C    (0.2) oder vielleicht kann ich den Würfel verzaubern?
         or maybe I can bewitch the die?

6    Ax_,,,down_____
     Cx_A_____
     Hx_A_____
C    (----------2.4----------) meinst du das hilft ↑auch
         do you think that helps ↑too
```

For a player to win this particular game, they must roll the die to the picture of the basket. In line 1, H suggests that another picture of a basket could be glued onto the die to improve the chances of winning. Towards the end of this turn, H selects A as the next speaker through gaze, and C also subsequently gazes at A. However, in spite of A's nomination as speaker, the design of H's turn does not explicitly project for any particular form of subsequent response from a co-participant. This ambiguity appears to be made relevant in line 2 through C's rephrasing of H's prior turn: note that C appends the tag 'ne', using rising intonation, which more strongly generates an expectation for A to speak (more specifically, to provide subsequent agreement with this turn; cf. Pomerantz, 1984). However, A does not react at all, and accordingly C's next move projects for a subsequent *disagreement* response from A through use of 'oder' ('*or*'). At this instant, A gazes down and stops grinning, and after a further non-response, C asks an explicit question following which A provides another minimal nod. However, unlike in Extract 1 this is not received as adequate, as C then produces 'ja' ('*yes*') with rising intonation before ultimately providing another or-prefaced candidate response.

Here, C's conduct indicates that Alexandra is expected to speak, yet does not. 'Ne' has previously been described as making relevant a recipient reaction (Selting, 1994; Selting et al., 2009; Simsek, 2012). Accordingly, when Alexandra does not respond to a 'ne' turn, C indicates this to be problematic in her subsequent turn. Unlike in Extract 1, Alexandra's non-verbal response is then treated as inadequate by C, as indicated by the subsequent (unsuccessful) attempt to project for a verbal contribution from A. Both adult participants frequently generate similar expectations for A to speak, and in Extract 3 it is H who initiates and pursues this course of action. The below occurs immediately following H's entry into the session. C announces to H that she and A have already played one round of a game, and H responds as follows:

Extract 3 [24:19]

```
1       Ax_down_____
        Cx_game_____,,,A__
        Hx_A_____
    H   und wer hat gewonnen,
        and who won,

2       Ax_down_____
        Cx_A_____
        Hx_A_____
        (------------0.7-------------)

3       Ax_,,,C___,,,down_____
        Cx_A_____
        Hx_C_____,,,A_____
    C   mh wer hat denn gewonnen
        mh who won

    H                   oder gibt's da gar nichts zu gewinnen.
                        or is there nothing to win.
                                ((A shrugs her shoulders))
```

4 Ax_down_____„,C__„,down_____
   Cx_A_____
   Hx_A_____
 C eigentlich hast (.) haben wir beide gewonnen ne
   actually you **we both won right**
      ((C shrugs shoulders))

At the end of H's question in line 1, A is nominated by gaze as the next speaker by both adults. Epistemically, C has the resources herself to answer H's question given that she too was a participant in this game. C's status as a possible answerer is arguably reflected by her own pursuit of a response from A's through repeating H's question (line 3). However, unlike in Extract 2, C does not then pursue a verbal response from A. In this case, she ultimately produces a turn in a sequential position where A remains nominated by gaze as the next speaker but appends 'ne' ('*right*') to project for A's subsequent agreement. That C, as non-selected speaker, appears to respond on behalf of A is in line with findings about multi-party talk in mundane context, where 'the preference for an answer overrides the preference for a response from the selected speaker' (Stivers & Robinson, 2006, p. 384).

Given A's diagnosis of SM, it is arguably unsurprising that her responses across Extracts 1–3 are either non-verbal or apparently absent. However, in marked contrast to the above, several sequences during the session were identified in which Alexandra more actively participated through non-verbal, and even verbal, means. As will now be demonstrated, these sequences are notably characterised by the *absence* of any expectation to speak being generated by adult turns prior to A's moves. For example, consider Extract 4, prior to which the three participants have been playing the game described above:

Extract 4 [39:10]

1 Ax_piece_____
   Cx_piece_____„,H_____
   Hx_piece_____
 H °kriegt der auch noch Füße°
   °*he gets feet as well*°
   ((A takes puzzle piece))

2 Ax_piece_____
   Cx_piece_____
   Hx_piece_____
   ((A places puzzle piece on board game))
   (---------------------6.5----------------------)

3 Ax_basket_____
   Cx_basket_____„,H_____
   Hx_basket_____„,A_____

```
         ((A puts die into basket, pushes towards H))
    H                     döt döt döt (.) dankeschö:n=
                          dut dut dut    thank you:
                          ((A smiles............))
                          ((H takes die))

4   Ax_basket_„,H____
    Cx_H_„,A_____
    Hx_A_____
    A    ((laughs))
    H         ((laughs))
```

In line 3, A non-verbally initiates a brief but clearly joint interactional project: instead of merely passing the die to H, A places it into the basket and slowly moves it towards her co-player. This renders the physical action of passing the die as being particularly salient, as reflected by H's production of 'sound effects' in co-ordination with A's movement of the basket. These vocalisations appear to occasion a smile from A, and following H's taking of the die, A produces a laugh (which is reciprocated by H) along with a flash of otherwise-rare eye contact.

This sequence is different to those discussed above in Extracts 1–3 for two reasons. First, although Alexandra again does not produce talk here, she does successfully initiate and sustain a spontaneous joint trajectory of action with a co-participant. Second and importantly for our arguments, A's non-verbal initiation of this joint project follows a turn by H which does *not* generate an expectation for A to speak subsequently, either by turn design or via use of gaze. Extract 5, which follows the above, demonstrates how A's participation becomes more pronounced as this episode develops:

Extract 5 [39:44]

```
1   Ax_basket_____
    Cx_basket_____
    Hx_basket_____
    ((A puts die into basket, moves it towards H))
    A                    ↑put ↑pu::t
                         ↑put ↑pu::t
    H                         ↑dö ↑dö ↑dö:::t
                              ↑do ↑do ↑do:::t
                                   ((A smiles.........))

2   Ax_basket__„,H___„,die_____
    Cx_A____„,H___„,die_____
    Hx_basket__„,die_____
    H   dankeschö:n¿
        thank you:¿
        ((A smiles...........................))
             ((H throws die))
```

```
3        Ax_die__„board_____
         Cx_die„,A_____
         Hx_die„,A_____
    H      ↑oh
           ↑oh
             ((A laughs))
             ((H laughs))
```

In line 1, A restarts the joint activity presented in Extract 4. This time, however, she mirrors H's prior conduct by producing the 'sound effects' herself, an action to which H aligns through her own co-ordinated vocalisations. Following H's *dankeschön* ('thank you'), A then briefly gazes at H, before H throws the die and subsequently produces a high-pitched sound (line 3). This appears to occasion shared laugher, which seemingly is initiated by A.

The reciprocal co-ordination between speakers that is evident in Extract 5 demonstrates that the A-initiated basket-moving project is a shared and joint one. Laughter is used affiliatively, with the embedded activity of die-passing and throwing jointly established and then collaboratively developed by the co-participants. In the above, Alexandra even produces playful vocalisations of her own. Given her diagnosis of SM, A's conduct here is notable as it indicates a willingness and ability to participate in *certain* forms of interactional sequence. And again, at no point in Extract 5 does an adult deploy a question form or a response-mobilising design, meaning that A is at no point positioned to *have* to produce a turn.

This last observation is strikingly demonstrated on the few occasions where Alexandra does contribute verbal talk to the interaction. Consider Extract 6 which occurred before those above; here, Alexandra has been chosen to initiate the board game:

Extract 6 [37:20]

```
1        Ax_straight„,die_____„,pear____
         Cx_A____„,die___„,A_____„,die___„,pear____
         Hx_C_____„,die___„,A_____„,die____„,pear____
    C    okay. bitteschön.
         okay. here you go.
                ((C gives A die))
                            ((A throws die, takes pear))

2        Ax_pear_____
         Cx_pear_„,game_____
         Hx_pear_„,basket_____
    A      Bi:rni
           pea:ri
               ((A puts pear in basket))
```

3   Ax_board_____
        Cx_game_____
        Hx_basket_____
   H   nicht <u>schlecht</u>
       not <u>bad</u>
       ((A takes pear))
       ((H takes basket))

4   Ax_pear_____
        Cx_H hand_____,,,A_____
        Hx_dice_____,,,A_____,,,board__
       ((A moves pear to mouth)) ((A pretends to eat pear))
       ((H moves hand towards dice)) ((H moves hand back))
   A   <u>hang hang hang</u>
                 hang hang hang

5   Ax_basket_____
        Cx_,,,H_____
        Hx_board_____
   H   bin ich überhaupt <u>dran</u> ja. ne¿
       is it my <u>turn</u>   **yes. right¿**
       ((A puts pear into basket))

In line 2, A contributes verbally following a turn by C that does not generate an expectation for her to speak: she first states which fruit she has just
collected, before pretending to eat it. While H does produce an acknowledgement in line 3, the adults do not shift gaze towards A, and therefore there
continues to be no expectation for A to talk further subsequently. In line 4,
A then produces another verbal turn. This time, however, H and C do gaze at
A, thereby nominating her as the next speaker. Notably, A then rapidly desists
her pretence, producing no further verbal turns within this sequence.

Finally, in Extract 7, Alexandra can be seen to produce a verbal contribution which renews and extends the context established by another participant's
prior turn. Unlike her verbal moves in Extract 6, in this instance A herself
projects more strongly for a subsequent response from her co-speakers:

Extract 7 [42:58]

1   Ax_C's basket_____
        Cx_C's basket_____
        Hx_C's basket_____
   C   °ich krieg° den <u>Ra:benschna:bel</u> hier
       °I get° the <u>ra:ven's bea:k</u> here

2   Ax_C's basket_____
        Cx_C's basket_____
        Hx_C's basket_____
       ((A laughs ...))
         ((H laughs ......))
           ((C laughs ......))
             ((A takes card))

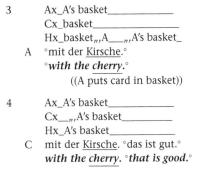

3   Ax_A's basket_____
        Cx_basket_____
        Hx_basket,,,A___,,,A's basket_
  A °mit der <u>Kirsche</u>.°
      °***with the cherry.***°
        ((A puts card in basket))

4   Ax_A's basket_____
        Cx___,,,A's basket_____
        Hx_A's basket_____
  C mit der <u>Kirsche</u>. °das ist gut.°
      with the <u>cherry</u>. °that is good.°

In line 1, C produces a turn with an intonation contour suggestive of it being designed to be heard as 'amusing'. Fittingly, A then initiates a laugh sequence involving both co-participants, thereby building affiliation with this humorous turn (cf. Glenn, 2008), before uttering *mit der Kirsche* ('with the cherry'). The omitted grammatical subject within A's turn indicates that it is to be heard as a continuation of C's prior description, and therefore serves to contribute to a joint trajectory of talk. Such verbal conduct by A is evidently notable for the co-participants themselves, as is reflected by C's subsequent repetitious receipt and explicitly positive assessment.

Clinical relevance

The findings presented above indicate two overlapping themes. First, the interactional conduct of *all* participants, including A, in this session varied depending upon sequential context. Second, the behaviours by A that could be described in terms of SM appeared directly responsive to the nature of pre-ceding adult turn. These variations are clearly demonstrated when comparing Extracts 1–3 to 4–7. The former three extracts illustrate more interactional problems (pauses, repair initiations, and so on) and less examples of child-initiated sequences and co-ordination between participants. As argued above, we attribute this to the presence of adult turns that generate an expectation for A to speak subsequently. Conversely, in Extracts 4–7 no such expectation to speak is generated by adult turns, and in following sequential positions Alexandra non-verbally and verbally initiates and/or aligns with joint courses of action involving her co-speakers. In spite of her diagnosis therefore, the child discussed in this chapter demonstrates a clear willingness to interact in specific interactional circumstances, a finding which provides initial empirical support for the conjecture that interlocutor conduct may shape the problem behaviours associated with SM.

These findings also begin to shed new light on previous work on SM; for instance, Nowakowski et al.'s (2011) investigation of joint attention in parent–child dyads. As discussed earlier in the chapter, the authors of that study reported that during structured discussion tasks, significantly fewer parent-initiated joint attention episodes occurred in dyads involving children diagnosed with SM. However, it is intuitive that during adult-led structured discussion there would be considerable likelihood that the child participants would frequently be positioned in sequential contexts where there is a requirement to speak, due to the often asymmetric nature of adult–child interaction. Conversely, Nowakowski et al. (2011) reported no significant group differences with regard to joint attention during unstructured tasks. Again, based on intuition, it would follow that unstructured tasks would be based around play or other joint endeavours, and therefore may be characterised by fewer turns that generate an expectation to speak. Thus, the findings of Nowakowski et al. (2011) may be consistent with our thesis that the behaviours associated with SM may, on a moment-by-moment basis, be less clearly observable where no expectation to speak has been established in prior talk. This may represent an important area for future research into SM.

But what is the clinical relevance of these findings? We suggest that there are two such relevancies implied by the above CA account, each relating to a different aspect of clinical practice involving children diagnosed with SM. The first relates to issues of diagnosis and SM's classification within psychiatric nosology. As discussed earlier in this chapter, there remains controversy around whether SM is most appropriately classified as an anxiety disorder or as a social phobia (Sharp et al., 2007). However, the findings presented above appear consistent with Omdal and Galloway's (2008) hypothesis that SM is *not* a general phobia, and at least in the case of this child, it does not appear to be a simple phobia of expressive speech either: Alexandra demonstrated that she is willing to interact, including through verbal means, although only in a restricted range of sequential positions. Hence, at least in the case of the child presented above, SM may reflect the behavioural manifestation of highly *context-specific* anxiety, associated with particular forms of interactional demands within particular relational contexts. While examination of interactions involving other diagnosed children is required to explore this suggestion further, these initial findings indicate that CA perspectives on SM may be of significant practical value, both in terms of differential diagnosis and the assessment of individual children's profiles.

The second point of clinical relevance arising from our account relates to issues of treatment. From an SLT perspective, there are broadly two means through which SM can be addressed clinically: by working directly with the child to increase their individual capacity to tolerate increasingly challenging communication environments (e.g. through behaviourist desensitisation therapy, as demonstrated in the analysed session) or alternatively by intervening upon individuals *around* the child (e.g. family members) with the aim of modifying their communication behaviours in order to reduce the systemic demands presented by a specific communication environment. While there are stark differences in the philosophical and epistemological commitments underpinning such individualised versus environmentalist therapy approaches, findings of the nature presented in this chapter could be applied to develop robust forms of either. If utilising a behaviourist desensitisation-based approach for instance, a CA-based modelling of relationships between various forms of adult turns and the subsequent expectation to speak could enable specification of nuanced hierarchies of interlocutor turns, ranging from those that generate no such expectation at one end to those that necessitate a response (e.g. open direct questions) at the other. Desensitisation treatment could then involve a gradual guided progression through such a hierarchy. Conversely, if adopting an environmentalist intervention approach, findings such as those reported above could inform tightly specified family training approaches regarding the kinds of communication behaviours that might be more or less facilitating for a diagnosed child's interactional participation. The value of CA-based coaching has already been demonstrated with other clinical populations, including adults with acquired language difficulties following stroke (e.g. Lock et al., 2001). For a simple summary of the implications for practice, see Table 27.1.

Summary

This chapter has presented a CA analysis of a treatment session involving a child diagnosed with SM, indicating one significant extra-individual factor that may influence the manifestation of problematic individual communication behaviours. Our findings interface intriguingly with existing theory about SM. For instance, the clinical features of SM are reported to vary significantly across children, ranging from apparent silence to the production of a few words, but in a monotone or otherwise-altered voice (e.g. McHolm, Cunningham, &

Table 27.1   Clinical practice highlights

1. The fine-grained CA analysis presented in this chapter demonstrates that a child diagnosed with selective mutism (SM) shows clear willingness to interact at certain points in a clinical interaction.
2. The variation in her interactional behaviour appears to depend upon the extent to which adults have, in prior turns, established an expectation for her to speak. Clinicians may wish to consider such issues when observing and assessing adult–child dyads involving children who are reluctant speakers.
3. Modelling a speaking hierarchy of turns that establish different levels of expectation to speak may be helpful when planning and implementing desensitisation therapies.
4. Minimising the occurrence of adult turns which require the child to speak subsequently may facilitate the interactional participation of children diagnosed with SM.

Vanier, 2005; Omdal & Galloway, 2008; Scott & Beidel, 2011; Shipon-Blum, 2007; Wong, 2010). However, our analysis indicates that one *individual* child demonstrates different ways of communicating within the *same* interaction: while appearing passive and avoidant in some sequences, Alexandra can be seen to contribute actively, and to initiate new sequences of talk non-verbally and verbally, at other junctures. Moreover, this variation appears systematic, depending in part upon whether an expectation to speak was established through immediately preceding interlocutor turns.

Evidently, analytic findings based upon one interaction with one child cannot be generalised with validity to all children with the diagnosis of SM, and further work is needed to investigate if interactions involving such children in different social contexts proceed comparably. However, as discussed above, the initial findings presented in this chapter have potential practical relevance for the clinical diagnosis, assessment, and treatment of children who behave in a manner consistent with SM's criteria. As such, our analysis reinforces the theoretical and practical value of CA work which aims to reposition ostensibly atypical communication behaviour as situated and reactive, rather than merely endogenous.

Appendix: Transcription conventions for gaze

x_____ = maintenance of gaze
„, = gaze shift

Each participant has her own gaze line. The initial at the beginning of the line indicates the participant (in this case A, C, or H). The first 'x_' on each line indicates the commencement of a gaze episode, and the word on the line indicates the direction of this gaze. Gaze can be directed at other participants (A, C, or H), an object (e.g. game), or more spatially (e.g. straight).

e.g. Ax_straight_____„,game_____

Here Alexandra first gazes straight ahead, before shifting gaze to the game.

References

American Psychiatric Association (2000). *Diagnostic and statistical manual of mental disorders* (4th edition). Washington, DC: American Psychiatric Publisher.

Bergman, R. L., Piacentini, J., & McCracken, J. T. (2002). Prevalence and description of selective mutism in a school-based sample. *Journal of the American Academy of Child and Adolescent Psychiatry, 41*(8), 938–946.

Carbone, D., Schmidt, L., Cunningham, C., McHolm, A., Edison, S., St Pierre, J., & Boyle, M. (2010). Behavioral and socio-emotional functioning in children with selective mutism: A comparison with anxious and typically developing children across multiple informants. *Journal of Abnormal Child Psychology, 38*(8), 1057–1067.

Damico, J. S., & Simmons-Mackie, N. N. (2002). The base layer and the gaze/gesture layer of transcription. *Clinical Linguistics & Phonetics, 16*(5), 317–327.

Elizur, Y., & Perednik, R. (2003). Prevalence and description of selective mutism in immigrant and native families: A controlled study. *Journal of the American Academy of Child and Adolescent Psychiatry, 42*(12), 1451–1459.

Garcia, A. C. (2012). Medical problems where talk is the problem: Current trends in conversation analytic research on aphasia, autism spectrum disorder, intellectual disability, and alzheimer's. *Sociology Compass, 6*(4), 351–364.

Glenn, P. (2008). *Laughter in interaction.* New York: Cambridge University Press.

Hellwig, B., & Somasundaram, A. (2011). *ELAN – Linguistic Annotator.* Retrieved November 25, 2011 from http://www.mpi.nl/corpus/manuals/manual-elan.pdf.

Hepburn, A., & Bolden, G. (2013). The conversation analytic approach to transcription. In T. Stivers & J. Sidnell (Eds.), *The Blackwell handbook of conversation analysis* (pp. 57–76). Oxford: Blackwell.

Jefferson, G. (2004). Glossary of transcript symbols with an introduction. In G. Lerner (Ed.), *Conversation analysis: Studies from the first generation* (pp. 13–23). Philadelphia: John Benjamins.

Johnson, M., & Wintgens, A. (2001). *The selective mutism resource manual.* Milton Keynes: Speechmark Publishing Ltd.

Katz-Bernstein, N. (2011). *Selektiver Mutismus bei Kindern* (3rd edition). München: Ernst Reinhardt Verlag.

Lock, S., Wilkinson, R., Bryan, K., Maxim, J., Edmundson, A., Bruce, C., & Moir, D. (2001). Supporting partners of people with aphasia in relationships and conversation (SPPARC). *International Journal of Language and Communication Disorders, 36*(Supplement), 25–30.

McHolm, A. E., Cunningham, C. E., & Vanier, M..K. (2005). *Helping your child with selective mutism.* Oakland, CA: New Harbinger Publication.

Muskett, T., Perkins, M., Clegg, J., & Body, R. (2010). Inflexibility as an interactional phenomenon: Using conversation analysis to re-examine a symptom of autism. *Clinical Linguistics and Phonetics, 24*, 1–16.

Nowakowski, M. E., Tasker, S. L., Cunningham, C. E., McHolm, A. E., Edison, S., Pierre, J. S., Boyle, M. H., et al. (2011). Joint attention in parent-child dyads involving children with selective mutism: A comparison between anxious and typically developing children. *Child Psychiatry and Human Development, 42*(1), 78–92.

Omdal, H., & Galloway, D. (2008). Could selective mutism be re-conceptualised as a specific phobia of expressive speech? An exploratory post-hoc study. *Child and Adolescent Mental Health, 13*(2), 74–81.

Pomerantz, A. (1984). Agreeing and disagreeing with assessments: Some features of preferred/dispreferred turn shapes. In M. Atkinson & J. Heritage (Eds.), *Structures of social action* (pp. 57–101). Cambridge: Cambridge University Press.

Sacks, H., Schegloff, E., & Jefferson, G. (1974). A simplest systematics for the organisation of turn-taking for conversation. *Language, 50*(4), 696–735.

Schegloff, E., & Sacks, H. (1973). Opening up closings. *Semiotica, 8*(4), 289–327.

Scott, S., & Beidel, D. (2011). Selective mutism: An update and suggestions for future research. *Current Psychiatry Reports, 13*(4), 251–257.

Selting, M. (1994). Konstruktionen am Satzrand als interaktive ressource in natürlichen Gesprächen. In B. Haftka (Ed.), *Was determiniert wortstellungsvariation? Studien zu einem interaktionsfeld von grammatik, pragmatik und sprachtypologie* (pp. 299–318). Opladen: Westdeutscher Verlag.

Selting, M., Auer, P., Barth-Weingarten, D., Bergmann, J., Bergmann, P., Birkner, K., & Al, E. (2009). Gesprächsanalytisches Transkriptionssystem 2 (GAT2). *Gesprächsforschung, 10*, 353–402.

Sharp, W. G., Sherman, C., & Gross, A. M. (2007). Selective mutism and anxiety: A review of the current conceptualization of the disorder. *Journal of Anxiety Disorders, 21*(4), 568–579.

Shipon-Blum, E. (2007). 'When the words just won't come out' – Understanding selective mutism (pp. 1–8). Houston. Retrieved November 25, 2014 from http://www.selectivemutism.org/resources/library/SM General Information/When the Words Just Wont Come Out.pdf.

Simsek, Y. (2012). *Sequenzielle und prosodische aspekte der sprecher-hörer-interaktion im Türkendeutschen.* Münster: Waxmann Verlag.

Stivers, T., & Robinson, J. (2006). A preference for progressivity in interaction. *Language in Society, 35*, 367–392, doi:10.1017/S0047404506060179.

Stivers, T., & Rossano, F. (2010). Mobilizing response. *Research on Language & Social Interaction, 43*(1), 3–31.

Ten Have, P. (2007). *Doing conversation analysis* (2nd edition). London: Sage.

Vecchio, J., & Kearney, C. (2005). Selective mutism in children: Comparison to youths with and without anxiety disorders. *Journal of Psychopathology and Behavioral Assessment, 27*(1), 31–37.

Wilkinson, R. (2008). Conversation analysis and communication disorders. In M. J. Ball, M. R. Perkins, N. Müller, & S. Howard (Eds.), *The handbook of clinical linguistics* (pp. 92–106). Oxford: Blackwell Publishing.

Wong, P. (2010). Selective mutism: A review of etiology, comorbidities, and treatment. *Psychiatry, 7*(3), 23–31.

Yeganeh, R., Beidel, D., & Turner, S. (2006). Selective mutism: More than social anxiety? *Depression and Anxiety, 23*, 117–123.

Recommended reading

- Elizur, Y., & Perednik, R. (2003). Prevalence and description of selective mutism in immigrant and native families: A controlled study. *Journal of the American Academy of Child and Adolescent Psychiatry, 42*(12), 1451–1459.
- McHolm, A. E., Cunningham, C. E., & Vanier, M. K. (2005). *Helping your child with selective mutism.* Oakland: New Harbinger Publication.
- Muskett, T., Perkins, M., Clegg, J., & Body, R. (2010). Inflexibility as an interactional phenomenon: Using conversation analysis to re-examine a symptom of autism. *Clinical Linguistics and Phonetics, 24*, 1–16.
- Sharp, W. G., Sherman, C., & Gross, A. M. (2007). Selective mutism and anxiety: A review of the current conceptualization of the disorder. *Journal of Anxiety Disorders, 21*(4), 568–579.
- Yeganeh, R., Beidel, D., & Turner, S. (2006). Selective mutism: More than social anxiety? *Depression and Anxiety, 23*, 117–123.

28

'I'm Happy with Who I Am': A Discursive Analysis of the Self-Characterisation Practices of Boys in 'Behaviour' Schools

Linda J. Graham

Introduction

Research in the psychological and medical fields has consistently found that children with behavioural disorders exhibit 'positive illusory bias' in their perceptions of self (Hoza, Pelham, Dobbs, Owens, & Pillow, 2002). Positive Illusory Bias (PIB) is 'operationally defined as a disparity between self-report of competence and actual competence' (Owens, Goldfine, Evangelista, Hoza, & Kaiser, 2007, p. 336) leading to 'an artificially inflated level of self-esteem' (Gresham, MacMillan, Bocian, Ward, & Forness, 1998, p. 405). Prior studies have compared the self-reports of behaviourally disordered boys relative to un-referred controls either on performance tasks or in relation to parent, peer, or teacher reports. Findings are varied; however, there is some evidence that boys with social, emotional, academic, and behavioural difficulties overestimate their competence, particularly in the domains in which they are weakest (Hoza et al., 2004; Jiang & Johnston, 2014).

As self-enhancement has been linked to poor social skills and psychological maladjustment over time, researchers suggest that 'accurate appraisals of self and of the social environment may be essential elements of mental health' (Colvin, Block, & Funder, 1995, p. 1152). Views as to treatment are mixed with some researchers arguing for the 'need to raise children's perceptions of competence' (Hoza et al., 2002, p. 277), and others advocating 'humility training' to bring these children's self-perceptions into line with reality (Gresham et al., 1998). The latter strategy could well be damaging however, as evidence exists that self-enhancing bias may serve a self-protective function; one that children may employ 'to mask feelings of inadequacy' (Ohan & Johnston, 2002, p. 239).

While extensive, the psychological and medical literature is overwhelmingly quantitative, disappointingly deficit-based, and risks exacerbating already impoverished and limiting views of these young people. It also neglects to consider the powerful role that language plays in shaping perception, for example in how children understand their own behaviours when associated with conditions such as ADHD (see e.g. Bradley & Butler, Chapter 26, this volume). Post-structural theories of language offer a different lens through which to view positive self-characterisations – not as an exercise in self-delusion by character-impaired young people – but as a means to both make and remake identity: a means by which one can punch out the limits of the identity that has been assigned. To better understand the complex links between language and identity (re)formation, this research draws on a Foucauldian approach to discourse analysis (Arribas-Ayllon & Walkerdine, 2008) that has been informed by post-structural and literary theory (Graham, 2007a, 2011). Such an approach is concerned less by what is *said* and more by what those words *do*: how they function in a constitutive sense and the political effects that flow from that act of constitution (Graham, 2007b). A Foucauldian approach, however, is not sufficient on its own to theorise acts of discursive resistance.

Post-structural approaches that draw on Foucault have been criticised for collapsing the constitution of objects with the formation of subjects *and* with the development of individual subjectivities, resulting in a nihilistic cul-de-sac with little room for agency (Yates & Hiles, 2010). Judith Butler resolves this problem by combining the discursive constitution of the subject with a theory of linguistic agency. She draws on Althusser's theory of interpellation together with Foucault's concept of discursive power to theorise the constitution of the subject, but extends this to describe how the process of discursive constitution holds within it an opportunity for re-inscription. Butler (1997a) described this by saying:

> One is not simply fixed by the name that one is called. In being called an injurious name, one is derogated and demeaned. But the name holds out another possibility as well: by being called a name, one is also, paradoxically, given a certain possibility for social existence, initiated into a temporal life of language that exceeds the prior purposes that animate the call. (p. 2)

Therefore, while 'a name tends to fix, to freeze, to delimit' (Butler, 1997, p. 35), being spoken into existence also opens a space for linguistic return; an opportunity for the subjected to subvert. This right of reply to address provides radical opportunities for the marginalised to speak themselves differently and, in so doing, engage in purposeful resistance.

Project overview

In this chapter, I draw on post-structural theories of language to examine the self-characterisation practices of 33 boys attending special schools for students with disruptive behaviour. During a semi-structured interview, each boy was asked to describe his personality and then to choose from a selection of positive/negative word pairs. The objective was to determine whether these young people would characterise themselves in positive or negative ways. Participants were then asked if there was anything they would change about themselves if they could. Responses were analysed and compared against a discourse model developed from media reports and interviews with their principals. Findings suggest that while discourse may well 'form the objects of which it speaks' (Foucault, 1972, p. 49) in the eyes of teachers, principals, psychiatrists, and paediatricians, it also offers a means through which the constituted subject can re-author itself in a more positive frame (Butler, 1997a).

In the first half of this chapter, I examine the ways in which boys who have been enrolled in special schools for challenging behaviour are described, first in the media and then by their principals, to gain a sense of such boys as 'discourse objects', an approach that could be described as Foucauldian. I then empirically examine the responses of these boys to a series of interview questions that ask them to describe their personality and whether there is anything that they would change about themselves if they could. Each boy also completed a personality word sheet devised of positive/negative word pairs.

The aim of my approach – comparing these boys' self-characterisations to a Foucauldian discourse model – is to determine whether their self-characterisations match the ways in which they have been or are characterised by others and, if not, whether any discrepancy is positively or negatively skewed. If negatively skewed, this could provide support for the theory – found in both positivist and poststructuralist analyses – that labelling and stigma negatively affect developing self-concepts (Yang, Wonpat-Borja, Opler, & Corcoran, 2010). If positively skewed, these characterisations could either provide support for the 'positive illusory bias' thesis described earlier or, from a Butlerian perspective, that these boys are employing discursive strategies to reframe negative characterisations that they know exist.

In other words, rather than indicate a lack of self-awareness or inflated self-esteem, positive self-characterisations may instead reflect the recognition that identity is discursively constructed and, as such, it can be reconstructed using language that depicts 'troubling' characteristics in a more accurate and positive light. As post-structural analysis has also been criticised for being theoretically dense, empirically opaque, and low in rigour (O'Farrell, 2006), the approach to discourse analysis that I take in this chapter could best be described as a hybrid: one that employs empirical methods of organising and reporting data but which applies a completely different epistemological lens in the process of analysis. The chapter begins by drawing on media reports to examine the discursive constitution of boys in behaviour schools as a particular discourse object: one that is not only dysfunctional but dangerous.

Mad, bad and, above all, dangerous

A story on the increased use of special schools for students with disruptive behaviour was published in Sydney's *Daily Telegraph* in 2011. Making full use of the sensationalism for which Rupert Murdoch's tabloids are renowned, the journalist – Bruce McDougall – paints a lurid picture of the danger that lurks within government schools in the state of New South Wales:

> MORE than 500 of the state's most dangerous students are now being taught in special schools because they are such a menace in mainstream classes – and the number is growing at 57 a year. But more than half the students, called 'the mad and the bad', never return to a regular school because principals are often too frightened to have them back.
>
> (McDougall, 2011, np)

McDougall neglects to mention that the students enrolled in these special schools are some of the most disadvantaged in Australian society, nor does he mention the overrepresentation of Indigenous children and children in out-of-home care (Graham, 2012). He also neglects to mention the very high percentage of students with language disorders or the role that school markets have in incentivising their exclusion from the mainstream (Granite & Graham, 2012).

Instead of portraying these young people with sensitivity and concern, *The Daily Telegraph* conjures images likely to increase the pressure to exclude. Phrases like 'filled with' and 'growing in number' give the impression that these special schools are bursting at the seams with many more 'dangerous' students still in mainstream. The use of words like 'frightened' with reference to school principals is particularly powerful. School principals are figures of authority but, according to McDougall, growing numbers of 'mad' and 'bad' children are terrorising even these powerful adults.

Later in the story, McDougall buttresses these claims with a statement that behaviour schools are 'filled with children with serious mental health issues' who are 'thought too dangerous to return to mainstream'. The word 'dangerous' is repeated five times in this short article and three of those uses are in the title, the opening paragraph, and in the caption beneath a darkened photograph of a boy caught in a menacing 'standover' position, which reads:

> **Dangerous:** An explosive report obtained by The Daily Telegraph reveals 14
> out of 35 behaviour schools are filled with children suffering serious mental
> health issues.

Further enhancing this discourse of 'dangerousness' is his claim that 'many
of the schools have locks on nearly every door, staff often carry walkie-talkies
and police are on call "to provide immediate assistance if required"'. Credibility is sought via reference to an 'explosive report' commissioned by the NSW
Department of Education and Training (DET) and 'obtained by the *Daily Telegraph* under Freedom of Information Laws'.[1] Whether true or not, the inclusion
of this salacious information serves to accentuate the paper's scoop, while also
suggesting that DET was reluctant for its contents to be made public.

As I mentioned earlier, the aim of Foucauldian discourse analysis is not so
much to analyse what is said but instead what those statements do: what
are the constitutive and political effects? In the case of boys in NSW government behaviour schools, we have a clear object of discourse: a dangerous and
unpredictable group of future mad-men who are terrorising the principals of
mainstream schools and putting other (innocent) children at risk. Readers of
the *Daily Telegraph* are left in no doubt. Of course it is necessary to segregate
these ticking time bombs: why wait until it is too late? But, how accurate are
these descriptions? As tabloid newspapers are given to exaggeration, I turn now
to examine how these young people were described by the members of the
public who would know them best: their principals.

Grotty, smelly, snotty, rude, obnoxious, aggressive...

In 2011, I met with the principals of five case-study behaviour schools to discuss their participation in the project. The aim of these meetings was to explain
the research and to learn more about their schools, as well as the young people within them. It was important to do this for a number of reasons: first,
to reassure the principals, each of whom had been warned not to participate
by principals of other government behaviour schools in NSW; and secondly,
to ensure that the research instruments were appropriate for the young people with whom we would be working. Over the coming months, four of these
principals – some with their deputies – participated in a series of interviews.[2]
The interviews ranged from two to three hours in duration and focused on
a wide range of issues, including teaching and discipline practices in mainstream feeder schools, the effects of social disadvantage, the aims and structure
of each behaviour school, the principal's own teaching background and leadership challenges, as well as the characteristics of our 33 student participants –
each of whom was attending one of the five participating behaviour schools.

Two of the principals were strongly influenced by a therapeutic perspective
and this drove both their approach and understanding of the young people

attending their schools. One of these principals, Lara, had once worked in a special school within an acute adolescent psychiatric unit and this strongly influenced how she perceived and responded to her students' behaviour. Rather than confronting distressed youngsters and escalating an already volatile situation, she would talk calmly to them using a language of emotion that was being taught in the school to help these young people recognise, understand, and manage their emotions. Despite McDougall's depiction of 'behaviour' schools in lock-down with walkie-talkies and the police on speed-dial, Lara said that she had never placed her hands on a student in all her 17 years of being in this field and if they needed time to chill out, they could have it. She had a very philosophical approach which was based on her understanding of behaviour as a language, not an innate character flaw:

> All of the students have, like in any setting, I suppose...they want to know where they fit in, and sometimes kids use violence and threatening behaviour and physical aggression to establish their place. In most cases, they'll shake hands, and that's over and done with. The problem with systems that immediately involve suspension, or other consequences, is that they lock in a particular moment of impulse, into what the student is like, and they're not. Where, if you give it the time and let the tide come out, there's a different outcome and we have to work on...bridging the gap between their desire to react, and the reaction.

Matthew, with a background in outdoor education, had a similarly gentle manner. With a calm, melodic voice and an impish sense of humour, he explained how his school aimed for a positive, strengths-based approach:

> I think they're great kids. I think they've all got – there's something in them, and if you look hard enough, you can find it. Sometimes we think that they're getting tougher, but then we find the way into the kids and they connect to the school, and suddenly that toughness disappears. It doesn't go away, it's just not the feature of the day. But we talk to the kids, well, certainly with the staff about a certain emotional flood. You know, you've got two brains – that rational brain and the emotional brain, and you need both of them. But when the tide's in on the emotional brain, that's not the time to be dealing with the issues. You've really just got to let things smooth out and then address them.

David was a little more critical. He noted that there was diversity in the system and that some behaviour schools, like Lara's, enrolled higher percentages of students with mental health issues than others. What set *his* school apart from hers, he said, was the level of criminality in the local area and the 'criminal element' within the school. He was interested in learning more about student trajectories and where they ended up after leaving his school because he was convinced that his school had two distinct types of 'bad': boys who would get into trouble because they were impulsive and boys who were just 'born criminals'.

> There's you know, I think about 47% of our kids from when I looked at the data from 1998 to 2009, about 47% had been incarcerated after they've left and about 44% of those had been for some kind of violence – but all to do with mental health issues. I mean a lot of our kids can't handle authority ... they'll punch a copper, punch a police officer.

David was sympathetic towards his boys and it was obvious that he cared about them, even if he was a little gruff in the way that he expressed it. One of the challenges he raised was the negative perceptions of others and how these might act as a self-fulfilling prophecy:

> I've got one paediatrician who tells some of my boys 'You're either going to be a millionaire or a serial killer. Get out of my office. I don't want to see you again. You're never going to be any good. You're just a criminal'.

Joan and her deputy, Mel, were the most negative of all the behaviour school principals, describing their 'typical clientele' as the kind who have 'difficulty with human relationships':

```
Joan:   These kids are not the kind who - they're not generally the
        kind of kids that other kids are going to...
Mel:    Flock to.
Joan:   Nurture [laughs]. But not even the adults are going to
        nurture them! When you get down to it, basically we don't
        have society's pretty kids; we have society's kids that
        everybody walks down the street and they cross the road!
Mel:    We have the kids that most teachers say, why would you want
        to work with him?
Joan:   You know, can be grotty, smelly, snotty...
Mel:    Rude.
Joan:   ...rude, obnoxious...
Mel:    Aggressive.
```

These principals' accounts, along with that of the media, paint a rather negative picture dominated by the repetition of particular words: dangerous, rude, and bad. Despite the kinder framing by Lara and Matthew, overall the public image of these boys is negative. Before even meeting them, readers of this chapter will have gained the impression that they are likely to be violent, impulsive, emotional, depressed, aggressive, untruthful, rude, unclean, dangerous criminal menaces.[3] But, again, how accurate are these descriptions? And, how – given the chance – do these young people characterise themselves? The remainder of this chapter engages with these questions by empirically examining the self-characterisation practices of boys from the special schools of which McDougall and these principals speak.

What do the boys have to say?

Thirty-three boys, aged between 9 and 16 years of age (mean 12.5 years) and who were enrolled in special schools for students with disruptive behaviour, each participated in a semi-structured interview. Each of the 33 participants was asked to tell the interviewer about themselves, starting with how they would describe their personality. Eleven boys (one third of participants) either could not or would not respond to the question. Six of these 11 boys either said nothing or replied 'I dunno', two responded by saying that they *could* describe themselves but were choosing not to, two provided answers about what they liked to *do* (e.g. 'watch videogames', 'ride my bike') as opposed to describe what they are *like* as a person, and one mistook personality to mean emotional state (e.g. 'happy').

Two-thirds (22) of the participants correctly interpreted the question and were willing to provide a response. Three of these 22 boys – Max, Aiden, and Charlie – characterised themselves in wholly negative terms. Interestingly, despite the question specifically asking how *they* would describe their personality, each referred to their perception of how *others* perceive them in order to describe themselves. For example, 13-year-old Max, who has received four mental health diagnoses (ADHD, Oppositional Defiance Disorder, anxiety and Asperger syndrome) and who has spent time in acute psychiatric care, replied, 'Well, I know everyone hates me and can't love me'. Aiden, aged 12, was equally bleak:

> Angry. I think some people would call me pathetic while they didn't know that I was there. Annoying, most of the time. Probably - some of them, scary. I don't want to be friends with that guy, stuff like that.

Thirteen-year-old Charlie also pointed to the perceptions of others; however, unlike Aiden, he appeared to draw something positive from being feared:

> People said my personality was quite a few things. Well, because my cousin has got a reputation for himself in my area, people like just call me a drug addict and everything and since I hang out with people from gangs and stuff, no-one really like, everyone respects me and yeah.

The negativity of these responses is perhaps not surprising, given the educational histories of these young people. What *is* surprising, however, is that the majority (19) of the 22 participants who could or would respond to the question provided answers that were mainly positive. While some might dismiss this finding as evidence of the limited self-awareness that is sometimes attributed to children with behavioural disorders (Owens et al., 2007) or a natural reluctance to acknowledge less positive personality traits (Colvin et al., 1995), an interesting pattern emerged upon further analysis.

Ten of the 19 boys who provided a positive response described one or two characteristics, such as being 'funny', 'open', 'nice', or 'friendly', without elaborating much further and without observing any negative qualities. The other nine, however, framed relatively frank self-assessments with a positive statement (see Table 28.1).

The statements of these nine boys demonstrate not just that they are self-aware but also that they are willing to acknowledge the existence of less positive personality traits. Indeed, some of these boys were remarkably candid

Table 28.1 Students who prefaced or framed negative remarks with positive statements

Pseudonym	Positive Framing
Mark	I'm nice *but if people aren't nice to me I can be an arsehole.*
Darrin	Um ... when I want to be, I'm nice. (pause) *I get in a lot of trouble.*
Scott	Uh, I reckon I'm pretty funny. Always have a joke. Uh ... can get along with people pretty good. *I do have a bit of anger but that's like a fuse – got a fuse in my head or somethin' that just goes off...* Pretty easy to get along with. And ... yeah. I don't bag people out, uh, I'm always the one to stand up for someone else. That's how I got into my first fight at this school. Standing up for someone else.
Chris	Um, I'm a lover, not a fighter. I'll put it like that ... Um, I'm easy-goin' *as long you don't get on the wrong side of me. Get on the wrong side of me ... watch out.*
Joseph	I don't know – I can be a good guy *when I want to be.* I'm good at helping people out.
Corey	Sometimes, I can be a bit funny. Um ... Fast. Yeah. *Sometimes, a little bit clumsy!*
Andy	Oh ... easygoing, *can be aggressive* ... I like to joke around. And respectful.
Kyle	I'm a really nice person, *but, like, sometimes, I get kind of angry easily. Yeah, and I get distracted easily. Like, say if I'm talking to someone, like, really important and one of my friends goes 'look over here', I'll look straight away.*
Tristan	*It depends on what type of mood I'm in. Depends who it is.* Like, most of the time, like, sometimes I might be caring *or sometimes might be grumpy.*

in relation both to this question and to other personal questions throughout the interview. However, given the potential for oral language competence to affect whether and how participants' responded (Snow & Powell, 2012), it was clear that this question could not be asked in isolation.

Personality word task

As receptive and expressive language difficulties are common in this population (Ripley & Yuill, 2005), it was anticipated that some boys would not be able to answer an abstract question that asked them to describe their own personality.[4] To address this problem and to enable comparability between participants, a series of word-pairs was devised. The aim was to determine whether participants would choose positive or negative self-descriptors and whether they would be more likely to choose words that aligned with *DSM* diagnostic criteria, such as 'hyperactive' and 'impulsive' (ADHD) or 'argumentative' (Oppositional Defiance Disorder). Additional words were selected from the descriptors that were used by the behaviour school principals to describe their students (e.g. 'rude', 'tough') and words commonly used to describe children with challenging behaviour (e.g. 'difficult', 'defensive'). Words that would too obviously tip the balance towards the negative, such as 'dangerous', 'aggressive', 'bad', or 'mad', were avoided and more subtle alternatives were chosen instead. Where corresponding negatives were not available, an antonym was chosen: for example, outgoing/shy. Words that are commonly used to describe personality such as 'kind' and 'outgoing' were included to provide an equal number of positive alternatives. The final word sheet had a total of 16 words printed in large font and arranged in no recognisable order.

The research assistant read the words aloud to each participant and explained their meanings. Each of the participants was then asked to select five words that they believed best represented their personality. To further check their understanding, each boy was asked why he had chosen that word. Thirty boys chose five words, one could only chose four, another chose three, and the last boy chose two. This boy volunteered another three words of his own: 'friendly', 'intelligent', and 'honest'. There were 159 selections from the personality sheet in total.

Analysis using frequency counts and relative risk ratios shows that the positive word was chosen more often from the eight positive/negative word pairs in all but one case (see Table 28.2).[5]

As shown in Table 28.2, 'determined' was only half as likely to be chosen as 'argumentative'; however, 'out-going' and 'easy-going' were almost twice as likely to be chosen as 'shy' and 'difficult', and 'cheeky' was twice as likely to be chosen than 'rude'. The words 'kind' and 'tough' were chosen almost equally as often, as were 'energetic' and 'hyperactive'. The most popular word-pair was

Table 28.2 Positive/negative word pairings frequency and relative risk

Word Pairs				
Positive	Frequency	Negative	Frequency	Relative Risk Ratio
Determined	6	Argumentative	12	0.50
Easy-going	12	Difficult	7	1.71
Protective	16	Defensive	11	1.45
Cheeky	12	Rude	6	2.00
Energetic	15	Hyperactive	12	1.25
Kind	13	Tough	12	1.08
Outgoing	9	Shy	5	1.80
Cautious	7	Impulsive	4	1.75

'protective' and 'defensive' with eight boys choosing both these words. It was clear from their explanations of why they chose each word that these boys placed a lot of value on being able to stand up for themselves and that they saw value in standing up for others, particularly their mothers. For example, when asked how he was protective, Jared replied, 'I protect my mum from junkies and all that stuff when they try to ask her for money.'

In total, positive descriptors such as 'kind' and 'easy-going' were chosen more often than words such as 'tough' and 'difficult'. Indeed, 'kind' was one of the three *most* popular word choices, along with 'protective' and 'energetic'. The *least* popular word choices were 'impulsive', 'shy', 'rude', and 'determined'. It goes almost without saying that these boys' self-characterisations differ markedly from the ways in which they are portrayed in the media and from the ways in which they were described by their principals. The question is why? Given that research has found that most people will self-enhance to at least some degree, a third interview question was posed. This question was deliberately framed to prompt consideration of negative characteristics.

What would you change?

Once participants had explained their reasons for the personality words they had chosen, they were asked whether there was anything that they would change about themselves, if they could. Only one-third of participants (11) said that there was. Their comments, which are outlined in Table 28.3, indicate again that these boys are both self-aware and willing to acknowledge less positive aspects of their personality. Interestingly, however, their comments also suggest that they tend not to view their emotions or circumstances as being within their control. This is not surprising, given their experiences. Chris, for example, was removed from his alcohol- and drug-addicted mother – whom he

Table 28.3   Characteristics participants said they would change about themselves

Pseudonym	'Is there anything that you would change about yourself if you could?'
Dallas	Stop being bad.
Mark	Just misbehavin' that much.
Darrin	My ADHD, my learnin', readin' and all that 'cos I'm dyslexic. I'd like to learn more, get a job and that. Help mum out. Yeah.
Scott	I'd stop tryin' to argue with everyone if I could. My anger management. If I could, like, take out all the anger and just throw it away or something I would, yeah. Throw it in the bin and burn it.
Aiden	My anger. Because it's the main problem with me.
Chris	I wish that I could go back live with my mum and everything went back, so we didn't have any trouble. Just have a fresh start.
Corey	Err…Clumsy! Because I'm going to get in trouble.
Cameron	My record.
Kyle	If I didn't really get annoyed, like, about anything, that easily. More calm, if the situation is, like not as big of a deal. Like, if I was just more calm…
Emerson	Be polite. (*Anything else?*) Nuh.
Jayden	Change? [Pause] I don't know really, just stop being naughty. (*Why is that?*) Because it gets me nowhere.

still sees and clearly loves – when he was just four years old. Obviously, this is not something that is within his power to change. Neither is Cameron's 'record' or Corey's clumsiness. Further, research has pointed to learned helplessness as a relatively common outcome of special education referral and diagnosis, which might explain why some boys nominate diagnoses as aspects of themselves that they would change 'if they could'.

The majority (22) of the boys however said that they would *not* change anything. Most were resolute in response, providing a firm 'Nuh!' in reply. Relatively few gave more expansive answers, which is not surprising given that a significant number experience expressive language difficulties. The responses of those who did expand in their reply to this question, however, provide a rare insight into why these boys may be choosing positive rather than negative self-descriptions. As per their answers to the very first question asking them to describe their personality, these boys tended to focus on qualities that *they* believed to be of worth, even if those qualities were few or not appreciated by others.

Fifteen-year-old Corbin, for example, who had chosen the words 'argumentative', 'cheeky', 'hyperactive', 'energetic', and 'impulsive' in the personality word test, responded by saying that he 'wouldn't really change anything'. He explained this by pointing to personality traits that he knew were negatively perceived and for which he was constantly getting in trouble at school (being 'cheeky' 'impulsive' and 'hyperactive') but which he believed also helped him to be something that *he* (and his friends) valued: 'funny'.

> I wouldn't really change anything. I love – like, if I change being cheeky, then, like... Basically the only thing that will get people laughing is when they make a comment you're going to say something about, if you're not fast – cheeky fast – if you wait, like, five or ten seconds to say it, then, like, you've lost it. You *have* to be impulsive and cheeky and hyperactive to be funny!

Fourteen-year-old Israel was even more emphatic. After choosing 'argumentative', 'cheeky', 'kind', 'tough', and 'outgoing' from the personality word list, Israel said that being kind was the best part of his personality. When asked if there was anything about himself that he would change if he could, he shook his head and said, 'I'm happy with who I am'.

Correcting the record

This study examined the self-characterisation practices of 33 boys in behaviour schools from three angles. First, the boys were asked to describe their personality. Of the 22 boys who were able to respond to this question, only three were wholly negative in response. The remainder responded with self-characterisations that were mainly positive in nature. Almost half of these 19 boys did acknowledge the existence of negative personality traits but framed these within an overall positive statement. While the latter tendency could be interpreted as support for the assertion that behaviourally disordered boys self-enhance, it does not indicate that they lack self-awareness or that they have an artificially inflated level of self-esteem. Findings from the personality word test cast even more doubt on these hypotheses.

In the personality word test, the positive word was more likely to be chosen from the positive/negative word pairs in all but one case, where 'determined' was half as likely to be chosen than 'argumentative'. Given the value the boys attached to being able to stand up for themselves (and others) elsewhere in the interview, it is debatable as to whether they even view 'argumentative' as a negative word. Also, as 'determined' has more than one meaning, it is questionable as to whether these boys understood what it meant in the context of personality. This may be why it was one of the least chosen words, along with 'impulsive', 'rude', and 'shy'.

Interestingly, impulsive and rude were two of the words used by some of the behaviour school principals to describe their students and these words reflect how these young people tend to be perceived by their teachers. As discussed earlier, much research in this area uses the perspectives of others (including teachers, as well as parents and peers) to judge positive illusory bias, but this research raises the question as to who might be biased and whether the informants know all that is important to know? In other words, while the majority of the participants in this study know that there are some things that they have difficulty with (e.g. anger management, compliance), they also believe that they have some qualities of worth that are worth talking about.

Human beings are complex and multidimensional and boys with challenging behaviour are no different. Yet, they are invariably seen though a one-way prism that is clouded by past impressions, rumour, and prejudice. During interview sessions, I was struck by inconsistencies between what I had been led to expect[6] and the boys' actual behaviour. For example, at the beginning of the interview (typically when asked why they did not like school) a number of boys asked if they could swear.[7] When offered refreshments during the interview and again when we provided them with a movie ticket as compensation for their time, the majority said 'yes please' and 'thank you'. Some would use the word 'pardon' when they didn't hear a question properly. The boys also recounted numerous examples of when they had 'stood up' for someone – often to their own detriment – and how they would try and defend themselves from bullying or from being picked on by their teachers. In short, I saw and heard plenty of evidence to suggest that there was more to these boys than their reputations.

A strong theme across the interview transcripts, however, was that these boys did not feel valued or well-liked by others. It was clear that much of their anger had its roots in the perceptions of others (particularly teachers) and how those perceptions appeared to affect the way that they were treated. Corbin, who achieved a standard score of 117 on the Peabody Picture Vocabulary Test (PPVT) and who was the most articulate of all the participants, encapsulated this general frustration in his account of why he began to dislike school:

> It was just – it was basically a time when I realised that, again, it's not me being bad so much as them kind of pointing you out. If one of the kids, basically, I'm just going to call them the dumb kids, like me and the others, as [the teacher] would call it, if we talked to one of the smart kids in the class, they would suddenly – like, if we talked to *each other*, they wouldn't care but if we talked to one of the smart kids they'd say, 'Stop it. Don't rub any of you onto them.' And they'd basically try to split you up as much as they can. And so the dumb kids just stay back in the corner of the room and we don't matter.

This sense of being disliked was underpinned by a strong sense of injustice and an almost visceral determination on the part of these boys to both value and like themselves. This may well explain why so many said that there was nothing they would change about themselves and why they focused on characteristics they valued. These strategies, however, are negatively perceived in the bulk of the research literature as the capacity to accept one's flaws is viewed as a pre-requisite to change. If, however, we were to analyse what young people are saying and how or why they might be saying it – instead of simply compar-ing whether their perceptions align with the (potentially biased) perceptions of others – we may learn a lot more about these young people, as well as identify areas of common ground from which to build relationships. Currently, how-ever, the positive things that these young people have to say about themselves are discounted as naïve or dissembling.

This is where the value of an alternative lens comes into play. As I discussed earlier, post-structural theories of language have been used to theorise both the constitution of objects (as demonstrated in the first section of this chapter) and the constitution of subjects, which involves the moulding and shaping of individual subjectivity or a person's beliefs, desires, perceptions, and actions. In order to become an acceptable subject, however, those beliefs, desires, per-ceptions, and actions must be consistent with the discourses by which the acceptable subject is called into being (Butler, 1997b). If they are not, the discourse shifts to a language of deviance and the norms are applied with 'redoubled insistence' (Graham & Slee, 2008).

Take, for example, the suggestion that children with disruptive behaviour receive humility training to bring their perceptions into line with those of their teachers, parents, or peers (Gresham et al., 1998). The norms that these 'others' accept and see as being transgressed influence their impressions of the chal-lenging young person. These norms are also transmitted through discourse; a discourse which, in turn, is used to define and describe young people as deviant. Should the young person *resist* that discourse by characterising them-selves in positive rather than negative ways, the dominant view in psychology is that those views ought to be corrected to ensure that the young person becomes aware of the need to work on themselves. However, if viewed through a Butlerian lens, positive self-characterisations could instead be recognised as an opportunity and a starting point; a window into a young person's life and an offer to meet them halfway.

In attempting to correct the record, positive self-characterisations also show promise. First, these attempts suggest that the young person is willing to engage in the difficult discursive work of re-authoring themselves. Secondly, it suggests that despite everything they have experienced and all the negative things that have been said about them, they still believe that effort is worthwhile. Finally, their willingness to engage in relatively sophisticated discursive strategies sug-gests that on some intrinsic level these boys understand the power of language and that identity is shaped and can be reshaped by it. Otherwise, why protest? Why persist with a medium that is not your friend if you do not believe that words carry weight and that they can make a difference?

In my experience with these boys, those who did not have anything positive to say about themselves stand the least chance of remaking themselves in positive form. Those with the *most* chance are the ones who refuse to believe that the negative traits they know and admit they have are the sum of who they are or can ever be. Another important qualification for researchers in this area is that what someone says does not necessarily reflect all that they believe. Therefore, rather than reflect an illusory bias on the part of deluded young people, positive self-characterisations may instead reflect who the young person wants to be and the image of themselves that they wish others had. It is therefore important that professionals in health and education understand the effects that language can have on the ways in which young people are described, the way in which they describe themselves, and why they might insist that they are happy with who they are.

Clinical/Educational relevance summary

This chapter has important implications for the fields of education and mental health. It demonstrates, first and foremost, how the theoretical lenses that we employ shape not only our interpretation of data but also our understanding and treatment of others. The data examined here could easily be interpreted as evidence of positive illusory bias, providing yet more support for a well-mined theoretical concept, but one that positions behaviourally challenging young people as dysfunctional, conniving, and/or helpless. Through the application of an alternative lens – one that considers the strategic use of language as an act of political agency – this analysis shows how positive self-characterisations might instead be perceived as agentive acts; ones that speak to potential strengths and the desire to prevail. Recognition of these different viewpoints is critically important in the fields of education and health because what professionals in each field *think* and *do* will affect how young people are perceived and treated. If we believe that young people are dysfunctional, conniving or helpless, we will treat them with suspicion and condescension, disabling them further. If, however, we believe that young people are capable, agentive individuals who have reason to value themselves – whether they demonstrate behaviour to support that assumption or not – we are more open to perceiving strengths and much more likely to be treated to the opportunity of seeing them. For a simple summary of the implications for practice, see Table 28.4.

Summary

This chapter drew on post-structural theories of language, particularly Judith Butler's (1997a, 1997b) identity politics, as a lens to examine the self-characterisation practices of boys who have been referred to special schools

Table 28.4 Clinical practice highlights

1. Language plays a powerful role in shaping perceptions of self and others.
2. What we think affects both what we say *and* what we do.
3. The application of a post-structural lens helps us to think differently.
4. By thinking differently, we gain a more nuanced appreciation of others and can identify new or alternative ways forward.
5. Strengths-based approaches help to open up opportunities for better understanding and a means to build respectful and more equal relationships.

for disruptive behaviour. Analysis of student responses shows that even boys who experience significant language difficulties employ discursive strategies to reframe the negative characterisations that they know exist. For example, the head principal and her deputy at one participating behaviour school described their 'typical clientele' using words like 'grotty, smelly, snotty, rude, obnoxious and aggressive'. However, student responses to the personality word test reveals that of the 16 words available, 'rude' was one of the *least* popular word choices, along with 'determined', 'shy', and 'impulsive'. In direct contrast to the way in which these boys are portrayed by schools and in popular media, the most common words chosen by the boys were 'protective', 'energetic', and 'kind'. Furthermore, while those who were able to provide spontaneous self-descriptions gave fairly frank assessments of their own character, the majority of boys prefaced their remarks with a positive statement, for example, 'Um, I'm a lover, not a fighter. I'm easy-goin' as long you don't get on the wrong side of me. Get on the wrong side of me...watch out!' The use of such discursive strategies suggests that these boys are well aware of how they are perceived but that they also believe that they have some qualities of worth that are worth talking about.

Acknowledgements

This research was supported under the Australian Research Council's Discovery Projects funding scheme (DP110103093). The views expressed herein are those of the author and are not necessarily those of the Australian Research Council. Ethics approval was obtained from the Macquarie University Ethics Committee (Final Approval No. 5201000654) and the NSW Department of Education and Communities (SERAP No. 2011027).

Notes

1. I secured this report via email to the Department without recourse to a Freedom of Information application.

2. The fifth principal, from a non-government behaviour school, was interviewed but the interview was not recorded.
3. Note, this cocktail of characteristics is one that some researchers point to as early warning signs of psychopathy, prompting our earlier advocate of 'humility training' to coin the term: 'fledgling psychopaths' (Gresham, Lane, & Lambros, 2000).
4. This turned out to be a wise decision. In the larger project from which this data has been drawn, we had three participant groups: 33 students in behaviour schools, 21 students with a history of severely disruptive behaviour still in mainstream, and 42 mainstream students with no history of disruptive behaviour. In addition to participating in semi-structured interviews, each student completed the Peabody Picture Vocabulary Test (PPVT) and the Expressive Vocabulary Test (EVT). We found a significant difference between groups with our 33 behaviour school participants scoring significantly lower on the EVT than students in mainstream (averaging 1 SD below the mean). A core-subgroup of behaviour school participants scored significantly lower on both the EVT and PPVT. For example, three brothers achieved age-equivalent scores that were between six and four years below their chronological age. The majority of participants in this sub-group were unable to answer the question 'How would you describe your personality?' and a number provided either grunts or one-word answers to other questions throughout the interview (see Graham & Buckley, 2014).
5. Relative Risk Ratios (RRR) were calculated to determine whether the probability, or 'risk', of choosing a positive word was greater than choosing a negative word. An RRR of 1 means a positive word choice was equally as likely as a negative word choice. An RRR higher than 1 means that a positive word choice was more likely, whereas an RRR lower than 1 means that a positive word choice was less likely.
6. Particularly by senior public servants from whom I had to gain permission to conduct this research (see Graham & Buckley, 2014).
7. My answer was 'Sure!' I have no problem with swearing.

References

Arribas-Ayllon, M., & Walkerdine, V. (2008). Foucauldian discourse analysis. In C. Willig & W. Stainton-Rogers (Eds.), *The Sage handbook of qualitative research in psychology*, (pp. 91–108). London: Sage.

Butler, J. (1997a). *Excitable speech: A politics of the performative*. London: Routledge.

Butler, J. (1997b). *The psychic life of power: Theories of subjection*. Stanford, CA: Stanford University Press.

Colvin, C. R., Block, J., & Funder, D. C. (1995). Overly positive self-evaluations and personality: Negative implications for mental health. *Journal of Personality and Social Psychology, 68*(6), 1152.

Foucault, M. (1972). *The archaeology of knowledge* (A. M. Sheridan Smith trans.). London: Tavistock.

Graham, L. J. (2007a). Speaking of 'disorderly' objects: A poetics of pedagogical discourse. *Discourse: Studies in the Cultural Politics of Education, 28*(1), 1–20.

Graham, L. J. (2007b). (Re)Visioning the centre: Education reform and the 'ideal' citizen of the future. *Educational Philosophy and Theory, 39*(2), 197–215.

Graham, L. J. (2011). The product of text and 'other' statements: Discourse analysis and the critical use of Foucault. *Educational Philosophy and Theory, 43*(6), 663–674.

Graham, L. J., (2012). Disproportionate over-representation of Indigenous students in New South Wales government special schools. *Cambridge Journal of Education, 41*(4), 163–176.

Graham, L. J., & Buckley, L. (2014). Ghost hunting with lollies, chess and Lego: Appreciating the 'messy' complexity (and costs) of doing difficult research in education. *The Australian Educational Researcher, 41*(3), 327–347.

Graham, L. J., & Slee, R. (2008). An illusory interiority: Interrogating the discourse/s of inclusion. *Educational Philosophy and Theory, 40*(2), 277–293.

Granite, E., & Graham, L. J. (2012). Remove, rehabilitate, return? The use and effectiveness of behaviour schools in New South Wales, Australia. *International Journal on School Disaffection, 9*(1), 39–50.

Gresham, F. M., Lane, K. L., & Lambros, K. M. (2000). Comorbidity of conduct problems and ADHD identification of 'fledgling psychopaths'. *Journal of Emotional and Behavioral Disorders, 8*(2), 83–93.

Gresham, F. M., MacMillan, D. L., Bocian, K. M., Ward, S. L., & Forness, S. R. (1998). Comorbidity of hyperactivity-impulsivity-inattention and conduct problems: Risk factors in social, affective, and academic domains. *Journal of Abnormal Child Psychology, 26*(5), 393–406.

Hoza, B., Gerdes, A. C., Hinshaw, S. P., Arnold, L. E., Pelham Jr, W. E., Molina, B. S.,…& Wigal, T. (2004). Self-perceptions of competence in children with ADHD and comparison children. *Journal of Consulting and Clinical Psychology, 72*(3), 382.

Hoza, B., Pelham Jr, W. E., Dobbs, J., Owens, J. S., & Pillow, D. R. (2002). Do boys with attention-deficit/hyperactivity disorder have positive illusory self-concepts? *Journal of Abnormal Psychology, 111*(2), 268.

Jiang, Y., & Johnston, C. (2014). Co-occurring aggressive and depressive symptoms as related to overestimations of competence in children with attention-deficit/hyperactivity disorder. *Clinical Child and Family Psychology Review, 17*(2), 157–172.

McDougall, B. (2011). Principals shun dangerous students. *The Daily Telegraph*. Retrieved June 3, 2014 from http://www.dailytelegraph.com.au/principals-shun-dangerous-students/story-fn6e0s1g-1226068221363

O'Farrell, C. (2006). Foucault and postmodernism. *The Sydney Papers, 18*(3/4), 182.

Ohan, J. L., & Johnston, C. (2002). Are the performance overestimates given by boys with ADHD self-protective? *Journal of Clinical Child and Adolescent Psychology, 31*(2), 230–241.

Owens, J. S., Goldfine, M. E., Evangelista, N. M., Hoza, B., & Kaiser, N. M. (2007). A critical review of self-perceptions and the positive illusory bias in children with ADHD. *Clinical Child and Family Psychology Review, 10*(4), 335–351.

Ripley, K., & Yuill, N. (2005). Patterns of language impairment and behaviour in boys excluded from school. *British Journal of Educational Psychology, 75*(1), 37–50.

Snow, P., & Powell, M. (2012). Youth (in)justice: Oral language competence in early life and risk for engagement in antisocial behaviour in adolescence. *Trends and Issues in Crime and Criminal Justice, 435*, 1–6.

Yang, L. H., Wonpat-Borja, A. J., Opler, M. G., & Corcoran, C. M. (2010). Potential stigma associated with inclusion of the psychosis risk syndrome in the DSM-V: An empirical question. *Schizophrenia Research, 120*(1), 42–48.

Yates, S., & Hiles, D. (2010). Towards a 'critical ontology of ourselves'? Foucault, subjectivity and discourse analysis. *Theory & Psychology, 20*(1), 52–75.

Recommended reading

- Davies, B. (2006). Subjectification: The relevance of Butler's analysis for education. *British Journal of Sociology of Education, 27*(4), 425–438.
- Laws, C., & Davies, B. (2000). Poststructuralist theory in practice: Working with 'behaviourally disturbed' children. *International Journal of Qualitative Studies in Education, 13*(3), 205–221.
- Youdell, D. (2004). Wounds and reinscriptions: Schools, sexualities and performative subjects. *Discourse, 25*(4), 477–493.

Part VI
Child Mental Health Practice

29

Therapeutic Vision: Eliciting Talk about Feelings in Child Counselling for Family Separation

Ian Hutchby

Introduction

Based on a research project which involved the tape recording, transcription, and analysis of talk between counsellors and young children who were experiencing parental separation or family break-up, this chapter outlines the key discourse practices involved in what can be called the 'therapeutic vision' of child counsellors. Therapeutic vision is a variant of 'professional vision' (Goodwin, 1994), which refers broadly to ways of seeing and understanding events according to occupationally relevant norms. Professional vision tends to involve three types of practice: (1) highlighting certain features of a perceptual field as opposed to others; (2) coding those features according to given, professionally available knowledge schemas; and (3) producing material representations (such as diagrams, graphs, tables, or models) of the salient phenomena.

For example, child counsellors routinely seek to highlight those aspects of children's talk that can be heard to be relevant for family-related or feelings-related matters. The interpretations they produce can be seen as coding events according to specific, counselling-relevant frames or schemas, such as *parents should sort it out; it's not the child's fault; children often get caught in the middle of parents' fights; parents' fights can make children feel angry/sad/guilty*. Finally, the frames themselves are often represented in literary or graphical form, in the context of manuals for child counselling procedure (e.g. Geldard & Geldard, 1997; Sharp & Cowie, 1998), or alternatively, storybooks provided within counselling practices themselves for clients – both children and their parents – to view while waiting to see their counsellor.

Through therapeutic vision, child counsellors deploy what the clinical literature refers to as 'active listening' (Sharp & Cowie, 1998, p. 85), not simply to respond empathically to the child's talk but to formulate that talk in terms that

are therapeutically relevant (Hutchby, 2005). Thus, we find that child counsellors interpret many of the things children say in terms that: (1) refer issues
back to the child's standpoint, that is what they may *think* about given events;
(2) refer issues to the child's subjective experience, that is what *feelings* the child
may have or concerns they may harbour; and (3) refer issues to the child's
relationships, in particular their consequences for the child's association with
their parents. The institutional work of child counselling therefore involves
the counsellor bringing into play events in the child's private or intrapersonal
sphere (thoughts, feelings, emotions, experiences) and 'translating' them into
the public or interpersonal sphere of talk-in-interaction (see also Kiyimba &
O'Reilly, Chapter 30, this volume).

A central problem in that process derives from the fact that while child
counselling, like all forms of counselling and psychotherapy, relies upon
an 'incitement to speak' (Silverman, 1996), that incitement is by no means
straightforwardly complied with by children. As in most forms of counselling
(Peräkylä, Antaki, Vehviläinen, & Leudar, 2008), the willingness of the client
(the child) to produce the kind of 'feelings-talk' that counsellors encourage
varies widely. In the analysis that follows, three conversational sequences are
described that the project found to be significant in counsellors' exercise of
professional vision in this context: (1) formulations; (2) perspective display
sequences; and (3) proffering sequences.

Project overview

The project involved data collection in a British child counselling and family therapy practice where the emphasis was on work with children currently
dealing with the prospect of, or consequences of, their parents' separation. The
child counsellors in this practice did not deal with children exhibiting severe
behavioural problems, who would more likely be referred to clinical psychologists, or at risk of harm, in which case the child would more likely be assigned
to social workers. Instead, they tended to deal with children whose parents
felt that some sort of help was needed in getting the child to come to terms
with the decision they had made to separate. Children were therefore referred
for counselling on a voluntary basis, although the volition was more usually
that of the parents rather than the child. Parents attended an initial assessment
meeting together with their child, but subsequently children were seen on a
one-to-one basis by the counsellor for between four and six sessions. At the end,
another meeting was held in which parents, children, and counsellors would
all be present. At this meeting, the counsellor would present their assessment
of the child and recommendations for future actions by the family members.

Altogether 15 full sessions were recorded on audio tape, including sessions
conducted by both male and female counsellors. Counsellors usually worked

alone with the children, though in one case a male and female counsellor worked together with a group of siblings. The cases included single children and siblings, male–male and male–female. Depending on the counsellors' assessment of behavioural and age dynamics, sibling groups were seen either together or as individuals. The children's ages ranged from 4 to 12 years.

Informed consent (of both parents and children) was obtained by means of a letter, written in styles that differed according to broad age-group categories, which explained some basic facts about the research project, guaranteed that the tape would only be available to the researcher, explained that the participants had the right not to agree or to withdraw at any point, and finally invited them to sign to say that they had read and understood the letter. The process of obtaining the informed consent of parents and children was done prior to the counselling period actually starting at the initial assessment meeting.

The recordings were carried out with the tape recorder in full view of the participants, the procedure being to place the device (a small, battery-operated portable machine) on a table at the side of the room, and situate two small, flat multidirectional microphones in different parts of the room (e.g. one near the armchairs where participants would sit, and one near the toy cupboard from which children would choose games, often at the counsellor's invitation).

Analysis was conducted using the methodology of 'conversation analysis' (CA). CA investigates the sequential organisation of verbal and non-verbal communication as a means of collaboratively organising natural forms of social interaction. As a method, it has been widely applied not just to the analysis of conversation per se but to numerous forms of so-called 'institutional' interaction including medical consultations, interviews, educational work, and counselling (Drew & Heritage, 1992; Heritage & Maynard, 2006; Hutchby, 2007; Hutchby & Wooffitt, 2008; Peräkylä, 1995; Rendle-Short, 2006; Sacks, 1992).

A distinctive feature of this method is that instead of relying on participants' retrospective accounts of interactions derived from interviews, it gathers data of actual interactions as they unfold in real time using video- or audio-recording technology. Through the application of CA, new or previously unnoticed details of the organisation of talk and interaction in the setting can be identified, often at a very fine-grained level. This method is therefore appropriate to the aim to reveal the in situ management of counselling with children.

Analysis

Formulations

Formulations have been described as turns which summarise the gist or upshot of a previous speaker's turn (Heritage & Watson, 1979). This makes them similar to the features of active listening as described by Sharp and Cowie (1998), which centrally include 'summarising and reflecting back the accounts and narratives' of children (p. 85).

A sequential pattern observable in the data is what I will describe as the Question–Answer–Formulation (QAF) sequence. In this sequence, the counsellor begins by asking the child a question – either introducing a new topical focus or following up on a topic already in play – the child then answers the question and, in the third turn, the counsellor formulates the gist or upshot of the child's remarks. In the following extracts, arrows (q), (a), and (f) mark out the Question, Answer, and Formulation turns in each sequence.

Extract 1 is taken from a session involving four siblings: two young brothers and two older sisters. Early on in the session, the children have informed the counsellor (C) that they have recently seen their father for the first time in months and have visited his home:

Extract 1: C07/00.

```
1   q →  C:    How many times ev y'seen 'im since w'last met.
2   a → (D):   °°Two°° =
3   a → (P):   =°°Two°°
4   f →  C:    Twi:ce. Is it- So tw[o Sundays in a row.
5       (P):                       [°Yeh°
6              (1.0)
7        C:    An' er y' gonna see 'im this Sunday.
8              (.)
9        D:    Ah think [so,
10       P:             [Probally.
11             (0.8)
12       P:    Yah.
13             (0.5)
14    →  C:    So::, (1.8) What bits o' that do y' like an' what
15             bits o' that don't y' like.
16             (2.0)
17       P:    Erm, (0.4) the on'y bit I don't like is that cuz
18             we 'ave t'go- (.) cuz we go ev'ry Sunday sometimes we
19             miss out on doing things,
```

In line 1, C begins by asking how often the children's visits have taken place. Two of the children offer the same response in lines 2 and 3, and at the start of line 4, C marks the newsworthiness of this response: 'Twi:ce'. He then appears to embark on the production of a next question ('Is it-') before cutting off and producing a formulation which foregrounds a particular aspect of the news that there have been two visits since the last meeting. With 'So two Sundays in a row', the counsellor marks out the fact that, from no visits at all (in previous meetings a key topic has been the lack of contact the children have with their father), the situation has been transformed into one where, possibly, weekly Sunday visits are now on the agenda. Although there is no verbal acknowledgement or uptake of this formulation in the one second silence at line 6 (though note the agreement produced in overlap with C's turn at line 5), C pursues the newsworthiness marked out in his formulation in his next utterance (line 7), where he asks if the new weekly regime is likely to continue this week. C subsequently seeks to relate that news back to the children's *feelings* about events, when in lines 14–15 he asks, 'What bits o' that do y' like an' what bits o' that don't y' like.'

Thus, while the formulation 'So two Sundays in a row' perhaps seems innocuous, it does interactional work of particular salience for child counselling in the context of family separation, in that it highlights a change in contact arrangements between the children and their father; one in which from virtually no contact they have moved to a situation of potentially regular contact.

In Extract 2, two 'so'-prefaced formulations are used in the process of foregrounding a counselling-relevant piece of information. This time the issue under discussion has associations with imagery frequently deployed in materials aimed at helping children to 'deal with' the experience of parental separation, particularly their responses to the occurrence of arguments between parents, which are depicted as involving increased levels of shouting:

Extract 2: C19/99.

```
 1   q→   C:   How does your mum get your dad to hear what she
 2             wants to say.
 3   a→   J:   Oh she shouts:.
 4        C:   Does your dad hea[r her.
 5   a→   J:               [She shout really loudly cuz
 6   a→        she:'s a teacher and she shouts sort uv .hh she's
 7   a→        got thisuh really lou[d voice ((squeals))
 8        C:                    [.h A:hh.
 9   f→   C:   So she's good at sort of shou[ting] like, like she's=
10        J:                          [Yeh.]
11        C:   =be[ing a teacher.
12        J:      [But she doesn- she doesn't do it in such a high
13             pitched voice. .hh If I did it the building would
14             probably blow up.
15             (0.8)
16        C:   What cuz you've got a high pitched voice.
17             (.)
18        J:   N[o:. Because-    ]
19        C:    [O:r just like yo]ur mum.
20        J:   Because it's:: really loud.
21             (0.3)
```

```
22        C:  And your dad's learning to be a teacher.
23        J:  Mmm.
24        C:  So is he learnin:g to shout [loud too.
25        J:                              [No h(h)e d(h)oesn't
26            shout.
27            (.)
28  q→    C:  Is 'e gunna learn to shout d'y' think like other
29            teachers. (.) Or d'you think he'll always not shout.
30  a→    J:  I don't think he will shout.
31  f→    C:  So he'll be a kind of teacher that doesn't shout.
32            (2.3)
33        J:  He doesn't like telling them off.
34            (0.5)
35        C:  A::h.
36            (1.8)
37        C:  Does 'e shout at you.
38            (0.2)
39        J:  N:::ot mu::ch,
```

The first QAF sequence here comes in the context of a discussion about how J
perceives differences between her mother who, according to J, 'shout(s) really
loudly cuz she's a teacher' (lines 5–6) and her more quietly spoken father. J's
association between the 'membership category' (Sacks, 1972) of teachers and
the activity of shouting is produced in the course of her answer to C's question
in lines 1–2, inquiring about how the mother gets the father 'to hear what she
wants to say'. The association is picked up and formulated by the counsellor in
lines 9–11. We might notice here a slight shift from the child's description of
her mother shouting 'cuz she's a teacher' (line 6) to the counsellor's formulation in which the mother is described as shouting 'like she's being a teacher'
(lines 9–11). In this shift, C's formulation foregrounds the fact that the shouting
is taking place not in the school, where the mother may be acting straightforwardly *as* a teacher, but in the home, where she is characterised as acting *as if*
she is being a teacher.

   The teaching/shouting association is brought into play by the counsellor
again in line 24, where he suggests that learning to be a teacher might lead
to the father 'learnin:g to shout'. In line 25, notably, J declines to extend the
association to her father, focusing instead on his purported trait as someone
who 'd(h)oesn't shout'. At this point, then, C is orienting to shouting as a
'category-bound activity' (Sacks, 1972) for members of the category teachers,
a category-boundedness which J herself had introduced in relation to her
mother. However, in relation to her father, J seeks to establish a separation
between the category-boundedness of this activity and his actual practice as
a category member. In the three turns that follow, we find a QAF sequence

in which this ambivalence is pursued by the counsellor. His question both reiterates the earlier question in line 24 and yet allows for the possibility that the father's 'natural' propensity for non-shouting behaviour may exempt him from the category-boundedness of shouting. J's response takes up the latter possibility, and C's 'so'-prefaced formulation ('So he̱'ll be a kind of teacher that d̲oesn't shout') foregrounds the breakdown of the category-boundedness between teachers and shouting.

Again, then, 'so'-prefaced formulations are used in the course of topicalising an issue with particular salience in the child counselling setting: differences in behaviour – especially aggressive behaviours such as shouting – between a child's parents. In quite a subtle way, C's pursuit of the strength of the bond between 'being a teacher' and 'shouting' is complicit in the child's differential construction of her parents' personalities, with the father, at this stage in the process, being seen in a considerably more positive light than the mother. A final point to note is the way in which, following the second QAF sequence, the counsellor once again indexes the father's lack of shouting directly back to the child in line 37 ('Does 'e shout at yo̲u.').

Perspective displays

The 'perspective display' sequence (PDS), as described by Maynard (1991), involves one speaker seeking a perspective, or opinion, from another in order, in the third turn of the sequence, to produce an opinion of his or her own in such a way as to incorporate the other's viewpoint somehow. In ordinary conversation, this strategy appears best suited to interactions among people who are not closely associated, as it allows caution when expressing views that may turn out to be incongruent. In Maynard (1991), the sequence was described in clinical interactions between paediatricians and the parents of young children who potentially have developmental problems and about whom a diagnosis is about to be announced. The PDS enables the doctor to create an environment in which the eventual diagnosis – which may contain quite devastating news about a child's developmental prospects and future quality of life – can be announced in such a way that the parents' own perspective is to some degree co-implicated in the clinical assessment.

The present project showed that child counsellors also use a type of PDS to invite children's views on things such as the situation at home, having parents living in separate houses, and so on. However, unlike Maynard's study of paediatric consultations, these frequently do *not* lead to cooperative exchanges in which the child's viewpoint can be incorporated by the counsellor, primarily because children frequently resist giving a perspective.

In Extract 3, we see some of the features of the PDS as it characteristically occurs in the child counselling data. (There are four siblings in this session; it is the interaction between the counsellor, C, and the elder sister Amanda, A, that is of primary interest here.)

Extract 3: C07/00.

```
 1   →   C:    So what- what d'you think,
 2       D:    An' we're ha[vin' this teacher called-
 3   →   C:              [Amanda what j'think about goin' t'see
 4             yuh da[d.
 5       D:          [We're havin' this [new teacher called ( )
 6       C:                            [Yer bein' very >quiet=
 7             'old it<shh! (.) shush °a minute shush,°
 8             (0.2)
 9   →   A:    I don't mi:nd really.
10             (0.9)
11     (D):    ku[hh ((cough))
12   →   C:      [Mind really.
13     (D):    kuh hugh
14             (1.2)
15       D:    Mand[y:,
16   →   C:        [Mmm do I sense a bit uv, (.) I'm not so su:re.
17   →         (1.1)
18       D:    ( [   )
19       C:      [Some good bits (.) an' some not suh good bits.
20   →         (0.8)
21       D:    Ple:ase can I ha-
22   →         (1.6)
23       A:    No jus' the same as Pam really like- [(.) y'get=
24       D:                                         [Please c'n=
25       A:    =[t'miss on some-]
26       D:    =[I have a little] (Man[dy)
27       A:                          [.h No:.
28       C:    [[Yih get t' miss out on bit[s.
29       A:    [[Yih get t'miss out-        [Bu- Dan give me back
30             my jui[ce.
31       D:        [( [         )
32       C:          [Da:n, (0.3) Da:n,
33       A:    No::wu[h.
34       P:          [Dan.
```
((Talk continues regarding D's purloining of A's drink))

The counsellor invites A's perspective on the topic of 'goin' t'see yuh dad' in lines 1–4. However, after a short pause, A's response is brief and noncommittal: 'I don't mi:nd really' (line 9). Following this there is a longer pause of almost a second (line 10) before C produces a partial repeat of A's turn (line 12). That partial repeat notably performs a particular operation on the prior turn, recasting it in different terms by shifting the pattern of emphasis. Whereas A placed the emphasis on 'mind' in 'I don't mi:nd really' (line 9), C emphasises 'really' in 'Mind really' (line 12). The effect of this is to transform the perspective from one of mild indifference to one which potentially manifests scepticism or uncertainty about the topic of 'goin' t'see yuh dad'. In other words, C can be understood here to be proffering a version of A's perspective: an interpretation which she herself may or may not wish to go along with.

However, what follows is another silence during which A declines to expand on her viewpoint (line 14), and (leaving out of account for now Dan's interjacent utterances requesting some of Amanda's drink) C subsequently pursues his own perspective on A's feelings about seeing her father. Lines 16 ('Mmm do I sense a bit uv, (.) I'm not so su:re') and 19 ('Some good bits (.) an' some not suh good bits') seek to do this work in the environment of numerous long pauses (lines 14, 17, 20, and 22) during which this alternative perspective is not topicalised by A.

When Amanda does elect to speak again, she begins to produce the pursued expansion on her perspective (lines 23–25); yet, it is noticeable that she does not explicitly align with C's proffered version emphasising uncertainty, but instead with her sister Pamela's view (expressed in a previous exchange) that seeing their father merely means that they sometimes miss out on other weekend events. C then shifts position in an attempt to topicalise this view (line 28) but the line of talk is disrupted at that point by Amanda directing her attention towards Dan who, following his earlier unsuccessful requests (see lines 21, 24–26, and A's self-interruptive refusal in line 27), has taken Amanda's drink for himself. Others in the room, including the counsellor, now also turn their attention towards Dan's actions, and the perspective-display series is abandoned at that point.

Briefly, we see the same pattern in Extract 4.

Extract 4: C:23/99.

```
((P is drawing))
1        C:   What 'ave you written there.
2             (0.8)
3        P:   They're numbers tuh show how-wu- [s:how-
4        C:                                    [To show how far
5             away it is.
6             (6.6)
7    →   C:   What does it feel like havin' the houses so far
             apa:rt.
8             (1.9)
9    →   P:   Don't know,
10   →   C:   Does it feel like this picture?
11            (1.2)
12   →   C:   (It feels) that picture looks, (0.9) a bit sad.
```

```
13   →         (2.2)
14       C:   Does this face ever get happy,
15   →         (3.0)
16       C:   What makes that face happy.=
17       P:   =°(Don't know.)°
```

Here, we find the same components. C produces a perspective-display invitation: 'What does it feel like havin' the houses so far apa:rt' (line 7). Although this utterance refers to feelings rather than to what the child 'thinks' about a certain event (as in Extract 3), it can be understood nonetheless as inviting a perspective on the salient topic of the child's parents living in separate houses a long way apart. As in Extract 3, this is followed by a pause and a non-committal response (lines 8–9). C pursues a feelings-based perspective in line 10, trying to encourage P to conceive of his feelings in terms of the drawing he has made. Following the lack of a response at line 11, C puts forward his own interpretation of the feelings depicted in the drawing ('that picture looks ... a bit sad': see the following section on 'proffering' sequences). Again, however, in subsequent turns that perspective is not topicalised by the child (note the pauses in lines 11, 13, and 15, and the final 'Don't know').

The interactional work accomplished by inviting perspective displays is responsive to three clinical imperatives that inform the professional orientation of child counsellors. First, the aim to construct the counselling arena as a space in which the child should speak from their own subjective standpoint. Second, relatedly, to engage in discourse which avoids leading the child towards particular topics: to engage in active listening but without being *too* active in interpreting the child's talk. Third, somewhat conflictingly, to use their professional (therapeutic) vision, which encourages them to try and topicalise 'difficult' issues in order to help the child to appreciate alternative perspectives that may help them to understand what is going on in their lives. It is this latter imperative, coupled with the highly variable willingness of children themselves to engage in speaking about such issues from their own standpoint, that leads to the perspective display sequence functioning differently in child counselling than in clinical contexts involving adult clients.

Proffering

A phenomenon related to the above types of activity, but which instantiates a more topically directive role on the counsellor's part, is the proffering sequence. Here, child counsellors mention what they are 'thinking', using such phrases as 'I was just thinking' or 'Do you know what I'm thinking?' In many examples, this type of utterance is produced when children are engaged in work with 'materials' – that is drawing, playing with blocks, and so on: something that regularly happens in child counselling. Typically, the counsellor watches the child for a time, maybe discussing with them what they are doing, then initiates a turn with a phrase such as 'I was just thinking...'.

In one sense, such interactional moves seem to go against the professional ideal of the counsellor as a 'neutral conduit' – a facilitator of the client's 'communication' rather than a 'communicator' per se. They seem to introduce or topicalise the counsellor's own thoughts on some matter rather than the thoughts or feelings of the child. When we examine the way these types of utterance function in interaction, it turns out that the 'I was just thinking' type of proffering turn is used to topicalise not the counsellor's but the child's feelings. As in the above types of sequence, however, there are different levels of cooperation that children can accord such interactional moves.

Extract 5 comes from a counselling session with a six-year-old male child whose parents are separated. In this session, the counsellor discovers that the child had been told by one estranged parent that he was going to Disneyland, only to hear from the other parent that he was not allowed to go. At this point, later in the session, he is drawing a picture about his troubled family life:

Extract 5: C:23/99.

```
((Child, P, is drawing))
1      C:   Is that a picture of EuroStar.
2           (0.5)
3      P:   Yeah.
4           (4.5)
5      C:   Cuz I guess you would've gone on EuroStar tuh go to
6           Disneyland wouldn't you.
7      P:   Don't know,
8           (.)
9      C:   °°Mm°°
10          (4.2)
11  →  C:   I wz jus' thinkin' if I could've gone tu:h
12          Disneywor:ld an' someone told me I couldn't I might
13          feel really cross with them.
14          (1.1)
15     C:   .hh Might get really ma:d an' cross an' go RRRAAAAAH!
16          (2.1)
17     C:   It would certainly make me think it wasn't fai::r,
18          (1.8)
19  →  C:   I'd be thinkin' (('child' voice)) 'it'sh not fa::ir.'
20          (4.3)
21     C:   'jus' cuz my mum an' dad aren't ge"in' on um not
22          allowed to go:tuh Disneyworld.'
23          (3.6)
24          It's pretty unfa:ir isn' it.
25          (12.5)
26     C:   Looks like a r:eally good train that t-train's going
27          somewhere. (.) Fas:t.
```

Of note here are the series of utterances by the counsellor, C, from line 11 through to line 23. The first two questions in the extract refer to elements of the drawing P is producing (line 1) and attempt to relate his drawing to the problematic Disneyland issue (line 5). Having received minimal responses showing no uptake of the topic, C then initiates a topic himself (line 11) beginning with the words 'I wz jus' thinkin'...'.

This initiates a series of turns that observably seek to foreground an interpretation of the child's hitherto unspoken *feelings* about the Disneyland trip. That interpretation instantiates the kind of therapeutic vision outlined earlier. It highlights feelings of being 'mad' or 'cross' (line 15) and of things being 'unfair' (line 17, line 24). In lines 21–22, it codifies those feelings in terms of the child counselling frame 'children often get caught in the middle of parents' fights'.

However, the child declines to respond as this account unfolds from the original 'I was just thinking' phrase. This recalcitrance is in line with the general behaviour of this particular child in these sessions (Hutchby, 2002). However, it seems the counsellor seeks the child's concurrence with the version of events he has proffered, in line 24 ('It's pretty unfair isn't it?'). But once again, no response is forthcoming, and following a 12.5-second silence, C ultimately returns to his previous tactic of topicalising elements of the drawing P is producing (line 26).

The upshot here, then, is that while the phrase 'I was just thinking' clearly prefaces a turn, or turns, in which the counsellor presents his own 'thoughts' on the situation, the aim is to present those thoughts as touched off by something the child has done or said, and in such a way that the child is invited at least to go along with, if not actively display uptake of.

In other examples, we find evidence of the more cooperative use of 'I was just thinking' as a means of topicalising therapeutic matters.

In Extract 6, the child J has been engaged for some time in playing with a building game consisting of differently coloured plastic strips and blocks. The counsellor once more has been partly watching and partly engaging in a conversation with J about what she has earlier described as the 'muddle' that is emerging at home as her parents negotiate their separation. J then announces (line 11) that she has completed the model of the 'muddle' that she has been creating and shortly afterwards C embarks on proffering his 'thoughts' about the model.

Extract 6: C:19/99.

```
1  J:  That one was meant- to be on top'v all of them bu:t,
2      .h I put it rou:n:d all of them didn' I.=
3  C:  =So w- sometimes when things go round things they
4      actually, .h end up kind of: .h underneath between
5      an' r:round the sides.
```

```
 6      J:   Yeah.
 7           (5.6)
 8      C:   So sometimes it's kinda hard t'know (0.3) where the
 9           muddle's gonna be.
10           (0.8)
11      J:   There we are!
12           (0.3)
13      C:   Gre:at.='s a good muddle.
14           (1.2)
15      J:   I think I'll leave it at that.
16           (1.2)
17      J:   (I want to.)
18      C:   'S a good muddle.
19      J:   (We) can carry it like that.
20           (0.2)
21  →   C:   D'you know what I w'z jus' thinkin' 'bout this muddle.
22      J:   °hNo.° .h
23           (0.6)
24      C:   I w'z thinkin:: (1.4) there's red, in the middle,
25      J:   Y:ea:h?
26      C:   The smallest one,
27      J:   Yea:h,
28           (0.2)
29      J:   Me:, ehuh=
30      C:   =Jenny!
31           (2.2)
32      C:   Who:'s the purple one.
33           (.)
34      J:   Mum.
35           (0.4)
36      C:   N'who's the green one.
37      J:   Dad!
38      C:   Ye:ah.
```

Here, C uses a question format to package his proposed thoughts about the muddle: 'D'you know what I w'z jus' thinkin' 'bout this muddle'. This way of formatting the turn makes relevant a next turn response from J (line 22), which in turn acts as a means of topicalising C's 'thoughts'. In that way, it is similar to a common strategy used by children more generally in interaction with adults, where prefatory questions such as 'You know what Daddy?', which are designed to be followed by the recipient with 'What?', function to provide clear conversational floor space for the child in a third turn (Sacks, 1975).

Three points are of note in the ensuing sequence. First, the way in which what C is 'thinking' turns out to be a way of linking the representational activity of building a 'muddle' with plastic strips to concrete circumstances in the child's home life. Second, the way that this linkage is achieved incrementally (in a similar sense as the incremental unfolding of the 'imagined emotional state' in Extract 5). Third, most significantly, the way that the child herself is invited to collaborate in developing the linkage, to such an extent that it is, in fact, J herself who explicitly verbalises the connections.

C begins, in line 24, simply by picking out one of the colours and observing that it is 'in the middle'. In line 26, he adds another observation, 'the smallest one', which J initially responds to just with a continuer (an invitation to 'go on'). However, after the short pause in line 28, she makes the link between 'in the middle' and 'the smallest one' (two conventional means by which young children are often depicted in family stories) and positions herself as represented by the red strip (line 29, confirmed in the counsellor's latched next turn). From there, counsellor and child collaboratively use the membership categorisation device 'family' (which children are able to use from a very early age) to enable J to position the two colours on either side of red as representing 'Mum' and 'Dad'.

Slightly later in this session, C again uses a 'just thinking' turn, but this time to *contrast* with the possible family future that the child is imagining.

Extract 7: C:19/99.

```
1        C:   How would this muddle be:, (0.5) whe:n, (.) mum an'
2             dad aren't living together.
3             (1.9)
4        C:   What will happen.
5             (1.4)
6        C:   T'your pu[zzle
7        J:            [THIs:::::. (0.2) would happen.
8             (2.0)
9        C:   Aha: so who's tha[t. Who've you just taken over there.
10       J:                    [No no no wait. Wait wait this
11            wouldn't happen, (.) I'll show you what would (happen)
12            (0.8)
13            ((BANG))
14            (0.3)
15       J:   (Wait) I haven't finished yet,
16            (2.0)
17       C:   S:so there's, (0.6) Ah ↑ka:y.
18            (3.5)
19       C:   Pulling apart.
20            (0.7)
21       J:   (Yeh they) break hih!
22            (2.1)
23  →    C:   Ah that's your >worry<=.h=But I was jus' thinkin',
24            (0.3) that, maybe this would happen. (.) Instead of
25            them being pulling apart breaking, .hhh (.) I like
26            the way y'did this y'put, .hh that green one over
27            the:re, .h an' that purple one over there cuz they're
28            bo:th .hh having a- (0.3) some ti:me with Jenny,
```

```
29        .hh but instead a' pulling they're pushing like this.
30        (0.2) s'they push together.
31   J:   Why::?
32        (0.4)
33   C:   (S'we just move this.) So they're not breaking. But
34        they're both looking a:fter'n, hugging Jenny.
```

Having been asked to speculate on what may happen once her parents finalise their separation, J complies by moving the coloured plastic strips (lines 7–8). Noteworthy in C's response (line 9) is the continued anthropomorphisation, as he refers to 'who' has been moved 'over there'. J, however, changes her mind in the course of this turn and, as her talk in lines 10–21 demonstrates, produces a different manipulation of the pieces which is ultimately characterised by C as 'Pulling apart' (line 19) and, more strongly, by J, '(Yeh they) break hih!' (line 21).

It is at this point – that is, where a fairly strong imagined future of the family 'breaking' has been introduced by J – that C produces another 'just thinking' turn. This time, he uses the disjunct marker 'but' and a stress marker on the first-person pronoun ('But I was jus' thinkin'', line 23) to foreground that his own thought is somehow different from that propounded by the child. As we see, he subsequently goes on to produce an alternative possible future in which the two parents, rather than 'breaking', are 'both looking a:fter'n, hugging Jenny' – a far more positive imagined future that the child is thereby invited to consider.

In general, thought-verbalisations such as these seek to respond to activities such as drawing a picture or playing with plastic strips in a way that makes the activity 'representational' – that is a representation of some aspect of the child's life as seen through the lens of therapeutic professional vision.

As part of this, the thought-verbalisation often involves an intricate imputation and adoption of identities and stances involving the counsellor and mutually available features of the child's work with materials. For example, a drawing of EuroStar is associated with the disappointment of not being taken to Disneyland, and the thought-verbalisation involves the counsellor taking on the standpoint of the child. A muddle of differently coloured strips is associated with the muddle of a family in process of separating, and the thought-verbalisation involves the counsellor (and, in the above case, the child too) imputing familial identities to the different strips.

Clinical relevance summary

Counselling practitioners, especially perhaps those who work with children, are already sensitive through their professional training to the effects of different styles and tropes of language use. However, a conversation analytic approach can reveal to practitioners potentially useful knowledge about how their chosen

strategies for engaging the child and exploring the child's feelings actually operate within the interactional dynamics that characterise child counselling as a
form of institutional discourse.

The three sequence designs outlined above are among the most significant
ways that child counsellors in this data corpus manage the contingencies of
therapeutic communication in the particular context of their professional work.
That professional context is characterised by two interrelated tensions. First,
the talk produced by counsellors places children within the social category of
'child' in as much as they are viewed in relation to their parents, whose actions
have consequences for them *as* children; yet, children are also invited to speak
in ways that are *outside* the normative parameters of 'childhood' as it tends to
operate in the context of child–adult interaction, for example to speculate on
their parents' reasoning, articulate their own feelings and responses to their
parents' actions, or develop proposals for how their parents can improve the
situation.

Second, counsellors' professional ethos and training encourages them to
place the child's 'story' at the heart of their work and avoid leading the child
or judging their words or actions. However, this requirement is problematised
by the variability in children's willingness to speak in ways that communicate
appropriately about emotions, feelings, and concerns. By the same token, counsellors' orientation to the counselling session as an environment where such
communication should ideally occur provides grounds for their often seeking
to topicalise emotions, feelings, and concerns even where these are not overtly
topicalised in children's own talk.

In that context, conversation analytic studies can reveal hitherto unnoticed
details about the patterns and structures that function within counselling dialogues, and about how different types of turn design operate within sequences
to enable more or less successful counselling to take place. Although the study
reported in this chapter was comparatively small-scale, it is perfectly feasible for
large-scale studies to be undertaken that evaluate the relative success of forms
of talk in relation to professional outcomes across a wide range of different
practice settings.

In addition to their usefulness for individual practitioners, the findings of
conversation analytic studies can have relevance for the wider clinical community and for the authors of clinical literature and guidelines for good
practice. For example, we have seen how 'active listening' can, though discourse
practices such as formulation and perspective display invitation, function in
distinctive ways that go beyond the recommendation to engage in empathic
and responsive interaction with the child. And we have seen how practices
such as 'proffering', which initially seem to take the counsellor beyond the
recommendation to avoid leading or topically directing the child's narratives,
can function in a positive way to facilitate consideration of alternative realities.

Professionals may therefore use studies such as this one to reflect upon and potentially modify their guidelines for practice. For a simple summary of the implications for practice, see Table 29.1.

Table 29.1 Clinical practice highlights

1. Counsellors working with children can become more informed about how their discourse techniques operate in actual practice.
2. Counsellors working with children can understand more about the interactional dynamics that feature in their professional work.
3. Counsellors working with children can see in more detail how different turns and sequence-designs offer responses to interactional problems that may occur in their professional work.
4. Clinical professionals and the authors of counselling literature can benefit from the analysis by using it to reflect upon, and possibly modify, their guidelines and/or professional vision.

References

Drew, P., & Heritage, J. (Eds.) (1992). *Talk at work*. Cambridge: Cambridge University Press.

Geldard, K., & Geldard, D. (1997). *Counselling children: A practical introduction*. London: Sage.

Goodwin, C. (1994). Professional vision. *American Anthropologist, 96*, 606–633.

Heritage, J., & Maynard, D. (Eds.) (2006). *Communication in medical care*. Cambridge: Cambridge University Press.

Heritage, J., & Watson, D. R. (1979). Formulations as conversational objects. In G. Psathas (Ed.), *Everyday language* (pp. 123–162). Mahwah, NJ: Lawrence Erlbaum Associates.

Hutchby, I. (2002). Resisting the incitement to talk in child counselling: Aspects of the utterance 'I don't know'. *Discourse Studies, 4*, 147–168.

Hutchby, I. (2005). Active listening: Formulations and the elicitation of feelings-talk in child counselling. *Research on Language and Social Interaction, 38*, 303–329.

Hutchby, I. (2007). *The discourse of child counselling*. Amsterdam/Philadelphia: John Benjamins.

Hutchby, I., & Wooffitt, R. (2008). *Conversation analysis* (2nd edition). Cambridge: Polity Press.

Maynard, D. (1991). Interaction and asymmetry in clinical discourse. *American Journal of Sociology, 97*, 448–495.

Peräkylä, A. (1995). *AIDS counselling: Institutional interaction and clinical practice*. Cambridge: Cambridge University Press.

Peräkylä, A., Antaki, C., Vehviläinen, S., & Leudar, I. (Eds.) (2008). *Conversation analysis and psychotherapy*. Cambridge: Cambridge University Press.

Rendle-Short, J. (2006). *The academic presentation: Situated talk in action*. London: Ashgate.

Sacks, H. (1972). On the analysability of stories by children. In J. Gumperz & D. Hymes (Eds.), *Directions in sociolinguistics* (pp. 325–345). New York: Holt, Rinehart and Winston.

Sacks, H. (1975). Everyone has to lie. In B. Blount & M. Sanchez (Eds.), *Sociocultural dimensions of language use* (pp. 57–80). New York: Academic Press.

Sacks, H. (1992). *Lectures on conversation.* Oxford: Blackwell.
Sharp, S., & Cowie, H. (1998). *Counselling and supporting children in distress.* London: Sage.
Silverman, D. (1996). *Discourses of counselling.* London: Sage.

Recommended reading

- Heritage, J., & Maynard, D. (Eds.) (2006). *Communication in medical care.* Cambridge: Cambridge University Press.
- Hutchby, I. (2007). *The discourse of child counselling.* Amsterdam/Philadelphia: John Benjamins.
- Hutchby, I., & Wooffitt, R. (2008). *Conversation analysis* (2nd edition). Cambridge: Polity Press.
- Silverman, D. (1996). *Discourses of counselling.* London: Sage.

30
Parents' Resistance of Anticipated Blame through Alignment Strategies: A Discursive Argument for Temporary Exclusion of Children from Family Therapy

Nikki Kiyimba and Michelle O'Reilly

Introduction

In this chapter, we utilise a discourse perspective to explore ways in which parents manage therapeutic alignment in family therapy. As therapy is an activity which relies heavily on the use of language (McLeod, 2001), we use a language-based analytic approach to explore child mental health, particularly as discourse analysis is most appropriate for looking at family therapy processes (Roy-Chowdhury, 2003). In this chapter, we present a case for the deliberate *temporary* exclusion of children in the initial stages of a series of therapeutic sessions. The purpose of this temporary exclusion is to provide opportunities for therapists to engage in active solution-focused alignment with parents in order to provide a foundation and set boundaries for later work with the whole family. We also argue that while this initial session with parents is taking place, the child could be otherwise engaged in a session of their own so that the child's perspective and expectations are also managed effectively.

We argue that the benefits of temporary exclusion relate to mitigation of the potentially competing accounts of parents and their children. First, it offers a forum where parents have the opportunity to provide an overview of family life from their perspective. Second, it affords the occasion for the therapist to negotiate therapeutic boundaries of appropriate conversational topics and ways of communicating when children are present. Third, it allows therapists to develop a framework with the parents which moves away from a 'blaming culture' towards helping them to work together to facilitate a solution to the family's problems.

In order to achieve this, we have drawn upon three of our own empirical papers to illustrate the benefits of purposeful temporary exclusion early on. In that research we worked with naturally occurring family therapy data and use this as the basis of our argument. Data used illustrated that where unreflective inclusion occurred, parents typically positioned the child as the problem in a negative and derogatory way, they 'gossiped' about their children as if they were absent and talked about topics which could be considered developmentally inappropriate for children to hear.

Family therapy

Any family is a unique interpersonal system with its own family identity, with external relationships connecting the family to the outside world (Dallos & Draper, 2010). However, families are not straightforward institutions and the family can experience problems (Hutchby & O'Reilly, 2010). Family therapy has been developed as a way of intervening in troubled family systems in order to mediate difficulties and collaboratively promote solutions. It provides valuable ways of treating children and their families together. Family therapy was built on a foundation of systemic thinking which understands the family as an interrelated system affected by all of its members (Ruble, 1999). From its historical roots which began converging in the 1950s, family therapy has evolved into a number of different strands which have been influenced by a range of psychological approaches (Goldenberg & Goldenberg, 2008). Modern family therapy is not a homogenous approach to treating families and broadly speaking three main types of family therapy exist – psychodynamic, behavioural, and family systems (Nichols, 1996). Notably, a social constructionist approach to family therapy has also became popular in contemporary culture (Goldenberg & Goldenberg, 2008).

A primary difficulty for families is a breakdown of communication between members, and therapy can be a useful way to mitigate modes of dysfunctional communication (Foley, 1974). Often the main challenges for family therapy are the initial engagement of all family members and maintenance of that engagement to avoid dropout. This is particularly true in the first meeting where there may be a sense of apprehension (Dallos & Draper, 2010). There are therefore four important considerations for therapists when engaging families; first, there is the potential for upsetting existing coping strategies; second, families may feel blamed; third, there may be a fear of change and; fourth, therapy can often be experienced as an additional stress (Withnell & Murphy, 2012). Young people in particular are especially difficult to engage in the therapeutic process (Thompson, Bender, Lantry, & Flynn, 2007) and children and adolescents dropping out of services are a significant problem (Kazdin, Holland, & Crowley, 1997). Typically, it is assumed that once the family members are initially engaged in the therapeutic relationship, they will continue to engage in therapy successfully; however, the process of engagement necessitates continual efforts on the part of the therapist (Withnell & Murphy, 2012). In addition, it is also often assumed that because children are present, they are engaged. However, their physical presence does not necessarily guarantee their participation and without this participation the family therapy may be ineffective (O'Reilly & Parker, 2013a).

Including and excluding children from therapy

Nonetheless, the literature on family therapy highlights that there is a general consensus that it is preferable to include children in family therapy sessions. One of the main reasons for this is that the therapist can witness the family dynamics which provide an understanding of the family processes (Hutchby, Chapter 29, this volume; Miller & McLeod, 2001). In addition, relationships within the family are an important source of engagement in the process with the development of connections between children and their parents facilitating their ability to solve problems (Thompson et al., 2007). An important part of family therapy is to gather the perspectives of all family members in relation to the topic in question. The format of therapy provides an environment that encourages children to communicate things that may be difficult to discuss (Hartzell, Seikkula, & von Knorring, 2009). Therapists are therefore in a better position to assess the full extent of the difficulties including the child's perspective (Hawley & Weiss, 2003). However, the other side of the argument presented in the literature has been that the inclusion of children in therapy has the potential to harm the child as the content of sessions may be inappropriate (Miller & McLeod, 2001).

This has led to an unhelpful polarisation of the argument between either including or excluding children from therapy. Including or excluding children from therapy should not be viewed as an 'either/or' option for therapists. Instead, therapists should make active decisions regarding whether or not to involve children, how long that inclusion may last, and how to effectively engage the child when present (Johnson & Thomas, 1999). These decisions need to be handled sensitively, as particular exclusions may be in the best interests of the child where the content of the discussion is of more of an adult nature (Ruble, 1999). However, family therapy research has not fully explored the specific issues that warrant the exclusion of children from therapy (Sori & Sprenkle, 2004). We therefore sought to engage with this question empirically in order to provide guidance for therapists regarding making these sensitive decisions. We note that the evidence that we present builds a case for the temporary purposeful exclusion of children from the initial family therapy session, but it is beyond the scope of the data that we have based our analysis on to make any broader claims about other points in therapy where exclusion may also be appropriate.

Thus, we explored the occasions on which it appeared to be more appropriate to temporarily exclude children from initial sessions for the benefit of overall progression. Our argument is that from the empirical evidence we have provided, it appears that when children are present, attempts by parents to perform blaming or accounting can be done in ways that are inappropriate or possibly damaging for children to witness. We contend therefore that there is potentially some benefit to temporary initial exclusion so that parental concerns can be expressed fully and that the therapist is afforded an opportunity to explain how therapy works and demarcate some boundaries and 'rules' about what might be appropriate to talk about when the children are included. This strategy has the potential to manage expectations and reconcile potential misconceptions. Additionally we argue that in order that the child is not made to feel that their exclusion is punitive, there may be benefits to organising a parallel child session.

Project overview

There is a tendency of researchers to objectify participants through research designs that are founded upon professional rhetoric which locates families as passively receiving treatment, rather than being actively engaged in the process (Roy-Chowdhury, 2003). In this chapter, we have moved away from such models of research to explore in detail how families made sense of their social worlds. We acknowledge that all members in family therapy were actively engaged in the complex process of generating meaning (Roy-Chowdhury, 2003) and we explored how this was achieved interactionally.

The setting

The data drawn on for this chapter were a corpus of video-recorded, naturally occurring family therapy sessions, which were provided by a family therapy centre based in the United Kingdom. These video-recorded sessions were routinely collected as part of reflective practice. Approximately 22 hours of therapy were recorded for research purposes, from four families who provided informed consent. Their details are provided in Table 30.1. The family therapy team were practising systemic family therapists who specialised in working with families of children with mental health difficulties, and two therapists were included in the data set. The data were transcribed in accordance with the analytic method, with Jefferson guidelines being followed (Jefferson, 2004). Our sampling was in accordance with the discursive epistemology and issues related to saturation are not intrinsic to the analytic method (O'Reilly & Parker, 2013b).

Table 30.1   Family demographics (pseudonyms applied)

Family	Description
Clamp family	The **Clamp** family consisted of two parents, **Daniel** and **Joanne**; one male uncle, **Joe**; and three children: **Phillip** ('special needs'[a]), **Jordan** ('handicapped'), and **Ronald** (learning difficulties).
Niles family	The **Niles** family consisted of two parents, **Alex** and **Sally**, and four children (one with a pending diagnosis): **Steve** (undiagnosed...suspected ADHD), **Nicola**, **Lee**, and **Kevin**.
Webber family	The **Webber** family consisted of two parents, **Patrick** and **Mandy**, and four children (one with a diagnosed disability): **Adam**, **Daniel** ('special needs'), **Patrick**, and **Stuart**.
Bremner family	The **Bremner** family consisted of the **mother**, the **grandmother**, and two children: **Bob** (autistic spectrum disorder) and **Jeff** ('mentally handicapped').

[a] Terms describing the children (e.g. handicapped) are the terms used by the families themselves.

The analytic approach

We recognised that there are a range of qualitative approaches suitable for the study of family therapy, and that process research is important in the field, as process research is concerned with understanding the process of how therapy works, as opposed to focusing on outcomes (Dallos & Draper, 2010). Of particular relevance is the qualitative method of discourse analysis, as this is methodologically congruent with family therapy theory and practice (Roy-Chowdhury, 2003). Thus, discourse approaches are well suited for the research questions that are of interest to practising family therapists (Tseliou, 2013).

In the broad field of discourse analysis, there are a number of different discourse approaches, and we align with Discursive Psychology as pioneered by Edwards and Potter (1992), as it has the benefit of drawing upon conversation analysis for a micro-analysis of the details of talk-in-interaction. This form of discourse analysis is committed to studying talk as social action and has an emphasis on the rhetorical organisation of language (Potter, 2004). The advantage of utilising a discursive approach which leans on the principles of conversation analysis is that it allows the analysis of the moment-by-moment accomplishment of interaction. We drew upon the discursive approach aligned with conversation analysis as it provides a framework to interrogate the detail of social actions as they occur in context and provides the opportunity for an investigation of process and function rather than merely content (Parker & O'Reilly, 2012).

The core advantage of using discourse and conversation analysis is that it provides an opportunity for a systematic analysis of the sequential aspects of the interaction within the family therapy setting (Parker & O'Reilly, 2012).

This is accomplished by the preference in discourse and conversation analytic studies for detailed transcription of sessions (Avdi & Georgaca, 2007). This therefore enables the analyst to view the contribution of each participant within the therapy from their respective positions (Roy-Chowdhury, 2006).

By examining the actual talk of family therapy sessions, it is possible to examine the adequacy of existing theoretical accounts regarding the therapeutic relationship (Roy-Chowdhury, 2006), and this is consistent with the epistemological requirement to utilise naturally occurring data for examination (Potter, 2004). Researchers engaging with discourse approaches seek to analyse the talk of participants in their actual settings and talk is seen as performative in generating and attempting to sustain versions of reality through appeals to rhetorical devices such as facts, common sense, or natural order (Edwards & Potter, 1992; Roy-Chowdhury, 2003). Thus discursive approaches are valuable tools in the exploration of constructive subjective realities that are locally produced within institutional settings (Tseliou, 2013).

Ethics

The project followed the deontological approach to ethics: respect for autonomy, provision of justice, ensuring beneficence and non-maleficence (Beauchamp & Childress, 2008). In practice, the core ethical principles were applied. Informed consent was collected from managers, therapists, and families, the right to withdraw was ensured, and the data were respected and protected.

Analytic overview

Through analysing the process of family therapy, we explored what may constitute an appropriate rationale for the purposeful temporary exclusion of children at specific junctures. It appeared that the institutional framework of family therapy predisposed parents to hold an expectation of being called to account for the behaviour of their children. Not least because historically parents (particularly mothers) have been blamed for the 'deviant' mental health–related behaviour of their children (Sommerfeld, 1989). From our data it was evident that family members used a number of discursive strategies in what appeared to be a social action of resistance against possible blame. First, resistance to blame and accountability were demonstrated by attempts at alignment with the therapist and active dis-alignment from the problems their children were reported to have. One of the ways that parents performed dis-alignment was through the use of derogatory descriptions of their children. It seemed that the institutional frame of therapy shaped the trajectory of the conversations which occurred within it and influenced the nature of reciprocity between the family and the institution. Second, resistance was demonstrated by the children in particular through passive and active disengagement with the therapeutic process. Third, resistance was managed in an iterative and reflexive way by the therapist through collaborative inclusion of all members as they moved towards therapeutic goals.

Alignment/(dis)alignment: negative descriptions of children

In the absence of having an initial session between the parents and therapist where the children are excluded, research has illustrated that parents regularly took responsibility for putting forward their perspective regarding reasons for attendance in therapy. Problematically, this presentation of reasons tended to be addressed in front of the children in a way that was potentially counterproductive. Additionally, parents used negative descriptions of their children in front of them including derogatory terms, negative dispositional inferences, and ascriptions of biomedical aetiology.

Extract 1: Niles family (from O'Reilly & Parker, 2013a, p. 498)

```
Steve:      I want to ↓go
Dad:    No (.) we're 'ere to get you sorted out kid (0.2) I
        reckon bo:ot (.) >boot camp< will sort you out
```

Extract 2: Webber family (from O'Reilly, 2013, p. 6)

```
Mum:   He's a slob (.) 'e's a com↑plete slob >an' 'e's a
       kleptomaniac<
```

Extract 3: Niles family (from Parker & O'Reilly, 2012, p. 464)

```
Dad:   <we've got t' sort> (.) o:r get some medication or somet
       t' calm 'is temper ↓down (.) cuz 'e's ↑schizo
```

In Extract 1, the child was clearly constructed as the problem. Positioning the child as 'needing sorting out' is a discursive strategy for managing parental identity as non-blameworthy (O'Reilly & Parker, 2012). In effect, this mitigated against potential inferences that the child's behaviour was more resulting from 'nurture' rather than 'nature'. Extract 2 exemplified parental ascriptions of children's problems as dispositional rather than behavioural. Again this positioned the problem as intrinsic to the child themselves, rather than a learned behaviour resulting from a parental inability to sanction socially unacceptable behaviour. There is a normative social and cultural expectation that parents have a moral accountability to discipline the deviant behaviour of their children to ensure that they fulfil a productive and acceptable role in society. Thus any failure of children to conform to those expectations is seen as a reflection of parental failure. In therapy, therefore parents appealed for pharmacological treatments for what they constructed as a biomedical problem. In

Extract 3, the parents positioned the child as 'schizo', a colloquial reference to schizophrenia which is understood to be an organic mental illness. In doing so, they worked to provide some distance between their parenting skills and the child's behaviour by implying the need for 'some medication or somet' as the appropriate intervention.

Clinical relevance

While it is recognised that these social actions are bound up with an over-all appeal to accountability and responsibility, the impact on the child has the potential to be quite damaging. This may put the therapist in a difficult position because if they challenge the parental ascriptions, there is a risk of undermining the parental authority, which may impinge on the systemic ethos of the family unit. Additionally, there is a risk of rupture and potential disengagement of parents from the therapeutic process if therapists are to overtly challenge. However, if they fail to address derogatory comments about the co-present children, the therapist in effect colludes with parental accounts and risks not only disengaging the child but also reinforcing the negative message.

We propose that a solution to these difficulties would be an initial temporary exclusion of the child from the first session. While in some ways this process may seem to be counter to the ethos of family therapy, we argue that some of the significant difficulties that we have observed empirically when children are present in the first session could be avoided. There are several potential benefits of allowing the opportunity to consult separately with parents and child(ren) initially. First, it provides space for addressing and managing parental expectations. This is particularly important as some parents may have unrealistic expectations and may disengage if they feel those expectations are not being met (Withnell & Murphy, 2012). Second, it affords a chance for the therapist to manage and allay any fears of being blamed that the parents may feel. Third, this allows the talk to unfold, providing an opportunity for the therapist to reconceptualise the problems in a more systemic way. Finally, it allows the therapist to set some parameters for future sessions regarding the appropriateness of the content of particular topics. Thus, it is important that a safe culture is created for parents and children, with parents feeling respected (Rober, 1998). While this first session is ostensibly parent focused, we argue that the child should be also afforded the opportunity to express themselves freely without their parents. This is particularly important as research has indicated that children find the presence of their parents inhibiting (Strickland-Clark, Campbell, & Dallos, 2000). Thus a session alone with the child could be beneficial, and this could take place concurrently with another member of the family therapy team. An advantage of offering parallel sessions would be that the potential problematisation of the child which may be incurred through mere exclusion could be avoided.

Engagement and resistance

It has been recognised that it is important to agree the framework of the trajectory, parameters, and expectations at the start of therapy. However, due to the dynamic nature of family interactions during the course of family therapy, the issue of engagement should be treated iteratively. Thus, while separate first sessions for the parents and children can be facilitative in managing expectations and blame, in isolation this is unlikely to be sufficient as therapy is a reflexive and iterative process. It is important to additionally manage ongoing sessions and for therapists to recognise and deal with indicators of resistance and disengagement, before they lead to therapeutic rupture and dropout. In our data, we noted that this tended to occur incrementally: with children initially demonstrating signs of resistance to parents' versions; then with children passively disengaging from the therapeutic conversation; and ultimately with children actively disengaging from the therapeutic session.

Extract 4: Niles family (from Parker & O'Reilly, 2012, p. 466)

```
Dad:      ↑Show Joe yer arm then
7 lines omitted
Steve:    there's nothin' th::ere
Mum:      Don't tell lies
Dad:      Looks like 'e's tried t' scratch the name o::r somethin'
          in 'is arm
Steve:    NO I ain't
```

Extract 5: Clamp family (from O'Reilly & Parker, 2013a, p. 396)

```
Dad:          I don't think Jordan understands what you're
              on about either (.) to be honest
FT:       Yeah
Dad:      I think Phil[lip(        )
Ron:                 [Heh h[eh hehheh ((Ron is jumping))
Jordan:              [hehhehhehheh ((Jordan is
          jumping))
Dad:      ↑Will you stop jumping
```

Extract 6: Bremner family (from O'Reilly & Parker, 2013a, p. 499)

```
FT:    ↑So (.) will you >come back again< (.) and see me again
       in fo:ur weeks?
Bob:   No
FT:    ↑Oh I think ↑so
Bob:   I will not
```

Extract 4 demonstrated that at this point in the therapy the child was an active participant, fully engaged in the therapeutic conversation. However, when an accusation was made by his parents that he deliberately scratched his arm, the child resisted the accusation that he would engage in this type of 'deviant' behaviour, stating 'there's nothing there'. When the parents made a direct allegation that he was telling 'lies', this precipitated a direct denial in order to mitigate blame and responsibility 'NO I ain't'. The first line of Extract 5 indicated that at this point in the session the children were being talked about rather than included in the conversation. Notably, during this talk the children were passively disengaged from the session evidenced by the fact that they were laughing and jumping on the chairs. However, if left unaddressed, children's disengagement may become more active as demonstrated in Extract 6. At this point, the child verbalised a clear message of not wishing to return to therapy 'No' and 'I will not'.

Clinical relevance

In studies of talk-in-interaction, it is argued that resistance is realised through acts such as disagreement, disconfirmation, rejection of advice, or the failure to provide a relevant answer (Muntigl, 2013). Our analysis has shown that children used these same discursive strategies in the context of family therapy as ways of resisting and disengaging. This resistance and potential disengagement was illustrated through various social actions. First, a challenge for therapists was the problem that there were competing versions of events from different family members. Second, it was typical for parents not to engage children in a constructive way by dis-aligning with them. Third, although children may be present in the session, they were not necessarily engaged with the therapy. Finally, children may disengage if the therapist failed to form or maintain a therapeutic alliance with them.

We propose that our assertion to temporarily exclude children from the initial session is a pragmatic solution to these difficulties. This affords the opportunity for the therapist to initiate an alliance with the parents in a way that facilitates the development of trust between them. This has the potential to promote a sense of security for the parents which may alleviate to some extent their anxieties about being judged. With careful attention on the part of the therapist to the concerns of the parents, there is a higher probability that parents will feel less undermined when attempts to fully engage the child's perspective are taken. This is because their version of events has already been heard and validated. This should lessen the likelihood of parents competing with children's versions when children are reintroduced into the therapy. This process could be facilitated by the therapist agreeing in the initial session with the parents that children's versions should be listened to without criticism. Despite this effort, ruptures in those later sessions may still occur and require careful management to re-establish therapeutic alliance. It is useful therefore if therapists are attuned to the subtle indications of potential ruptures and take some initiative when a suspected rupture has occurred (Safran, Muran, Samstag, & Stevens, 2001). Where potential ruptures are not appropriately attended to, they are likely to remain unresolved which may lead to premature treatment termination (Aspland, Llewelyn, Hardy, Barkham, & Stiles, 2008).

Progression in family therapy

The process of family therapy is gradual and incremental in terms of positive changes within the family system. It is important to recognise that families are expected to engage with the therapeutic process over a particular agreed time period. It is possible therefore that some of the parental insecurities and concerns of being accountable for the behaviour of their children will re-emerge during this process. It is important that therapists have an appreciation that the initial exclusion of the child will be insufficient to ensure that parental alignment is consistently maintained. As reflective practitioners therefore, therapists should remain mindful of the need for revisiting issues of alignment and alliance throughout the therapeutic process. This should lead to a reduction in blame behaviour, the development of collaborative goals, and eventually to the acknowledgement of positive systemic change.

Extract 7: Niles family (from O'Reilly, 2013, p. 6)

```
Dad:    'e can be good and 'e can be bad (.) it's jus' the <fact
        that> (.) we've got t' get t' the bottom of ↑why he's
        like the incredible Hulk because 'e changes like the
        wea↓ther
```

Extract 8: Clamp family (from O'Reilly, 2013, p. 7)

```
Dad:    but Phillip >I think< (.) needs somebody t' talk to cuz
        if we are doin' things wrong we'd like t' ↑know
```

Extract 9: Clamp family (from O'Reilly, 2013, p. 12)

```
Dad:    Against because of er you know tryin' t' get things
        sorted
FT:     Right
Dad:    We jus' wanna be a happy family
```

Extract 10: Clamp family (from O'Reilly, 2013, p. 16)

```
Mum:   there's another situation I wuz gonna say as well Joe
       (0.6) that (.) that Phillip (.) and Jordan (.) was
       beginin' t' love each other
Dad:   ↑Yeah th[ey were they were gettin' on as well
FT:            [Right
```

In Extract 7, the use of the inclusive pronoun 'we' demonstrated a systemic understanding of shared responsibility for identifying the causal explanation for the child's difficulties. This is arguably a move forward to goal setting in the therapy, and while parents may revert back to a blaming discourse, this type of shared responsibility could be encouraged and fostered by the therapist. Extract 8 demonstrated that the parents displayed openness to accepting the possibility that it may be beneficial to their child if they changed their parenting practices or behaviour 'if we are doing things wrong we'd like to know'. The willingness to change was thus interrelated with goal setting in terms of ascertaining what the therapeutic goals may be. Extract 9 illustrated this in the systemic goal oriented to by the father 'we just wanna be a happy family'. The way in which the goal was phrased has a colloquial idiomatic quality which demonstrated that it was offered from a personal perspective and was non-institutional. This is a subjective qualitative desired outcome of the therapy. The effectiveness of therapy for this family therefore was illustrated in Extract 10 in the explicit behaviour of 'love' between the siblings, which was argued to have had an emotional 'love' and behavioural 'getting on' aspect to the improvement.

Clinical relevance

Arguably, an initial session with just the parents present has the potential to build a therapeutic alliance which dissipates parental anxiety about blame and accountability. This in turn may facilitate receptiveness to the possibility of a systemic reframing of the child's difficulties. This is particularly important given the context of family therapy which is aimed at discussing the family unit as a whole with actions being treated as carrying a moral charge (Hutchby & O'Reilly, 2010). When this reframing happens, parents move from blaming the individual child to developing a more inclusive familial acceptance of shared responsibility, acknowledging that as parents they may play a role in therapeutic outcomes. Notably, this is not a straightforward process. Family systems theory recognises that resistance to change is expected and unavoidable (Nichols & Schwartz, 1991). Change inevitably has a degree of uncertainty and with uncertainty there is likely to be anxiety. It is the blaming discourse that inoculates against potential parental stake (see Potter, 1996), in the sense that blaming the child manages the parental anxiety that they will be held accountable. Thus, the therapist's efforts to promote therapeutic alliance with the parents work towards softening resistance to therapeutic change. Notably, indicators of reductions in resistance are marked by the systemic reframing from the parents and active systemic goal setting. In therapy this marks a turning point in the attribution of blame and agency, with an acceptance of shared responsibility.

Clinical relevance summary

While it is recognised in family therapy that it is important and necessary to elicit all members' viewpoints, the mechanisms for achieving that goal are variable. In some cases, children and parents are all present in the initial meetings where goals and perspectives of family members are gathered. However, we have argued that dominant parental perspectives tend to be negatively conveyed which in part is due to issues of managing accountability and blame, and the assignation of personal agency. This is problematic as children may grow and develop with the belief that these 'problems' are characteristic of their identity and are irrefutable (Turns & Kimmes, 2014). Furthermore, research has shown that children can be affected by parental problems (Sori & Sprenkle, 2004). Additionally, it is likely that children will be inhibited in expressing their views in front of their parents. Therefore, the benefit of temporary early exclusion is that the therapist and parents can work together to externalise the problem, so that when children are included the difficulties have already begun to be reframed.

We have suggested that one potential solution to this complexity is to temporarily exclude the child from the first session. The purpose of this temporary exclusion is for the therapist to set the frame of reference for how subsequent therapy sessions may unfold and to allay parents' anxieties about being blamed or being judged. It is important that therapists establish a strong therapeutic alignment with parents early on to facilitate progress and to help prevent premature termination (Hawley & Weiss, 2003). Furthermore, it may be advantageous to also spend time with the child/children separately from their parents as a way of eliciting the child's perspective in a safe environment. This may also be pertinent to particular age groups whereby the therapist may choose to use participatory techniques with younger children to facilitate engagement and for older children an environment to freely express their opinion without needing to attend to competing versions from other family members. Notably, however, this has not been a primary focus of research in family therapy and we argue that it would be valuable to undertake further research in this area.

An advantage of separate parent and child initial sessions is that expectations on both sides can be addressed more fully and both children and their parents have more freedom to speak openly about anxieties regarding potential censorship. Additionally, the institutional context of the therapy setting demands a particular kind of conversational style which is unlike the normative rules of everyday mundane talk. Parents are unlikely to be familiar with these expectations and may need guidance regarding issues such as the appropriate topics for discussion in front of children, appropriate terms of reference, and the ethos of a therapeutic non-blaming environment. We propose that while this is a good practice approach to initial sessions, there may be divergence from the agreed parameters as the therapeutic sessions unfold in which case it may be helpful to have a mid-therapy recalibration session to reiterate some of these fundamental principles. We recognise that many therapists do work in this reflexive way and we encourage and advocate this practice.

Ultimately, we recognise that it is advantageous that the majority of the family therapy sessions are held with adults and children together and Korner and Brown (1990) illustrated two core reasons why inclusion is promoted. First was that the therapist is likely to have a better understanding of the family system where more members are present in the setting. Second, children often bring a straightforward perspective that is not constrained by societal norms or expectations, which may facilitate change. However, this collaborative style requires careful management and monitoring. Children may not be used to having their views validated and therefore the therapist should be clear that their point of view is welcomed (Cooklin, 2001). Children are less likely to resist participation in the therapy if the therapist provides a safe environment for them to express their views (Lund, Zimmerman, & Haddock, 2002).

Traditionally goal setting has been viewed as a task assigned to the initial session of therapy. However, in the context of family therapy where there are frequently competing versions of events, there may also be a discrepancy of goals between parents and their children. This is especially likely given that children are taken to therapy by their parents as being the 'identified patient' (Berg & Steiner, 2003) and children are often involuntary participants in therapy (White, 2007). During initial sessions, it may become apparent that parents and children have opposing views regarding the problem but also polarised ideas about what they believe successful goals should be. This is particularly an issue for families bringing older children or teenagers to therapy whereby the task of goal setting is more complex. Achieving a consensus is particularly challenging and can complicate the therapist's identification of the target problems (Hawely & Weiss, 2003). We suggest therefore that collaborative familial goal setting is itself a goal of therapy and one that is emergent during the process of therapy as goals become less divergent and more

Table 30.2   Clinical practice highlights

1.	The inclusion of children in family therapy requires careful consideration and management.
2.	The purposeful temporary exclusion of children from the initial session can be beneficial.
3.	A single session with only the child present can be beneficial in obtaining the child's perspective.
4.	Alliance and engagement with individual family members can be better facilitated where there is a mixture of individual and family sessions.
5.	Collaborative goal setting is an ongoing process in family therapy.

systemically defined. For a simple summary of the implications for practice, see Table 30.2.

Recommendations

For the purposes of clarity, we present our final recommendations in three broad areas. First, we propose a number of recommendations, which fall within the rubric of structural considerations. Structural considerations are those pragmatic and organisational decisions that therapists can make to facilitate engagement and alignment. Second, we propose a number of recommendations, which fall within the rubric of content considerations. Content considerations are issues, which can be addressed within the structural environment that has been facilitated. Finally, we propose a number of recommendations which fall within the rubric of reflexive considerations. These are aspects of reflective practice which may be helpful for therapists to consider at different stages of therapy; that is, those issues that the therapist needs to be reflexively aware of during the ongoing process of therapy. These are outlined in Table 30.3.

We acknowledge that this table is a heuristic aid and the contents within it are not discrete elements or intended to be rigidly chronological. It is offered as a synopsis of the detailed contents of the chapter and argument presented within it.

Summary

In this chapter, we have explored some of the long-standing arguments regarding the inclusion and exclusion of children from family therapy. Drawing upon three of our empirical papers, we have proposed the argument that the initial temporary exclusion of children is beneficial. We have acknowledged that it is common for parents to talk about their children in negative ways in front of them, which may (at least to some extent) be avoided through this practical strategy. Additionally, we have recognised that parents perform alignment

Table 30.3 Structural, content, and reflexive considerations

Structural Considerations	Content Considerations	Reflexive Considerations
Separating children and parents for the initial meeting.	Managing expectations. Clarifying how therapy works. Setting boundaries. Eliciting different versions of the difficulties.	Different linguistic frameworks between mundane and institutional talk. Anxieties regarding accountability and blame.
Encourage all members of the family to participate in the majority of sessions.	Managing alignment shifts between different family members. Allow space for child contributions. Develop and form family goals. Decide on indicators of change and success.	Consideration of how resistance manifests. Asserting therapeutic authority without undermining parental authority. Reflecting on their skills for engaging children in multi-generational sessions.
Additional parental meeting later in the progress.	To re-visit agreements regarding managing the framework of therapy. To re-visit and reinforce systemic understanding of the family difficulties.	Sensitivity to discursive markers which indicate softening of resistance. Recognition of the progress towards systemic goals.

work with the therapist which has potential for children to disengage from the therapeutic process. We argued that this alignment work could be more appropriately managed in an initial parent only session. Finally, we noted that the process of goal setting and therapeutic progress is ongoing and requires careful management by the therapist as the formation of an agreed upon family goal may in itself be a valid outcome.

References

Aspland, H., Llewelyn, S., Hardy, G., Barkham, M., & Stiles, W. (2008). Alliance ruptures and rupture resolution in cognitive-behaviour therapy: A preliminary task analysis. *Psychotherapy Research, 18*(6), 699–710.

Avdi, E., & Georgaca, E. (2007). Discourse analysis and psychotherapy: A critical review. *European Journal of Psychotherapy and Counselling, 9*(2), 157–176.

Beauchamp, T., & Childress, J. (2008). *Principles of biomedical ethics* (6th edition). Oxford: Oxford University Press.

Berg, I., & Steiner, T. (2003). *Children's solution work*. New York: W. W. Norton and Company.

Cooklin, A. (2001). Eliciting children's thinking in families and family therapy. *Family Process, 40*(3), 293–312.

Dallos, R., & Draper, R. (2010). *An introduction to family therapy: Systemic theory and practice* (3rd edition). Berkshire: Open University Press.

Edwards, D., & Potter, J. (1992). *Discursive psychology*. London: Sage.

Foley, V. (1974). *An introduction to family therapy*. New York: Grune and Stratton.

Goldenberg, H., & Goldenberg, I. (2008). *Family therapy: An overview* (7th edition). USA: Thompson Brooks.

Hartzell, M., Seikkula, J., & von Knorring, A. L. (2009). What children feel about their first encounter with child and adolescent psychiatry. *Contemporary Family Therapy, 31*, 177–192.

Hawley, K., & Weisz, J. (2003). Child, parent, and therapist (dis)agreement on target problems in outpatient therapy: The therapist's dilemma and its implications. *Journal of Consulting and Clinical Psychology, 71*(1), 62–70.

Hutchby, I., & O'Reilly, M. (2010). Children's participation and the familial moral order in family therapy. *Discourse Studies, 12*(1), 49–64.

Jefferson, G. (2004). Glossary of transcript symbols with an introduction. In G. H. Lerner (Ed.), *Conversation analysis: Studies from the first generation* (pp. 13–31). Amsterdam: John Benjamins.

Johnson, L., & Thomas, V. (1999). Influences on the inclusion of children in family therapy: Brief report. *Journal of Marital and Family Therapy, 25*, 117–123.

Kazdin, A. E., Holland, L., & Crowley, M. (1997). Family experience of barriers to treatment and premature termination from child therapy. *Journal of Consulting and Clinical Psychology, 65*(3), 453–463.

Korner, S., & Brown, G. (1990). Exclusion of children from family psychotherapy: Family therapists' beliefs and practices. *Journal of Family Psychology, 3*(4), 420–430.

Lund, L., Zimmerman, T. S., & Haddock, S. (2002). The theory, structure, and techniques or the inclusion of children in family therapy: A literature review. *Journal of Marital and Family Therapy, 28*(4), 445–454.

McLeod, J. (2001). *Qualitative research in counseling and psychotherapy*. London: Sage.

Miller, L., & McLeod, E. (2001). Children as participants in family therapy: Practice, research, and theoretical concerns. *The Family Journal: Counseling and Therapy for Couples and Families, 9*(4), 375–383.

Muntigl, P. (2013). Resistance in couples counselling: Sequences of talk that disrupt progressivity and promote disaffiliation. *Journal of Pragmatics, 49*(1), 18–37.

Nichols, M. (1996). *Treating people in families: An integrative framework*. New York: Guildford.

Nichols, M., & Schwartz, R. (1991). *Family therapy concepts and methods*. Needham Heights, MA: Allyn and Bacon.

O'Reilly, M. (2013). 'We're here to get you sorted': Parental perceptions of the purpose, progression and outcomes of family therapy. *Journal of Family Therapy*, doi: 10.1111/1467-6427.12004.

O'Reilly, M., & Parker, N. (2012). 'She needs a smack in the gob': Negotiating what is appropriate talk in front of children in family therapy. *Journal of Family Therapy*, doi: 10.1111/j.1467-6427.2012.00595.x.

O'Reilly, M., & Parker, N. (2013a). You can take a horse to water but you can't make it drink: Exploring children's engagement and resistance in family therapy. *Contemporary Family Therapy, 35*(3), 491–507.

O'Reilly, M., & Parker, N. (2013b). Unsatisfactory saturation': A critical exploration of the notion of saturated sample sizes in qualitative research. *Qualitative Research, 13*(2), 190–197.

Parker, N., & O'Reilly, M. (2012). 'Gossiping' as a social action in family therapy: The pseudo-absence and pseudo-presence of children. *Discourse Studies, 14*(4), 457–475.

Potter, J. (1996). *Representing reality: Discourse, rhetoric, and social construction.* London: Sage.

Potter, J. (2004). Discourse analysis as a way of analysing naturally occurring talk. In D. Silverman (Ed.), *Qualitative research: Theory, method and practice* (pp. 200–221). London: Sage.

Rober, P. (1998). Reflections on ways to create a safe therapeutic culture for children in family therapy. *Family Process, 37*, 201–213.

Roy-Chowdhury, S. (2003). Knowing the unknowable: What constitutes evidence in family therapy? *Journal of Family Therapy, 25*, 64–85.

Roy-Chowdhury, S. (2006). How is the therapeutic relationship talked into being? *Journal of Family Therapy, 28*, 153–174.

Ruble, N. (1999). The voices of therapists and children regarding the inclusion of children in family therapy: A systematic research synthesis. *Contemporary Family Therapy, 21*(4), 485–503.

Safran, J., Muran, C., Samstag, L., & Stevens, C. (2001). Repairing alliance ruptures. *Psychotherapy, 38*(4), 406–412.

Sommerfeld, D. (1989). The origins of mother blaming: Historical perspectives on childhood and motherhood. *Infant Mental Health Journal, 10*, 14–24.

Sori, C. F., & Sprenkle, D. (2004). Training family therapists to work with children and families: A modified DELPHI study. *Journal of Marital and Family Therapy, 30*(4), 479–495.

Strickland-Clark, L., Campbell, D., & Dallos, R. (2000). Children's and adolescents' views on family therapy. *Journal of Family Therapy, 22*(3), 324–341.

Thompson, S., Bender, K., Lantry, J., & Flynn, P. (2007). Treatment engagement: Building therapeutic alliance in home-based treatment with adolescents and their families. *Contemporary Family Therapy, 29*(1/2), 39–55.

Tseliou, E. (2013). A critical methodological review of discourse and conversation analysis studies of family therapy. *Family Process, 52*(4), 653–672.

Turns, B., & Kimmes, J. (2014). 'I'm NOT the problem!' Externalising children's 'problems' using play therapy and developmental considerations. *Contemporary Family Therapy, 36*(1), 135–147.

White, M. (2007). *Maps of narrative practice.* New York: Norton.

Withnell, N., & Murphy, N. (2012). *Family interventions in mental health.* Berkshire: Open University Press.

Recommended reading

- Avdi, E., & Georgaca, E. (2007). Discourse analysis and psychotherapy: A critical review. *European Journal of Psychotherapy and Counselling, 9*(2), 157–176.
- Dallos, R., & Draper, R. (2010). *An introduction to family therapy: Systemic theory and practice* (3rd edition). Berkshire: Open University Press.

- Edwards, D., & Potter, J. (1992). *Discursive psychology.* London: Sage.
- Parker, N., & O'Reilly, M. (2012). 'Gossiping' as a social action in family therapy: The pseudo-absence and pseudo-presence of children. *Discourse Studies, 14*(4), 457–475.
- Tseliou, E. (2013). A critical methodological review of discourse and conversation analysis studies of family therapy. *Family Process, 52*(4), 653–672.

31

Parentification: Counselling Talk on a Helpline for Children and Young People

Susan Danby, Jakob Cromdal, Johanna Rendle-Short, Carly W. Butler, Karin Osvaldsson, and Michael Emmison

Introduction

This chapter investigates counselling interactions where young clients talk about their experiences of taking on family responsibilities normatively associated with parental roles. In research counselling literature, practices where relationships in families operate so that there is a reversal of roles, with children managing the households and caring for parents and siblings, is described as parentification (see also Hutchby, Chapter 29, this volume; Kiyimba & O'Reilly, Chapter 30, this volume, for further discussion on family therapy talk). Parentification is used in the counselling literature as a clinician/researcher term, which we 'respecify' (Garfinkel, 1991) by investigating young clients' own accounts of being an adult or parent and by showing how counsellors orient to these accounts. As well as providing understandings of how young people propose accounts of their experiences of adult–child role reversal, the chapter contributes to understanding how children and young people use the resources of counselling helplines, and how counsellors can communicate effectively with children and young people.

The concept of parentification

This chapter investigates young clients' accounts of their experiences of caring for family members, roles normatively associated with being an adult or parent, and how counsellors orient to their accounts. The counselling sessions occurred in phone and web chat sessions on Kids Helpline, an Australian national helpline for children and young people.

The concept of parentification refers to the reversal of roles within family relationships, with children managing the household and caring for parents and siblings (Jankowski, Hooper, Sandage, & Hannah, 2013; McMahon & Luthar, 2007). These practices can be found in contexts of dysfunctional family dynamics, or in response to a family crisis such as a parent with a serious illness (McMahon & Luthar, 2007). Parentification can involve two kinds of children's support for parents, instrumental and emotional, and sometimes both come into play when caring for parents and younger siblings, looking after the household and generally acting as an adult (Leon & Rudy, 2005; McMahon & Luthar, 2007). Parentification is often described as the blurring of parent–child roles, breaking down the 'generational hierarchy', with the consequence of children and young people coping with stress and anxiety (Hooper, 2008). The concept of parentification is not to be confused with children helping to some extent with household chores or other family responsibilities, such as caring for younger children, activities that are considered to develop concepts of altruism and positive self-identity (McMahon & Luthar, 2007).

Many studies of parentification have relied on retrospective reports by adults about their childhood, although more recently children are being directly asked about their experiences (Earley & Cushway, 2002). In a study that reported on how adolescent daughters managed when the mother had been diagnosed with breast cancer, the daughters reported that they had difficulties attending to the needs of their mother as well as their own needs (Stiffler, Barada, Hosei, & Haase, 2008). They struggled to isolate themselves from the new responsibilities by escaping from the situation, but also not wanting to leave their mothers without support, recognising that their mothers required increased care and someone else to be responsible for everyday household tasks (Stiffler et al., 2008). In a study of parental conflict, Leon and Rudy (2005) found that young children of parents who had higher levels of parental conflict drew pictures of themselves more often taking on an adult caring role in the family. A Belgian study of children aged 7–14 years focused on the children's experiences of caring for parents with depression. The children reported not feeling recognised for the care they gave, and they also reported protecting their parents by not disclosing their emotional states or discussing sensitive or difficult topics (Van Parys & Rober, 2013).

Parentification can affect children's mental health outcomes. In their study of almost 800 university students in the United States, Jankowski and colleagues (2013) examined the negative effects of parentification. The students in their study described a sense of unfairness and injustice due to their own needs being unmet, including negative emotions such as stress. Parentification has been found to have a negative effect on children's long-term well-being well into adulthood, where they experience 'imposter' feelings of being an adult: feeling 'like a child trying to behave like an adult' (Castro, Jones, & Mirsalimi, 2004, p. 209). More recent studies show how parentification can lead to some positive short- and longer-term benefits for children, such as an increased sense of competence, particularly when children reported that the responsibilities they undertook were recognized by the parents (Hooper, 2008; Jankowski et al., 2013).

While parentification is a well-known concept in clinical research, to date there are relatively few empirical studies to specifically explore the phenomenon (Leon & Rudy, 2005; McMahon & Luthar, 2007). There is even less written about therapeutic support and intervention (see however, DiCaccavo, 2006). By investigating how young clients talk about their experiences of acting like an adult, undertaking roles such as caring for family members, including parents and siblings, we show the 'when' and 'how' experiences of parentification introduced by young people and discussed within a therapeutic context of a helpline dealing specifically with young clients. Considering the strategies that young people use to talk about their experiences has application for understanding the concerns that they might face and ways to support them in these situations.

Project overview

The data reported here consists of a subset from 50 audio-recorded telephone calls and 100 logged synchronous web chat counselling interactions on Kids Helpline (Danby, Butler, & Emmison, 2009; Harris, Danby, Butler, & Emmison, 2012). Older clients are more likely to use web chat and email, whereas younger clients are more likely to use the phone (BoysTown, 2013). As well, more serious issues such as self-harm and suicidal thoughts are more likely to be discussed within an online environment (BoysTown, 2013). Approval was sought and given to use the transcripts of the audio-recorded phone calls and chat/email logs. Names and other identifying information were replaced with pseudonyms.

In 2013, almost 250,000 young people aged 5–25 years made contact with Kids Helpline, with mental health concerns the main reason for contact (BoysTown, 2013). The counsellors are paid professionals and have university qualifications in counselling, psychology, or social work. In line with the helpline philosophy 'We care, we listen', counsellors encourage and support the children and young people (Danby, Baker, & Emmison, 2005; Emmison & Danby, 2007).

Counsellors on Kids Helpline support young clients to tell their troubles and they respond with counselling support. They provide interactional space for children and young people to design their own way into the counselling sessions and respond with displays of sensitivity that draw on the skill of the counsellors to actively listen across a mode of modalities, including counselling through telephone calls, web chat and email modalities (Danby et al., 2005; Danby et al., 2009). The counsellors use strategies such as questions designed to help clients consider their own situations and to come up with solutions, to empower clients through promoting self-directedness, an essential component of the Kids Helpline philosophy (Butler, Potter, Danby, Emmison, & Hepburn, 2010). Another strategy that counsellors use is where they volunteer detailed models, described as script proposals, of what clients could say to a third party, such as friend or a parent (Emmison, Butler, & Danby, 2011).

In this chapter, we use ethnomethodological and conversation analysis approaches to 'respecify' (Garfinkel, 1991) the term 'parentification' by beginning with how the young people and counsellors 'produce and exhibit...logically, reflexively accountable orderliness' (p. 17) as they talk about experiences of caring for their family members in ways that clinicians or researchers might identify as parentification. One focus of our analysis is the use of category work by client and counsellor; that is, how each orient to the clients' family relationships drawing on common-sense and normative categorisations that orient to distributions of family rights and responsibilities, and normative lifespans. Through analysis, we show how membership categorization is 'an *activity* that is carried out in particular local circumstances' (Hester & Eglin, 1997, p. 22). It is in the '*use* of these categories that culture is constituted' (Hester & Eglin, 1997, p. 20). Here, family and stage-of-life (SOL) devices are made relevant by young clients and counsellors to make sense of family relationships and to validate clients' accounts of their experiences.

Analysis

Below, we present extracts from two counselling sessions of children and young people presenting accounts of their experiences of caring for family members.

Extract 1 is of a young caller using phone counselling, and Extract 2 is from a web counselling session.[1] These two extracts show the phenomenon under discussion.

Extract 1 (PC150508_1807) (phone)

```
1   Cou:    .HH [ So you- y:-     ]ou::::: you left ho::- left =
2   Call:       [ °Mum (thought)-]
3   Cou:    = schoo:l, t- tih help out at ho:me?,
4           (0.2)
5   Cou:    .hh [and s'you could be there tah help out mum a =
6   Call:       [°°Ye:ih.°°
7   Cou:    = whole lot more?
8           (0.3)
9   Call:   °Ye:ah.°
```

Extract 2 (WC103531) (web chat)

(Counsellor)	how do you feel about your mum being like that?	16:22	18:39	00:55
(Client)	um it kinda sucks. i would like her to look after me for once. and nurse me when im not well or hug me when i cry etc. but i have come to accept thats not gonna happen so i just have to lump it.	16:23	19:20	00:41
(Counsellor)	it's good that you look after you mum like that and take such a big responsibility. Do you ever put your own needs first?	16:24	20:57	01:37
(Client)	she is mainly my first priority. she is so needy and when i was younger i couldn't help and just had to watch and now i can i feel and obligation to protect her.	16:25	21:43	00:46

In Extracts 1 and 2, counsellors and clients did not directly refer to parentification practices, although they did talk about young people undertaking caring roles in looking after parents. As researchers, we rely on clients' own accounts of their experiences and emotions, and how the counsellors respond to these descriptions, but we do not name them as instances of parentification.

In the two cases we discuss below, beginning with Extract 3, we show how members describe situations, problems, and experiences as a way of producing and recognising particular activities that go along with the category of being a young person (Atkinson, 1980), and the difficulties they report acting within a social category usually associated with adults and parents. The membership categorisation devices of family and SOL are used by counsellors and clients alike in counselling on Kids Helpline (for discussion of SOL device, see Cromdal, Danby, Emmison, Osvaldsson, & Cobb-Moore, forthcoming). We show how membership categories referring to lifespan and family categories are 'inference rich' (Sacks, 1972) in that a great deal can be understood through drawing on common-sense understandings located within the categories.

Some days where I'm the mum

We focus in depth on one call where Hayley, a teenage caller, initially calls to talk about fighting with her sister and later reports that her mother is ill, and that she has extra responsibilities at home. Extract 3 starts just after the call begins, with Hayley describing how she has been fighting with her younger sister earlier that day, before adding that she fights often with her mother. Up until now, she is reporting on family relationships. The extract begins with the counsellor's formulation of the client's current circumstances.

Extract 3 PC150508_1807

```
10  Cou:   .hhh HA:::↓ha::=sounds like yeah it's bee::n (0.6) °m- m-°
11         bit hard for you this afternoon¿=And- (0.2) prob'ly I'm
12         guess:ing: (0.6) been (.) bit hard over the last (0.3)
13         while as well?
14         (0.5)
15  Call:  Yeah cos mum's been sick.
16         (0.5)
17  Cou:   °tch oka:y,° (.) yeah? .h (0.5) wha- what's mum dealing
18         with at the m[o:ment;
19  Call:               [Um fibromya:lgia.
20  Cou:   .HHH f:::ibromy↑algia?
21         (0.2)
22  Call:  °Yea::h° =
23  Cou:   =did I say that right?=.HH [You're] you're gonna haf to=
24  Call:                            [ Yeih ]
25  Cou:   =educa:te me on what that i:s, is[::
26  Call:                                  [Um it's whe:re all
27         your muscles and (0.2)
28  Cou:   mh:m
29  Call:  stuff don't they just give u:p and you have days whe:n
30         ~you can't walk an~ ((voice breaking))
31  Cou:   .HH ri::ght oh: wow that must be such a challenge for
32         he:r¿
33         (.)
34  Cou:   Ye:a:h.
35         (0.2)
36  Call:  °yeih°
37         (.)
38  Cou:   tch a::nd I- I gue:ss:- (0.3) it's a challenge for he:r
39         >b't it's< a:lso a challenge for a challenge for ↑you
40         guys as we:ll.
41         (0.2)
```

In this extract, Hayley presents her initial reason for her call to Kids Helpline, that of fighting with her sister. The counsellor initially responds to the specific incident that led to the call with an empathic formulation that it has been a 'bit hard' for the caller this afternoon (line 11), and he 'guesses' that it's been a hard for a while (lines 12–13). Hayley offers an agreement and expands by disclosing that her mother has been sick (line 15). The counsellor quietly receipts this and follows with an upwardly rising 'yeah?' and he follows up with a question specifically asking what her mum has been 'dealing with at the moment' (lines 17–18). She names the illness, fibromyalgia (line 19). After repeating this, the counsellor asks her to 'educate' him about it (lines 23 and 25), which works to put the client into an epistemic position of being more knowledgeable about the disease than the counsellor. This early work in the call by the counsellor is designed to encourage the caller to talk about her experiences and to minimise the inherent asymmetry found in professional–lay consultations (Heritage & Sefi, 1992) and in adult–child interactions (Speier, 1973). This move by the counsellor also works to elicit the caller's perspective.

As the client describes her mother's condition, she stops when her voice starts breaking, and the counsellor responds in lines 31–32 with a loud in breath, drawn out receipt (ri::ght) and an assessment that it must be 'such a challenge for her' (line 38). After a slight pause, the counsellor reinforces his perspective with an extended and downward falling 'Ye:a:a:h' (line 34) to which the caller quietly agrees. The counsellor's slightly troubled delivery suggests attending to the delicacy of this matter (Silverman & Peräkylä, 1990), and he returns the focus of the conversation from the mother back to the caller in the form of 'you guys' (lines 39–40), which provides the interactional space for the caller to talk more about her family situation, including how her father had to work two jobs (not shown in the transcript). The counsellor's turn invokes the family device, treating her account as a warrantable troubles telling. He now pursues Hayley's experiences of the family's challenges and its impact on her (Extract 4).

Extract 4 PC150508_1807

```
42   Cou:   tkl I mean I 'ink .hhh I think e:very fa:mily has::
43          .hhhh y'ow their differen↑c:es and y'ow their
44          °d-° (.) cha:llenges that they fa:ce, b't .hhh (i')
45          (you're) dealing with an illness like tha:t¿ (0.3)
46          ye:ah is som:e some cha:llenges: that no:t that many
47          families °↓have to deal with:↓. Ye:[ah.
48   Call:                                    [Mm
49   Cou:   M:m tk .HHHH and you as the oldest daughter¿
50          (0.5)
51   Cou:   Do you end up feeling pretty stressed?=Because of mum's
52          illness?
53   Call:  ~°Oh:° there's some days where I'm the mum.~
54          (0.4)
55   Cou:   Ye[ah?
56   Call:    [~Which I find hard because sometimes they're the days
57          that me and my sister are fi:ghting.~
58          (0.4)
59   Cou:   .HHH O:kay, .h so some b- <YEAH when y- sometimes
60          you have to be the mu:m,
61          (.)
62   Cou:   Wh[ich means you: haf tah be the respo:nsible o:ne¿,
63   Call:    [°°Yeah°°
64   Cou:   .hhhh and the:::y're- °o-° >that< (.) of'en leads to
65          you an your si:ster fighting.
66          (0.5)
67   Call:  °°Y[ip°°
68   Cou:     [Ye:ah¿
```

```
69            (.)
70   Cou:     O:kay,
71            (0.3)
72   Cou:     .hh °f-° ↑what's it like for you tuh be: .h a f- (.)
73            you 'ow fif:tee:n: (0.2) an:d having tu(h)h be
74            m↑um↑ sometimes: hh[.hh
75   Call:                      [°W'll: (0.7) like- (0.2) my mu:m's
76            giving me most stu' t'do cause I'm turning sixteen in
77            a little bit¿
78            (0.3)
79   Cou:     [Yeah?
80   Call:    [B't- (0.2) and she's giving me more an more
81            responsibilitie:s for it- (0.2) th'n- (0.2)
82            [b't (.) yet she- (0.5) sez I'm the reason the=
83   Cou:     [Mm
84   Call:    =reason the family's falling apa:rt an- (0.6)
85   Cou:     Y[:- You:'re the reason
86   Call:     [Cos I:'m the one (-------------)
87            (0.6)
88   Cou:     O↑↓h::.
89            (1.3)
90   Cou:     ↑Tha- that must be hard for you to hear from mu:m
91            °he:y.°
```

In lines 42–47, the counsellor invokes the 'family' device – first, by describing what 'every family' goes through, and the challenges they face, but then the challenges faced by the caller's family as extreme – beyond the norm. The counsellor picks up on the parts of the account to do with the 'challenging stuff' that the caller has had to deal with in her family situation. In line 42, the counsellor introduces the institution of family, shifting from the generic 'every family' (line 42) to 'your' family (lines 45–47) dealing with a particularly difficult situation. The counsellor's turn works to upgrade the magnitude of what the caller has identified, that the circumstances are beyond 'normal', and to elaborate on what it means for her family ('some challenges not too many families have to deal with', lines 46–47). By identifying the caller's situation as 'not normal', the counsellor affiliates with the caller by assessing her situation as difficult. This summarising formulation gives the gist of what has been said so far, as well as ratifying her account (Antaki, 2008; Heritage & Watson, 1979).

The counsellor extends on this version of the hardship faced by Hayley, not only as a member of a family with specific and extreme challenges but as the 'older daughter' (line 49). This device brings common-sense understandings about the roles, rights, and responsibilities attributed to different roles/categories within the device family. He has zoomed in on the caller's particular and personal challenges by virtue of her membership as the 'older daughter', invoking a distinct set of rights and responsibilities. In this way, he validates and ratifies the caller's reported 'stress' as a consequence of having additional responsibilities due to her membership as 'oldest'. The categorisation of 'older sister' invokes age, but it is relative to the ages of other children in the family – a relational category.

The counsellor continues by asking about 'feeling pretty stressed', said quickly and latching to the reason, mum's illness (lines 51–52). In this way, the counsellor invokes the type of ranking and related distribution of family responsibilities, that the oldest daughter has greater responsibilities and can be expected to be more stressed than younger members of the family. The counsellor's categorisation is crucial in ascribing a problem to the caller and shows how the counsellor draws on common-sense cultural knowledge to preset his understanding of the client's telling.

After the counsellor offered a formulation of the caller's emotional state and (justifiable) reason for this, via a categorisation, the caller's 'oh-prefaced' response (Heritage, 1998) indicates her epistemic primacy over the relevance of the category ascription. Hayley then maps herself into a different category: 'some days I'm the mum'. With this self-categorisation, she upgrades the extent of her responsibilities as resulting in activities predicated to the 'mother' category. This draws on common-sense understandings, providing a snapshot of the situation, highlighting the scale of the problem she is having to cope with; as a teenager, her life situation has resulted in her having to act as the mother – 'there's some days where I'm the mum' (line 53). This works to upgrade the counsellor's categorisation in terms of workload and domestic responsibilities. The role reversal serves to categorically upgrade the trouble description, still by drawing on the same kind of 'what anyone knows' type of knowledge of the moral orders of families indexed by the counsellor's previous turn.

After a slight pause and the counsellor's encouraging 'yeah?' to continue (line 55), Hayley provides an account of the consequence of having to be the mum, of assuming a position of authority and responsibility in the family, she gets into fights with her younger sister. The counsellor receipts Hayley's account with a formulation (lines 59–60), elaborating on the trouble of being in both categories of mum and sister. He formulates an explanation of why the sisters are fighting, working from the common-sense understanding of relationships between older and younger sisters. He provides the upshot of this formulation in line 62 by beginning to unpack the implied shared cultural knowledge in the identity claim – that being 'the mum' means being responsible, to which the caller agrees.

Following this agreement, the counsellor seeks her perspective about what it is like to be aged 15 and to be the mother (lines 72–74). The implication in the counsellor's description of being 15 and having to be the mum (lines 72–73) suggests that the troubles she is experiencing might belong more rightfully within the category of being an adult rather than being associated with her chronological age. Here, the counsellor has introduced the SOL device, with the categories of older sister and mum being mutually exclusive. The younger sister treats her as a sister, but Hayley is doing things via her membership as mum.

Using a quantifiable measure such as age can offer 'a routine, reliable and objective way of describing someone,... [and] can also function as a bottom-line argument, as the last resort for checking someone's "objective" age-category' (Nikander, 2002, p. 75). As Sacks (1995) pointed out, 'the age class a person is in turns on the topic with respect to which they're being talked of' (Vol. 1, p. 754). In this way, we can see the counsellor suggesting to the client that there has been some breach of 'some normative or common sense notions of age' (Nikander, 2002, p. 149).

In this extract, we saw how the counsellor made the family device relevant through introducing the category of the 'oldest daughter' and displayed care and sympathy through his counsellor's affiliative stance and alignment that encouraged more talk from the client. This display of affiliation involved drawing on everyday understandings of social and cultural life, through the use of membership categories. Moreover, the counsellor's questions worked to demonstrate his awareness of the authenticity of the client's own stated feelings.

'I feel like the adult'

In the previous extract, the counsellor initiated the family device through his introduction of the category of 'oldest daughter'. In the web session below (Extract 5), it is the client who initiates the SOL device by referring to herself: 'I feel like the adult'. Up until this point, the client had provided an account of how she was living in a family where her father had lost his job and was drinking a lot, that her mother had depression and that her elder sister was leaving home so that the client would be left alone to deal with the family situation. (The transcript is produced exactly as was shown in the web page; see Endnote 1 at the end of this chapter for an explanation of how to understand the time stamps on the transcript.)

Extract 5 WC103531

(Counsellor)	sounds like you're almost having to cope with this by yourself. Would am right in saying that?	16:13	09:36	01:43
(Client)	yeh its pretty daunting. i feel like the adult in the situation trying to work out what to do.. how to look after everything. pay for stuff. its just overwhelming.	16:14	10:19	00:43

(Counsellor)	that would be overwhelming. Overal, how do you feel you're coping?	16:15	11:19	01:00
(Client)	yeh not very well. hahe. i would lvoe to say i am brilliant and coping just fine. but its not that case. i am in over my head dealing with it all	16:15	12:02	00:43
(Counsellor)	sorry to hear that, sounds full on!	16:17	13:28	01:26
(Counsellor)	what are you wanting to do about the situation? How much longer do you think you can go on like this?	16:17	13:56	00:28
(Client)	i dunno. i just dont know what to do. i wanna move out, but its kinda a race between me and my sister who goes first gets to go kinda thing. because whoever loses has to look after mum. and my sister already has plans to move so i lost. i just dont know.. im stuck	16:19	15:04	01:08
(Counsellor)	that makes it difficult when you're wanting to look after your mum. Sounds like you really care for her	16:20	16:07	01:03

Extract 5 begins almost 10 minutes into the web chat session. At 09:36 (see the middle column that indicates the length of time that the web chat has been going; see Endnote 1), the counsellor provides an upshot of the family situation that the client has described. He brings the focus back to her and how she has to cope with her family situation by herself. The client agrees, adding that it is 'daunting', and introduces the SOL device, saying 'i feel like the adult' (10:19). She follows with a gloss of what this entails, and finishes with an assessment using an extreme case formulation (Pomerantz, 1986) of 'just overwhelming' that the counsellor picks up on, repeating this formulation, which works to show acceptance of the client's account.

The counsellor now shifts into a counselling sequence where he asks a series of questions. The first question is used to assess the client's assessment of her current situation from her personal perspective: 'how do you feel you're coping?' (11:19). The client's next posting displays the intensity of her feelings through the contrasting positions she proposes: 'i would lvoe to say i am brilliant and coping just fine. but its not that case. i am in over my head' (12:02). The client's idiomatic expression of being in 'over her head' is unchallengeable (Drew & Holt, 1988) and provides the counsellor with straight forward evidence from the client that she is not coping well at all, delivered as an upshot, or punchline (Sacks, 1995, Vol. 2), as evidenced by the counsellor's next short post, proffering an empathic stance (13:28).

The client's previous turn now provides sufficient warrant for the counsellor to shift to a future-oriented action, 'what are you wanting to do about the situation? How much longer do you think you can go on like this?' (13:56). The client's reply takes over a minute to post. She describes how she has lost the race with her sister (to move out of home) and that she will have to look after her mum as a consequence. Interspersed within this description, she provides a sequence of talk that begins with 'i dunno' to 'I just don't know' to 'I just don't know..im stuck'. This is category-resonant with 'being in over one's head'. The final 'im stuck' sums up the situation for the client and provides the jumping off point for the counsellor. The counsellor selects, from all this talk of wanting to move out, a formulation that recognises that it is a difficult situation and it 'sounds like' the client 'really cares for her mum' (16.07).

Extract 6 continues 9 minutes after Extract 5, and both counsellor and client continue their discussion related to SOL categories.

Extract 6 WC103531

(Counsellor)	I'm sorry to hear that you feel that way. A concern is how this is affecting your life and your future. How do you feel this is affecting you?	16:29	25:30	01:27
(Client)	um well i feel like i am like thirty. old before my time. i feel so run down. and just on the verge of my own personal breakdown. ya know. and i just want it to be someone elses problem and run away.. i also get scared that i will end up like her, bitter and resentful to everyone.	16:31	27:15	01:45
(Counsellor)	it's understandable to feel that way considering what how you've been supporting her.	16:32	28:10	00:55
(Counsellor)	It sounds like you're wanting more your life. What would you like to be doing?	16:32	28:44	00:34
(Client)	i dunno being 18, clubbing, having friends, achieving stuff. going places becoming famous. hahe. just normal stuff minus becoming famous.. hahe. tho that would be nice.	16:33	29:31	00:47

(Counsellor)	that's normal and what most people want. To enjoy life. Do you feel that you could take a backward step with things at home, still help out a bit but then also do the things you want to be? Because when people aren't happy and helping themselves then they aren't usually in a position to best help others. Do you know what I mean that?	16:36	32:40	03:09

The counsellor begins a series of questions seeking the client's perspective about her life situation. At 25:30 minutes into the web chat, the counsellor asks about she feels the family situation at home is affecting her. The client begins with a 'thinking token' before going on to say that she feels 30 years old, and old before her time (27:15). Following empathic receipt from the counsellor and a formulation that 'it sounds like' she is wanting more for her life, he continues his pursuit of her own perspective on her life, and asks a forward-projecting question about what she would 'like to be doing' (28:44). The client's next post presents herself as being 18 (almost half the age she previously indicated she felt), and she produces a list of activities associated with common-sense norms of what teenagers like to do (clubbing, going places, and so on). She finishes with what she'd like to do is 'just normal stuff', meaning the sorts of normal stuff that teenage girls like to do (29:31).

In the client's presentation of herself, she established chronological age as a way of indicating a mismatch between what she is currently doing and what she would like to be doing as a teenager. Making chronological and lifespan categories relevant is one way to 'describe and account for our own and others' actions' (Nikander, 2009, p. 864). As Atkinson (1980) points out, in ' "growing up" ... one moves "from" childhood "into" adolescence and "then into" adulthood' (p. 33). Growing up is part of the lifespan, but growing old before your time is not. She juxtaposes the mismatch of her 'natural lifetime schema' (Atkinson, 1980) of what it is like for her now, using the age device of being almost double her current age, with the normative teenage experiences she would like to be doing. In this way, she makes immediately recognisable some features of normative and non-normative orders of the lifespan (Cromdal et al., forthcoming).

The counsellor picks up the client's account of what she would like to be doing, 'that's normal' (32:40), which works to confirm and validate the client's perspective as not only her individual perspective but one that 'most people' would want. Built into the counsellor's response is an acceptance and validation of the client's perspective that being a teenager has a normative distribution of rights and activities, such as clubbing. His post recognises and accepts the cultural knowledge embedded within the category of being a teenager. Once both counsellor and client agree on the shared category of being a teenager and associated normative activities, the counsellor works to explore ways that the client could manage the situation at home.

Discussion

In this chapter, we have shown extracts from a phone call and from a web chat session. In each, we presented an analysis of how clients reported their experiences of being in situations where they felt that, even though they were teenagers, they were acting as adults and having to take on adult-like roles of caring for parents and other family members. The focus on membership categories of family and SOL highlight the normative assumptions that underpin everyday common-sense assumptions, made relevant by clients and counsellors as a way of talking about and making sense of the client's experiences.

We saw how the family and SOL devices were made relevant as a way to describe the situation and to confirm and validate young people's accounts of their emotional states. Invoking categorisation, counsellors oriented young people's circumstances to confirm, commend, validate, or seek more information by probing and asking questions and to produce formulations that displayed affiliation with the caller's stance, orienting to the caller as someone with feelings who is responding to an abnormal situation. The clients drew on family and SOL devices to build cases that displayed the extent of the problems they were facing at home, and the counsellors took up or introduced family and SOL devices to display an empathic hearing that displayed shared affiliation and validation of the clients' accounts.

We saw that counsellors and clients drew on the social structures and cultural norms as part of making sense of the young clients' experiences and situations. Clients invoked categorisations to describe what they were doing in the family context and how they felt about that. Counsellors invoked norms to validate the clients' experiences and also to address the client as an individual. In this way, common-sense family and SOL categorisations are useful ways to understand normal development. Additionally, and shown through the counselling talk here, membership categorisations provide opportunities to show mismatches between cultural norms and the clients' own individual experiences and situations and to use these understandings as a way of talking about the clients' own situations.

This chapter contributed to exploring ways in which analysis of membership categories and sequence organization can jointly explicate how communication occurs on a children and young people's helpline. With our analytic focus on children and young people's accounts of their own experiences, we drew on how they made sense of the family and SOL categories. In other words, it was not the researchers' description but rather how the participants themselves organised their talk to draw on such category work.

Clinical relevance summary

The chapter has focused on how young clients provide accounts of their activities and responsibilities within family life. Family issues are one of the main reasons that young callers contact Kids Helpline, and counselling sessions open up possibilities for discussing norms of family life, roles and relationships of family members, and family responsibilities. The analytic approach of membership categorisation has shown that clients' accounts and experiences of disconnect between normative and actual responsibilities can be understood through common-sense category membership work. When there is mismatch within category membership (family and SOL), these descriptions might inform clinicians of how clients discuss activities within 'dysfunctional' families and provide clinical possibilities for how to explore these matters with clients.

Family situations are complex and require sensitivity and understanding of the diversity of family contexts. Children and young people use counselling support across different modalities to report their troubles. Counsellors draw on a range of communication strategies to support the young clients to understand their particular contexts of family life. Counsellor strategies are designed to confirm, commend, validate, and seek more information from the client by probing and asking, and clients are oriented to as someone with feelings and with the competence and resources to understand and find solutions to their troubles.

Clients and counsellors draw on common-sense knowledge of family relationships with their normative distribution of rights and responsibilities. This type of cultural knowledge is a built-in feature of membership categories, and it is shown here within the categorisation work of both clients and counsellors as they make sense of the situation. Within this chapter, we showed that the counsellors went beyond a sense of shared understanding to proceed to expand the categorical relations that were invoked by clients and counsellors. The shared understanding develops into displays of affiliation that are oriented directly to the institutional philosophy of Kids Helpline: 'we care, we listen'.

The research findings of this chapter investigating how counsellors and young people interact on helplines address a broader research agenda. By taking these findings into account, this chapter argues against the use of prepared agendas or templates for counsellors to use when interacting with young people in helpline contexts. Building the clinician-client relationship requires a sensitive and skilled approach where counsellors actively listen to the clients' presentations of their concerns. This relationship is built moment by moment through the clinician–client interaction as it unfolds. The counsellor's work

Table 31.1 Clinical practice highlights

---

1. Family situations are complex and require sensitivity and understanding of the diversity of family contexts, taking into account differing family, cultural and religious practices.
2. Counselling sessions can open up possibilities for discussing family life, roles and relationships of family members, and family responsibilities.
3. Invoking category membership (family and SOL) descriptions might provide clinical possibilities for how to explore family matters with young clients.
4. Listening for mismatches between category membership can provide important clues for how clients discuss family matters and provide a strategy for clinicians for how to explore these concerns.
5. The clinician-client relationship is fragile and built moment by moment through the interaction as it unfolds. It would not be possible to achieve this quality interaction by following a prepared service guideline template on how to respond to the clients. An interactive approach is required where the clinician is able to expand, with the client, their shared understandings of family categories and relationships as they unfold.

---

is to expand, with the client, shared understandings of family categories and relationships as they unfold. If the counsellor were constrained by service guidelines that require following a template or stepped procedures on how to respond to the client, then no template could ever fully consider the multiplicities of possibilities that the client might want to talk about, or the unfolding of the counselling session. For a simple summary of the implications for practice, see Table 31.1.

Summary

Analysing client–counsellor talk as it unfolds moment by moment shows how counsellors work within an institutional philosophy of empowering the young caller by recognising their competence and capacity to resolve their issues and to come up with resolutions to their problems. Analysis shows how both client and counsellor make family and age categories relevant, with openings for both commendation and acknowledgement (the counsellor's response to how the client has managed) and complaint (the client's account of missing out on the SOL of being a teenager). The chapter contributes to understandings of service provision for children and young people, such as how counsellors can communicate effectively with children and young people.

Acknowledgements

This project was supported by an Australian Research Council Discovery grant (Project ID: DP0773185) with ethical approval from Queensland University of

Technology. We thank Kids Helpline and BoysTown, and the counsellors and clients who took part in the study.

Note

1. The web chat session transcripts consist of three time slots: For example, in the first line in Extract 2, the first slot [16:22] refers to the actual time of the interaction on a 24-hour clock. The second time slot refers to when the turn was posted [00:50], occurring 50 seconds into the interaction. The third time slot [00.41] identifies the lapse of time since the last posting, 41 seconds. We use the second slot to identify the line being discussed.

References

Antaki, C. (2008). Formulations in psychotherapy. In A. Peräkylä, C. Antaki, S. Vehviläinen, & I. Leudar (Eds.), *Conversation analysis and psychotherapy* (pp. 26–42). Cambridge: Cambridge University Press.

Atkinson, M. A. (1980). Some practical uses of 'a natural lifetime'. *Human Studies, 3*, 33–46.

BoysTown (2013). *Giving voice: BoysTown annual report 2013*. Brisbane: BoysTown.

Butler, C. W., Potter, J., Danby, S., Emmison, M., & Hepburn, A. (2010). Advice implicative interrogatives: Building 'client centred' support in a children's helpline. *Social Psychology Quarterly, 73*(3), 265–287.

Castro, D. M., Jones, R. A., & Mirsalimi, H. (2004). Parentification and the imposter phenomenon: An empirical investigation. *The American Journal of Family Therapy, 32*, 205–216.

Cromdal, J., Danby, S., Emmison, M., Osvaldsson, K., & Cobb-Moore, C. (forthcoming) "umm basically..its the usual whole teen girl thing": Stage of life categories on a children and young people's helpline.

Danby, S., Baker, C., & Emmison, M. (2005). Four observations on opening calls to *Kids Help Line*. In C. D. Baker, M. Emmison, & A. Firth (Eds.), *Calling for help: Language and social interaction in telephone helplines* (pp. 133–151). Amsterdam/Philadelphia: John Benjamins.

Danby, S., Butler, C. W., & Emmison, M. (2009). When 'listeners can't talk': Comparing active listening in opening sequences of telephone and online counselling. *Australian Journal of Communication, 36*(3), 91–113.

DiCaccavo, A. (2006). Working with parentification: Implications for clients and counselling psychologists. *Psychology and Psychotherapy: Theory, Research and Practice, 79*, 469–478.

Drew, P., & Holt, E. (1988). Complainable matters: The use of idiomatic expressions in making complaints. *Social Problems, 35*, 398–417.

Earley, L., & Cushway, D. (2002). The parentified child. *Clinical Child Psychology and Psychiatry, 7*(2), 1359–1045.

Emmison, M., Butler, C., & Danby, S. (2011). Script proposals: A device for empowering clients in counselling. *Discourse Studies, 13*(1), 3–26.

Emmison, M., & Danby, S. (2007). Troubles announcements and reasons for calling: Initial actions in opening sequences in calls to a national children's helpline. *Research on Language and Social Interaction, 40*(1), 63–87.

Garfinkel, H. (1991). Respecification: Evidence for locally produced, naturally account-able phenomena of order, logic, reason, meaning, method, etc. in and as of the essential haecceity of immortal ordinary society, (I) – an announcement of studies. In G. Button (Ed.), *Ethnomethodology and the human sciences* (pp. 10–19). Cambridge: Cambridge University Press.

Harris, J., Danby, S., Butler, C., & Emmison, M. (2012). Extending client-centered support: Counselors' proposals to shift from email to telephone counseling. *Text and Talk, 32*(1), 21–37.

Heritage, J. (1998). Oh-prefaced responses to inquiry. *Language in Society, 27*, 291–334.

Heritage, J., & Sefi, S. (1992). Dilemmas of advice: Aspects of the delivery and reception of advice in interactions between health visitors and first-time mothers. In P. Drew & J. Heritage (Eds.), *Talk at work: Interaction in institutional settings* (pp. 359–417). Cambridge: Cambridge University Press.

Heritage, J., & Watson, D. (1979). Formulations as conversational objects. In G. Psathas (Ed.), *Everyday language: Studies in ethnomethodology* (pp. 123–161). New York: Irvington Publishers.

Hester, S., & Eglin, P. (1997). Membership categorization analysis: An introduction. In S. Hester & P. Eglin (Eds.), *Culture in action: Studies in membership categorization analysis* (pp. 1–23). Washington, DC: International Institute for Ethnomethodology and Conversation Analysis & University Press of America.

Hooper, L. M. (2008). Defining and understanding parentification: Implications for all counsellors. *The Alabama Counselling Association Journal, 34*(1), 34–43.

Jankowski, P. J., Hooper, L. M., Sandage, S., & Hannah, N. J. (2013). Parentification and mental health systems: Mediator effects of perceived unfairness and differentiation of self. *Journal of Family Therapy, 35*, 43–65.

Leon, K., & Rudy, D. (2005). Family processes and children's representations of parentification. *Journal of Emotional Abuse, 5*(2–3), 111–142.

McMahon, T. J., & Luthar, S. S. (2007). Defining characteristics and potential conse-quences of caretaking burden among children living in urban poverty. *American Journal of Orthopsychiatry, 77*(2), 267–281.

Nikander, P. (2002). *Age in action: Membership work and stages of life categories in talk.* Helsinki: The Finish Academy of Science and Letters.

Nikander, P. (2009). Doing change and continuity: Age identity and the micro-macro divide. *Ageing & Society, 29*, 863–881.

Pomerantz, A. (1986). Extreme case formulations: A way of legitimizing claims. *Human Studies, 9*, 219–229.

Sacks, H. (1972). An initial investigation of the usability of conversational data for doing sociology. In D. Sudnow (Ed.), *Studies in social interaction* (pp. 31–74). New York: The Free Press.

Sacks, H. (1995). *Lectures on conversation* (G. Jefferson, Trans. Vol. I and II). Oxford, UK: Blackwell.

Silverman, D., & Peräkylä, A. (1990). AIDS counselling: The interactional organisation of talk about 'delicate' issues. *Sociology of Health and Illness, 12*(3), 293–318.

Speier, M. (1973). *How to observe face-to-face communication: A sociological introduction.* Pacific Palisades, CA: Goodyear Publishing Company.

Stiffler, D., Barada, B., Hosei, B., & Haase, J. (2008). When Mom has breast cancer: Adolescent daughters' experiences of being parented. *Oncology Nursing Forum, 35*(6), 933–940.

Van Parys, H., & Rober, P. (2013). Trying to comfort the parent: A qualitative study of children dealing with parental depression. *Journal of Marital and Family Therapy, 39*(3), 330–345.

Recommended reading

- Atkinson, M. A. (1980). Some practical uses of 'a natural lifetime'. *Human Studies, 3,* 33–46.
- Hester, S., & Eglin, P. (1997). Membership categorization analysis: An introduction. In S. Hester & P. Eglin (Eds.), *Culture in action: Studies in membership categorization analysis* (pp. 1–23). Washington, DC: International Institute for Ethnomethodology and Conversation Analysis & University Press of America.

32

'And You? What Do You Think Then?' Taking Care of Thought and Reasoning in Intellectual Disability

Marilena Fatigante, Saverio Bafaro, and Margherita Orsolini

Introduction

Intellectual disability is a frequent but still ill-defined condition. This encompasses individuals with heterogeneous cognitive profiles who are included in a same diagnostic category, on the basis of (1) an IQ score below approximately 70 and (2) a clinical assessment of deficits in adaptive functioning. In the new *Diagnostic and Statistical Manual of Mental Disorders (DSM-5)*, the core deficits of *intellectual functioning* include the ability to reason, plan, solve problems, think abstractly, comprehend complex ideas, learn academic skills, and learn from experience. Although each of these complex abilities could be analysed through both specific neuropsychological tasks and qualitative ecologically sound methods, the assessment of intellectual functioning is still centred on IQ. In recent years, there has been considerable debate as to whether a general mental function such as intelligence can map onto measures drawn from statistical analyses on the subtests concurring to IQ or, whether it should be better grounded onto the findings of developmental psychology and neuropsychology. It has been pointed out that there are several types of intelligent thinking (Gardner, 1999; Sternberg, 1985), and that acting purposefully, monitoring behaviour, planning and organising activities, are important cognitive functions that are not tapped by psychometric intelligence tests (Ardila, 1999; Fiorello et al., 2007; Greenspan & Woods, 2014; Salvador-Carulla & Bertelli, 2008).

An IQ denoting even a mild intellectual disability has been traditionally conceived as an index of *thinking* disability. Based on this assumption, it has been long anticipated not only that thinking, reasoning, and solving novel problems are core cognitive deficits of individuals with intellectual disability but also that such deficits cannot be modified by training or rehabilitation practices. Arguing against such a claim, we highlight that reasoning processes (including for instance the ability to logically operate upon reality, link results to antecedents, build and verify conjectures, reflect upon experience etc.) are supported by specific cognitive functions (e.g. working memory) which, while they are often impaired in individuals with intellectual disability (Danielsson, Henry, Rönnberg, & Nilsson, 2010; Edgin, Pennington, & Mervis, 2010; Karmiloff-Smith, 1998; Schuchardt, Maehler, & Hasselhorn, 2011; Van der Molen, Van Luit, Van der Molen, & Jongmans, 2010), they can nonetheless improve as a consequence of training in individuals with mild intellectual disabilities (Orsolini, Latini, Salomone, & Melogno, 2014; Söderqvist, Nutley, Ottersen, Grill, & Klingberg, 2012; Van der Molen, Van Luit, Van der Molen, Klugkist, & Jongmans, 2010). Whether or not improvements in each of these individual functions can transfer onto more comprehensive reasoning skills is still an open research issue; yet, the claim that reasoning and thinking cannot be treated in individuals with intellectual disability is not empirically grounded.

We should also consider that the ascription of limitations in overall reasoning abilities leads parents, teachers, and clinicians to create simplified communicative environments that 'inhibit the acquisition and display of pragmatic language skills' (Hatton, 1998, p. 80). Thus, emergent reasoning skills in children diagnosed with a mild intellectual disability are likely to be hidden by conversational practices aimed at simplifying discourse, rather than soliciting cognitive and pragmatic skills.

In this study, we take the adult–child discourse as an interactive context in which reasoning, far from being the property of a single individual's 'mind', can emerge as a social activity and a co-regulated process, being locally managed by both the adult and the child, who are mutually responsible for its development and its consequences (cf. Wootton, 1997). Analysing discourse in such a context allows the psychologist to observe verbal reasoning as emerging within what Vygotsky (1934/1978) called 'zone of proximal development', that social and interactional environment in which the adult engages with the child in the process of meaning-making, and enhances the child's attempts to make sense out of his own (and the interlocutor's) utterances and speech, by progressively explicating and explanating (see also Schäfer & Muskett, Chapter 27, this volume).

By complementing the cognitive assessment this way, we can promote a change in the perspective of developmental assessment overall, in line with what Bruner mentioned for typically developing children: 'given an appropriate, shared social context, the child seems more competent as an intelligent social operator than she is as a "lone scientist" coping with a world of unknowns' (Bruner & Haste, 1987, p. 1).

Bruner's claim is particularly suited for children who have been diagnosed with intellectual disability; due to their experiences of failure in complex tasks and academic performance, the lack of reciprocity they suffer in interaction with peers (Tipton, Christensen, & Blacher, 2013), and the simplified forms of discourse they are addressed, these children often construct their self-image as individuals with low intelligence, with dramatic effects on their self-esteem (Heiman, 2001).

As psychologists who take discourse as an essential arena to observe and assess the emergent interpretive abilities of the child, our efforts aim to identify *in* discourse those elements that are conceived as traditional ingredients of thought and reflection. Vygotsky and the cultural-historical approach to thought and learning is certainly the main framework inspiring this enterprise, given the strong premise of this school that higher psychic functions stem from internalisations of social actions and are *mediated* by language as a primary symbolic artefact (Cole, 1996; Vygotsky, 1934/1978).

Reflecting in discourse: A Deweyan approach

Besides the Vygotskian approach, we were also greatly supported by the tradition represented by the work of John Dewey and his inspiring formulation of 'Reflective Thought'.

From his elaboration, we sorted criteria to identify and collect instances of talk, which more than others can be offered as evidences of the reasoning competences of a child, diagnosed with an intellectual disability. In 1933, Dewey defined the notion of reflective thinking as 'active, persistent, and careful consideration of any belief or supposed form of knowledge in the light of the grounds that support it and the conclusion to which it tends' (p. 9).

In contrast to a surface approach to learning and thinking, which relies on a procedure of 'trial and error' and brings about 'unassimilated knowledge' (p. 143), reflective thinking implies at a basic level conceptualising a relation between 'what we try to do and what happens in consequence' (p. 143). Questioning how (i.e. the thinking procedures by which) this relation is produced is the further step of a reflective searching for the detailed connections between activities and consequences. Bearing the uncertainty, a sense of concern and emotional involvement in what is going on, a questioning orientation, taking the risk of not arriving at certain conclusions, are all constitutive elements of reflective thinking.

Project overview

Relying on the conceptual background so outlined, this chapter explores emergent reasoning and reflective thinking in a boy who has been diagnosed with intellectual disability. We analyse how the boy interacts with a psychologist who *engages* in the child's spontaneous narratives and supports his attempts to explain and make sense out of what happens to him.

Data and context

The data we analyse are drawn from a series of individual assessment and treatment sessions between one senior psychologist (Margherita, Director of the Centre) and Davide (pseudonym), 14.2 years old at the time of the diagnostic consultation (started in October 2009) in the University clinical centre. The conversations we examined include both the (overall) four meetings, which constitute part of the assessment protocol and a total of 10 adult–child sessions, extracted from the first 1.5 year of the treatment (up to September 2011).

To this corpus, we added a set of six peer (dyadic) sessions of therapy involving Davide and a peer who was also diagnosed with a mild intellectual disability. In both these sessions two junior psychologists worked with the adolescents on semi-structured activities, including emotion recognition of visual stimuli and writing and discussion of self-narratives. The sessions took place in a laboratory of the university clinical centre and were video-recorded (except for the peer sessions, which were only audio-recorded). Informed consent for the recording and the treatment of the material was gained in written form by the children's parents and obtained orally by the adolescents. The conversations of all sessions were first transcribed as regards content only; a detailed Jeffersonian transcription was applied to those extracts that were sorted as instances of reflective and argumentative episodes for the paper (please see preface of the handbook for a table of symbols).

A Vygotskian framework for assessment and treatment

Professionals and trainees working in the centre assume conversation as a privileged observational platform to qualitatively analyse how the child is able to take into account the other's perspective, construct and make sense of the experiences in which s/he is involved, engage in explanations, and label and interpret emotions.

Discourse also has a central role in cognitive treatment of children with mild intellectual disabilities. In this clinical centre, treatments involve *core trainings* of specific cognitive functions (particularly of verbal working memory as reported in Orsolini et al., 2014), and *strategy trainings*, which are implemented through adult-child verbal interaction inspired by the Feuerstein approach (Feuerstein, Falik, Rand, & Feuerstein, 2006). Phrases such as 'according to you', 'how this thing *appears* to you' or 'what did this *mean* to you', 'why do you think this happened?' are examples of adult language stimulating explanatory discourse and self-reflection. Eliciting and discussing narratives about the child's personal experiences is the *warming up* stage of each training session. For some children, including Davide, narrative talk becomes an activity 'per se', and may result particularly useful to report problematic events that have caused him some suffering and 'urge', then, further elaboration.

Davide

Davide is a boy who arrived at the centre with a diagnosis of intellectual disability of a non-specific aetiology, whose IQ ranged between 60 and 70 in different tests across the elementary and junior school years. At age three, Davide was given a diagnosis of global developmental disorder. After two years of treatment within a small group of children, his communication skills increased in a remarkable way and the diagnostic label changed to that of a specific language impairment (evidenced in receptive language, verbal dyspraxia, phonological disorder), associated with difficulties in emotion regulation. Davide then attended speech therapy and entered the primary school one year later than expected, assisted by a special educator who, according to the Italian law, helps the children with special needs for a varying amount of time (according to the severity of their impairment) in regular classes.

In the initial interview in our university clinical centre, Davide's parents reported that their son had always been very concerned with the difference in achievement he could detect between himself and his peers at school. Since his referral to our centre when he was 14 years old, he showed his interest in sorting his experiences and difficulties as objects of explicit narrative and shared discussion. He eventually decided to accept to attend the therapy in our centre telling us: 'you can perhaps change my life'. We then clarified that we could not 'change life' but only train skills and give support to his attempts to improve. To the question 'What are the directions in which you would like to change?', Davide answered that his desire was to communicate more with his schoolmates. After an initial treatment that was centred on working memory and executive functions (see Orsolini, 2012), Davide participated in weekly meetings first with another teen and then with a small group of peers with intellectual disabilities.

Analysis

Reflecting on others' behaviour

We examine an extract from the second meeting of the assessment stage. Davide refers to a scout excursion, focusing on the arrival of a new peer. We focus on the argumentative abilities of the boy displayed in the narrative, his inquiry attitude, his involvement in searching for reasons and his sharing of this journey with the interlocutor. Our extracts show the original Italian, and the translation to English (in italics and bold).

Extract 1a. 30 November, 2009

Participants: Davide (D.), T (senior psychologist)

```
1   D:    poi c'era questo compagno nuovo,
          then there was this new mate,
2   T:    ah si? è un nuovo entrato?
          uh really? is he a new entrance?
3   D:    si, non è: no:: è l'unico. so:: so entrati altri nuovi
          compagni.
          yes, he's no:t no::t the only one. o:: other mates have
          entered.
4         però questo, è un compagno nuovo (.) che è russo.
          but this one, he is a new mate (.) who is Russian.
5   T:    addirittura russo! ma parla italiano?
          Russian seriously! but does he speak Italian?
6   D:    si. parla italiano, (0.6) si chiama Pavel.
          yes. he speaks Italian, (0.6) his name is Pavel.
7   T:    mh!
8   D:    e questo ragazzo::: non è bravo.
   →      and this gu::::y is not good.
9         eh::: mh:::: perché::: c'ha dei problemi grossi.
   →      eh::: mh:::: becau:::se he has big problems.
10  T:    ho capito.
          I see.
11  D:    e:::: dà:: dà fastidio a tutti.
          and:::: he:: he annoys everybody.
12        per esempio, ieri sera,
   →      for instance, yesterday evening
13  T:    mh=mh
```

Sequence 1–13 shows how Davide progressively builds the boy of the narrative as a *special* character, anticipated as *different* (a *new* mate, *Russian*) even before he is properly identified and given a name: Pavel (line 6). In this way, Davide establishes in discourse the sense of uncertainty and challenge due to the observation of another's 'strange' behaviour, and anticipates his inquiry into the *reasons*, which motivated the event.

After having identified the boy, Davide makes a straight evaluative claim: *he is not good* (line 8), immediately supplementing the description of the *dispositional* state of the child with behavioural evidence: *he annoys everybody* (line 11), to account for his claims. After line 13, Davide provides an example of how the child behaved; the telling of the actual episode is used as a source of evidence for his original assessment of the child. We then observe displayed in this discourse a basic argumentative structure, characterised by evaluative claims that are accounted for by descriptive or narrative evidence.

Extract 1b.

```
18   D:    e questo ragazzo:, imitava:: la voce degli ↑a:nimali,
           and this guy:, imitated:: the voice of the ↑a:nimals,
19   T:    ma dai! (.) e come mai?
           you kidding! (.) and how come?
20         glielo avete chie:sto come mai faceva così?
           did you ((pl.)) a:sked him why he was doing this?
21   D:    no. purtroppo:: i versi degli animali, si!
           no. unfortunately:: the verses of the animals, right!
22         eh::: dei cani, si! no. non l'ha chiesto.
           eh::: of the dogs, right! no. *hasn't asked.
23         però↑ i miei compa:gni hanno avuto m↑o::lto fastidio e:::
           bu↑t my ma:tes have been much ann↑o::yed and:::
24         e poi si so' inca↓zzati.
           and then they got ↓pissed.
25   T:    p↑ure!
           r↑eally!
*((null subject in the original))
```

As storyteller, Davide is oriented towards providing as much minute details as he can about how the character's odd behaviour affected those present in the episode. In so doing, he presents himself as a valid and reliable observer of the social world he inhabits. The therapist intervenes before the story reaches a conclusion: she casts, by her assessment (line 19), the child's behaviour as exceptional, thus substantially aligning with Davide's stance in reporting the facts he witnessed. Also, the therapist proposes a scenario for which Davide and the other peers could inquire after the *reasons* underlying the boy's behaviour (lines 19–20). Interventions like these ones show the therapist's attempts to have the child engage in questioning motivations and ascribing causes to events (and behaviours), which can be made explicit and debated in interaction. Reflecting upon the unconventional, bizarre behaviour and attitudes manifested by the child in the narrative, fostered Davide to question further the 'reasons' behind such an oddity.

Extract 1c.

```
60   D:         però io non ri↑esco a capire qu↑esto:: sa parlare bene.
           →    still I c↑an't understand th↑is:: he talks well.
61              ma- si comp↑orta da bambino piccolo!
                But- he beh↑aves like a small child!
62   T:    mh!
63   D:    non si capi:: non riesco::: a cap↓ire:::
           →    it's not unde: I canno::t underst↓a::nd
```

64		perché qu↓esto si comporta, (.) da bambino p<u>i</u>ccolo.
		why this g↓uy behaves, (.) as a small ch<u>i</u>ld.
65		non cap↓isco.
	→	*I cannot underst↓and*
66	T:	eh! c'avrà le sue difficoltà probabilmente.
		eh! he will probably have his difficulties.
67	D:	e::: io capivo meglio se:: i::
	→	*e::: I would have better understood if:: I::*
68		io capivo m<u>e</u>:glio,
	→	*I would have b<u>e</u>tter understo<u>o</u>:d*
69		se questo non sapeva par<u>la</u>:re, allora
		if this guy could not <u>ta</u>:lk, then
70	T:	eh!
71	D:	però non riesco a cap<u>i</u>:re, (.) perché questo c'ha dei probl<u>e</u>mi gr<u>o</u>ssi,
	→	*but I cannot underst<u>a</u>:nd, (.) why this has such b<u>i</u>g pr<u>o</u>blems,*

The extent and limits of understanding or *not* understanding is what Davide wonders about. In the overall sequence, Davide manifests his struggle to make sense of a peer's strange behaviour that appears as not justified by any language impairment. It is easy to recognise how the report of Pavel's trouble constitutes a chance for Davide to make relevant his own difficulties (cf. Finlay & Lyons, 2000 on the specific reference to others' *bad behaviour* as a tool to develop downward comparisons), linked to, among other things, articulatory problems. That is, as he interrogates the social and interpersonal reality *around* him, Davide also questions his own abilities, i.e. what happens *inside* him.

Explicitly questioning the *reasons* why certain events happen, and why people behave in certain ways, is the first evidence of Davide's reflective attitude developing in these learning therapy sessions, which attracted our attention. While reporting, Davide addresses the therapist – as well as himself – with questions regarding the underlying forces or, dispositions, motivating a person's behaviour, displaying as deeply aware of the normative grounds for which certain demeanours are regarded as 'plain' and conventional, whereas others are rejected as problematic and bizarre.

'Visible to oneself': Davide's reflections on his own intellectual abilities

In this section, we analyse a set of different instances, which show how Davide applies a questioning stance and reflection to himself and examines in a critical light his 'impaired' abilities too. In (2a), Davide reports to the senior therapist a narrative, implying quite a severe derogatory judgement on his way of thinking:

Extract 2a. 4 March 2010

```
1    D:    i:o certe volte quando parlo:: (    ) (.) ho detto una
           stupidaggine.
           I: sometimes when I spea::k (    ) (.) I told silly
           things.
2          quand'↑era il mio compleanno:: o:: (.) il °compleanno di
           mio fratello°,
           when it w↑as my birthda::y or:: (.) °my brother's
           birthday°,
3          ↑ah si:! il mi:o compleanno, (0.4) sono venu:ti dei mie:i
           nonni,
           ↑uh ye:s! my: birthday, (0.4) my gra:nd parents ca:me,
4          che abitano a:: xxx= di xxxx
           they live in:: ((name of a town))=from ((revision of the
           name of the town))
5          >gli ho ↑detto ai nonni di xxx<
           >and I ↑told the grandparents from < ((home village of
           the grandparents))
6          vole:te che vi facciamo porta::re la ↓torta. a casa?
           would yo:u like that we gi::ve you some ↓cake. to bring
           home?
7          mia z↑ia, (.) mi ha detto. ↑↑sei sicu:ro di quello che
           stai dice:ndo?
           my a↑unt, (.) told me. ↑↑are you su:re of what you are
           sa:ying?
8          perché gu↑arda che la t↑orta non la puoi portare. ↑sennò
           si squaglia!
           for l↑ook you ↑can't bring the cake. ↑otherwise it will
           melt!
9          (1.0)
10         se fosse stato mio- mio cugi:no, (0.4) non la diceva mai
           if my- my cou:sin would be there, (0.4) he would have
           never told
11         una sciocchezza del genere.
           a foolish thing like this.
12         e nemm↑↑e::no un mio comp↑agno di ↓scuo:la. (.) h.=
           nor ↑↑a:ny of my sch↑oolmate ↓wou:ld. (.) h.=
13   T:    =dici che è stata una grande sciocche:zza?
           you say that it was such a big foo:lishness?
14   D:    s:i.
           y:es.
```

Here, Davide portrays a negative characterisation of his conduct: he tells 'silly things'.

Note how he builds this assertion. He reports that something, namely *telling silly things*, happens *sometimes* during a routine activity such as, talking. As such, the formulation is manufactured as a 'script formulation' (Edwards,

1994), which, despite not occurring on a regular basis (such as, *always* or, *habitually*) can still be forecasted as possible, and non-exceptional.

In much the same format as he built the description of Pavel in two steps, first ascribing him a stable characteristic (he is 'not good', and 'he has big problems') and then adding the evidential basis for this ascription, Davide reports in detail an exemplary instance, which accounts for the characteristic he mentioned. The narrative of the cake, including the animation (Goffman, 1981) of the aunt's voice challenging the boy's plan, works as an argumentative resource in support of the original claim *'when I talk I tell silly things'*.

Attending to his report, the therapist first offers the boy a chance to consider again his claim (line 13) then, she powerfully expresses a stance upon the matter, casting Davide's formulation as extreme (*goodness!*), and envisioning an exception to the script that he laid out.

Extract 2b.

```
15   T:   ↑m̲a̲mma m↑↑i̲::a =
          ↑my ↑↑goo̲:ddess=
16        a ↑ME non mi ↑se̲:mbra tu̲tta sta' grande sciocche̲:zza.=
          Da̲vide.
          I ↑DOn't SEE: it su̲ch a bi̲g foo̲:lishness.=Da̲vide.
17        ↑m̲a̲mma mia! non ↑h̲ai pensa̲:to che magari forse la to̲:rta,
          ↑my̲ goodess! you did ↑no̲t thi̲:nk that perhaps the ca̲:ke,
18        si sarebbe potuta rovina̲:re.
          could have ru̲:ined.
19        ci stanno delle ↑t̲orte che ↑m̲i̲ca si ro↑vi̲:nano.
          there are ↑c̲akes that do ↑n̲o̲t ru̲:in at all!
20   D:   la torta si può rovina̲:re, <da quando> è stata ↓presa.
          the cake can ru̲:in, from the time when it has been
          ↓bought.
21        loro abitano lo- lonta̲:no, e e- giustame̲:nte,
          they live fa- far awa̲:y, and and- o̲:bviously,
22        dopo tanto t- te̲mpo, (0.6) si può squa̲:g°liare°.
          after a l- long ti̲me, (0.6) it can me̲:°lt° .
23        ma i̲o n- (0.4) non c'ho pen°° ↓sato. (.) a questo fatto.°°
          but I̲ did n- (0.4) not con°° ↓sider. (.) this fact.°°
24   T:   non ↑↑è così gra̲:ve.= Davide se̲nti. non E' cosI' grave!
          it's ↑↑not tha̲:t serious. = Davide li̲sten. it's NOT
          thAt serious!
25   D:   non ↑è grave.
          isn't ↑it serious.
26   T:   ↑no::. non ↑è gra:ve Da'. ↑eh! ch↑e è.
          ↑no::. it's ↑not serious Da'. ↑eh! for ↑gooddess sake.
27        non ↑so:' cose gravi. =senti ti dico una cosa Davide:.
          they're ↑no:t serious things.= listen I tell you
          something Davide:.
```

28		se <↑tu stai più ca:lmo.> e ri<fletti con ca:lma>,
		if <↑you 're more rela:xed.> and <reflect pa:tiently>,
29		senza avere sempre questo pre:side dentro, (.)
		without always having this schoo:l dire:ctor inside, (.)
30		che ti pa:rla, (.) >che ti dice sempre< prendi ci:nque,
		who talks to you, (.) >who always tells< you get fi:ve,
		((=poor school grade)),
31		sba:gli, dici le sciocche:zze,
		you make mista:kes, you tell foo:lishness,
32	D: →	ma- ma ↑perché. que:lla non è una sciocche-zza?
		but- but ↑why. i:sn't it a foolish-ness then?
33	T:	cosa?
		what?
34	D: →	il ↑fatto della to:rta! (.) non è una sciocchezza?
		th↑at of the cake! (.) isn't it a foolishness?
35	T:	↑no! = non E' una sciocchezza! è una cosa normale, una
		↑cosa che-
		↑↑no! = it's NOt a foolishness! it's a normal ,
		s↑omething that –
36	D: →	è una cosa sTU:pida!
		it's something sTU:pid!
37	T:	ma ↑↑non è: una cosa stupida! perché prima di tutto,
		but it's ↑↑no:t a stupid thing instead! 'cause first of
		all,
38		ci stanno delle torte che non si rovinano.
		there are cakes that do not ruin. (.)
39		anche se le fai stare tanto tempo fuo:ri. (.)
		even if you take them out for a lo:ng time.
40		di↑pende dal tipo di torta.
		it de↑pends on the type of cake.
41		è una cosa a cui un raga:zzo potrebbe non averci pensato.
	→	*it's something which a gu:y could even not think about.*

The strength of argumentation, which Davide uses to account for the rationality and legitimacy of his position, is such that the adult therapist enters a series of vigorous attempts to remove the severe moral judgement that Davide has laid out on his performance, explicitly entering a disagreement with him (e.g. lines 15–16). The therapist provides an alternative explanation for his presumed failure, casting his belief as something that might instead have proved true, given different conditions (e.g. for cakes that do not ruin). Yet, Davide re-enacts his argument (lines 20–23), better articulating the deductive reasoning upon which his judgement relies: (cakes) *can ruin* (after a certain time), grandparents *live far away*, (that) *cake can melt* (due to the length of time of the travel).

Whereas Davide is oriented towards getting confirmation for his explanation of the episode the therapist actively contests the moral implication that Davide drew from it, halting his inclination towards a negative judgement upon his thoughts. The therapist's turns stem as urgent attempts to provide Davide with a less incriminating framework to interpret his actions: she opposes Davide's assertions, and eventually (line 27) she explicitly turns to an instructional frame, advising him – by using the metaphor of a school director wired *into* the boy – to ban this critical attitude once and for all (lines 29–31). The therapist's advice (which remains uncompleted) does not stop Davide from questioning the validity of his claim (see lines 25, 32, 34, 36), thus demonstrating his ongoing engagement in the reflexive interrogation about how his mind operates. At this point the therapist elaborates an alternative account with more detail, which is able to withstand comparison with the original one (lines 37–41).

The provision of a new reading of the story acknowledges the *situated* and mobile character of the interpretations, and the chance to account for events in varied ways: whereas Davide had adhered 'irreflexively' to the aunt's interpretation of the event, the therapist prompts him to consider other routes and 'tentative hypotheses', and see how he may take a different, more self-advantageous seat, as an observer of his thinking procedures. To this aim, also noticeable is the therapists' shift (line 41) from reference to Davide as the subject of the narrative to reference to a general 'guy'. This also works to cast Davide and his reasoning attitude as non-exceptional but, rather, similar to all kids of his age.

Rapley, Kiernan, and Antaki (1998) showed that adults with intellectual disabilities – contrary to common assumptions about their tendency to deny ascriptions of incompetence and disability – were able to orient relevantly to their self-categorisations, exhibiting themselves as acute observers of their (varied, indeed) identities. Here, Davide pushes this orientation further, making all the more visible his own self-ascription of intellectual impairment. At the same time, he also makes visible – and open to be acknowledged by the therapist and discussed with her – his determined and extensive interrogation about how his mind works.

Reflecting upon 'stupidity'

We also examined instances of talk developing during more structured activities, led by two junior therapists in the peer setting. Here, Davide and a peer (Lelio) were presented with material (photographs or sketches) meant to solicit emotion identification and labelling. In one of these, Davide is asked to comment upon a picture, portraying a child who sits at a desk with a notebook, with a fist tightened and the other hand pushed on his head.

Extract 3a. 12th May 2010

Participants: t1 and t2 (junior psychologists), D (Davide), L (Lelio, 13.5 years old)

```
1    t1:   che sta pensando?
           what is he thinking?
2          (2.2)
3          che si sta dicendo?
           what is he telling himself?
4    D.:   fo:rse sta pensando a:, ma quanto sono stupido.
           ma:ybe he is thinking of:, but how stupid I am.
5    t1:   dici?
           you say so?
6          (2.0)
7          può: esse:re.
           it may be:.
8          dice (.) ma quanto so: stupido!
           he says (.) but how stu:pid I am!
9          non mi sta riuscendo questo compito
           I'm not succeeding in this task
10         [può: esse:re
           [it ca:n be:
11   D.:   [si!
           [yes!
12   t2:   [come mai pensa che? ((to D))
           how come he thinks like that?
13   L.:   [giusto!
           right!
14   t2:   co:me mai pensa che sono stupido? ((to D))
           how co:me that he thinks I am stupid?
15   L.:   °(    ) tutte le cose che diceva lui° però=vabbè
           °(    ) all the things he said° but =ok
16   t1:   lo ha detto bene lui
           he said it the right way
17   L.:   e=vabbè quindi giusto
           well =right so it's fine
18   t1:   che dici Saverio?
           what do you say Saverio ((=t2's name))?
```

The child here interprets the angry emotion shown by the child's posture as associated with an ascription of incompetence (stupidity). Similarly to the senior therapist in line 13 of Extract 2a, the junior psychologist here (t1) addresses the child with the question: *you say X?*, thus casting the child's response as open to revision (Schegloff, Jefferson, & Sacks, 1977). Not receiving any response from Davide, the therapist accepts his contribution; still, she qualifies it as something uncertain, hypothetical (line 7); while doing so, she embodies the version given by the child and animates the character accordingly (lines 8–9). Having acknowledged the interpretation suggested by Davide (who confirmed, line 11), t2 pushes the inquiry a step further, soliciting Davide (lines 12, 14) to consider (in a virtual scenario) what kind of evidences can motivate the (presumed) ascription of stupidity that the child in the picture does to himself. See how a general preference towards considering any claim as open to justification and debate, inspires also the question that t1 addresses to t2 (line 18), as she asks for his opinion. Yet, t2's response orients more to the institutional aim of soliciting reflection from the adolescents, rather than displaying his own stance, and thus sends back the question again to Davide:

Extract 3b.

```
19   t2:     dico. come mai pensa che sò stupido?
       →     I say. how come he thinks I am stupid?
20   D.:     ah: è perché è una cosa che faccio io
             uh: it's because it's something I do
21   t1:     ah
             uh
22   D.:     allora:: penso: che: questo si sta dicendo che è
             stupido.
       →     so:: I think: that: this one is telling himself that he
             is stupid.
23   L.:     io me lo dico de:ntro. >però non faccio così<
             I tell myself inside. >but I don't do like this<
```

Different to *why*, the question *come mai* (how come) indexes a quality of problematisation and critique to the object it addresses. Interestingly, Davide begins his response (line 20) with a change of state token (Heritage, 1984), as if revealing that he acknowledged a new meaning of that question. We cannot tell whether the child would have referred to himself – as someone self-ascribing a judgement of 'stupidity' – if the testing, *plus* the therapists' questions had not made that particular self-categorisation relevant in talk (Finlay & Lyons, 2000). However possibly solicited by the ongoing activity, we maintain our focus on the reflective activity in which Davide engages: he takes his experience as a target of description and formulation (*because it's something I do*) and from there, he draws implications on the way he *thinks upon* the world he is presented.

See at this point, how he now frames the response he gave as *relative* to his own thought (*I think*, line 22), taken as an object of judgement *per se*. The twist that Davide does from responding about the picture to referring about himself, attracts Lelio too, who discloses his own habit of internally – and hiddenly – ascribing himself a judgement of stupidity (line 23).

The analyses conducted up to now lead us to consider how the conversational context arranged in this therapeutic setting may foster, support or (as in Davide's case) enhance the likelihood that the child takes his own experience as an object of 'cognitive' and reflective exploration.

Further, the setting such as the one examined also solicits the adolescents to orient to the identities and labels made *salient* (Finlay & Lyons, 2000, 2005) in that particular context, related to the intellectual impairment. In this regard, the following section discusses how Davide engages in a reflection on categorisations of normality attributed to himself.

Reflecting upon being 'normal'

In this session, Davide is referring to a boy, who does therapy in the centre, who reported that he suffered exclusion in the school.

Extract 4a. 28 September 2011

```
1   D   mi dis- piace: che: si tro- ma- a scuola,
        I am so- rry: tha:t he fe- ba- at school, ((meaning: he feels
        bad at school))
2       perché s- se lui <arrivasse in questa scuola
        'cause I- if he <would get in this school ((meaning, the same
        school as Davide's))
3       si sentirebbe meglio forse>
        he would feel better perhaps>
4   T   e::h. forse sì.=
        e::h. perhaps he would. =
5   D   =perché qua ci so-no ta:nti ragazzi::
        =for here there a- re ma:ny boys::
6       tipi come lui (2.0) e anche un pochettino come °me°.
        guys like him (2.0) and a little bit like °me° too.
7       ((D smiles))
8   T   h. eh. eh.=un pochettino come tutti
        h. eh. eh.= ((she smiles)) a little bit like everybody
9       forse. (.) ↑pure ah ah .h
        perhaps. (.) ↑too uh uh .h
(...)
```

Talking about the school where he is, highly known for including several pupils with learning or other kind of disabilities, Davide applies a categorisation process (line 6), which first assigns the boy he is reporting a certain class (*guys like him*) then, with some hesitancy (signalled by the pause, decreased voice volume, smile), it includes also himself (*like °me°*). The label 'like me' helps the boy not to target specific (likely problematic) features of the salient category of 'learning disabled individuals' (cf. Finlay & Lyons, 2000, 2005; Rapley, 2004; Rapley, Kiernan, & Antaki, 1998), leaving them implicit. The design of Davide's formulation at line 6, exhibits the child's struggle between competing orientations: one that proposes his *difference* from a set of typified others (*guys*), to which his (rejected) companion would belong, and the other that proposes instead a *similarity* between him and the companion. The problematic quality of the assertion is indexed by the smile, which accompanies the statement and acts as remedial action (as light laughter often does; cf. Potter & Hepburn, 2010). According to this interpretation, Davide's smile would signal to the interlocutor that he is repairing what he has just said, and acknowledging that his previous formulation (attributing a certain identity to his friend only, *guys like him*) was partially incorrect or imprecise.

The therapist affiliates with Davide by smiling back to him; still, she actively intervenes with a reformulation of the category, up to the point as to make it applicable to all individuals (including herself, cf. the plural *we*); in so doing she rejects Davide's proposal, offering him a normalising, 'lightened' alternative (see laughter at lines 8, 9) to the creation of a special category that would include himself and his friend. The therapist continues, compelling Davide to openly account for the category to which they both have alluded.

Extract 4b.

```
17   T   perché dici.= bè- ci stanno raga-
         why do you say.= well - there are gu-
18       è una scuola .h dove si troverebbe bene.
         it's a school .h where he would feel well.
20   D   forse sto:: più verso:: (.) i:: ragazzi:: (4.0)
         maybe I a::m closer:: (.) to:: guys:: (4.0)
21       quelli norma:li ↑forse.
         the no:rmal ones ↑maybe.
22   T   <e chi è proprio °normale secondo te°?>
         <and who 's someone who is °just normal according to you°?>
23       <che vuol di:re. essere normali.>
         <what does it mea:n. to be normal.>
24   D   bè:: (.) avere: mh: (2.0) problemi:: eh:: mh: (.)
         we::ll (.) to ha:ve mh: (2.0) pro::blems:: eh:: mh: (.)
25       non ↑gravi, problemi:: h:: (.) che si possono:: risolvere.
         not ↑serious, problem:: h:: (.) that ca::n be solved.
26   T   risolvere ec[co!
         solved ri[ght!
27   D             [e questi sono: °[prob-
                   [and these are °[prob-
28   T                             [problemi che s- si possono
         attenu[a:re
                                   [problems that m- may attenu[a:te
29   D   [e:: sì.
         [e:: right.
(...)
```

Asked by the therapist to account for the reasons why he suggested his school as a more suitable context for his friend, Davide makes an explicit reference to the category of 'normality'. His several delays and other markers of uncertainty, punctuating his response, make evident the sensitive and dispreferred nature of making a judgement about normality (that is obviously forced in a dichotomous view as opposed to abnormality) explicit. The complex task that Davide has to solve is to reconcile the ascription of himself to the category of 'normal' guys, with the acknowledgement of the difficulties that led him to attend the therapy sessions for learning disabilities. He solves the problem through a sophisticated compromise (lines 24–25), in which errors are mentioned and acknowledged as part of normality, as long as they can be acted upon and eventually changed. Upon this 'creative' solution the therapist agrees with the child and further elaborates his turn (lines 26, 28), stressing the modifiable character of the problems, which – with regards to the learning difficulties – plays a central role in the theoretical and methodological framework shared by the therapists in the centre.

Clinical relevance summary

Inspiring the overall treatment is the idea that cognitive processes can be modified, that certain impairments can be mitigated, and that, to this aim, having a child practising and discovering their thinking and argumentative skills is pivotal. Besides soliciting argumentative thinking, that is, establishing explicit links between what s/he observes/claims and the procedures by which s/he draws observations, the conversational setting arranged in the therapeutic sessions we have examined appear also to solicit the child (or adolescent) to tirelessly engage in a more benevolent consideration of how their mind works. Testing (and improving) the strategies by which the child makes sense of events means, within this perspective, also testing the *stances* s/he takes on them, assessing – and possibly changing – the extent to which s/he positions her/himselves as hostile, helpless, or discouraged towards her/his reasoning abilities. For a simple summary of the implications for practice, see Table 32.1.

Summary

In this study, we analysed how thinking and reasoning were displayed by an adolescent with an intellectual disability, within conversations with a psychologist who supported a varied range of conversational activities such as making evaluative claims, talking about norms (e.g. disputing what should and should not be done or said), and explicitly discussing categorisations and ascriptions of 'normality'. We observed in Davide a genuine interest in formulating hypotheses, accounting for evaluative statements, and disputing and challenging the psychologist's different point of view. We have analysed discursive sequences in which Davide revealed he was able to identify problems, raise questions, and have his demands responded to, including demands upon himself, his ways of interpreting the world, and his own difficulties.

Table 32.1 Clinical practice highlights

1. Always include a conversational section in the therapy setting as a way to offer the child/adolescent opportunities to narrate, share, and 'test' interpretations about their own everyday experience. This will foster the child's 'natural' abilities of reasoning and reflection and make her/his questions 'how the world around them works', public and open to debate.
2. A pivotal role of the child-therapist's explicit discussion and debate is about how the child/adolescent accounts for their learning mechanisms and failure, where a self-attribution of incompetence or 'stupidity' surfaces in the child's narratives and performance. Dedicate a special time to work with it.
3. Always ask yourself whether the ability (and pleasure) of cognitive exploration of the child/adolescent may be halted and impaired also by his consideration of 'how his mind works'.
4. Engage and share with the child/adolescent a sense of concern and emotional involvement in what is going on, and support a questioning orientation, bearing the risk of not arriving at certain conclusions (constitutive elements of reflective thinking).
5. Display to the child/adolescent that therapy means not only to improve learning, or 'compensate' certain difficulties, but also to take a good 'care' of his thought. Include the analysis of changes of the child's *stance* towards her/his own way to think and reason.

Davide's narratives and descriptions analysed in this chapter display a main concern, related to the question of whether or not he thought differently and 'abnormally' from others. Davide made this question relevant as he reported about the annoying conduct of his peer Pavel, who had behaved in a problematic way and had been blamed by his peers (Extract 1). In this conversation, Davide's reflective thinking was somehow focused on the heterogeneous nature of *abnormal* behaviour: someone can have skilled language but immature conduct. In the second extract, Davide validated the criticism he was given by his aunt, engaging in argumentative discourse that supported the criticism and solicited a debate with the therapist regarding a self-ascription of foolishness. We also observed in Extract 3 how the child and a peer made relevant, in manners that were also coherent with the scenarios evoked by the therapists, their reflections about stupidity and self-criticism.

The final extract showed the therapist addressing Davide with the crucial question 'what does it mean to be normal according to you?', obtaining a version from him that reconciles the acknowledgement of problems and normality, and that also displays his awareness of the therapeutic work relevant here, which is aimed to remedy and lessen the impairing outcomes of the child's learning disability. In the local environment of the interaction, Davide demonstrated the ability to build an argumentative and reflective context for demands that are relevant to him; he was able to discuss and evaluate uncertain situations, to search for connections between events, persons, and thought in order to understand social scenarios; and he also displayed himself as keen to question his own way of thinking and acting in the world. This data also unearthed how Davide engaged in self-categorisations and made relevant social comparisons to others (Finlay & Lyons, 2000). Due to the acknowledgement of the topic as worthy of an in-depth analysis, we opted for deferring this examination to another place, while advocating instead the need for psychologists to consider these instances as indicators of the child's reasoning abilities unfolding in conversation, which can add valid information to testing results on cognitive performances.

References

Ardila, A. (1999). A neuropsychological approach to intelligence. *Neuropsychological Review, 9*(3), 117–136.

Bruner, J., & Haste, H. (1987). *Making sense: The child's construction of the world.* Law Book Co of Australasia.

Cole, M. (1996). *Cultural psychology: A once and future discipline.* Cambridge, MA: Harvard University Press.

Danielsson, H., Henry, L., Rönnberg, J., & Nilsson, L. (2010). Executive functions in individuals with intellectual disability. *Research in Developmental Disabilities, 31*, 1299–1304.

Dewey, J. (1933). *How we think: A restatement of the relation of reflective thinking to the educative process* (1910, revised edition). Boston: Heath.

Edgin, J. O., Pennington, B. F., & Mervis, C. B. (2010). Neuropsychological components of intellectual disability: The contributions of immediate, working, and associative memory. *Journal of Intellectual Disability Research, 54*, 406–417.

Edwards, D. (1994). Script formulations: An analysis of event descriptions in conversation. *Journal of Language and Social Psychology, 13*(3), 211–247.

Feuerstein, R., Falik, L., Rand, Y., & Feuerstein, Ra. S. (2006). *Creating and enhancing cognitive modifiability: The Feuerstein Instrumental Enrichment program.* Jerusalem: ICELP Press.

Finlay, W. M. L., & Lyons, E. (2000). Social categorizations, social comparisons and stigma: Presentations of self in people with learning difficulties. *British Journal of Social Psychology, 39*(1), 129–146.

Finlay, W. M. L., & Lyons, E. (2005). Rejecting the label: A social constructionist analysis. *American Journal of Mental Retardation, 43*(2), 120–134.

Fiorello, C. A., Hale, J. B., Holdnack, J. A., Kavanagh, J. A., Terrell, J., & Long, L. (2007). Interpreting intelligence test results for children with disabilities: Is global intelligence relevant? *Applied Neuropsychology, 14*, 2–12.

Gardner, H. (1999). *Intelligence reframed.* New York: Basic Books.

Goffman, E. (1981). *Forms of talk.* Philadelphia: University of Pennsylvania Press.

Greenspan, S., & Woods, G. W. (2014). Intellectual disability as a disorder of reasoning and judgement: The gradual move away from intelligence quotient-ceilings. *Current Opinion in Psychiatry, 27*, 110–116.

Hatton, C. (1998). Pragmatic language skills in people with intellectual disabilities: A review. *Journal of Intellectual & Developmental Disability, 23*, 79–100.

Heiman, T. (2001). Depressive mood in students with mild intellectual disability: Students' reports and teachers' evaluations. *Journal of Intellectual Disability Research, 45*(6), 526–534.

Heritage, J. (1984). A change of state token and aspects of its sequential placement. In J. M. Atkinson & J. Heritage (Eds.), *Structures of Social Action* (pp. 299–345). Cambridge: Cambridge University Press.

Karmiloff-Smith, A. (1998). Development itself is the key to understanding developmental disorders. *Trends in Cognitive Sciences, 2,* 389–398.

Orsolini, M. (2012). La valutazione e l'intervento con D: Disabilità intellettiva o disturbo generalizzato dell'apprendimento? *Psichiatria dell'infanzia e dell'adolescenza, 79,* 178–194.

Orsolini, M., Latini, N., Salomone, S., & Melogno, S. (2014). *Treating verbal working memory in children with intellectual disability or borderline intellectual functioning.* Paper presented at the 10th European Conference on Psychological theory and research on Intellectual and Developmental Disabilities. Linköping, Sweden, June 12–14.

Potter, J., & Hepburn, A. (2010). Putting aspiration into words: 'Laugh particles', managing descriptive trouble and modulating action. *Journal of Pragmatics, 42*(6), 1543–1555.

Rapley, M. (2004). *The social construction of intellectual disability.* New York: Cambridge University Press.

Rapley, M., Kiernan, P., & Antaki, C. (1998). Invisible to themselves or negotiating identity? *Disability and Society, 13*(5), 807–827.

Salvador-Carulla, L., & Bertelli, M. (2008). 'Mental retardation' or 'intellectual disability': Time for a conceptual change. *Psychopathology, 41,* 10–16.

Schegloff, E. A., Jefferson, G., & Sacks, H. (1977). The preference for self-correction in the organization of repair in conversation. *Language, 53,* 361–382.

Schuchardt, K., Maehler, C., & Hasselhorn, M. (2011). Functional deficits in phonological working memory in children with intellectual disabilities. *Research in Developmental Disabilities, 32,* 1934–1940.

Söderqvist, S., Nutley, S. B., Ottersen, J., Grill, K. M., & Klingberg, T. (2012). Computerized training of non-verbal reasoning and working memory in children with intellectual disability. *Frontiers in Human Neuroscience, 6,* 1–8.

Sternberg, R. J. (1985). *Beyond IQ: A triarchic theory of human intelligence.* New York: Cambridge University Press.

Tipton, L. A., Christensen, L., & Blacher, J. (2013). Friendship quality in adolescents with and without an intellectual disability. *Journal of Applied Research in Intellectual Disabilities, 26,* 522–532.

Van der Molen, M. J., Van Luit, J. E. H., Van der Molen, M. W., & Jongmans, M. J. (2010). Everyday memory and working memory in adolescents with mild intellectual disability. *American Association on Intellectual and Developmental Disabilities, 115,* 207–217.

Van der Molen, M. J., Van Luit, J. E. H., Van der Molen, M. W., Klugkist, I., & Jongmans, M. J. (2010). Effectiveness of a computerised working memory training in adolescents with mild to borderline intellectual disabilities. *Journal of Intellectual Disability Research, 54,* 433–447.

Vygotsky, L. (1934/1978). *Mind in society: The development of higher psychological processes.* Cambridge, MA: Harvard University Press.

Wootton, A. J. (1997). *Interaction and the development of mind.* Cambridge, MA: Cambridge University Press.

Recommended reading

- Finlay, W. M. L., & Lyons, E. (2000). Social categorizations, social comparisons and stigma: Presentations of self in people with learning difficulties. *British Journal of Social Psychology, 39*(1), 129–146.
- Rapley, M. (2004). *The social construction of intellectual disability*. New York: Cambridge University Press.
- Rapley, M., Kiernan, P., & Antaki, C. (1998). Invisible to themselves or negotiating identity? *Disability and Society, 13*(5), 807–827.
- Wootton, A. J. (1997). *Interaction and the development of mind*. Cambridge, MA: Cambridge University Press.

33

'You Just Have to Be Cheerful Really': Children's Accounts of Ordinariness in Trauma Recovery Talk

Joyce Lamerichs, Eva Alisic, and Marca Schasfoort

Introduction

This chapter investigates children's accounts of ordinariness when they talk about trauma and trauma recovery. We were struck by the fact that children when they were invited to talk about their experience with trauma would say things like 'It was just a coincidence' or 'I just started to forget it'. Our first thoughts were that the particle 'just' seemed so 'ill-fitted' to the description of these serious, painful events, as they seemed to construct what happened as 'nothing out of the ordinary'. Why would children describe potentially life-altering occurrences like losing a sibling or a parent in ways that suggested it had only had a minor impact on their lives? These initial questions prompted us to explore what these mentions of 'just' do in children's tellings of trauma and trauma recovery.

Studies have shown that many children are exposed to traumatic events in their lives, such as natural disasters, domestic violence, accidents, death, or war. At least 14% of primary-school-aged children in the Netherlands are exposed to trauma; general population studies in peacetime have shown that exposure to traumatic events is fairly common in childhood, with prevalence rates ranging from 14% to more than 65% (Alisic, Boeije, Jongmans, & Kleber, 2011; Alisic, Van der Schoot, Van Ginkel, & Kleber, 2008). The consequences of such traumatic experiences for children are often severe and debilitating. The literature that looks at how children experience trauma, largely focuses on post-traumatic stress disorder (PTSD) as the most prevalent disorder after exposure to trauma. One of the problems with this construct is that it was designed to describe adult symptoms while several studies have indicated that children's reactions to trauma are different from adult's reactions (see Alisic et al., 2008 for an overview). Moreover, it has also been pointed out that children have different cognitive abilities for processing traumatic events than adults (Salmon & Bryant, 2002).

The identification of these theoretical gaps about studying children's experience of trauma and trauma recovery has marked the beginning of a qualitative interview study that was undertaken by one of the authors (EA, see for an overview Alisic et al., 2011; see also Van Wesel, Boeije, Alisic, & Drost, 2012). The study aimed to take a child-focused approach in highlighting children's perspectives on trauma and trauma recovery: children were encouraged to talk about how they processed the trauma in their own terms, focusing on 'how they perceive the traumatic event, how they tell its story, and how they make sense of the trauma and why it occurred' (Urman, Funk, & Elliott, 2001, p. 405).

This is not to say that taking such an approach in inviting children to talk about their experiences then becomes easy. Many handbooks have laid out the difficulties professionals may encounter when considering which questions are most appropriate or how to best summarise the child's answers before moving on to the next topic (cf. Geldard & Geldard, 2008). Studies that have examined the actual talk in institutional settings in which children are involved have also begun to detail the dilemmas professionals face in settings when they talk with children about possibly sensitive topics, such as how to maintain progressivity in investigative interviews about sexual abuse (Fogarty, Augoustinos, & Kettler, 2013) or how to interactionally manage 'I don't know' answers provided by children in child counselling sessions (cf. Hutchby, 2007; see also Lamerichs, Alisic, & Schasfoort, under review).

In this chapter, we present a secondary analysis of the interview materials that were collected as part of the aforementioned qualitative study by EA, in which we consider the interview as an interactional accomplishment. Taking such an interactional approach which is based on the principles of conversation analysis (CA) and discursive psychology (DP), means that we refrain from considering children's tellings about these traumatic experiences as a window on cognitive schemas that reside in their heads or as indicative of mental states. We take as our focus that tellings of trauma and trauma recovery are talked into being. In line with this view, we consider the way people describe themselves, and how they are taken as members of certain identity categories as interactional matters to be avowed, ignored or rejected and therefore essentially discursive projects (cf. Antaki & Widdicombe, 1998). Our study can be placed within the broader tradition of studies in CA and DP that are empirically oriented and adopt a fine-grained perspective on language as interaction (cf. Te Molder and Potter, 2005). Work in this domain with a focus on children's tellings about trauma (recovery) is still scarce (but see Bateman, Danby, & Howard, 2013 on facilitating trauma talk in the context of the preschool classroom; Bateman, Danby, & Howard, Chapter 22, this volume; also Lamerichs et al., under review).

Project overview

The data for our secondary analysis are psychological research interviews with children conducted by a trained psychologist and co-author of this article (EA). On the basis of this collection of idiographic accounts, the study aimed to inquire how children themselves experienced a traumatic event. In its narrative detail, the accounts go beyond the information that can be collected by questionnaires and surveys (cf. Urman et al., 2001). Since the results of this study are reported elsewhere (see Alisic et al., 2011 for a full account), we do not engage in a detailed presentation of the findings. Rather, we highlight a broader conversational practice in these interviews in which children describe what happened to them and their efforts to cope as 'get(ting) on with life as usual' and 'picking up normal routines' (2011, pp. 487–888); a practice that in more than 200 instances in our data includes mentions of 'just'. For this chapter, we explore a subset of 33 mentions of 'just', where children seem to downplay the importance of the trauma by providing 'accounts of ordinariness' when asked to provide advice to a hypothetical peer who experienced a similar traumatic event or to offer help. We set out to further investigate how it is that these mentions of 'just' play a role in resisting the action agenda embedded in the interviewer's questions.

A recent review study (Van Wesel et al., 2012) has shown how orienting to 'ordinariness' is overwhelmingly present in children's tellings about trauma (recovery). The study documented how children describe the trauma and their recovery as a staged process in which 'normalcy' and 'difference' are descriptive categories that are drawn upon interchangeably in their tellings, for example when they talk about whether life had changed and in what way (p. 6).

These 'accounts of ordinariness' can also be linked to a range of interactional studies such as Harvey Sack's early observations in CA that if we experience something and tell others a story, 'we first look for its 'ordinary' aspects and search for 'ordinary' explanations. So if we can, we will report that we first heard what turned out to be gunshots as a car backfiring or tell about a plane hijack as a joke' (Sacks, 1992, p. 220). In studying peoples' accounts of their experience with the paranormal, such as sightings or an invisible presence, Wooffitt (1992) has demonstrated one particular way in which 'doing being ordinary' might be accomplished and how mentions of 'just' play a part in this. Wooffitt shows how people frequently employ a two-part format: 'I was just doing X . . . when Y occurred'. In this format, X is used to describe the speaker's (mundane) activities at the time, while Y describes the first awareness of the phenomenon (1992, p. 18; see also earlier work on this format by Sacks, 1984 and Jefferson, 2004a). Wooffitt shows how the two-part format acquires a particular rhetorical force when the speaker's credibility for reporting something highly unusual is at stake; it creates a contrast between the mundane, inconspicuous circumstances that are initially considered, and the untoward state of affairs that is presented in what comes next. Juxtaposing these images of 'normality' and 'strangeness' enable the speaker to defuse potential sceptical claims about the veracity of the account and the speaker's reliability (Wooffitt, 1992).

In our data, the credibility of the children is not at stake as the veracity of their trauma tellings is far from being contested. However, children produce accounts of ordinariness in our interview data and these accounts are coupled with mentions of the particle 'just', for example when they describe the trauma (e.g. 'I'm just having bad luck it's with me') and when they talk about the process of recovery (e.g. 'you just have to do fun things'). The children's accounts seem to bear a strong resemblance to Jefferson's observations about the 'At first I thought X, then I realised Y' sequence, where people report on their experience of accidents and trauma as merely unremarkable or as 'an incidental occurrence' (Jefferson, 2004a, p. 145). In our analysis we will examine in greater detail what the mentions of 'just' do in children's tellings about trauma.

Mentions of 'just' have been considered for the ways in which they express a 'depreciatory meaning' (Lee, 1987) or function as 'grammatical limiters' (e.g. 'I just stabbed him', Rymes, 1995). They can be used to intensify or de-intensify meaning depending on the context in which they are used (Quirk & Greenbaum, 1973). In this way, mentions of 'just' may work to mitigate agency for violent behaviour rather than heightening it ('I got mad and I just hit 'im', Rymes, 1995, p. 505) or may downgrade a 'polite directive' ('If you'd just sign there for us', Lee, 1987, p. 383). In what comes next we will further unpack how mentions of 'just' are produced in our psychological research interviews, but first we will turn to the data.

Data

Our data consist of 14 psychological research interviews with children between 8 and 12 years of age who experienced single-incident trauma (e.g. involving sudden loss, violence or accidents with injury). Examples are presented from a subset of 33 instances in which the child is asked to provide advice for his peers, to elaborate on the advice and indicate ways of helping, and uses mentions of 'just' to do so. We set out to explore the identity work these descriptions with 'just' accomplish here.

The children who took part in the interviews were no longer receiving mental health care at the time of the interview and were asked to talk about their traumatic experience and factors they considered helpful or hindering for recovery. The study protocol was approved by the Medical Ethics Committee of the University Medical Centre Utrecht (the Netherlands). Children and parents were contacted through the University Medical Centre Utrecht and the families approved of secondary use of the interview materials for educational purposes.

The data are transcribed following the transcription conventions developed by Gail Jefferson (2004b) and we present a two line transcription that represents the Dutch original along with an idiomatic English gloss (in italics and bold) to capture the local interactional meaning (Hepburn & Bolden, 2013).[1]

Analysis

Our analysis is guided by the observation that when asked to give advice to hypothetical peers experiencing trauma, children employ the particle 'just' in different ways. One way is to offer advice about what is helpful to recuperate in mundane and generalised terms (e.g. as something everyone would suggest in these circumstances). Our exploration of the data has shown how children's replies to the interviewer's request initially start with dispreference markers, suggesting that they have difficulties answering this question. After these initial interactional difficulties, and in response to further questions by the interviewer, children overwhelmingly present the 'advice' they come up with as ordinary and as a routine matter (e.g. 'you just have to do fun things'). The answer is thereby constructed as a 'minimal suggestion' that makes available the implication that to recover from trauma can be routinely achieved and lies in doing 'normal' things. It also casts the advice as 'the kind of thing anyone could come up with'. In what follows we want to analyse two examples to demonstrate how these 'just' mentions are part of a larger set of practices by which the child resists being ascribed the identity category of someone who has special knowledge on trauma (recovery). Extract 1 is from an interview with an 11-year-old girl whose mother was killed by her father and occurs towards the end of the interview. In the extracts, I is interviewer, C is child.

Extract 1 [K23] father has killed mother (K23/723-816/25:34)

```
1   I:    ja.
          yes.
2         (1.8)
3   I:    °oke°,
          °okay°,
4         (5.1)
5   I:    en uhm:
          and uhm:
6   I:    (.) stel nou dat een vriendinnetje van jou zoiets zou meemaken,=
          (.) what if a friend of yours would go through something
          like this,=
7   I:    =>wat jij hebt mee gemaakt;<
          =>what you have been through;<
8   I:    (.) wat zou je der dan voor t↑ips geven;
          (.) what kind of t↑ips would you give her;
9   C:    •pt ↑weet ik niet,
          •pt I don't know,
```

```
10  C:      maa:r ehh ja nou ehm: (2)
            but ehh yes well ehm: (2)
11  C:      nou ik zou der--
            well I would her--
12  C:      •h ehh er was een vriendinnetje
            •h ehh there was a girlfriend
13  C:      >maar dat was niet zoiets<;
            >but that wasn't something like this<;
14  C:      >nja;<
            >mwell;<
            ...............
            ((20 lines omitted in which the child elaborates on what happened
            with her friend from school whose mother also died. She tells how
            the head teacher called her at home to ask if she wanted to get in
            touch with the girl to offer her some support))
            ...............
15  C:      en ja (1.5) tip uhm: (1) ja;
            and yes (1.5) tip uhm: (1) yes;
16  C:      ik bedoel je moet weer ↑doo:r;
            I mean you have to move ↑o:n again;
17  C:      dus •h het heeft geen zin om; (.)
            so •h it is of no use to; (.)
18  C:      alsmaar te gaan huilen en--
            keep on crying and--
19  C:      •h en het niet achter je >te laten<;=
            •h and not to leave it behind you;=
20  C:      =>want ik bedoel<;
            =>because I mean<;
21  C:      =je moet t op een gegeven moment wel >achter je laten<;=
            =you do need to leave it behind you at some point;=
22  C:      =•hh want je kan er niet heel je leven aan ↑vast zitte',
            =hh because you can't remain ↑stuck to it your whole life,
23  I:      hmhm;
            hmhm;
24  C: →    dat is gewoon- (.)dat helpt niet=
       →    that is just- (.) that doesn't help=
25  C:      =dat heeft geen zin.
            =that's no use.
26  I:      °oke.°
            °okay.°
27  C:      >dan heb je ook geen fijn leve'<.
            >then you don't have a nice life either<.
28  I:      °hm°
            °hm°
29  I:      (.)•h en als die vriendin dan zegt van ja maar hoe doe ik dat,
            (.)•h and if that friend then says like yes but how do I do that,
30  I:      dat achter me laten;
            this leaving it behind me;
31  I:      zou je dan: (.) advies hebben?
            would you then: (.) have advice?
32  C: →    je moet er gewoon aan werken;
       →    you just have to work on it;
33  C:      >nouja< •h bijvoorbeeld (.)
            >well yes< •h for example (.)
34  C:      met pleegzorg
            with foster care
35  C:      of jeugdzorg of sp-
            or youthcare or pl-
36  C:      of speeltherapie;
            or play therapy;
```

```
37   C:      of of dit >zeg maar<,=
             or or this >say<,=
38   I:      [hmms
             [hmms
39   C:      [•h over praten;
             [•h talk about it;
40   C:      nou ja (.)
             well yes (.)
41   C:      •hh dat-
             •hh that-
42   C:      <het lucht ↑op;>
             <it brings relief;>
43   I:      ja.
             yes.
44   C:      (1.5) dan heb je het tegen iemand gezegd.=
             (1.5) then you have told someone.=
45   C:      =>en dan< (1) voel je je f(h)ijner.
             =>and then< (1) you feel b(h)etter.
46   I:      (0.8) °oke.° (3)
             (0.8) °okay.° (3)
47   I:      en zijn er nog andere dingen die je zou ra- eh als tip zou geven;
             and are there other things you would adv- eh give her as a tip;
48   C:      ehm: (5) °ehm:°. (7)
             ehm: (5) °ehm:°. (7)
49   C:      naja •h zeg maar ehm:. (2)
             wel yeah •h like ehm:. (2)
50   C:      ja het is heel erg dan;=
             yes it is really awful then;=
51   C:      =m↑aar, (.) >ik bedoel< je moet er doorh↓een;=
             =b↑ut, (.) >I mean< you have to get through it;=
52   C: →    =dus je moet •h >je moet gewoon leuke dingen doen<;
         →   =so you have •h >you just have to do fun things<;
53   I:      hmm
             hmm
54   C:      >om het te vergeten.<
             >to forget it.<
55   I:      ↑ja.
             ↑yes.
56   I:      (2)
             (2)
57   I:      °oke.° (1.8)
             °okay.° (1.8)
58   I:      en wat zou je zelf doen als die vriendin dat had meegemaakt;
             and what would you do if that friend would have experienced that;
59   C: →    eh: (.) haar steune' en gewoon; (4)
         →   eh: (.) support her and just; (4)
60   C:      uhm: zeggen (.) het was ut-
             uhm: tell her (.) it was ut-
61   C:      nja >het was vroeger<;
             hm well <it used to be<;
62   C:      >je kan er •h niks meer aan verandere';
             >you can't do •h anything to change it;
63   C:      >en je-je moet niet heel je leven der aan vast zitte';
             >and you-you should not remain stuck to it your whole life;
64   C:      •h j-je hoe bot het ook klinkt;=
             •h y-you how rude that may sound;=
65   C:      =[je moet weer •h doo:r dus-
             =[you have to •h move on so-
```

```
66   I:      = [hmhm;
             = [hmhm;
67   C: →    en gewoon;
        →    and just;
68   C: →    nou ja je moet gewoon eigenlijk vro:lijk zijn.
        →    well yes you just have to be cheerful really.
69   I:      ja,
             yes,
70   I:      (2.5) °oke.°
             (2.5) °okay.°
71   I:      ((snuift))
             ((sniffs))
```

In lines 5–8, the interviewer poses a hypothetical question that inquires what the girl would give as 'tips' if one of her friends would experience something similar. Referring to the traumatic event with 'something like this' and 'what you have been through' (lines 6 and 7) offers room for the child to consider different types of situations in her reply, while its unspecified character also highlights its delicacy. Casting the child's response in terms of 'tips' not only suggests that there might be 'ready-to-be-formulated' tips to provide for dealing with these kinds of experiences, but also positions the child in the category of advice giver. It addresses her as someone in the position to formulate tips for a friend who also experienced a traumatic event on the basis of her special expert knowledge and experience.

In line 9, the child claims to have no knowledge and retracts for the first time, followed by a but-prefaced continuation of her turn coupled with markers of dispreference. Her reply suggests she is having trouble answering, but also displays how she actively considers what might be an appropriate response. We see further illustrations of her attempts to do so in the restart in line 11 where the child orients to what would possibly be relevant advice and again in lines 12–14, where she displays thinking as to whether an experience she had with a friend at school would count as an appropriate reply here. She seems to dismiss this story then first in line 13 but after a 'mwell' in line 14 that displays doubt, she does go on to recall the episode in great detail (data not shown here).

In line 15, the child comes back to the notion of 'tips' in the interviewer's initial question and displays further thinking on how to answer it ('and yes (1.5) tip uhm: (1) yes;'). She seems to dismiss the category of 'tips' by pre-emptively accounting for why they are not useful: 'I mean you have to move ↑o:n again' (line 16), which shows her ongoing concern with the interviewer's category of 'tips' as a problematic term for answering. The child formulates an upshot in line 17 that speaks against its desirability and usefulness ('so •h it is of no use') by presenting an extremised scenario of *not* being able to cope with what happened ('keep on crying and-- •h not to leave it behind you', cf. Pomerantz, 1986). She continues with a similar pre-emptive account that further explains and recycles what she mentioned before in lines 18–19 in even stronger and idiomatic terms (line 21, 'you do need to leave it b<u>e</u>hind you at some point', and line 22 'you can't remain ↑<u>stuck</u> to it your whole life'). Both descriptions of what is needed to recover from trauma are presented as universal truths (cf. Drew & Holt, 1988) while the modal verbs of necessity and a negative form of 'can' cast it as a moral obligation and express impermissibility. After a continuer by the interviewer (line 23) she offers a list which seems to do closing work (Atkinson, 1983). The first list item starts out as a judgement about the course of action she just described and is then aborted and repaired ('that is just- (.) that doesn't <u>help</u>', line 24). After her repair she continues with the second list item that is presented as a stronger description and attends to the futility of the scenario she previously described ('that's no <u>use</u>', line 25). The third list item is produced after a continuer by the interviewer and is hearable as a generalised list completer ('>then you don't have a nice life either<', line 27). It receives a confirming, hearably softer produced uptake by the interviewer in line 28.

The interviewer pursues with a hypothetical follow up question (see also lines 6–7). By conveying a question the previously mentioned friend could ask and using reported speech to echo the girl's own words mentioned in line 21 ('yes but how do I <u>do</u> that, this leaving it behind me', lines 29–30), she requests more precise advice from the girl on behalf of the hypothetical friend; the girl is asked to specify the working of the strategy she adopted to overcome the trauma ('leaving it behind me'). In line 32 the child comes with a straightforward answer: 'you just have to work on it', which again suggests a moral obligation, this time to work on your recovery. It stresses the need to put in an effort to get better but at the same time, the recommendation is also constructed as common knowledge; as a clichéd expression that is universally available (see the use of the generic 'you', Bull & Fetzer, 2006; also Wiggins, 2004). As a clichéd expression, it is presented as not particularly tied to the girl's experience with the trauma, and as such we can begin to see how it enables her to resist the suggestion that on the basis of her experience alone, she is able to provide 'special' advice or explain the working of her coping strategy.

In lines 33–37 the child offers a three-part list of examples to exemplify the kind of work that may be required. In doing so, she displays extensive knowledge of how the health system works and its available therapies. Listing them also suggests a sense of completeness of possible options. Including the current interview ('or or this >say<', lines 37 and 39) highlights 'talking' as a legitimate part of the work needed for recovery. The extensive listing of 'remedies' might also indicate an accommodating reply to the interviewing psychologist and expert. In lines 40–45 the importance of talking and its generic benefits are emphasised again more strongly ('>and then< (1) you feel b(h)<u>e</u>tter'), which constructs 'feeling better' as something that is worth striving for (see also line 27).

In line 47 the interviewer continues with a further question inquiring whether there are any other tips the girl could give. What follows is a hesitant start with pauses and reformulations (lines 48–49), which displays the child's uneasiness of being addressed in the role of advice giver once more. In lines 50–52 she again seems to dismiss the category of 'tips': the reply starts off with a qualifying response ('yes it is really <u>awful</u> then') only to retract and pre-emptively account that tips might not be sufficient and something more 'encompassing' is needed ('>I mean< you have to get through it', line 51). While subtly taking issue with the terms of the interviewer's question here again (see also lines 15–16) she stresses the importance of a particular 'mindset' that is presented as more important for recovery. In lines 52 and 54 'so you have •h >you just have to do fun things to forget it<' is presented as a logical consequence of such a mindset. Its obligatory character is reinforced while the particle 'just' mitigates the effort needed to execute 'fun things'. It presents what is needed for recovering as something that can be achieved without great difficulty and as nothing special. It also constructs it as a generic and clichéd form of support; not as something that requires special expertise or experience but as universally available and knowable. Line 54 also draws the sequence to an end.

We see a confirming response by the interviewer in line 55, a pause and an 'okay', which is followed by another pause of 1.8 seconds and a further question that inquires what the child herself would do (as opposed to what anyone would do) if a friend had been through a trauma (referred to here with 'that', treating it as unspecific and possible delicate). The child hesitantly produces the general clichéd expression 'support her' which is followed by a mention of 'just'. After a four second pause she continues by pointing out what she would say to her friend, ('you can't do •h anything to change it', line 62) which presents her statement as a matter of fact and as not up for further discussion. What starts with indirect reported speech (line 60) is then formulated as two 'generic lessons' that are worded in a similar fashion as the previous 'lessons' (see lines 21–22). The child accounts for the bluntness of what she constructs as another universal 'lesson' that everyone knows and should apply ('you have to •h move on so-', line 65). Ultimately she presents her conclusion as a 'generic summary' (Drew & Holt, 1998) in which she recycles her previous 'advice' (line 52) with a slight reformulation ('you just have to be cheerful really', line 68) that attends to its straightforwardness. With the mention of 'just' she constructs this as the essence of her advice (see also line 24) and the child's turn also is closure implicative in an environment where the interviewer is pursuing an elaboration. The interviewer confirms and brings the sequence to a close with a 'yes', followed by a pause and an okay' in line 70 that is hearably softer produced.

To conclude, fragment 1 shows how the interviewer, after her initial hypothetical question that inquires whether the child could provide advice, pursues by asking a question for clarification (lines 29–31) and again by two more follow up questions (lines 47 and 58). We have seen how the child retracts and restarts her initial answer in response to the interviewer's first question, displaying resistance against the notion of 'tips'. We want to argue that the child's replies throughout this fragment demonstrate how she does not present herself as someone with special knowledge on trauma (recovery) and does not mobilise that specific knowledge in her answers. What she does is quite the opposite almost: by stressing the importance of generic, almost innocuous lessons (i.e. through listing, idiomatic expressions, and presenting generic clichéd summaries), she constructs her 'advice' not only as uncomplicated and matter of fact, but also as the kind of advice anybody would give in those circumstances ('just do fun things'). In doing so, the child resists the category bound features that are made available by the interviewer that she, as someone who has experienced trauma, holds a special epistemic position on the basis of which she is able to provide advice to her peers or knows how to offer help overcoming the trauma. Sequentially, with the mentions of 'just' as part of the child's 'generic clichéd summaries' the child also attempts to bring the sequence to an end while the interviewer continues to seek a further response (see lines 52–54, 68 but also 24–27).

Mentions of 'just' occur more often when children attempt to provide completeness. In fragment 2 we see an example of how mentions of 'just' occur in an environment where the interactional pressure for the child to provide (additional) answers is rising. Extract 2 is taken from an interview with a 10-year-old boy who lost his sister in a train accident. The segment we present occurs eight minutes into the interview. At this point in the interview, the interviewer has reformulated her initial question about giving advice twice while the child has demonstrated great difficulties answering (see also Lamerichs et al., under review for an analysis of this fragment and other instances in which children make use of 'I don't know' answers).

Extract 2 [K10] loss of sister due to train accident (K10/362-404/08:02)

```
1   I:      >•h en als een< vriendje (.) van jou iets mee zou ma↑ken;
            >•h and if a friend< (.) of yours would go through something;
2   I:      (.)wat- wat zou je dan zelf doe:n.
            (.)what-what would you d:o then.
3           (1,3)
4   C:      hh
            hh
5   C:      °jah;°
            °yeah;°
6   C:      (2)
7   C:      ehm;=
            ehm;=
8   C:      =>wee k nie.<=
            =>don't know.<=
9   C: →    =>nou gewoon< •hh
       →    =>well just< •hh
```

```
10  C:      (1)
11  C:      ehm s:teunen °ofzo°,
            ehm offering support °or something°,
12  I:      hmh mm?
            hmh mm?
13          (0,8)
14  I:      >en h↑oe zou je dat doen?<
            >and h↑ow would you do that?<
15          (1,5)
16  C:      mmm da >wee k nie<.
            mmm I >don't know<.
17  K:      •h
            •h
18          (0,8)
19  I:      m:ag e↑ffe nadenk↑e;
            ta:ke your time to think;
20          (0,7)
21  C:      >em•h<
            >em•h<
22          (1,4)
23  C:  →   gewoo:n;
        →   ju:st;
24  C:      w (vanwe) (.)
            w (vanwe) (.)
25  C:      •hhh
            •hhh
26  C:      (.)
27  C:      hm
            hm
28  C:      naem vaak n- kaa↑rtje st-m-st↑uren;=
            weuhm often n-se-m-sending a postcard;=
29  C:      =en naar ze toegaan,
            =and going to see them,
30  I:      hm mm;
            hm mm;
31          (1,5)
32  I:      °•h ok↑ee°;
            °•h ok↑ay°;
33          (4,6)
34  I:      ((smakt)) en waarom zou je kaartjes sture.
            ((smacks)) and why would you send postcards.
35          (2)
36  C:      na;=
            well;=
37  C:      =•hhh hhhh•
            =•hhh hhhh•
38  C:  →   gewoon;
        →   just;
39  C:      mm da ie:dereen krijgt dan (.) kaartjes °van° [(.)(e]h) van mensen.
            mm everybody gets postcards then (.) °from° [(. e)h] from people.
40  I:                                                   [hm mm]
                                                         [hm mm]
41  I:      jha;
            yha;
42          (4,8)
43  I:      ok↑ee:,
            ok↑ay:,
```

In line 1, the interviewer inquires with an and-prefaced hypothetical question what the child would do if a friend would go through 'something'. The traumatic event itself is glossed over, as it was in the initial question by the interviewer (e.g. as 'something really bad', data not shown here) displaying her orientation to its sensitive character (see also fragment 1). With pauses and perturbations in lines 3–7 and with a claim to no knowledge in line 8, the child displays difficulties in answering the question. The 'I don't know' occurs in turn initial position and is quickly followed by '>well just<.hh (1) ehm offering support °or something°', which constructs the clichéd expression 'offering support' as a mere guess (see also the post-positioned 'or something' in line 11 that is hearably softer produced and with rising intonation). It also suggests that this is what one would usually or routinely do; it is presented as a generic form of help.

After a continuer and a pause, the interviewer pursues with an and-prefaced follow up question in line 14 that inquires in greater detail how the child would offer support. After a 1.5 second pause the child produces another claim to having no knowledge, and after another small pause in which the child does not produce anything further, the interviewer prompts the child to take some time to think (line 19). The interviewer's receipt demonstrates that she does not treat the child's previous answer as a sufficient reply; she both offers the child a pause to think and an instruction to do so. After a pause, a disrupted TCU and an in-breath that suggests further uneasiness, the child continues with a turn interspersed with perturbations. With a turn initial 'just' in line 23, the child constructs what comes next ('often sending a postcard', line 28; 'going to see them', line 29) as a generalised routine activity everybody would consider doing in these circumstances. The turn is also closure implicative.

After a continuer ('hm mm'), and a pause in lines 30–31 in which the child does not elaborate, the interviewer produces a softly spoken 'okay', which seems to indicate she considers this reply to be sufficient for now. However, after a long 4.6 second pause in line 33 and while the child has not said anything further, the interviewer continues by asking another follow up question ('and why would you send postcards', line 34). The question does not ask for more specific advice but requests an account for the child's previous response. In lines 36–37 the child's reply is prefaced with 'just' after markers of dispreference that display ongoing difficulties answering. They are also hearably softer produced, which might display the child's orientation to the sensitive nature of accounting for his answer and not offering anything more here at the time. In what comes next, the child offers a basis for his previous answer: stressing its general truth value with the use of an extreme case formulation (ECF) works to present this course of action as what anybody would legitimately do in these circumstances (Pomerantz, 1986). The situation of the hypothetical peer is thereby presented as nothing out of the ordinary and as not requesting any special attention. The extremised gloss that is offered also does closing work and we see that after the initial question in lines 1–2 and 3 follow-up questions, the interviewer offers an affiliative response, produced partly in overlap, in lines 40–41. After a long pause in which the child does not offer anything further, the interviewer proffers an 'okay' that marks the point of moving on to another question in the interview.

The fragments we have presented occur in an environment in which the interviewer asks the child to come up with 'tips' for a hypothetical peer (fragment 1) or inquires what the child would do if a friend would have experienced something similar (fragments 1 and 2). In both instances, the interviewer seeks further specifications from the child (fragment 1, lines 29–30 and fragment 2, line 14), whereas in fragment 2 the child is prompted to elaborate (line 19) and even asked to account for his previous answer (line 34). Parallel to what we have seen in fragment 1, the child in fragment 2 also recycles his 'just' prefaced answer in order to present his advice (offering support, sending postcards) as 'routine business' and with the clichéd summaries, also attempts to do closure work.

However, when the interactional pressure rises as is the case in fragment 2 (see the 'I don't know' answers, the child's perturbated turns and the interviewer's prompt in lines 14–27) the child seems to take on a more defensive position. In this environment, the 'just'-prefaced answer in lines 37–39 appear at the beginning of the turn and not only show how the child resists to be positioned as an advice giver, but also seems to resist further questions. As such, fragment 2 provides a more overt illustration of how the generic clichéd forms of support that are suggested by the child seem to stand in contrast to the 'special status' the interviewer seems to invoke and reinforce with her repeated questions.

Clinical relevance summary

We want to stress two points we consider important for professionals. Based on the analysis of our data, we think that professionals could benefit from perceiving trauma talk with children about sensitive matters as a collaborative interactional achievement. When doing so, they may be stimulated to reflect on their footing positions, the effects of the structural properties of their question formats, the answer categories that are implied, and the question's action agenda in terms of ascribing particular identity membership categories to the child. To be more attentive to these aspects can make professionals more aware of the (initial) difficulties and reluctance children seem to display when they are addressed in the position of an expert and asked to provide advice for their peers or be knowledgeable about how to help another child. It may also increase professionals' awareness of how the use of repeated follow-up questions, prompts and why-questions suggest that to be able to come up with advice is very important here. Constructing this as particularly significant may result in increasing interactional pressure for both participants, especially when the interview is already progressing with some difficulties. Extract 2 shows how 'just' prefaced answers may work as rather 'minimal' responses where invitations to elaborate are little successful; towards the end of Extract 2 they might even be hearable as a defensive move to resist the interviewer's why-question.

This brings us to our second point which has to do with the 'performative' qualities of interview settings (cf. Seale, Charteris-Black, MacFarlane, & McPherson, 2010). We have started to detail some of the identity work that children attend to in these interviews, when resisting the category of someone with a special experience or knowledge. We want to highlight how accommodating responses from the child might constitute another aspect of such identity work (cf. Fogarty et al., 2013; Van Nijnatten, 2013). We have glossed the child's demonstration of having detailed knowledge about therapies as an accommodating response (i.e. the child is 'doing being a good pupil' in the presence of an expert). We want to suggest that displaying a strong normative orientation to proposing possible routes for recovery as straightforward and as a felt obligation might also be considered as such (see Seale et al., 2010, p. 603 on 'presenting a moral front'). We think professionals will benefit from an enhanced understanding of approaching trauma recovery talk as an interactional matter and as an important locus for identity work. Adopting such a perspective might be a useful tool to highlight potential problem areas, such as inviting accommodating answers and the action agendas embedded in the interviewer's questions that children then seem to resist. While we have demonstrated how children may display subtly ways of countering these identity categories and may find ways to close off a line of questioning, we have also seen that a subtle balance needs to be struck between eliciting a detailed and individual perspective from a child on these matters, and prompting children to elaborate, so as not to run the risk of the interview stalling. Finding out which questions would be more 'successful' to get at the detailed accounts that are needed to improve our understanding of how children cope with traumatic events in their lives is a topic that deserves further study. For a simple summary of the implications for practice, see Table 33.1.

Summary

This chapter has explored how children talk about trauma and trauma recovery. What started out with an initial gloss of how 'just' plays a role in 'accounts of ordinariness' has resulted in a more detailed view of how mentions of

Table 33.1 Clinical practice highlights

1. Interviewer's questions carry important action agendas that make available and elicit particular identity categories.
2. We have begun to show that children may resist the category bound features that suggest they are a person with a special experience and hence special knowledge that might legitimately place them in the position of giving advice to others about trauma recovery.
3. Resisting particular identities can become even more problematic when the interview is progressing with some difficulties and mentions of 'just' (which can be part of generic summaries that do closing work) might be used to end a line of questioning.
4. Professionals could benefit from an awareness of their own footing position in interviews and the 'identity work' children attend to in answering, such as resisting the 'expert-role', emphasising the normative character of their advice or offering accommodating replies.

'just' operate in answering a particular type of question in psychological research interviews where children are asked to offer advice and help to a hypothetical peer.

Our analysis has begun to show how with these mentions of 'just' both children find ways to resist the category-bound implications that are part of the interviewer's questions: that on the basis of their experience with a traumatic occurrence, the children have special knowledge or expertise to draw upon and to give advice or help a friend who has experienced a similar traumatic event. We have demonstrated how mentions of 'just' are used to construct the advice that children eventually provide as nothing special and as the kind of universally knowable, generic suggestions everybody would offer.

As such, children's self-characterisations emphasise 'doing being ordinary' by structurally employing three-part lists, generic constructions, and idiomatic clichéd expressions that contain modal verbs. They work as generic summaries to underpin their accounts and as attempts at closing a sequence while the interviewer pursues further elaborations to get at more specific advice.

Note

1. As we can only guess how to capture interspersed laughter particles, stress or pitch movements within words in the English translation, we have not attempted to include these detailed notations in the translated gloss (except where words remain identical, e.g. 'tips'). We have however added hearable aspiration, pauses, unit final intonation, volume and speed of delivery in the translation where the translation (e.g. word order) allows. The reader has access to the full transcription details in the Dutch original that is placed immediately above the translation in English.

References

Alisic, E., Boeije, H. R., Jongmans, M. J., & Kleber, R. J. (2011). Children's perspectives on dealing with traumatic events. *Journal of Loss and Trauma: International Perspectives on Stress and Coping, 16*(6), 477–496.

Alisic, E., Van der Schoot, T. A., Van Ginkel, J. R., & Kleber, R. J. (2008). Looking beyond posttraumatic stress disorder in children: Posttraumatic stress reactions, posttraumatic growth, and the quality of life in a general population sample. *Journal of Clinical Psychiatry, 69*, 1455–1461.

Antaki, C., & Widdicombe, S. (Eds.) (1998). *Identities in talk.* London: Sage.

Atkinson, M., (1983). Two devices for generating audience approval: A comparative study of public discourse and texts. In K. Ehrlich (Ed.), *Connectedness in sentence, text and discourse* (pp. 199–236). Tilburg, the Netherlands: Tilburg Papers in Language and Literature.

Bateman, A., Danby, S., & Howard, J. (2013). Everyday preschool talk about Christchurch earthquakes. *Australian Journal of Communication, 40*(1), 103–121.

Bull, P., & Fetzer, A. (2006). Who are we and who are you?: The strategic use of forms of address in political interviews. *Text and Talk, 26*(1), 1–35.

Drew, P., & Holt, L. (1988). Complainable matters: The use of idiomatic expressions in making complaints. *Social Problems, 35*(4), 398–417.

Drew, P., & Holt, L. (1998). Figures of speech: Figurative expressions and the management of topic transition in conversation. *Language in Society, 27*(4), 495–522.

Fogarty, K., Augoustinos, M., & Kettler, L. (2013). Re-thinking rapport through the lens of progressivity in investigative interviews into child sexual abuse. *Discourse Studies, 15*(4), 395–420.

Geldard, K., & Geldard, D. (2008). *Counselling children: A practical introduction.* London: Sage.

Hepburn, A., & Bolden, G. (2013). The conversation analytic approach to transcription. In J. Sidnell & T. Stivers (Eds.), *The handbook of conversation analysis* (pp. 57–76). Oxford, England: Wiley-Blackwell.

Hutchby, I. (2007). *The discourse of child counselling.* Amsterdam, The Netherlands: John Benjamins.

Jefferson, G. (2004a). 'At first I thought'. A normalizing device for extraordinary events. In G. Lerner (Ed.), *Conversation analysis: Studies from the first generation* (pp. 132–167). Amsterdam, The Netherlands: John Benjamins.

Jefferson, G. (2004b). Glossary of transcript symbols with an introduction. In G. Lerner (Ed.), *Conversation analysis: Studies from the first generation* (pp. 13–31). Philadelphia, PA: John Benjamins.

Lamerichs, J., Alisic, E., & Schasfoort, M. (under review). Managing claims to knowledge when eliciting trauma talk: A sequential analysis of children's I don't know answers in psychological research interviews.

Lee, D., (1987). The semantics of just. *Journal of Pragmatics, 11*, 377–398.

Pomerantz, A., (1986). Extreme case formulations: A way of legitimizing claims. *Human Studies, 9*, 219–229.

Quirk, R., & Greenbaum, S., (1973). *A university grammar of English.* London: Longman.

Rymes, B. (1995). The construction of moral agency in the narratives of high-school drop-outs. *Discourse & Society, 6*(4), 495–516.

Sacks, H. (1984). On doing 'being ordinary'. In J. M. Atkinson & J. Heritage (Eds.), *Structures of social action: Studies in conversation analysis* (pp. 413–429). Cambridge: Cambridge University Press.

Sacks, H. (1992). *Lectures on conversation* (Vol. 1 and 2), edited by Gail Jefferson, with an introduction by Emmanuel A. Schegloff. Oxford: Blackwell Publishers.

Salmon, K., & Bryant, R. A. (2002). Posttraumatic stress disorder in children: The influence of developmental factors. *Clinical Psychology Review, 22*, 163–188.

Seale, C., Charteris-Black, J., MacFarlane, A., & McPherson, A. (2010). Interviews and Internet forums: A comparison of two sources of qualitative data. *Qualitative Health Research, 20*(5), 595–606.

Te Molder, H. F. M., & Potter, J. (Eds.) (2005). *Conversation and cognition.* Cambridge, England: Cambridge University Press.

Urman, M. L., Funk, J. B., & Elliott, R. (2001). Children's experiences of traumatic events: The negotiation of normalcy and difference. *Clinical Child Psychology and Psychiatry, 6*(3), 403–424.

Van Nijnatten, C. (2013). Downgrading as a counterstrategy: A case study in child welfare. *Child and Family Social Work, 18*, 139–148.

Van Wesel, F., Boeije, H., Alisic, E., & Drost, S. (2012). I'll be working my way back: A qualitative synthesis on the trauma experience of children. *Psychological Trauma: Theory, Research, Practice, and Policy, 4*, 516–526.

Wiggins, S. (2004). Good for 'you': Generic and individual healthy eating advice in family mealtimes. *Journal of Health Psychology, 9*, 535–548.

Wooffitt, R. (1992). *Telling tales of the unexpected: The organization of factual discourse.* Hemel Hempstead: Harvester Wheatsheaf.

Recommended reading

- Peräkylä, A., & Vehviläinen, S. (2003). Conversation analysis and the professional stocks of interactional knowledge. *Discourse & Society, 14*(6), 727–750.
- Antaki, C., (Ed.) (2011). *Applied conversation analysis: Intervention and change in institutional talk.* Basingstoke: Palgrave MacMillan.
- Brom, D., Pat-Horenczyk, R., & Ford, J. D. (Eds.) (2009). *Treating traumatized children: Risk, resilience and recovery.* East Sussex: Routledge.

Glossary

Accountability: The discursive process of justifying or explaining actions as a result of perceived blame.

Address term: A word, phrase, name, or title (or some combination of these) used in addressing someone.

ADHD: Attention deficit hyperactivity disorder is one of several labels used to describe a collection of characteristics, behaviours, and 'symptoms'.

Affiliation: Displays support of, and endorses, the conveyed stance of the other person.

Alignment: A particular position adopted by one person in attempting to take the same side or perspective as the other person.

Applied Behaviour Analysis: A form of intervention based on the principles of operant conditioning.

Argumentation: Type of discourse which includes an assertion or claim accompanied by evidences and warrants which validate it.

Asperger's syndrome: An autism spectrum disorder (ASD) that is characterised by significant difficulties in social interaction and non-verbal communication, alongside restricted and repetitive patterns of behaviour and interests.

Attributional bias: A tendency for individuals to attribute cause or blame in a biased way depending on their own viewpoint and access to information.

Autism (ASD): A lifelong neurodevelopmental disorder, characterised by a triad of impairments including difficulties in social interaction, difficulties in communication, and limited flexibility in thinking (see Wing, 1981).

Behavioural therapy: A term that refers to a variety of approaches and methods most often used to therapeutically address behavioural needs and/or challenges.

Behaviourist desensitisation approach: Treatment approach aiming to reduce anxiety responses acquired through classical conditioning, where patients are gradually exposed to the anxiety-inducing stimulus while applying relaxation techniques.

Body distress: This is a term that is sometimes used by critical scholars as a less medicalised and pathologising alternative to 'eating disorders'. It refers to negative feelings and experiences associated with body weight, shape, and/or size.

Body image: Slade and Brodie (1994, p. 32) define 'body image' as 'the mental picture of the size, shape and form of the human body; and to the feelings concerning these characteristics and the constituent body parts'.

Body modification: The deliberate altering of the body via practices such as dieting, self-starvation, piercing, tattooing, and so on. This may be done for a variety of reasons, including for aesthetic reasons, to enhance sexual attractiveness, to display group membership/affiliation or self-expression (among others).

Causal attributions: Explanations about the causes of an event or phenomenon.

Cognitive theories: Explanations relating to mental processes.

Conversation analysis: A qualitative approach to analysis used to examine the sequential nature of talk-in-interaction. Those practicing conversation analysis use naturally occurring data and explore the social actions produced in talk.

Dialogic polarity: Two kinds of explanation that are set up as mutually exclusive.

Discourse: The written or spoken language and/or interaction.

Discourses: Discrete ways of understanding, communicating, and responding to human aspirations and challenges, such as well-being, mental health, and mental illness.

Discourse analysis: A qualitative research method used to analyse verbal interactions, speech, and texts. The most common distinction made by writers is between a 'micro'-level analysis that is orientated closely to the positions taken by speakers as demonstrated in the talk itself, as opposed to 'macro'-level analyses which seek to situate the talk within social and institutional structures. Talk is seen as performative and the analyst is less interested in the truth or falsity of accounts than the accounts themselves and what is done with the talk in presenting versions of reality and negotiating meaning.

Discursive action: A performative action that is accomplished through discourse; for example, making or undermining a truth claim, and constructing a version of events.

Discursive contexts: The contexts in which discourse takes place. Contexts can be 'local' (situated in conversation and place) and also in 'wider' culture or society.

Disruptive mood dysregulation disorder: A disorder that has been newly described and included in *DSM-5*, published in 2013. DMDD is famously characterised by 'temper outbursts' and has an age onset before 10 years, and diagnosis can be made between 6 and 18 years.

DSM: *Diagnostic and Statistical Manual of Mental Disorders* used in diagnosing mental health conditions.

ELAN (EUDICO Linguistic Annotator): An annotation tool for video and audio data, developed at the Max Planck Institute for Psycholinguistics, Nijmegen, The Netherlands.

Extreme case formulations: Linguistic expressions used to represent events and phenomena as extreme examples (see Pomerantz, 1986).

Forum: A site on the Internet where people can post messages and engage in discussion.

Genetic theories: Explanations relating to genes and heredity.

Hedging: Refers to hesitant and stilted utterances in discourse.

Indexical: Referring to how the meaning of a word is determined by the context of its use.

Initiating Action: See: sequence initiating action.

Interpretative repertoire: A key theoretical concept of critical discursive psychology referring to a patterned collection of commonly used expressions and ways of explaining and interpreting the social world.

Lexical choices: Words or phrases that are used in interaction.

Medicalisation: The interpretation of social phenomena in medical terms. Managing human conditions, concerns, and struggles as medical conditions; using medical studies, definitions, diagnostic procedures, prevention modalities, or designing and conducting treatments according to medical traditions.

Moderator: A person that monitors the forum and ensures that people post online with the forum rules, and amends or removes comments that fail to conform to the rules.

Moral imperative: A prescriptive social recommendation to act, think, or feel in a particular way.

Multimodal analysis: An analysis of multimodal interaction, in particular an analysis that is sensitive to the participants' use of the difference modes that are present.

Neurobiological theories: Explanations relating to the physical nervous system and the brain.

Non-normative communication: Communication approach and patterns that fall outside the norm.

Operant conditioning: The term was introduced by the American psychologist B. F. Skinner (1904–1990) to refer to the form of learning in which the likelihood of a certain behaviour being produced depends on the previous consequences of producing that behaviour under similar conditions.

Positivist-empiricist: Approaches to research that emphasise the verification of 'truths' or facts about phenomena via observation and scientific testing.

Post-structuralism: Used to refer to a body of work produced in the mid-20th century, mostly by French and continental philosophers and critical theorists. This work emphasises the centrality of language in the creation of human cultures and is sceptical of the search for explanatory structures underlying social phenomena (as is characteristic of structuralism).

Post-traumatic stress disorder (PTSD): PTSD has been identified as the most prevalent disorder after traumatic exposure. PTSD is characterised by symptoms of intrusion (e.g. recurrent distressing memories, nightmares), avoidance and numbing (e.g. avoiding conversations about the experience, losing interest in former hobbies), and hyperarousal (e.g.

irritability, concentration difficulties). The disorder can impair children's development in emotional, social, academic, and physical domains.

Predicates of ADHD: Attributes defining the category.

'Pro' sites: Websites which encourage talk about, or the activities of, self-harm, eating disorders, and other forums of self-injury; usually, these are not moderated.

Pseudomedicine: This refers to those treatments that claim to be working within the concepts of traditional medicine that either they have no objectively verifiable benefit or they are incompatible with the current state of knowledge in science (Psiram, 2013).

Psychosocial: Relating to both psychological and social factors.

Randomised controlled trial: This is an experimental method for assessing outcomes of healthcare treatments which, when applied to assessing the effectiveness of a psychotherapy, controls for non-treatment variables through such methods as manualised treatment to increase therapist conformity to the model, as well as the use of a non-treatment control group.

Reflection: According to Dewey, reflection is both an intellectual and emotional enterprise implying the establishment of connections and ties between elements of experience (e.g. antecedents and consequences). It includes subsequent steps: a sense of a problem with the related attitude of bearing the uncertainty, a sense of concern and emotional involvement in what is going on, a questioning orientation, and the acknowledgement of thinking as 'an actual part of the course of events and designed to influence the results' (Dewey, 1980, p. 154).

Repair: Mechanisms through which parties in conversation deal with problems in speaking, hearing, or understanding.

Response mobilising design features: Linguistic and paralinguistic features of a speaker's turn that 'increase the recipient's accountability for responding' (Stivers & Rossano, 2010, p. 4).

Responsive action: An action that responds to a prior sequence-initiating action.

Selective mutism: A childhood psychiatric diagnosis assigned to children who pervasively do not speak in certain social situations where speaking would be expected.

Sequence initiating action: An action (such as a question or a request for an object) that launches an interactional sequence through making a responsive action (such as an answer or the passing of an object) relevant.

Single-incident trauma: Single-incident trauma is an acute event that did not occur in the context of chronic abuse, chronic maltreatment, or war.

Social cognition: An approach within social psychology which seeks to explain a range of individual and interpersonal behaviours by examining the mental encoding, storage, retrieval, and processing of social information.

Social constructionism: There is no single position of social constructionism, but broadly speaking, it takes a critical position against the taken-for-granted knowledge produced in daily interactions (see Burr, 2003). Social constructionism gives emphasis to language and narrative and sees reality as a product of social processes.

Speaking hierarchies: A desensitisation approach specific to selective mutism, where patients are led through various interactional situations involving increasing demand for spoken communication.

Speech and language therapy: An allied health profession whose remit is the assessment, diagnosis, and treatment of disorders of speech, language, communication, and swallowing in adults and children.

Stake and interest: A personal reason for taking a particular viewpoint.

Subject position: A key theoretical concept of critical discursive psychology meaning an identity position taken up or attributed in discourse while explaining and interpreting the social world.

Superflex curriculum: A cognitive behavioural curriculum designed by Stephanie Madrigal and Michelle Garcia Winner to assist children in their thinking, perspective-taking, and social behaviours.

Talk-in-interaction: A general, all-encompassing term for conversation.

Therapeutic alliance: The secure non-judgemental relationship with clients that therapists seek to foster.

Therapeutic rupture: Points of difficulty or break down of communication or therapeutic alignment.

Thread: A series of posts on an Internet site – an online 'conversation'.

Transparent representation: A representation that allows a view of a reality beyond itself.

Trouble sources: Trouble sources may be defined as a problem of speaking, hearing, or understanding (Schegloff, 1992). In identifying trouble sources as such, it is important to scrutinise the orientation of speakers themselves to these conversational moments that appear to be problematic. Typically, attempts will be made to correct a 'defective utterance' by the speaker (self-initiated repairs) or the listener (other-initiated repairs). Where the speaker does not accept the listener's invitation to correct the prior problematic utterance, the listener will typically undertake the repair herself in a subsequent turn (other corrections).

'Truth will out' device: A linguistic device used to validate a truth claim by showing how an outcome proves an initial claim (see Gilbert & Mulkay, 1984).

Turn construction unit: Basic units of talk that comprise of a turn. Unit types include lexical, clausal, phrasal, and sequential.

Zone of proximal development (ZPD): For Vygotsky, it refers to the distance between the actual development level as performed by the child independently and the level of potential development, which the child can reach under adult guidance or in collaboration with more capable peers. In this study, the discourse between the adult therapist and a boy with intellectual disability was analysed in terms of the ZPD metaphor, creating a supportive environment for reasoning skills.

References

Burr, V. (2003). *Social constructionism* (2nd edition). London: Routledge.

Dewey, J. (1916/1980). Democracy and education. In J. A. Boydston (Ed.), *The middle works, 1899–1924* (1916, Vol. 9). Carbondale and Edwardsville: Southern Illinois University Press.

Gilbert, G. N., & Mulkay, M. (1984). *Opening Pandora's box: A sociological analysis of scientists' discourse.* Cambridge: Cambridge University Press.

Pomerantz, A. M. (1986). Extreme case formulations: A new way of legitimating claims. *Human Studies, 9,* 219–230.

Psiram (2013). *Pseudomedicine.* Retrieved 12 August 2014 from http://www.psiram.com/en/index.php/Pseudomedicine.

Slade, P., & Brodie, D. (1994). Body-image distortion and eating disorder: A reconceptualization based on the recent literature. *European Eating Disorders Review, 2*(1), 32–46.

Schegloff, E. A. (1992). Repair after next turn: The last structurally provided defence of intersubjectivity in conversation. *American Journal of Sociology, 97,* 1295–1345.

Stivers, T., & Rossano, F. (2010). Mobilizing response. *Research on Language & Social Interaction, 43*(1), 3–31.

Wing, L. (1981). Language, social and cognitive impairments in autism and severe mental retardation. *Journal of Autism and Developmental Disorders, 11*(1), 31–44.

Index

.